Molecular Genetics and Gene Therapy of Cardiovascular Disease

FUNDAMENTAL AND CLINICAL CARDIOLOGY

Editor-in-Chief

Samuel Z. Goldhaber, M.D.

Harvard Medical School
and Brigham and Women's Hospital
Boston, Massachusetts

Associate Editor, Europe

Henri Bounameaux, M.D.

University Hospital of Geneva
Geneva, Switzerland

ADDITIONAL VOLUMES IN PREPARATION

Molecular Genetics and Gene Therapy of Cardiovascular Disease

edited by

Stephen C. Mockrin

National Heart, Lung and Blood Institute
National Institutes of Health
Bethesda, Maryland

Marcel Dekker, Inc.   New York • Basel • Hong Kong

Library of Congress Cataloging-in-Publication Data

Molecular genetics and gene therapy of cardiovascular diseases /
    edited by Stephen C. Mockrin.
        p.   cm. -- (Fundamental and clinical cardiology ; 26)
    Includes index.
    ISBN 0-8247-9408-7 (alk. paper)
    1. Cardiovascular system--Diseases--Genetic aspects.
2. Cardiovascular system--Diseases--Genetic therapy. I. Mockrin,
Stephen C. II. Series.
    [DNLM: 1. Cardiovascular Diseases--therapy. 2. Gene Therapy.
3. Genetics, Biochemical. W1 FU538TD v.26 1996 / WG 120 M718 1996]
RC669.M64   1996
616.1'042--dc20
DNLM/DLC
for Library of Congress   95-25992
                                                                         CIP

*This book was edited by Dr. Stephen C. Mockrin in his private capacity. No
official support or endorsement by the National Institutes of Health is
intended and none should be inferred.*

The publisher offers discounts on this book when ordered in bulk quantities. For more
information, write to Special Sales/Professional Marketing at the address below.

This book is printed on acid-free paper.

Marcel Dekker, Inc.
270 Madison Avenue, New York, New York 10016

Current printing (last digit):
10 9 8 7 6 5 4 3 2 1

PRINTED IN THE UNITED STATES OF AMERICA

Series Introduction

Marcel Dekker, Inc., has focused on the development of various series of beautifully produced books in different branches of medicine. These series have facilitated the integration of rapidly advancing information for both the clinical specialist and the researcher.

My goal as editor-in-chief of the Fundamental and Clinical Cardiology series is to assemble the talents of world-renowned authorities to discuss virtually every area of cardiovascular medicine. In the current monograph, Stephen Mockrin has edited a much-needed and timely book. Further contributions to this series will include books on molecular biology, interventional cardiology, and clinical management of such problems as coronary artery disease and ventricular arrhythmias.

Samuel Z. Goldhaber

Preface

The rapid progress and recent developments in gene mapping and manipulation provide experimental and clinical strategies that have revolutionized biomedical research and medical practice. Examples include:

The production of increased amounts of therapeutic compounds that exist in scarce amounts

The design of novel, biologically active molecules

Localization of DNA-based markers of disease

The identification and isolation of disease-causing genes

Better understanding of both normal and pathophysiological mechanisms

A new generation of highly specific diagnostic tests for human disease in advance of symptoms

The development of genetically altered animals that allow the investigator to study the biological effects of specific genes at all levels of complexity, to understand how different genes interact with one another and with environmental factors to produce multifactorial diseases, to generate new models of human diseases, and to test new therapeutic modalities

The establishment of somatic cell gene therapy to restore or enhance normal cellular activities, to confer new cellular functions, or to prevent unwanted cellular activities

These new strategies offer unprecedented opportunites for improved diagnosis, detection, prevention, and treatment that will have far-reaching effects on the morbidity and mortality, and the enormous fiscal and emotional costs imposed on society by cardiovascular diseases. As a consequence, this volume has been assembled to provide the critical dimensions and selected highlights of these new technologies as they pertain to cardiovascular research and medicine. Written by investigators at the leading edge in their respective areas, the chapters provide the latest significant advances and trends within the context of the impact of pioneering research on medical practice and prevention. In addition, special overview chapters introduce principles, techniques, and paradigms to assist the reader who is unfamiliar with these specialized areas. By combining topics that illustrate key developments and trends with distinctive overview chapters that define the major concepts and strategies, this book should be valuable and practical to basic research and clinical scientists who have little experience in the field as well as specialists who desire a timely volume of the latest advances.

Stephen C. Mockrin

Contents

Contributors

W. French Anderson, M.D.   Director, Gene Therapy Laboratories, and Professor of Biochemistry and Pediatrics, Norris Cancer Center, University of Southern California School of Medicine, Los Angeles, California

Greg Boivin, D.V.M.   Assistant Professor, Department of Pathology, University of Cincinnati College of Medicine, Cincinnati, Ohio

Jan L. Breslow, M.D.   Frederick Henry Leonhardt Professor and Director, Laboratory of Biochemical Genetics and Metabolism, The Rockefeller University, New York, New York

François Cambien, M.D.   Director, DNA Bank for Cardiovascular Research, INSERM, Paris, France

Aravinda Chakravarti, Ph.D.   Professor, Department of Genetics, Case Western Reserve University School of Medicine, and University Hospitals of Cleveland, Cleveland, Ohio

Kenneth R. Chien, M.D., Ph.D.   Professor, Department of Medicine, Center for Molecular Genetics, and the American Association-Bugher Foundation Center for Molecular Biology, University of California, San Diego, School of Medicine, La Jolla, California

Marilyn Dammerman, Ph.D.   Assistant Professor, Laboratory of Biochemical Genetics and Metabolism, The Rockefeller University, New York, New York

Ronald J. Diebold, Ph.D.   Doctoral Assistant, Program of Excellence in Molecular Biology of Heart and Lung and Department of Molecular Genetics, Biochemistry, and Microbiology, University of Cincinnati College of Medicine, Cincinnati, Ohio

Harry C. Dietz, M.D.   Associate Professor, Departments of Pediatrics, Medicine, and Molecular Biology and Genetics, The Johns Hopkins University of Medicine, Baltimore, Maryland

Thomas Doetschman, Ph.D.   Associate Professor, Program of Excellence in Molecular Biology of Heart and Lung, and Department of Molecular Genetics, Biochemistry, and Microbiology, University of Cincinnati College of Medicine, Cincinnati, Ohio

Victor J. Dzau, M.D.   Arthur L. Bloomfield Professor of Medicine, Chairman, Department of Medicine, and Director, Falk Cardiovascular Research Center, Stanford University, Stanford, California

Michael Eis, Ph.D.   Postdoctoral Assistant, Department of Molecular Genetics, Biochemistry, and Microbiology, University of Cincinnati College of Medicine, Cincinnati, Ohio

Randy C. Eisensmith, Ph.D.   Assistant Professor, Department of Cell Biology, Baylor College of Medicine, Houston, Texas

Ingrid L. Grupp, M.D.   Professor, Department of Pharmacology and Cell Biophysics, University of Cincinnati College of Medicine, Cincinnati, Ohio

Richard O. Hynes, Ph.D., F.R.S.   Professor, Department of Biology, Howard Hughes Medical Institute and Center for Cancer Research, Massachusetts Institute of Technology, Cambridge, Mssachusetts

*Howard J. Jacob, Ph.D.**   Assistant Professor, Cardiovascular Research Center, Massachusetts General Hospital, Charlestown, Massachusetts

Mark T. Keating, M.D.   Howard Hughes Medical Institute, Eccles Institute of Human Genetics, University of Utah, Salt Lake City, Utah

Dean H. Kedes, M.D., Ph.D.   Howard Hughes Medical Institute and Departments of Medicine, Microbiology, and Immunology, University of California, San Francisco, California

Laurence H. Kedes, M.D.   Institute for Genetic Medicine, Department of Biochemistry and Molecular Biology, and Department of Medicine, University of Southern California School of Medicine, Los Angeles, California

---

Current affiliation: Associate Professor, Department of Physiology, Medical College of Wisconsin, Milwaukee, Wisconsin.

John H. Krege, M.D.   Howard Hughes Physician Post-Doctoral Fellow, Department of Medicine, University of North Carolina at Chapel Hill, Chapel Hill, North Carolina

José E. Krieger, M.D., Ph.D.   Assistant Professor, Heart Institute HCUSP, University of São Paulo, São Paulo, Brazil

Helena Kuivaniemi, M.D., Ph.D.*   Department of Biochemistry and Molecular Biology, Jefferson Institute of Molecular Medicine, Jefferson Medical College, Thomas Jefferson University, Philadelphia, Pennsylvania

Theodore W. Kurtz, M.D.   Professor, Department of Laboratory Medicine, University of California, San Francisco, California

Eric S. Lander, Ph.D.   Professor, Department of Biology, Massachusetts Institute of Technology, and Director, Whitehead Institute/MIT Center for Genome Research, Cambridge, Massachusetts

Fred D. Ledley, M.D.   Vice President, Medicine, GeneMedicine, Inc., and Associate Professor, Departments of Cell Biology and Pediatrics, Baylor College of Medicine, Houston, Texas

Jeffrey M. Leiden, M.D., Ph.D.   Departments of Medicine and Pathology, Unviersity of Chicago, Chicago, Illinois

Leslie A. Leinwand, Ph.D.   Professor, Departments of Microbiology and Immunology, Medicine, and Genetics, Albert Einstein College of Medicine, Bronx, New York

Richard P. Lifton, M.D., Ph.D.   Associate Professor, Howard Hughes Medical Institute, Boyer Center for Molecular Medicine, and Departments of Medicine and Genetics, Yale University School of Medicine, New Haven, Connecticut

Robert W. Mahley, M.D., Ph.D.   Director, Gladstone Institute of Cardiovascular Disease, and Professor, Departments of Medicine and Pathology, University of California, San Francisco, California

John J. Mullins, Ph.D.   Centre for Genome Research, University of Edinburgh, Edinburgh, Scotland

Linda J. Mullins, Ph.D.   Centre for Genome Research, University of Edinburgh, Edinburgh, Scotland

Elizabeth G. Nabel, M.D.   Professor, Department of Internal Medicine, and Director, Cardiovascular Research Center, University of Michigan, Ann Arbor, Michigan

---

**Current affiliation*: Associate Professor, Center for Molecular Medicine and Genetics, and Department of Surgery, Wayne State University School of Medicine, Detroit, Michigan.

Gary J. Nabel, M.D.   Professor, Departments of Internal Medicine and Biological Chemistry, Howard Hughes Medical Institute, University of Michigan, Ann Arbor, Michigan

Andrew Plump, M.D., Ph.D.   Post-Doctoral Associate, Laboratory of Biochemical Genetics and Metabolism, The Rockefeller University, New York, New York

Darwin J. Prockop, M.D., Ph.D.   Chairman and Professor, Department of Biochemistry and Molecular Biology, Jefferson Institute of Molecular Medicine, Jefferson Medical College, Thomas Jefferson University, Philadelphia, Pennsylvania

Howard A. Rockman, M.D.   Assistant Professor, Department of Medicine, University of California, San Diego, School of Medicine, La Jolla, California

Robert D. Rosenberg, M.D., Ph.D.   Professor, Department of Medicine, Harvard Medical School, and Beth Israel Hospital, Boston, and Department of Biology, Massachusetts Institute of Technology, Cambridge, Massachusetts

Robert S. Ross, M.D.   Assistant Professor, Department of Medicine, and the American Heart Association-Bugher Foundation Center for Molecular Biology, University of California, San Diego, School of Medicine, La Jolla, and Veterans Administration Hospital-San Diego, San Diego, California

Nicholas J. Schork, Ph.D.   Assistant Professor, Department of Genetics, Case Western Reserve University School of Medicine, and University Hospitals of Cleveland, Cleveland, Ohio, and Associate Research Scientist, The Jackson Laboratories, Bar Harbor, Maine

Christine E. Seidman, M.D.   Associate Professor, Department of Medicine; Associate Investigator, Howard Hughes Medical Institute; and Director, Cardiovascular Genetics Center, Brigham and Women's Hospital and Harvard Medical School, Boston, Massachusetts

Jonathan G. Seidman, Ph.D.   Professor, Department of Genetics, and Investigator, Howard Hughes Medical Institute, Harvard Medical School, Boston, Massachusetts

Marcia M. Shull, Ph.D.   Postdoctoral Assistant, Program of Excellence in Molecular Biology of Heart and Lung and Department of Molecular Genetics, Biochemistry, and Microbiology, University of Cincinnati College of Medicine, Cincinnati, Ohio

Michael Simons, M.D.   Assistant Professor, Department of Medicine, Harvard Medical School, and Beth Israel Hospital, Boston, Massachusetts

Louis C. Smith, Ph.D.   Professor, Department of Medicine, Baylor College of Medicine, Houston, Texas

Oliver Smithies, D. Phil.   Excellence Professor of Pathology, Department of Pathology, University of North Carolina at Chapel Hill, Chapel Hill, North Carolina

Florent Soubrier, M.D., Ph.D.   Professor, INSERUM U36, Collège de France, Paris, France

*Gerard Tromp, Ph.D.**   Research Assistant Professor, Department of Biochemistry and Molecular Biology, Jefferson Institute of Molecular Medicine, Jefferson Medical College, Thomas Jefferson University, Philadelphia, Pennsylvania

Denisa D. Wagner, Ph.D.   Associate Professor, Department of Pathology, and Senior Investigator, Center for Blood Research, Harvard Medical School, Boston, Massachusetts

Savio L. C. Woo, Ph.D.   Professor, Department of Cell Biology, Howard Hughes Medical Institute, Baylor College of Medicine, Houston, Texas

Stephen G. Young, M.D.   Scientist, Gladstone Institute of Cardiovascular Disease, and Associate Professor, Department of Medicine, University of California, San Francisco, California

---

Current affiliation: Assistant Professor, Center of Molecular Medicine and Genetics, Wayne State University School of Medicine, Detroit, Michigan.

1

Molecular Genetics and Its Application to Understanding Cardiovascular Disease

Dean H. Kedes
Howard Hughes Medical Institute
University of California, San Francisco
San Francisco, California

Laurence H. Kedes
Institute for Genetic Medicine
University of Southern California
Los Angeles, California

> *One will not be able to enter these organizations and work effectively from within unless he is scientifically competent, technically capable and skilled in the practice of his own profession.**

INTRODUCTION

The impact of molecular biology and genetic research on the discovery of the root causes of a wide variety of hereditary and acquired diseases has long been self-evident. As the later chapters of this volume testify, that impact is being increasingly felt in defining the molecular basis of a variety of cardiovascular disorders and their treatments. The purpose of this initial chapter is to provide a brief outline of basic molecular and genetic principles and techniques to enable the reader to more fully appreciate and understand the sophisticated, complex approaches and implications provided in the ensuing chapters. Clearly our effort has been neither encyclopedic nor profound. Our assumption is that the reader has a general knowledge of biochemistry and cell biology as taught in medical schools. We focus on events in mammalian cells and, when possible, use examples culled from cardiovascular-specific instances. We focus on the beauty of the biology rather than on the ugliness of the details but, as all researchers in these fields know too well, the critical messages lie

---

*Pope John XXIII. Encyclical, April 10, 1963.

in the minutiae. We have placed in **_boldface italics_** key words or concepts as they are first mentioned and defined.

Some Recommended Texts and Monographs

Fortunately, for those who desire to review background and supporting data in more detail, as well as obtain a deeper exposition of important details of current knowledge about molecular genetics, there are a number of excellent recent textbooks and monographs. The following is a very short list of publications that we can recommend after having reviewed them in preparation for this chapter. Absence from this list—and there are many other excellent publications—does not necessarily imply that any given book or monograph is not recommended, but *caveat emptor*!

> Molecular Biology of the Cell. Alberts B, Bray D, Lewis J, Raff M, Roberts K, Watson JD. New York: Garland Publishing, 1994.
> Biochemistry. Stryer L. New York: WH Freeman, 1995.
> Molecular Basis of Cardiology. Roberts R, ed. Boston: Blackwell Scientific Publications, 1993.
> Genes in Medicine: Molecular Biology and Human Genetic Disorders. Raskó I, Downes CS. London: Chapman & Hall, 1995.
> Recombinant DNA. Watson JD, Gilman M, Witkowski J, Zoller M. New York: Scientific American Books, 1992.

From Genes to Proteins to Phenotype

This is a book about diseases of the cardiovascular system—many heritable, others acquired—and about the potential of the application of molecular genetics through gene therapy to rescue the diseased tissues. The proximate molecular mechanisms responsible for essentially all the diseases discussed are altered or absent **_proteins_** that affect the **_phenotypes_** of the cells involved. And the promise of gene therapy lies predominantly in its potential to produce novel or normal proteins in the affected tissues, thus restoring in whole or in part the normal phenotypes.

DEFINING A GENE

Genes are heritable traits. Genes are also defined segments of DNA (**_deoxyribonucleic acid_**) that are passed to succeeding generations through **_germ cells_** (spermatocyte and oocyte progenitors) and are equally represented in every **_somatic_** (nongerm) **_cell_**. The eventual product of most genes is a protein whose function is responsible for the heritable trait and thus the phenotype of the cell, tissue, organ, or organism. How the information stored in the DNA is duplicated during cell replication and then processed and converted into proteins—at the correct time in the life of the organism, in the correct tissues, and in the correct amounts—lies at the heart (no pun intended) of understanding the essentials of molecular genetics. Each of these processes is itself regulated by scores of intracellular pathways defined by interactions among proteins; hence, there are genes that regulate other genes. Recombinant DNA technology and genetic engineering are laboratory methods that allow us to manipulate genes and alter the phenotype of cells, tissues, and organs. The power of these methods engenders the promise of gene therapy.

The Basic Building Blocks of DNA Structure

Macromolecules such as proteins and DNA consist of polymerized chains of subunits held together by covalent bonds. The highly specific interactions of macromolecules, whether in the binding of a polypeptide growth factor to its receptor, in the mechanical motion of myosin heavy chains sliding along an actin filament, or in the forces holding two DNA strands together, depend on noncovalent, weaker intermolecular bonds including—in descending order of their strength—salt bridges, hydrogen bonds, hydrophobic interactions, and Van der Waals forces. A major characteristic of the structure of such polymers is that they tend to form a linear helix—a conformation that minimizes the free energy associated with their contacts and also ensures that the relationship of adjacent monomers is essentially the same throughout the length of the polymer.

DNA is a polymer made up of *nucleotide* molecules linked covalently; thus, DNA is a *polynucleotide*. Nucleotides are made up of three components: a nitrogen-containing ringed **base**, a five-carbon **sugar**, and a **phosphate** group (Figure 1). The base is attached to the 1' carbon of the sugar to form a molecule called a *nucleoside*.* A string of three phosphate groups is covalently attached to the 5' sugar carbon of an individual nucleoside, forming a nucleotide. In a DNA polymer, the 3'-OH group of each nucleotide forms a link to the proximal phosphate group of its neighbor, and in the process the two distal phosphate groups are lost. Phosphate groups are covalently attached to the 5' and 3' sugar carbons, and the phosphates form the links between adjacent nucleotides in the DNA polymer. The bases face inward in the polynucleotide, and the phosphate–sugar linkages form an external *backbone*. Since the polymer has a 5' attachment site at one end and a 3' attachment

---

*Some of us always seem to have trouble remembering which has the phosphates attached, the nucleoside or the nucleotide. One simple mnemonic is to remember that the nucleoside has its side exposed (no phosphates). But this reminds us of the classic conversation aired on American radio by the comedians Bob and Ray in which the spelunker being interviewed by Wally Baloo describes how he remembers the difference between a stalagtite and a stalagmite: "a stalag*tite* holds *tight* to the roof of the cave and a stalag*mite* *might* not. Or is it the other way around?"

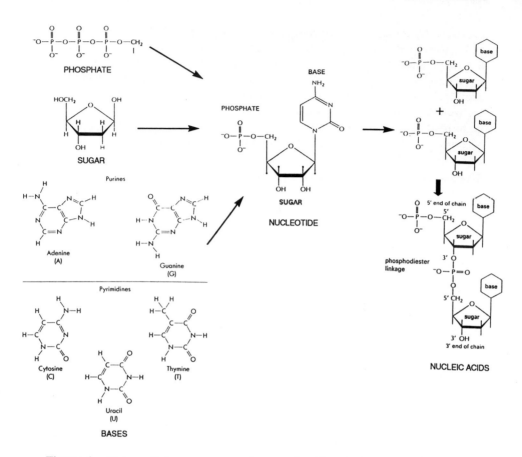

Figure 1   Major cellular components form nucleotides, the basic building blocks of nucleic acids. The sugar molecule in the illustration is a deoxy pentose. It has five carbon atoms. Each carbon on the sugar is conventionally numbered and followed by a prime mark; e.g. the labeled carbon in the illustration is the 5-prime carbon (5'). The phosphates are normally joined to the 5'C hydroxyl of the sugar. The bases, either pyrimidines or purines, are ring compounds containing nitrogen. A nucleotide is composed of a pentose sugar, a base, and at least one phosphate group. The base is linked to the sugar at the C1 (1') position. In the illustration, a cytosine base (a pyrimidine) has joined to form the monophosphorylated nucleotide, deoxycytosine (dCMP). Uracil replaces thymine when ribose rather than deoxyribose is the sugar used to form RNA. Phosphodiester linkages join the 5' and 3' carbon atoms of the nucleotide sugars to form nucleic acids.

site at the other, there is stereochemical *polarity* of the strand. By convention the DNA strand is represented as 5′ to 3′ from left to right.

The DNA of animal cells (*eukaryotic cells*) is double-stranded and located in the nucleus. RNA (*ribonucleic acid*), which is a copy of one of the two DNA strands, is a single-stranded polynucleotide *transcribed* (synthesized) and modified in the nucleus that, as part of its function for information transfer, enters the cytoplasm where it is involved in directing the synthesis of all proteins (Mitochondria, located in the cytoplasm, also contain double-stranded DNA that engenders single-stranded RNAs engaged in the production of mitochondrial-specific proteins, but these polynucleotides remain in the mitochondrial compartment.) This information flow from DNA to RNA to protein remains the central tenet—if no longer the dogma—of molecular genetics. However, in terms of molecular biology, some viruses whose genes are stored as single-stranded DNA or as RNA are important exceptions.

The five-carbon sugar present in the backbone of both DNA and RNA is *ribose*. The nucleosides used to synthesize DNA contain a ribose that has an –H rather than an –OH, hence *deoxy*ribonucleic acid versus ribonucleic acid. Only four bases are used in the DNA chain: *adenine*, *thymine*, *guanine*, and *cytosine*. These bases, and their nucleosides and nucleotides, are often abbreviated as *A*, *T*, *G*, and *C*. The other major distinction between the basic structure of DNA and RNA polynucleotides is that one of the bases used in DNA, thymine, is replaced by *uracil* (*U*) in RNA.

Base Pairing and Complementarity Explain Heredity

Adenine and guanine have a double-ring structure and are *purines*. The simpler single-ring structures of thymine (uracil in RNA) and cytidine classify them as *pyrimidines*. The stereochemistry of these bases in the polynucleotides enables them to form highly specific hydrogen bonds from one chain to the other: a purine can form *base pairs* with a pyrimidine in the opposite chain (Figure 2). Guanine and cytosine form three hydrogen bonds and pair only with each other, and adenine and thymidine pair by forming two such bonds. Accordingly, the bonds between G:C base pairs are stronger than those between A:T pairs. Because of the stereochemical structure of the oligonucleotide strands, these base pairings can form only when the two strands lie with opposing (*antiparallel*) orientations. Thus, one strand is represented 5′ to 3′ and its base-paired partner lies intertwined 3′ to 5′. Although the hydrogen bonds pairing opposing bases are relatively weak, the sum of these bonds extending in both directions along the double-stranded polynucleotide chain creates highly stable structures that must be relaxed during both DNA and RNA synthesis. The ability of polynucleotides to *hybridize* to each other with such a high degree of affinity and specificity provides the foundation for a

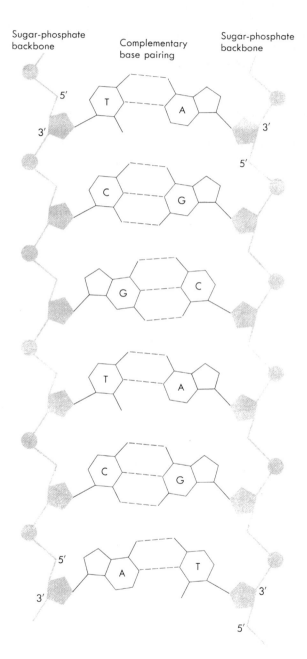

Figure 2   The base-pairing and anticomplementarity of two DNA chains. (Adapted from Watson JD, Gilman M, Witkowski J, Zoller M, eds, Recombinant DNA, 2nd ed. New York: WH Freeman, 1992.)

variety of tools used for the identification, purification, and manipulation of specific nucleic acid sequences.

The rigid stereochemical rules of base pairing (G with C and A with T) immediately explain the *complementary* nature of DNA double strands: the order of nucleotides on one strand predicts the order on its opposite strand. Creating a new strand from one of the two older strands exactly recreates the original duplex, thus explaining the structural basis for heredity. This elegant principle, including its implications for the *replication* (copying) of DNA and the basis for the stability of heritable DNA sequences, did not escape the notice of James Watson and Francis Crick in their seminal short papers in *Nature** on the structure of DNA as a base-paired, antiparallel, double-stranded oligonucleotide. Nor did it escape the notice of the Nobel committee.

DNA: A Lesson in Information Storage and Retrieval

The sequence of bases in DNA carries novel genetic information. The fact that there are only four symbols (A, G, C, T) in this nucleotide alphabet does not limit the infinite complexity of information that DNA can store. After all, the information in an encyclopedia, or all of human knowledge for that matter, can be represented by two symbols: a dot and a dash in Morse code, or a 0 and 1 in the binary code that drives computers. The nucleotide sequence in the DNA gives each gene its unique informational content. The linear nucleotide sequence encodes the information for the linear sequence of amino acids that make up novel polypeptides. That information, and additional nucleotide sequences that govern when, where, and how much of a protein is to be made, is the basic definition of a gene: a DNA segment that encodes the information for assembling the linear amino acid sequence for a polypeptide. In general, one gene codes for one polypeptide. Important exceptions to this rule are described later.

The amino acid sequence of the protein is *colinear* with the nucleotides of the gene that encodes them. As the DNA has polarity (5' to 3'), so then must the amino acid sequence. By convention, the DNA strand represented 5' to 3' and from left to right in a duplex DNA is the *coding strand* (see Figure 3). The sequences at the 5' end of the coding strand encode the amino terminus of the encoded protein and the sequences at the 3' end encode the carboxy terminus of the protein. DNA and RNA polynucleotide synthesis both proceed from 5' to 3'. The polymerization of amino acids into polypeptides similarly

---

*"It has not escaped our notice that the specific pairing we have postulated immediately suggests a possible copying mechanism for the genetic material." Watson JD, Crick FHC. Molecular structure of nucleic acids: a structure for deoxyribonucleic acid. Nature 1953; 171:737-738.

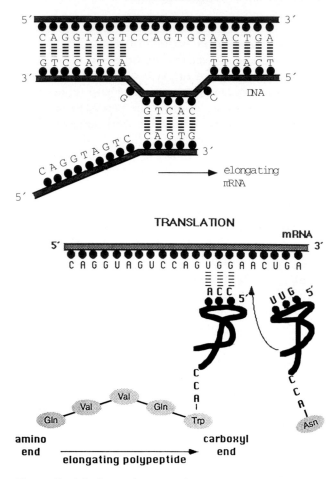

Figure 3   The basic elements of genetic information storage and retrieval. (Top) The various bases (filled circles) and the phosphate sugar backbones (shaded lines) in the double-stranded DNA form a base-paired template (short dashes). A short section of the DNA is shown as looping out to schematically illustrate the formation of an elongating messenger RNA copy by transcription. The transcription process also involves base-pairing. The sequence of the DNA bases on the upper strand, read in the 5′ to 3′ direction (the "coding" or "sense" strand by convention), can be interpreted as an amino acid sequence. The actual physical template for mRNA production, however, is the "antisense" or "noncoding" DNA strand. The resulting single-stranded mRNA thus contains the "sense" information and orientation. In the cell, the RNA template contains uracil (U) rather than thymine (T) bases. (Bottom) The encoded information on the mRNA is stored as a unidirectional (5′ to 3′) but unpunctuated triplet code. Once the specific site of initiation of peptide synthesis is identified by the translation machinery,

begins by assembling the amino terminus based on the 5′ sequences of the informational RNA and terminates at the carboxy end near the 3′ end of the RNA.

MUTATIONS: A CHANGE IN THE INFORMATION STORED IN DNA

*To err is human, but to really foul things up requires a computer.**

Replication of DNA, while elegant in the simplicity of concept, is a highly complex process involving a large number of biosynthetic steps, each regulated by the interactions of many enzymes and other proteins. Although errors are very infrequent, and additional complex "proofreading" mechanisms have evolved to correct them, errors do occur and provide a direct explanation for the occurrence of many kinds of *mutations*. An A is placed rather than a T, or a G is incorporated rather than a C. Sometimes one or a few extra bases are inserted or one or a few skipped. The correction mechanisms do not always know which was the original strand and which the new complementary strand, and correction, if it takes place at all, can lead to a "certification" of the new strand as accurate. This in turn leads to the creation of a perfectly paired duplex DNA that has a different sequence from its immediate ancestor. If such errors occur in germ cells, then the error can be passed on to succeeding generations. If the errors occur in somatic cells, the daughter cells and their progeny carry the new sequence.

---

*Anonymous. Printed on our coffee mug.

---

the bases in the RNA chain are used to base-pair sequentially to specialized transfer RNA molecules (tRNAs), each carrying a specific amino acid. While the tRNAs are involved in base-pairing, the translation machinery also creates a new polypeptide bond to add an additional amino acid to the growing polypeptide chain. In the illustration, a tRNA carrying a tryptophan has a 5′-CCA-3′ anticodon sequence that recognizes the 5′-UGG-3′ triplet codon on the mRNA while an asparagine-bearing tRNA is about to dock at its codon site. In the example, the triplet codes on the mRNA were CAG (glutamic acid), GUA (valine), GUC (also valine), CAG (glutamic acid). Following the incoming asparagine is the triplet UGA. No tRNAs correspond to this triplet, consequently referred to as a "nonsense" codon. The failure of the three nonsense codons (UAG and UAA are the other two) to base-pair with an aminoacylated tRNA leads to the dissociation of the translation machinery from the mRNA and terminates the polypeptide elongation process. Such "termination" codons are critical to defining the carboxyl end of all naturally synthesized proteins.

Entirely different kinds of mutations involving gross rearrangements of chromosomes also occur frequently and are often extensive enough to be visible by cytogenetic analysis of chromosomes by light microscopy. Losses of chromosomes (or pieces of chromosomes), switching of large segments of chromosomes, and duplications of chromosomes (or of pieces) have all been reported and are often associated with specific phenotypes or disease entities. Similar rearrangements, but on a smaller scale, take place even more frequently. Many of these events, when they occur during *meiosis* (the chromosomal events leading to production of germ cells with their complement of only one of each pair of chromosomes—germ cells are said to be *haploid*), are an important natural process that ensures the mixing of genes in our gene pools. Since we each have a maternal and a paternal copy of each of our chromosomes (*diploid*), mechanisms have evolved to guarantee that germ cells end up with a mixture of maternal and paternal genes. For genes on different chromosomes that is not a problem, since the maternal and paternal copies of each chromosome sort randomly into the haploid germ cells. But to ensure a mixing of genes on the same chromosome, meiotic cells, whose chromosome pairs line up in parallel, undergo a series of reciprocal breaks and rejoinings such that a resulting chromosome is a mosaic of genes derived from both maternal and paternal origins. These events, usually called *reciprocal crossing-over*, are naturally occurring examples of *genetic recombination*. The events occur randomly in chromosomes in different germ cells. The events also occur so frequently that genes located at opposite ends of a chromosome behave as if they are not at all *linked* to one another and *segregate* randomly. Genes very close to one another on a chromosome, on the other hand, often travel together through a family (i.e., have a low versus a high *recombination frequency*). The molecular events associated with crossing over, as well as other less well understood mechanisms, can go awry and lead to visible or microscopic rearrangement abnormalities.

The Role of Mutations in Natural Selection

Organic life, we are told, has developed gradually from the protozoon to the philosopher, and this development, we are assured, is indubitably an advance. Unfortunately it is the philosopher, not the protozoon, who gives us this assurance.[*]

Mutations, by their very nature, change the informational content of a gene and, in many cases, lead to changes in the amino acid sequence of the polypeptide the gene encodes. Such mutations are usually deleterious to the

---

*Bertrand Russell, *Mysticism and Logic*, 1917.

cells where the polypeptide acts, and to the organism. Occasionally a mutation will increase or add to the efficiency of a biochemical function and be advantageous to the organism. Such *variation* among the members of a species, fashioned much by chance, is itself the basis on which *natural selection* can allow the ascendancy of organisms with new phenotypes, better adapted to survival and procreation, and is the basis for much of *evolution*. Without DNA replication error (or several other error-inducing mechanisms), there would be no biological variation, no diversity among individuals of a species, and, in all probability, no species at all.

Human beings are the only evolved species that is capable of controlling many of the environmental factors that otherwise allow natural selection to work. While the rate of chance mutations in the human population has not changed measurably over the course of eukaryotic evolution (despite what some might infer from Ames' test results*), there are now limited conditions to which human beings are subject that might provide grist for the Darwinian mill. Putting aside arguments that *geotechnical* changes in the environment (changes brought about by our persistant meddling in the chemical and physical aspects of the earth) might yet lead to subtle changes in procreation or survival of humans, and hence of continuing selective (as opposed to random) evolution, it is more likely that the *biotechnical* changes epitomized by the premise and promise of this book will work *against* natural selection and impede any residual evolutionary vectors. This follows since advances such as gene therapy may allow the survival and probably enhanced procreation of some individuals afflicted with a wide variety of historically deleterious hereditary diseases and conditions. Rather than be shocked by this notion, upon reflection, we must see such goals—the very antithesis of the abhorrent historical concepts and practices of eugenics—as among the noblest aspirations of our species.

THE INFORMATION SUPERHIGHWAY: DNA TO RNA

Information, a crumb of information, seems to light the world.[†]

It is an RNA copy of the DNA coding strand that brings the information for polypeptide polymerization from the nucleus to the protein-synthesizing machinery in the cytoplasm. The initial polynucleotide RNA that is synthesized is colinear, base for base, with the coding strand. The process is called

---

*This test, described by University of California Berkeley professor Bruce Ames, measures microbial mutations as an index of the mutagenic potential of natural and environmental compounds.

†John Cheever, *The Journals*, "The Sixties," entry in 1966.

transcription. Accordingly, the DNA strand that serves as the template for RNA synthesis is the noncoding (or *antisense*), antiparallel strand. Only certain segments of a protein coding gene are actually transcribed into an RNA strand, and these usually encompass the segment of DNA encoding the polypeptide information as well as variable lengths of sequences adjacent to the 5' end (often called *upstream flanking sequences*) and the 3' end (*downstream sequence*). Specialized segments of the gene, interacting with other nuclear macromolecules, contain information about when, where, and how much RNA to synthesize. Such regulatory segments are usually located variable distances upstream from the transcription initiation site, but often enough can be found downstream or even within the transcribed segment. Additional sequences of the gene are involved in directing where on the polynucleotide sequence of the DNA to start and stop the RNA transcription process. The discovery of such features has refined the physical definition of a "gene" to extend far beyond the protein coding segments and to include all the DNA sequences of a functional transcription unit including upstream and downstream regulatory elements.

Processing RNA: Separating the Wheat from the Chaff

A novel feature of most eukaryotic protein coding genes is that the DNA nucleotide sequence that encodes a polypeptide is interrupted frequently by lengthy stretches of polynucleotides usually encoding nothing at all (Figure 4). Such segments, referred to as *introns*, punctuate and segment the coding regions of a gene, consequently called *exons*. Every intron is flanked by exons. When initially discovered, the existence of such disrupting *intervening sequences* also disrupted our stable presumption of colinearity of DNA and protein sequence, a notion that had been soundly based on extensive prior observations on bacterial genes and on early observations of eukaryotic *histone genes* (encoding proteins that package DNA into chromosomes), neither of which have introns. Histone genes and the genes for a few other proteins have emerged as exceptions to the presence of introns in most eukaryotic protein coding genes.

One of the reasons that the presence of introns came as such a surprise is that the oligonucleotide sequences of eukaryotic *messenger RNA* (*mRNA*) molecules were, in fact, colinear with the amino acid sequences rather than with the DNA sequences of the genes: the intron sequences were missing from the mRNAs. It quickly became clear that long molecules of RNA coding strands that contain intron sequences exist in the nucleus and that these represented initial transcripts from which intron sequences were subsequently excised. The complex process by which introns are excised is referred to as *splicing*, and the long transcripts represent mRNA precursors collectively referred to as *heterogeneous nuclear RNA* (*hnRNA*).

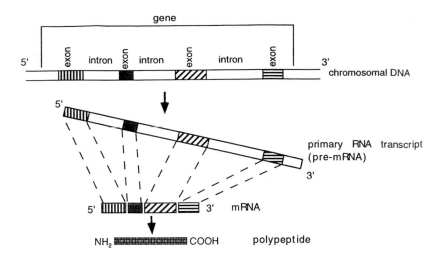

Figure 4 Genes encoding most proteins are discontinuous. The information stored in chromosomal DNA to encode a continuous polypeptide chain (protein) is usually interrupted. The coding information in the double-stranded DNA occurs in clusters, the exons (hatched and shaded boxes), interrupted by introns (open boxes). A copy of the coding DNA strand (the single-stranded pre-mRNA) is synthesized during transcription in the nucleus and includes RNA sequences corresponding to both introns and exons. The introns are carefully excised by a complex series of enzymatic splicing events regulated by sequences embedded near the intron–exon boundaries. The resulting mRNA is exported to the cytoplasm, where it can be translated into a polypeptide. In the figure, the 5′ and 3′ ends of the gene and the RNA products are designated and correspond to the amino (NH_2) and carboxyl (COOH) ends of the polypeptide.

Polynucleotide Cryptography: Information in mRNA is Read as Triplets

When mRNA molecules relocate to the cytoplasm, they engage the protein-synthesizing machinery of the cell and instruct it in the precise polymerization of a polypeptide chain of defined sequence (***translation***). Every identical copy of an mRNA from any one gene always encodes the identical amino acid sequence. Since there are only four bases used in the code (remember U rather than T in RNA) but 20 different amino acids as building blocks of polypeptides, some combination of bases must specify each amino acid. The minimum combination that could give a unique set of bases for each of 20 amino acids, sets of three, called a ***codon***, is in fact what has evolved (see Figure 3). Sets of two ($4^2 = 16$) could encode only 16 amino acids, whereas sets of three ($4^3 = 64$) can accommodate all amino acid coding requirements. The result of this arrangement is that the code is redundant—several combinations encode the same amino acid. The amino acid coding nucleotide triplets in the RNA sequence contain no punctuation between them, but there are three *nonsense* codons assigned to signal the *termination* of translation (the carboxy terminus of the polypeptide).

Nucleic Acid Complementarity and Polypeptide Synthesis

The principle of complementarity of antiparallel nucleic acid sequences lies at the heart of the ability of the mRNA strand to direct the sequential assembly of its encoded amino acids. Every cell contains a pool of the 20 amino acids, each linked by specific enzymatic reactions to the end of a distinct small *transfer RNA (tRNA)* molecule about 80 nucleotides long (see Figure 3). The role of the tRNA moiety of these *aminoacylated tRNAs* is to base-pair with the appropriate codon of the mRNA through the orchestration of an anticodon triplet and, in so doing, to bring its attached amino acid to the correct position within the peptide bond–forming machinery.

Equally important in this complex process is the role of *ribosomes* in providing the mechanisms for positioning of the mRNA codons and the aminoacyl tRNAs in the correct positions for formation of the peptide bonds, releasing the tRNAs and then translocating themselves down the mRNA in a 5′ to 3′ direction. The ribosomes recognize the *initiation codon* located a variable distance downstream of the 5′ end of the mRNA as the position to begin amino acid polymerization and form the amino terminal end of the polypeptide. Each triplet in turn is recognized by the ribosome as the polypeptide grows in length until the ribosome reaches a *stop (nonsense) codon* for which there is no complementary tRNA.

Ribosomes are complex organelles consisting of two different long RNA molecules enwrapped with a series of many small polypeptides to form a *ribonucleoprotein particle (RNP)*. The *ribosomal RNAs (rRNAs)*, the numerous tRNAs, and a series of other *small nuclear RNAs (snRNAs)*, some of which make up organelles that oversee the mRNA spicing reactions, are major types of RNAs transcribed from genes that do not code for proteins.

MISSING THE POINT AND OTHER GENETIC ERRORS

> *God . . . created a number of possibilities in case some of his prototypes failed—that is the meaning of evolution.*[*]

A number of kinds of mutations can affect the polypeptide coding region of a gene. Errors that lead to *substitution* of one base by another, often called

---

*Graham Greene, *Travels With My Aunt*, 1969.

point mutations, alter the nucleic acid sequence of the mRNA and the triplet codon. Since the triplet code contains many synonymous codons, substitutions, especially in the third base position of the codon, often lead to no change in the coding information and are usually referred to as *neutral mutations*. Most substitutions, especially those in the first or second base position of the codon, change the coding information and result in a substitution of an amino acid in the polypeptide product. Such nucleotide changes are examples of *missense* mutations. The impact of missence mutations on the function of the polypeptide is not always predictable, but can range from no measurable effect (equivalent to a neutral mutation) to serious consequences for cell and organism phenotype. One type of substitution mutation can also lead to the creation of a premature termination (stop) codon (TAG, TAA, or TGA), which aborts polypeptide chain elongation. Such *nonsense mutations* lead to the synthesis of an abbreviated peptide that is usually unstable and results in a loss of peptide function. Because the amino acid code is read in frames of sequential nucleotide triplets, *deletion* or *insertion mutations* of one or two bases would immediately cause a *reading frame shift mutation*. The consequences for the fate and function of polypeptides produced from such aberrant mRNAs is usually quite profound, but depend on how much of the amino terminus of the molecule is intact before the garbled amino acids begin toward the carboxy end of the protein. Most frame shift mutations lead to the chance creation of a termination codon a short distance downstream. If the number of bases deleted or inserted is a multiple of 3, then the reading frame will not be disturbed but the effect on the polypeptide will be the *deletion* or *insertion* of one or more amino acids. Similarly, point mutations at the sites of the specific bases involved in ensuring correct splicing out of introns can grossly alter the splicing pattern and lead to mRNAs with residual introns or intron segments. The effects of such *splice-site mutations* and the appearance of additional bases in the midst of the mRNA are variable, ranging from instability and destruction of the mRNA to creation of premature termination codons on the mRNA to insertion of stretches of unpredictable amino acids within the polypeptide.

One corollary of the existence of 1) numerous regulatory nucleotide sequences upstream, downstream, and within genes and 2) the presence of nucleotide sequence-specific signals for the starting and stopping of transcription is that mutations at such sites can also affect the amino acid sequence of the resulting polypeptides as well as their quantity. Hence, such mutations are as likely to affect the phenotype of a cell or organism as are mutations in the coding sequence of a gene. Examples of deleterious mutations in regulatory and splicing signals abound in many organisms including *Homo sapiens*.

The Impact of Mutations on Phenotype: Passive and Aggressive Outcomes

Biologically the species is the accumulation of the experiments of all its successful individuals since the beginning. *

The impact on phenotype by such mutations is not always predictable in people since, for genes located on the *autosomal* (non-sex-linked) chromosomes, each cell has two independent copies of each protein coding gene. In this regard we are a diploid species. In an individual *heterozygous* for a mutation, each cell carries both the mutant and normal (*wild-type*) copy of the gene. The impact of the mutant gene will depend on the role of the protein. For example, if the polypeptide is a complete enzyme, then having a mixture of normal and nonfunctioning enzymes may have limited or no impact on the cell unless the 50% reduction of enzymatic activity is critical. Such mutations are then said to be *recessive*. If, on the other hand, the polypeptide is a component of a larger complex—e.g., a cytoplasmic structure such as the myosin heavy chain that comprises a sarcomeric thick filament—then the presence of large numbers of abnormal molecules mixed into the multimolecular structure will adversely affect the structure and exert a *dominant* effect as turns out to be the case in familial hypertrophic cardiomyopathy (see Chapter 5).

Natural Variations and Linkage

The two copies of a gene are on *homolog* chromosomes and are referred to as *alleles* of each other. These alleles can be identical or can vary from one another because of mutations introduced in preceding generations. Such variations in the nucleotide sequences of genes are common in the population and are referred to as *allelic polymorphisms*. Hundreds of such polymorphisms have been identified in the human population, and each has been assigned with some precision to a specific region of a chromosome. Such polymorphisms may or may not be associated with changes in the phenotype of the organism and may exist as apparently *neutral mutations*; that is, they provide no basis for evolutionary selective advantage, and are detectable only as differences in DNA sequence between chromosomes and between individuals. Many such polymorphisms provide the laboratory basis for determining the mode of inheritance of disease genes in a family. For example, if polymorphism of type A is associated with a known genetic disease and testing reveals that the father's chromosomes have polymorphisms A and B and that the mother's chromosomes have polymorphisms C and D, then it is essentially

---

*H. G. Wells, *The Works of H. G. Wells*, Vol 9, 1925.

always the case that an offspring with polymorphisms A and D has the disease gene and inherited it from the father (Figure 5). Such polymorphisms need not be located directly within the gene but may be located nearby. Since large segments of chromosomes are inherited together, the location of polymorphism A near a gene carrying a disease-causing mutation allows the prediction that inheriting polymorphism A means that the individual has also inherited the disease gene. The polymorphism is said in this case to be **linked** with the disease, and the science of predicting such associations is referred to as **genetic linkage analysis**. By studying the linkage or lack of linkage of many different polymorphisms in the population with the inheritance of a specific disease, it has become possible to find specific linkages. Since the location of the polymorphism on the chromosomal map—its **locus**—is known, recombinant DNA technology makes it possible to isolate (**clone**) defined regions

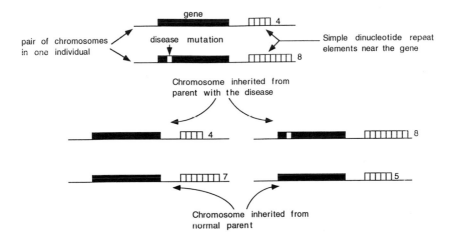

Figure 5 Naturally occurring DNA polymorphisms often serve as markers for inherited diseases. Simple repeating sequences such as dinucleotides (. . . GTGTGT . . .) or trinucleotides (. . . AGGAGGAGG . . .) are scattered throughout the genome with such frequency that they flank most genes. The lengths of these repeats (i.e., 5, 6, . . . 10, . . . 35 copies) at the same location are variable among members of the population but are usually heritably stable. The specific repeat elements are thus linked to the nearby gene and can be used to follow the inheritance of that particular copy of the gene from generation to generation. Thus, if a parent has an element with four repeats on one autosome and an element with eight repeats at the same site on the other autosome, as illustrated in the figure, the children will inherit either the chromosome with four repeats or the one with eight repeats. If the nearby gene carries a mutation that leads to a disease phenotype, then the presence of the linked repeat marker in a descendent also signals the inheritance of the mutated gene.

of the genome that contain the polymorphic marker and, by implication, the disease locus. The approaches used to find the disease gene in such a locus and to identify the mutation that leads to the disease are the subject of a number of the chapters in this volume.

Whereas the effect on phenotype of mutations located on autosomes is not predictable, mutations in genes on the X chromosome can always have a dominant effect on phenotype since in males, by definition, they are single-copy genes. The impact of mutations in the *dystrophin* gene, the X chromosome (*X-linked*) gene responsible for Duchenne muscular dystrophy and the resulting skeletal and cardiac muscle myopathies, is a clear example. Dystrophin is an extraordinarily large polypeptide involved through interactions with other proteins in stabilizing the membranes of skeletal and cardiac muscle cells. Mutations in the dystrophin gene take on many forms, from large deletions in which no mRNA is synthesized, and hence no polypeptide, to small deletions or altered polypeptide sequences. Having half the normal content of dystrophin apparently leads to minimal or no signs of muscle-cell instability since females who carry one normal and one mutant X chromosome are usually clinically unaffected, and normal males have only one copy of the gene. However, males who inherit the mutant chromosome and produce mutant dystrophin or no dystrophin at all are severely affected. Such mutations are *X-linked recessives*.

Some important autosomal protein coding genes are represented in multiple copies (up to several hundred), either scattered among many *chromosomal loci* or in one or a few *tandemly repeated clusters*. Often the proteins encoded by such genes are not strictly similar, and so these genes represent alleles without being homologs. Whether mutation in one or a few copies of such multiple-copy genes can lead to a phenotype depends on whether the products of the individual genes play different roles with respect to cellular function, whether they play the same role but are expressed at different times of development, or whether the presence of mutant polypeptides can interfere with the function of the normal polypeptide.

While we have focused predominantly on the impact of point mutations in DNA, a variety of other mutational mechanisms can also lead to the change of production or synthesis of an abnormal protein. Large-scale deletions of chromosomal segments can occur, chromosomal *translocations* or segmental *inversions* can break a chromosome in the midst of a gene segment, and changes in *gene dosage* accompany duplication or loss of either whole chromosomes (Trisomy 21 or Turner's syndrome—XO—are familiar examples) or a segment of a chromosome. Such cataclysmic karyotypic events, which usually occur during meiosis through *genetic recombination*, can all have effects on the production of abnormal amounts or kinds of proteins.

PROTEINS: "FORM EVER FOLLOWS FUNCTION"*

The same weak molecular interactions that are so critical for the structure of DNA duplexes play an equally critical role in the folding of polypeptide chains into their active configurations and their ability to form more complex structures with multiple polypeptide subunits. The specificity of the formation of functional multipolypeptide complexes is provided by the stereochemical configuration of the folded chains and the creation of interacting subdomains. Such subdomains can interact with a variety of cardiovascular intracellular and extracellular ligands, including atoms (such as sodium, potassium, or calcium ions), neurotransmitters (acetylcholine), hormones (epinephrine), structural proteins (actin), and polynucleotides (RNA polymerase).

Protein Flexibility: Allosteric Changes in Conformation

An important feature of proteins is that their interactions with their ligands can lead to a reversible conformational change in the structure of the folded protein. Such *allosteric* changes can affect binding affinities, such as the effect of hydrogen ions (pH) on the affinity of hemoglobin for oxygen through allosteric changes in the globin moiety. Similarly, such events can lead to the activation of key enzymes, to the opening or closing of membrane channels, or to the ability of regulatory proteins to induce the transcription of key genes.

In a myocardiocyte, for example, at the initiation of the electrical signal for contraction, divalent calcium ions (Ca^{2+}) are released into the cytoplasm from cellular stores and quickly saturate specific calcium binding sites on the troponin C polypeptide. Troponin C is one of three specifically intertwined peptides (the others being troponin I and tropinin T) that make up the troponin complexes lying at equal intervals along the actin thin filaments of the sarcomere. The presence of calcium in specific binding pockets on the surface of troponin C induces a change in its shape (an allosteric change). This change in turn affects the physical contacts between other domains of the same troponin C molecule and the adjacent troponin I protein. Following this gentle nudge, troponin I shifts its position, thus allowing the underlying cardiac α-actin monomers to make contact with the powerful heads of the cardiac myosin heavy-chain molecules. This interaction permits the power stroke that slides the actin filaments toward the center of the myosin bearing thick filaments and mechanically shortens the sarcomere, thus contracting the cell and the myocardium. As calcium is pumped back into its storage pools, the Ca^{2+}

---

*Louis Henry Sullivan, U.S. architect, "The Tall Office Building Artistically Considered," Lippincott's Magazine, March 1896.

molecules leave troponin C and the allosteric process is reversed, preventing the association of actin with myosin and allowing the sarcomere to relax.

Polypeptide Modifications Can Activate or Repress Function

Many polypeptide growth factors—for example, those that trigger smooth-muscle hyperplasia following vascular injury—interact with cell-surface receptors and initiate gene activity or mitosis through well-studied *signal transduction second-messenger pathways*. Alteration of the activity of a large number of intracellular signaling molecules, often by a *phosphorylation* or *dephosphorylation* event, is the mechanism that leads to the nuclear response. Phosphorylation is a common event that induces allosteric transitions in proteins and the signal transduction pathways often make use of this mechanism. The attachment of a phosphate group to an amino acid residue within a protein adds a double negative charge and can effect an allosteric transition by, for example, attracting positively charged neighboring side chains. The primary donor of phosphates in mammalian cells is ATP, and hundreds of related enzymes, the *protein kinases*, function to phosphorylate specific protein substrates while a smaller number of *protein phosphatases* selectively remove phosphate groups. The protein kinases are part of the complex intracellular signaling network in which many of the kinases are themselves activated (or inactivated) after being phosphorylated by a specific kinase.

Complex Folding Patterns

> *Everything is complicated; if that were not so, life and poetry and everything else would be a bore.* *

Proteins, like nucleic acids, are composed of repetitive components—the monomeric amino acids linked through repeating peptide bonds. Like all such repetitive polymers, proteins tend to form **secondary structures** to minimize the free energy of the contacts between adjacent amino acids. The polypeptides themselves tend to form regular hydrogen-bonded structures, predominantly β sheets and α helices (Figure 6A). Most globular proteins contain extensive regions of β-pleated sheets and short stretches of α helices. The distribution of various negatively and positively charged amino acid side chains and of hydrophobic and hydrophilic groups dictated by the **primary sequence** of the polypeptide chain leads to the formation of local domains in proteins. The final, complex, three-dimensional, **ternary** structure of a

---

*Wallace Stevens, *Letters of Wallace Stevens*, no. 336, 1967. *Author's note*: "everything else" includes molecular genetics.

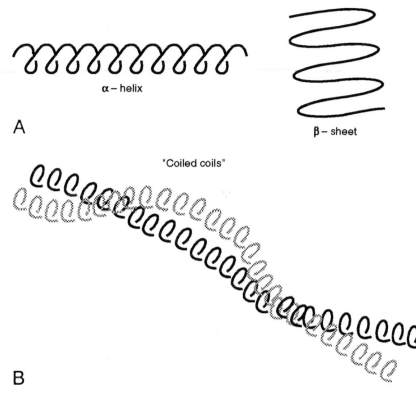

α – helix

β – sheet

A

"Coiled coils"

B

Figure 6 (A) Repetitive polypeptides form regular hydrogen-bonded secondary structures to minimize the free energy of the contacts between adjacent amino acids. (Left) α-helices form when the amio acids are hydrogen-bonded to neighbors in the *same* chain, leading to the formation of a rigid cylinder. (Right) β-sheets form when short stretches of polypeptide fold back and forth in an antiparallel fashion, producing a rigid structure held together by the hydrogen-bonding between amino acids on *adjacent* segments. (B) The final complex three-dimensional, ternary structure of a polypeptide chain is dictated by the folding of domains with each other. Two α-helices often wrap around each other to form coiled coils, a commonly encountered ternary protein structure.

polypeptide chain is dictated by the folding of domains with each other and, often, the formation of covalent links (such as sulphydryl bonds) between side chains (Figure 6B). While a detailed exposition of protein-folding mechanisms and structure is beyond the scope of this section, we stress again that the function of proteins as enzymes, motors, and structural components of cells depends on their shape. Their shape, and hence their function, is dictated by their primary amino acid sequence. And, as we have discussed earlier, mutations in the nucleic acid sequence that lead to an alteration in the shape of the resulting protein can have a major impact on the phenotype of human cells, tissues, and organs. Many of the chapters in this volume focus on the consequences of such mutations in the cardiovascular system.

Proteins usually fold into local domains along the polypeptide chain before the domains interact with each other. Related proteins that carry out similar functions, such as protein kinases or classes of DNA binding proteins, often share similar domains. During evolution, gene duplication events have allowed two genes to diverge in specific function while retaining components (domains) of specific motifs. Often, but not always, exons represent conserved protein domains. The sharing of such motifs can be used to construct a tree of relatedness among proteins (and their gene sequences) and has revealed the presence in the genome of large subsets of *gene families* and sets of families or *superfamilies*. Once a polypeptide is synthesized and forms its ternary structure, it is capable of interacting with other proteins. Many cellular activities are controlled by *multiprotein complexes* made up of many different monomeric protomers such as numerable enzymes (e.g., DNA polymerase), membrane channels (e.g., sodium-potassium pumps), or receptors (e.g., acetylcholine receptor). Some inactive protomers interact with themselves to form active *homodimers* (Figure 7A) or with closely related proteins to form active *heterodimers* (Figure 7B) (e.g., certain transcription-regulating DNA binding proteins). The assembly of subunits represents the *quaternary* level of protein structure.

An important principle, alluded to earlier, is that a mutation in a gene that engenders an altered protomer can have a negative impact on the function of a quaternary complex (Figure 8). In the simpler case of a homodimer, a mutant protomer can effectively disable the function of 75% of the complexes even in a heterozygous individual—all the dimers formed by self-assembly of the mutant protomer (25%) and all those formed in association with the wild-type protomer (50%) are affected. In the case of a complex structure consisting of many copies of the same protein subunit, the random assortment of mutant polypeptides into the structure affects the function of every structure (e.g., a mutant myosin heavy chain would affect the structure and function of every sarcomeric thick filament as in some cases of familial myocardopathy). Such mutations are often referred to as *dominant negative*. In a case in which a protomer functions as a heterodimer or as a single unit of a larger complex, half the complexes carry the mutant protomer and half remain wild-type. The impact of these structures on the phenotype of such heterozygous carrier individuals is usually recessive: minimal or no impact. However, they too can be dominant if the abnormal complex damages the environment of a cell (e.g., keeps an ion channel open) or irreversibly binds a critical substrate.

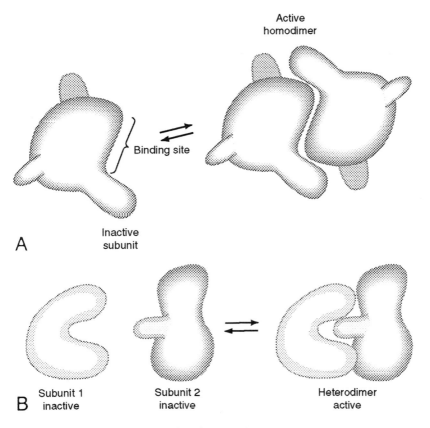

Figure 7 Inactive polypeptide subunits often combine to form protein complexes with enzymatic activities. Inactive subunits have often evolved to form specific dimers (or higher-order multimers) with themselves (A, homodimers) or with other polypeptides (B, heterodimers). Dimer formation may itself be regulated to control enzyme activity.

RNA AND PROTEIN SYNTHESIS: THE NITTY-GRITTY

Those "who wish to know about the world must learn about it in its particular details."

—*Heraclitus*

Transcription: A DNA Simulcast

Proteins are not synthesized directly from their respective genes in the DNA duplex since, in eukaryotic cells, proteins are synthesized in the cytoplasm whereas the coding DNA remains in the nucleus. Instead, the cytoplasmic

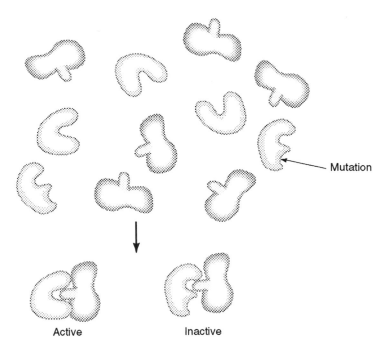

Active   Inactive

Figure 8 Mutations in a subunit can affect the activity of a multiprotein complex. When one copy of a gene is mutated, its protein product may be crippled and interfere with activation of a heterodimer complex. Since the normal gene product is also synthesized, half the complexes will be active. The phenotype of such a mutation depends on the function of the heterodimers.

machinery that assembles new proteins uses an RNA copy of the appropriate section of the DNA to provide the necessary genetic instruction. After some modifications (see below), the RNA travels from nucleus to cytoplasm. In this way, the RNA acts as an intermediary or messenger (mRNA). Production of the final mRNA is a complex process that begins in the nucleus with the synthesis of a colinear RNA representation of the DNA template in the process known as transcription. Transcription is directed by a protein complex that includes the enzyme *RNA polymerase II (pol II)* along with auxiliary transcription factors. *RNA polymerases I* and *III* synthesize other, nonmessenger RNAs that have structural or catalytic functions. The RNA pol II transcription complex attaches to a specific region at the beginning of the gene referred to as the *promoter* (see "How and Why Genes Are Regulated" below for more detail). Once bound to the promoter, the complex employs the specificity of base-pairing to accurately "read" the appropriate DNA strand, incorporating ribonucleotides into a faithful complementary copy of the DNA sequence, except for uracils (U) rather than thymines (T) in RNA. The DNA strand that serves as the template is, by convention, referred to as the *antisense* strand and RNA pol II proceeds along it in a 3′ to 5′ direction. Since RNA–DNA base-pairing dictates the choice of each ribonucleotide incorporated into the growing nascent RNA, the synthesis must occur in an orientation antiparallel from that of the antisense strand. Therefore, the 5′ end of the RNA emerges first and the 3′ end last. When the end of the gene is reached, the transcription complex dissociates, releasing the RNA, which, after modification into its final messenger form (see below for a further explanation of the processing of mRNA), is transported into the cytoplasm where a new set of reactions translate its ribonucleotide sequence into a protein.

RNA Processing: The Final Edition

Messages should be delivered by Western Union.*

As alluded to earlier, the RNA molecules destined to code for proteins initially emerge in a precursor form (or pre-mRNA) following their transcription off of their corresponding gene. Before their transport out of the nucleus and attachment to ribosomes, these initial transcripts must undergo a series of highly regulated modifications. Only after such processing is the transcript a mature mRNA capable of directing translation. One of the first modifications of the nascent RNA transcript is a chemical modification (*cap*) of the 5′ end. The cap consists of a modified guanosine residue linked to the 5′ end of the RNA in an unusual 5′ to 5′ linkage (Figure 9). The cap is thought to render the 5′ end of mRNAs more resistant to RNA degrading enzymes (*ribonucleases, RNases*) and in addition plays a role in the proper positioning of the ribosomes during initiation of translation. At least some of the enzymes necessary for attaching the cap are likely attached to the RNA pol II enzyme since the structure appears almost immediately after the initiation of transcription and the RNAs that are synthesized by either RNA pol I or III (non-mRNAs) lack the cap structure.

The next major processing event of the pre-mRNAs is splicing—the removal of the introns. In order for the mature mRNA to be colinear with the encoded protein, the excision of the introns must be precise. A mistake of even one nucleotide would disrupt the ribosomal reading frame, leading to a dysfunctional protein. The locations of the pre-mRNA cuts are dictated by certain nucleotide sequences at the junctions between introns and exons. These junctional sequences are similar among all an organism's primary RNA

*Samuel Goldwyn, quoted in Arthur Marx, *Goldwyn*, 1976.

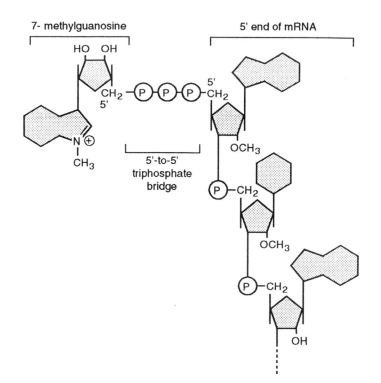

Figure 9 The mRNA cap structure. The cap consists of a 7-methylguanosine residue linked by its 5' carbon to the 5' end of the RNA via a triphosphate bridge. This contrasts with the usual 3' to 5' phosphodiester bonds between all the other adjacent RNA residues. The circled P represents a phosphate group. (From Alberts et al., Molecular Biology of the Cell. New York: Garland Publishing, 1994.)

pol II transcripts and are therefore called *consensus sequences*. At least two such sequences appear at these intron–exon junctions: the *5' splice site* or *donor site* lies at the junction between the 3' end of an exon and the 5' end of the adjacent downstream intron; the *3' splice site* or *acceptor site* lies at the opposite end of the intron where its 3' end joins the 5' end of the next (downstream) exon. After the removal of each intron, the flanking exons are joined reforming a seamless phosphate backbone (Figure 10).

The details of splicing are moderately well understood. The splicing machinery that catalyzes these reactions involves the orchestration of a number of RNA-protein complexes that together form the *spliceosome*. More specifically, each of the major splicesome components consists of a small nuclear RNA (snRNA) complexed with a number of proteins. Since these complexes remain in the nucleus, they are called *small nuclear*

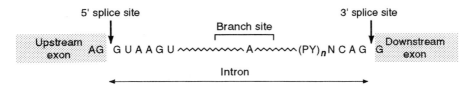

5' splice site

Branch site

3' splice site

Upstream exon AG G U A A G U ∿∿∿∿∿A∿∿∿(PY)$_n$N C A G G Downstream exon

Intron

Figure 10 Splicing consensus sequences. The sequences lying at the intron–exon junctions are similar among all an organism's primary RNA pol II transcripts. The average or consensus sequence appearing at the *5' splice site* lying at the junction between the 3' end of an exon and the 5' end of the adjacent downstream intron is shown. During the splicing reaction, the initial cleavage occurs invariably between the dinucleotides AG and GU, as indicated by the first vertical arrow. The sequence defining the *3' splice site* that marks the border between the intron and the next downstream exon is less exact in its requirements but consists of a string of pyrimidines [(Py)n] followed by any single residue (N), then usually CAG as shown. The 3' splice occurs after the AG dinucleotide in the intron as indicated by the second vertical arrow. The intron sequence defining the *branch point* or *site* (see text and Figure 11) is even less stringent but depends on an A residue 20–50 nucleotides upstream of the 3' splice site. (Modified from Stryer L, Biochemistry, 3rd ed. New York: WH Freeman, 1988.)

ribonucleoprotein particles, or *snRNPs* (pronounced "snurps"). Not unlike the RNA–RNA base-pairing found between the codon and anticodon in translation, the RNA component of some of the major splicesome snRNPs exploit transient base-pairing to align precisely the intron excision sites. For example, the RNA within one spliceosome component (the U1 snRNP) base-pairs with the 5' splice site within the pre-mRNA. Although a complete discussion of the extraordinary molecular events involved in splicing is beyond the scope of this book, the overall reaction mediated by the spliceosome can be summarized as follows (see Figure 11). In the first step, an A nucleotide just upstream of the 3' splice site attacks the first nucleotide of the intron (at the 5' splice site), thereby cleaving it from its upstream neighbor. This first step results in a covalent linkage between the 2'-OH group of the A and the 5' phosphate of the first nucleotide of the intron, creating a lariat-shaped RNA intermediate. The second step in the splicing reaction involves the ligation of the two exons, with simultaneous cleavage at the 3' splice site and release of the intron.

In the usual mode of splicing pre-mRNAs, adjacent exons are ligated with their intervening introns excised. This is known as *constitutive splicing* and a single mRNA results (Figure 12A). However, in many instances of splicing, one or more exons can be skipped as a result of the direct ligation of an upstream exon to another further downstream while jumping over exons in between (Figure 12B). Such aberrations from the conventional splicing are known as *alternative splicing* and can be as simple as this exon-skipping or

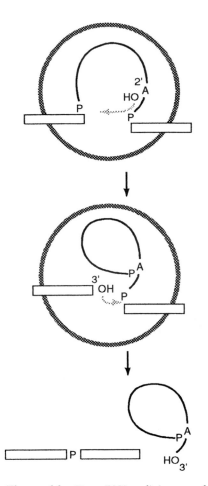

Figure 11 Pre-mRNA splicing can be divided into two steps. In the first, an A nucleotide just upstream of the 3′ splice site attacks the first nucleotide of the intron (top, curved arrow). This nucleophilic attack at the 5′ splice site cleaves the first intron nucleotide from its upstream neighbor and results in an unusual covalent linkage between the A's 2′-OH group and the 5′ end of the intron. Since the A nucleotide also retains its bonds with the nucleotides on its 3′ and 5′ sides, it forms a lariat-shaped intermediate (center). In this lariat, the three concurrent covalent linkages give the A nucleotide its branched structure. The second step in the splicing reaction involves the ligation of the free 3′ end of the upstream exon to the 5′ end of the downstream exon with simultaneous cleavage at the 3′ splice site. This results in release of the intron in its lariat form. Exons are shown as open rectangles and the flanked intron as a thin line. The shaded circle represents the spliceosome complex, which actually consists of a dynamic group of snRNPs and regulating proteins (see text). (Sharp PA, Science 1987; 235:769.)

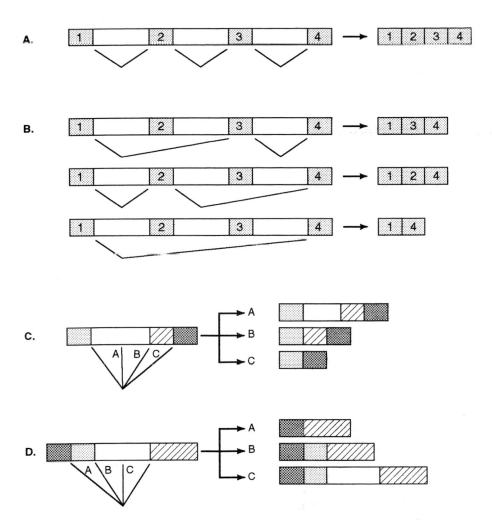

Figure 12   In constitutive splicing (A), all the tandemly arrayed exons (boxes 1 to 4) in a pre-mRNA are spliced directly together in a head-to-tail arrangement, resulting in a single mRNA designated 1-2-3-4. In one form of alternative splicing (B), one or more exons can be removed (or "skipped") along with the introns (open boxes). In the four-exon model shown, exon 2 or 3 or both can be spliced out along with their respective introns, resulting in the three mRNAs: 1-3-4, 1-2-4, or 1-4, respectively. In a variation on this theme, the splicing patterns can often reflect use of alternative 5' or 3' splice sites. In alternative 3' splice site selection (C), a single 5' splice site can be used in combination with any one of three tandemly arrayed 3' splice sites. Alternatively, in alternative 5' splice site selection (D) any of three possible 5' splice sites can be used in combination with a single 3' splice site. In either scenario shown, the number of mRNAs generated (A, B, and C) reflects the number of alternative splice sites utilized. A combination of both 5' and 3' splice site selection can add an even greater complexity to the possible outcomes. In panels C and D, different segments of the pre-mRNA are indicated by striped or shaded boxes for clarity in following the different splicing patterns.

can arise from more subtle changes such as the use of cryptic 5' or 3' splice sites that may normally lie within an exon itself. For example, a single 5' splice site could be ligated to either one of two or more tandemly arrayed 3' splice sites and vice versa (Figure 12C and D). In these latter cases it is clear, then, that the boundaries of exons and introns are not necessarily fixed along the length of a pre-mRNA. In some pre-mRNAs, a discrete set of splicing combinations can give rise to many different exon arrangements. For example, alternative use of exons in a pre-mRNA with four exons and three introns could, in theory, result in four different final mRNAs. In more complicated cases, such as the troponin T pre-mRNA, 64 possible mRNAs can be generated from the use of both constitutive and combinatorial splicing. Control of alternative splicing is usually (if not always) a concerted effort on the part of the cell since the patterns are not random but instead seem well ordered. These patterns can vary with a particular cell type, tissue, or developmental stage. The polypeptide products that arise from alternatively spliced mRNAs define one class of *protein isoforms* —highly related proteins that can differ in their final length, exon composition, and functional capability.

In addition to the modifications of capping and splicing, processing of pre-mRNA (with only few exceptions, such as the histone mRNAs) must include the addition of a polyadenylate tail to the 3' end (Figure 13). The poly(A) tail (100–200 nucleotides long) gives stability to the mRNA, protecting it from degradation at its 3' end, and may also play a role in proper translation initiation. These A residues are added by a specialized enzyme called *poly(A) polymerase* (not RNA pol II since they are not encoded by the DNA template). During transcription, RNA pol II incorporates the consensus sequence AAUAAA downstream from the protein coding region. This hexanucleotide sequence in the primary transcript signals another cellular enzyme to cleave the pre-mRNA approximately 10 to 30 nucleotides downstream, leaving a freshly generated 3'-OH group. It is at this downstream site that poly(A) polymerase adds its string of A residues. At this point the capped, spliced, and polyadenylated mRNA is ready for translation by the ribosomes.

Translation: "The Other Side of a Tapestry"*

Translation, formally the conversion of the information stored in an mRNA sequence into the synthesis of its corresponding protein, requires another large group of molecules including transfer RNAs (tRNAs) and ribosomes. Every cell contains a pool of the 20 amino acids, and a subset of this pool is linked by specific enzymatic reactions to the end of amino acid–specific small tRNA molecules, each about 80 nucleotides long. The function of the tRNAs is to

---

*Attributed to Miguel Cervantes: "Translation is the other side of a tapestry."

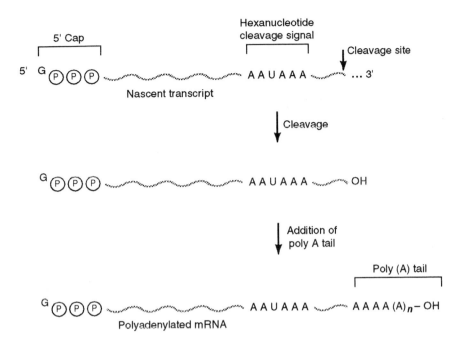

Figure 13 Polyadenylation of mRNA. During transcription, RNA pol II incorporates the consensus sequence AAUAAA downstream from the protein coding region. Then, as with splicing (see Figure 11), polyadenylation of nascent transcripts occurs in two sequential steps. First, a cellular enzyme cleaves the pre-mRNA approximately 10 to 30 nucleotides downstream from the consensus sequence (also called the ***hexanucleotide cleavage signal***). This cleavage leaves a 3'-OH group on the mRNA intermediate. Second, the enzyme ***poly(A)polymerase*** adds a string of 100 to 200 A residues [(A)n] at this downstream site. With this poly(A) tail, a 5' cap and spliced out intron(s), the mRNA is now fully ***processed*** and ready for export to the cytoplasm.

present single amino acids to the protein synthesis apparatus attached to the mRNA.

Each amino acid awaiting incorporation into protein is bound by its carboxy group to the 3' end of a single specific tRNA (Figure 14). A tRNA carrying alanine is designated tRNA^{ala}, and so on. Base-pairing, a theme commonly exploited in the flow of genetic information from DNA to protein, is essential for matching codon to amino acid in translation. This time, however, the complementary nucleotides are in the form of mRNA-tRNA base pairs. The three nucleotides in the tRNA that base-pair with the complementary mRNA codon are referred to as the ***anticodon***. Resolution of the problem of redundancy in the amino acid code results from a moderate degree of

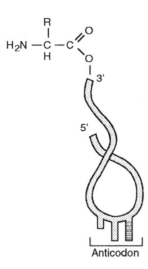

Figure 14 The transfer (t) RNA molecules carry specific amino acids to the site of protein synthesis. At its 3' end, each tRNA is attached to (*charged* with) a specific amino acid in an amino-acyl bond. The three-dimensional structure of tRNAs (shown here schematically as a shaded loop structure) allows three specific internal ribonucleotide residues (*anticodon*) to base-pair with three complementary residues (*codon*) on a mRNA. The first (5'-most) residue of the anticodon (indicated by rectangular blocks protruding from the looped tRNA) has less base-pairing fidelity than the second and third residues and is called the *wobble* base (horizontally striped block). The charged amino acid is shown as a chemical structure with the R designating any amino acid side group.

base-pairing infidelity by the anticodons. Whereas the second and third nucleotides of the tRNA anticodon must perfectly base-pair to the mRNA, the first (or *wobble base*) is often less choosy. In some cases, more than one species of tRNA can specify the same amino acid. Combined, these mechanisms can allow up to six different triplet combinations to code for the same amino acid. The three stop codons (UAA, UAG, and UGA) in the mRNA have no corresponding tRNAs. Instead, other proteins called *release factors* bind directly to stop codons and induce the release of the polypeptide chain from the tRNA and the dissociation of the translational machinery from the mRNA. The portion of the mRNA that lies downstream of the stop codon is the *3' untranslated region* (3' UTR) (Figure 15).

The tightly coordinated synthesis of proteins is catalyzed on ribosomes, which are complexes of RNA and proteins. Each ribosome is composed of one small and one large subunit. The small subunit binds both the tRNA and the mRNA, while the large subunit coordinates the covalent joining of amino

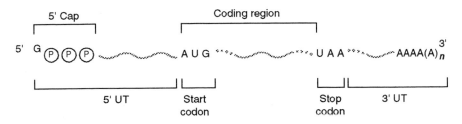

Figure 15 Fully processed mRNA. Mature mRNAs contain a 5′ cap structure (*cap*), followed by a 5′ untranslated region containing noncoding sequence (*5′ UT*), the coding region beginning with the requisite AUG methionine start codon and ending with a stop codon (UAA in this example), and finally a 3′ untranslated region (*3′ UT*) which includes the poly A tail.

acids. In protein synthesis, the polypeptide chain grows as amino acids are added stepwise to the carboxy terminus (Figure 16). This directionality originates from the ability of each successively added amino acid to use its free amino group (the carboxy group is bound to the tRNA) to form a covalent linkage with the carboxy group of the previously incorporated amino acid. In forming this linkage, called a peptide bond, the newest tRNA displaces the previously attached tRNA while transferring the growing peptide chain to itself. The released penultimate tRNA is then free to pick up another amino acid cargo for later use. This cycle of peptide chain growth continues until a stop codon is reached. In the end the polypeptide represents the colinear translation of its mRNA with the amino (N) and carboxy (C) termini corresponding to the 5′ and 3′ ends of the mRNA, respectively.

Since codons consist of three nucleotides each, three potential reading frames exist in an mRNA polynucleotide sequence. As a result, each cellular mRNA could potentially code for three different polypeptides yet only one is synthesized. For example, the mRNA sequence . . . GCACU . . . could be read in a codon frame of . . . **GCA**CU . . . G**CAC**U . . ., or . . . GC**ACU**. . . . Each would give rise to a completely different polypeptide. Similarly, if translation started anywhere within the mRNA, the resultant polypeptide(s) would have varied lengths. This is also not the case: From any one species of mRNAs, uniformity in polypeptide length is the rule. Such restrictions demand that ribosome initiation must be accurate to the exact nucleotide because unregulated frame selection would result in incorrect translations two out of three times. How does the translation machinery start at exactly the correct nucleotide? Although the basis for this precision is only incompletely understood, it clearly relies on a complex set of proteins interacting with a specialized tRNA^{met} (methionine-tRNA). One of the earliest events controlling this accuracy depends on a protein called *eukaryotic initiation factor-2* (eIF-2),

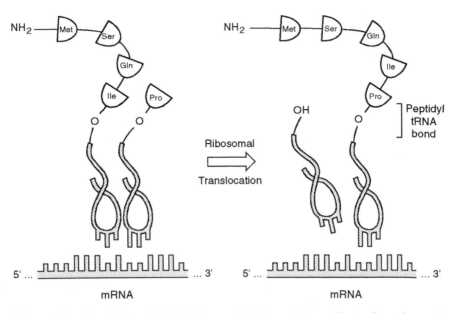

Figure 16 Translation of mRNA into protein. In protein synthesis, the polypeptide chain grows as amino acids are added stepwise to the carboxyl terminus. At left, the 5′-most tRNA has a four-amino-acid peptide chain attached to its 3′ end (by a *peptidyl-tRNA bond*), already having undergone translation initiation with a methionine residue (Met) three codons earlier. For the translation of the mRNA to continue, an incoming tRNA must first base-pair via its anticodon to the immediately adjacent downstream codon (in this case, coding for a proline). The transfer of the growing peptide chain to the newly base-paired tRNA^{Pro} is coordinated by the ribosome (see text) as it proceeds along the mRNA in a 5′ to 3′ direction. At right, the peptidyl-tRNA bond between the tRNA^{Ile} and the peptide chain has been broken and the peptide chain has been transferred to the tRNA^{Pro}. This results in the formation of a new peptide (Pro-Ile) bond and allows the penultimate tRNA^{Ile} (previously anchoring the peptide chain) to float off the mRNA, where it will pick up another amino acid (isoleucine in this case) for later use. The directionality of the growing of the peptide chain is maintained since each successively added amino acid can use only its free amino group (the carboxyl group is bound to the tRNA) to form the covalent linkage with the carboxyl group of the preceding amino acid. The cycle of peptide chain growth continues until a stop codon is reached (see text). As in Figure 14, the anticodons are indicated by rectangular blocks protruding from the looped tRNAs and, similarly, the complementary codons are indicated by rectangles at right angles to the long axis of the mRNA. Note that the short and long rectangles represent pyrimidines (C or U) and purines (A or G), respectively. (Modified from Mathews and Van Holde, Biochemistry. Bowen D, ed. Redwood City, CA: Benjamin-Cummings, 1990.)

which is essential for the binding of the tRNA^{met} to the small ribosomal subunit. One molecule of eIF-2 binds tightly to the initiator tRNA as soon as the latter acquires its methionine. This small ribosomal subunit–tRNA^{met} complex then recognizes and attaches near the cap structure on the 5' end of mature mRNA (see "RNA Processing" above for details). The complex then scans the mRNA beginning from its 5' end until reaching the first AUG codon (the one and only triplet that codes for methionine) called the *initiation*, or *start*, *codon*. Note that the 5' portion of the mRNA upstream of this first AUG is call the *5' untranslated region* (5' UTR). Once the scanning complex reaches the start codon, other initiation factors previously bound to the small ribosomal subunit dissociate, allowing the large ribosomal subunit to bind and translation to proceed as described above. In this way, only one species of polypeptide chain results from each mRNA. When mRNAs code for just one protein they are termed *monocistronic*.

DNA Repair

We are the products of editing, rather than of authorship.[*]

Mutations in DNA sequence can arise either from mistakes in DNA replication or from intracellular metabolic or extracellular environmental influences such as radiation and pollutants. As discussed earlier, if such changes in the DNA sequence were allowed to persist unchecked, the consequences would be disastrous for an organism's survival. Mutated DNA usually leads to a mutant protein that, if essential, would be lethal to the cell and, if present in the germ line (sperm or oöcytes), lethal to the organism's offspring. Thankfully, our cells, like those of most higher organisms, possess a sophisticated system of error correction referred to as *DNA repair*, which, when working properly, keeps the rate of permanent mutation to a mere trickle. For example, it is estimated that thousands of changes to DNA occur in each of our cells (each having 3×10^9 base pairs) every day, yet a germ line cell, for example, retains only 10 to 20 base-pair mutations per year.

DNA repair, in general, exploits the double-stranded nature of its substrate. When one strand of the double helix is damaged, repair enzymes use the information stored in the complementary bases on the opposite strand as a template for repair. The mechanisms behind the system of DNA repair are moderately well understood, with cells usually employing one of two major repair pathways (depending on the nature of the damage) to restore sequence to its native state (Figure 17). The first such pathway, *base excision* repair,

---

[*]George Wald, "The Origin of Optical Activity," Ann NY Acad Sci 1957; 69. (There is actually a rather recently discovered cellular process of RNA editing that does alter the readout of some mRNAs.)

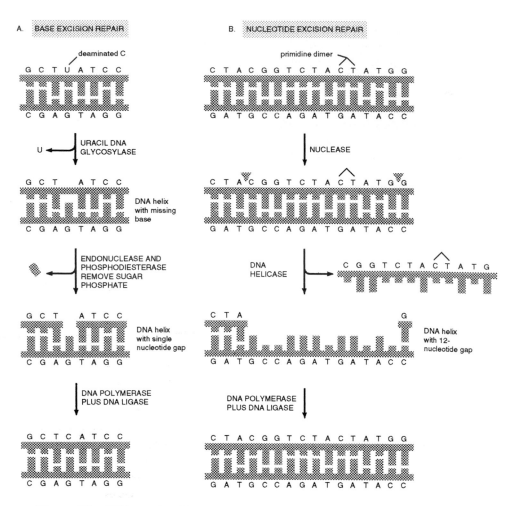

Figure 17 DNA repair. (A) *Base excision* repair involves the recognition of a mismatched base pair by an excision enzyme. For example, when a C is deaminated, converting it to U (deamination is a very common occurrence), yet remains situated opposite the original G, a small irregularity appears in the helix. The enzyme uracil DNA glycosylase then excises the uracil, leaving exposed a segment of the sugar phosphate backbone. A set of repair enzymes consisting of an endonuclease and a phosphodiesterase remove this sugar phosphate segment, leaving a single-stranded gap. DNA polymerase then uses the template strand to repair the gap with the correct complementary base, in this case a C. Finally, a fourth enzyme *DNA ligase* regenerates the continuity of the DNA backbone by reforming the broken phosphodiester bond. (B) *Nucleotide excision* addresses bulky lesions in the DNA that may arise from attachment of large chemical groups to one or more of the bases or inappropriate covalent bonding of adjacent bases on the same strand. The example of such a lesion outlined here is the pyrimidine dimer in which the nitrogen-containing ring structures

corrects base-pair mismatches. Such mismatches can result from a DNA polymerase proofreading error or a chance chemical modification of a base after DNA replication. These mismatches can lead to a small irregularity in the helix, allowing their recognition by an excision enzyme. The enzyme removes the mismatched base, leaving exposed a segment of the sugar phosphate backbone. A set of repair enzymes remove this sugar phosphate segment, leaving a single-stranded gap. DNA polymerase then uses the template strand to repair the gap with the correct complementary base. Finally, a fourth enzyme, **DNA ligase**, regenerates the continuity of the DNA backbone by reforming the broken phosphodiester bond.

The second major repair pathway, **nucleotide excision**, addresses bulky lesions in the DNA, which may arise from attachment of large chemical groups to one or more of the bases or inappropriate covalent bonding of adjacent bases on the same strand. A common example of such a lesion is the pyrimidine dimer in which the nitrogen-containing ring structures of adjacent pyrimidines (T or C) become directly bonded to each other in addition to their normal phosphodiester linkage. Such covalent dimerization can be introduced into DNA from exposure to ultraviolet radiation (such as sunlight). The cell in such cases takes a more radical approach to repair. A series of enzymatic steps leads to removal of a segment of single-stranded DNA (usually a 27- to 29-nucleotide oligomer) containing the damaged area. DNA polymerase then transcribes the unaffected template strand, thereby filling in the gap, and DNA ligase reseals the phosphodiester backbone.

Despite these impressive repair mechanisms, the system is not flawless and mutations do persist, albeit at low levels. However, nature has cleverly engineered even this slight imperfection—without it, natural selection and continued evolution would cease. For individuals who lack the usual safety net provided by the normal DNA repair system, however, particular mutational troubles await, for example, when an individual retains a mutation in a repair enzyme itself. Their inability to repair spontaneously occurring DNA damage during development and—if they survive long enough—during their life almost always results in a premature death, most often from cancers. Other mutations can also

---

of adjacent pyrimidines (in this case a C and a T) become directly bonded to each other. First, a multienzyme complex (**exonuclease**) recognizes the damage, and its nuclease component makes two widely spaced cuts in the phosphodiester backbone on either side of the lesion (indicated by downward solid arrowheads). Next, a DNA helicase unwinds and removes the cut segment containing the damaged DNA (here a 12-nucleotide oligomer). Finally, DNA polymerase transcribes the unaffected template strand, thereby filling in the gap, and DNA ligase reseals the 3' end of the phosphodiester backbone to complete the repair. (From Alberts et al., Molecular Biology of the Cell. New York: Garland Publishing, 1994.)

be missed by the repair system, either through its inherent error rate or by its inability to recognize them as errors. An example of the latter would be an inherited mutation or a spontaneous one that was appropriately base-paired.

HOW AND WHY GENES ARE REGULATED

He that would govern others, first should be the master of himself.[*]

From the outset of this chapter we have stressed that the unique phenotypes of our disparate tissues and organs are based predominantly on differences in the ability of various cell types to synthesize different sets of proteins. Clearly cardiac cells synthesize proteins that order the assembly of sarcomeres, while red blood cell precursors synthesize globin and proteins that form its unique membrane; cardiac cells don't synthesize globin and red cells don't assemble sarcomeres. Since nearly all cells of an individual contain identical sets of chromosomes and genes, the tissue and cell-specific proteins must be promulgated by *tissue-specific patterns of gene expression*. However, many, if not most, of the proteins synthesized by different cell types (and hence the genes expressed) are actually identical and provide, for example, for basic metabolic pathways, for events of the cell cycle, for DNA replication and repair, for protein translation, and for maintenance of normal intracellular ionic environments. The genes for such proteins are often referred to as *housekeeping genes*, and many of them are not regulated in response to environmental signals and express *constitutively*.

Gene and Transcript Complexity Implies Gene-Regulation Mechanisms

Best recent estimates place the total number of human genes at about 60,000. Estimates vary on the numbers of different genes that are expressed (i.e., represented as unique mRNAs or protein molecules) in a single cell type. Combining the results of protein-separation techniques, massive *cDNA* (synthetic DNA complementary to an mRNA) nucleotide-sequencing efforts, and hybridization assays that can estimate the numbers of kinds (complexity) of mRNAs or proteins that turn up has led scientists to conclude that most cell types express about 10,000 different genes. Most of these genes are expressed in many other cell types (housekeeping genes); others are confined to one or a few cell types and the proteins they encode define the novel phenotype of the cell. Not all genes are expressed to the same degree in a given cell. Given that the basic RNA polymerase II transcription machinery is used by all cells and by all active genes in a given cell, there clearly are many layers of regulatory

---

[*]Philip Massinger (1583–1640), English dramatist, *The Bondman*, Act 1, Scene 3, 1624.

control to determine which genes are expressed in which cells and, for a specific cell type, to determine the rate of expression of different genes.

Regulation of Information Flow at Every Step

Regulatory mechanisms have evolved for essentially every step in the flow of information from DNA to RNA to protein. Among the major mechanisms are:

Transcribed genes in a chromosome tend to be in a different configuration than those that are not being transcribed. Chromosomal DNA, even in interphase chromosomes, is mostly tightly packaged as **chromatin** by reiterative coiling into higher-order structures of supercoils entwined with a variety of nuclear proteins, most notably the histones. Chromatin is thus sequestered from access by the transcription machinery. Genes destined for expression tend to be more loosely packaged, thus allowing the formation of transcription complexes.

Even when there are multiple genes in regions of the chromosome that become loosely packaged in chromatin, only one or a few may engage the transcription machinery. This second level of control over gene expression is predominantly regulated by the presence of tissue-specific combinations of nuclear proteins that act as activators required to bolster transcription or repressors that specifically bind to inactivate genes. The workings of such proteins and their interactions with DNA elements are discussed in more detail later.

Mechanisms that control the rates of mRNA elongation, 5' capping, polyadenylation, and splicing can affect the rate at which specific transcribed sequences are transported to the cytoplasm and hence affect the rate of synthesis of the encoded proteins.

The notion that mRNA base sequences intrinsically carry instructions to deliver themselves to specific subcytoplasmic locales remains controversial. However, it is clear that, following the start of translation and while still attached to the ribosomal translation machinery, those mRNAs encoding polypeptides destined for secretion are translocated to the membranes surrounding the Golgi apparatus. The hydrophobic **signal peptide** at the amino terminus of such secreted protein precursors is responsible for this translocation.

mRNAs are degraded at different rates. Some mRNA species undergo random degradation and have a steady half-life. Others can be quite stable and then undergo rapid degradation following some intracellular signal. In very general terms, the number of molecules of a specific mRNA per cell is directly related to the rate of synthesis of the encoded polypeptide. Thus, any difference in the rate of turnover of mRNAs from two genes transcribed at the same rate would be reflected in differences in the rates of synthesis and the concentrations of the proteins. For polypeptides whose concentrations are rate-limiting, such mechanisms can regulate cell phenotype.

Finally, proteins turn over at very different rates. Thus, the phenotype of the cell can be affected by events that destroy or stabilize specific polypeptides.

Tissue-Specific Patterns of Gene Expression in Muscle and Heart

Skeletal and cardiac muscle provide excellent examples of tissues subject to differential gene-expression events. The primary components of the cardiac and skeletal muscle sarcomere comprise thick filament proteins (including myosin heavy chain, the alkali myosin light chains, and the regulatory myosin light chains) and thin filament proteins (including α-actin, α and β tropomyosin, the three proteins of the troponin complex, and tropomodulin).

Several of the contractile proteins are encoded by members of complex muscle gene families, many of which contain separate striated muscle fiber-type isoforms (cardiac or skeletal), smooth-muscle isoforms, and nonmuscle isoforms. For example, there are at least six different human actin isoform genes, including the two striated muscle isoforms, cardiac and skeletal α-actins; two smooth-muscle isoforms, smooth α- and γ-actins; and two nonmuscle isoforms, cytoplasmic β- and γ-actins. During skeletal and cardiac muscle differentiation, muscle-specific isoforms are induced and nonmuscle isoforms are suppressed.

The diversity of other contractile proteins is based on a more limited set of genes, or even single genes, whose alternative protein products are generated by a multiplicity of posttranscriptional splicing events. Still other muscle-specific proteins, including both contractile proteins of the sarcomeres as well as tissue-specific enzymes such as muscle-creatine kinase, are identical in both the heart and skeletal muscle and are the products of the same gene.

Off and On, Genes Are Flipped On and Off

Transcription factors are nuclear proteins that play a pivotal role in regulating the expression of protein coding genes. These are a class of proteins characterized by their ability to activate (or sometimes inhibit) transcription being carried out by the *general transcription complex* that includes RNA polymerase II and its associated proteins (Figure 18). Transcription factors bind to specific elements in the DNA known as enhancers and promoters.

Enhancers are often located at a significant distance from the start of transcription, but as genetic elements they are essentially always linked to the same segment of DNA as the gene being regulated. Accordingly, they are

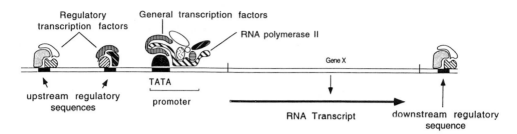

Figure 18   RNA transcription involves many protein complexes and highly regulated polypeptide interactions as well as specific DNA–protein interactions. The figure illustrates a region of a chromosome (a transcription unit) in which a gene (gene X) is being actively transcribed. A common set of general transcription factors, including RNA polymerase II, is present on most active protein coding transcription units and assembles in a sequential manner. In many cases, one of the components directly contacts the DNA at an AT-rich site known as the TATA box and delineates the location of the transcription complex approximately 25 base pairs upstream (5′) of the site of transcription initiation. Additional sites on the DNA interact with other complexes of proteins and can act to enhance or even inhibit transcription.

referred to—along with promoters that are also linked—as *cis regulatory sequences*, as opposed to the factors themselves, which operate in *trans*. Although most enhancers are located upstream of transcription units, they have also been found within genes (in introns) and downstream. This *position-independent* behavior is a major characteristic that defines enhancer elements. Additionally, many such elements are *orientation-independent* as well; that is, reversal of their 5′ to 3′ polarity has no effect on their activity (Figure 19). The reasons that enhancers are orientation-independent is that many consist of short palindromic sequences with twofold symmetry. Such symmetry suggests that the transcription factors themselves have structural symmetry, and this has been borne out by the isolation and structural analysis of many enhancer-binding transcription factors.

Transcription Factors Have DNA Binding and Activation Domains

Enhancer-binding proteins often have distinct polypeptide domains with many of the features of proteins described earlier. Often, these transcription factors are symmetrical homodimers or heterodimers of closely related protomers (Figure 20). The *DNA binding domain* is often an α-helix with positively charged amino acid residues that fit nicely into the major groove of the double helix at nearly a right angle to the long axis of the DNA. Many factors have a *dimerization domain* that allows the factor to become a dimer

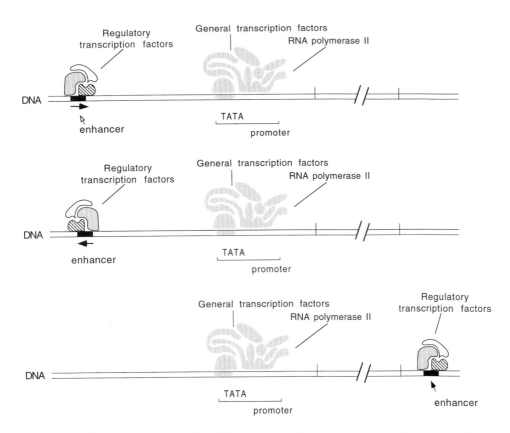

Figure 19   Enhancer sites that bind transcription regulatory proteins or protein complexes can be located close to or at great distances from the transcription initiation site. The figure illustrates the observation that the same regulatory complexes can participate in the activation of transcription of different genes. Furthermore, the enhancer sites can be located either upstream or downstream of the gene, and the 5′-3′ polarity of the enhancer usually does not affect its activity.

such that the two helices can insert symmetrically into the major grooves on opposite sides of the DNA.

A third domain, often comprising many positively charged amino acid residues, acts as a *transactivator* that interacts with the polymerase transcription complex or with other transcription factors, or both. Enhancer elements and their bound factors do not normally act alone without the RNA polymerase complex, which is bound at the *basal promoter* site. Like other activating proteins, transactivating factors often act as the final transducers of signals from the cell environment. Some transcription factors, such as myocyte

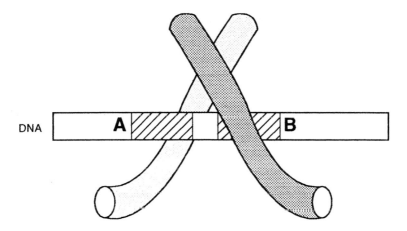

Figure 20 Many enhancers are symmetrical DNA sites that interact with the DNA binding domains of a symmetrical protein or pair of proteins.

enhancer factor 2 (**MEF-2**) and myogenic regulatory factors (**MRFs**) (discussed later), are phosphorylated by cellular kinases that cause allosteric modifications that enable DNA binding or activation. Other transcription factors are nuclear hormone receptors. For example, when thyroid hormone binds to its receptor, the receptor can then bind to a *thyroid-responsive element* (TRE) in the enhancer of cardiac muscle myosin heavy-chain genes and greatly modulate their expression. Since the enhancer elements and their binding proteins are often far upstream of the basal promoter (hundreds to thousands of base pairs away) and since the transcription factors appear to make contact with other proteins located at or near the RNA polymerase complex, it is believed that the intervening DNA between the enhancer and basal promoter loops out so that these sets of proteins can make contact and initiate transcription (Figure 21).

The thyroid hormone receptor and the MRFs are two sets of transcription factor families that act as dimers and exemplify the fact that most transcription factors can be classified on the basis of their structural motifs. The thyroid hormone receptor is a typical member of the class of *zinc finger proteins*, which also includes steroid, vitamin D, and retinoic acid receptors (Figure 22, top). These ligand-activated proteins have either two pairs of cysteines or a pair of cysteines and a pair of histidines clutching a zinc ion with the polypeptide between these pairs looping out into a finger shaped appendage. The fingers are usually tandemly repeated three or more times. The MRFs are characterized by a *basic helix-loop-helix* (bHLH) domain (Figure 22, bottom). The four-member MRF multigene family includes MyoD, myogenin, MRF4, and myf5,

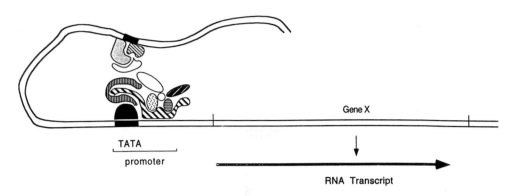

Figure 21 Looping out of DNA may explain how enhancer binding complexes located at great distances from the transcription initiation site can interact with and regulate the activity of the general transcription factors.

which share a nearly identical bHLH polypeptide domain. This region of the protein consists of a cluster of positively charged amino acids immediately adjacent to two α-helical stretches separated by a loop. The HLH segments are responsible for forming the heterodimers with a closely related but non-cell-specific bHLH protein. The basic regions of the two polypeptide chains contact the DNA on opposite sides of the short recognition sequence $\frac{CANNTG}{GTNNAC}$, often called an E-box. Other motifs such as *helix-turn-helix* or *leucine zippers* (repeating leucine residues that can allow inter- or intramolecular attractions) characterize other important transcription-factor families.

The discovery of transcription factors in eukaryotic cells has not completely explained tissue-specific patterns of gene expression because most transcription factors are active in many different cell types. Only a few are highly restricted (see Chapter 16). For example, many recent observations point to the myogenic regulatory factors as primary activators in establishing and maintaining the myogenic program in skeletal muscle cells. The lack of the MRFs in the heart might be the cause of the differences observed in the patterns of expression of muscle-specific genes in heart and skeletal muscle cells. However, some genes, such as cardiac α-actin, depend on an intact MRF binding site for their transcription in both heart and skeletal muscle cells. Most likely, the combination of multiple transcription factors and mechanisms that make specific regions of chromatin accessible to transcription machinery are in large measure responsible for establishing tissue-specific patterns of gene expression.

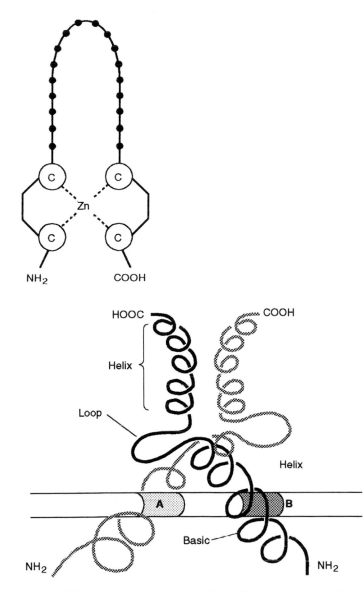

Figure 22 Schematic representations of two important classes of transcription regulatory proteins. (Top) The zinc-finger class of proteins includes many transcription factors. Here, four cysteine residues form a zinc-binding pocket and generate a finger from the intervening polypeptide (filled circles). Several fingers in tandem are common and can interact directly with specific sites on the DNA. (Bottom) Basic helix-loop-helix proteins can form symmetrical homo- or heterodimers and interact with symmetrical DNA sequences.

MASTERING NATURE: RECOMBINANT DNA MOLECULES

*Man masters nature not by force but by understanding. This is why science has succeeded where magic failed: because it has looked for no spell to cast over nature.**

Just as the cell continually exploits the complementary nature of nucleic acids and the uniqueness of their sequence, so do molecular biologists in their attempts to manipulate genes in vitro. Base-pairing provides the basis for piecing together or *recombining* new segments of DNA, and sequence specificity of specialized enzymes, namely *restriction endonucleases*, allows stretches of DNA to be cut at precise locations. Other enzymes allow pieces of DNA to be pieced back together (*ligases*) or degraded on their ends (*DNA exonucleases*) or throughout their length (*DNA endonucleases*). Numerous other enzymes are also available that can chemically modify pieces of DNA, allowing them to be followed in different biological experiments or stages of purification. Finally, precisely defined DNA and RNA molecules can be synthesized in a test tube employing purified enzymes. Taken together, these procedures of genetic engineering, combined with the ability to place the engineered DNA molecules back into bacterial or animal cells, are known as *recombinant DNA technology*.

Restriction Enzymes

Restriction-modification endonucleases, better known as restriction enzymes, are a class of bacterially derived enzymes that recognize specific sequences of DNA and then cleave both strands of the phosphodiester backbone at or at a fixed distance from that site. These cuts can be either straight across both strands, giving a *blunt* end, or asymmetrical, leaving short overhanging regions of single-stranded DNA (Figure 23). Since the latter will readily base-pair with other complementary regions of single-stranded DNA, they are often referred to as *sticky* ends. The stretch of sequence recognized by a restriction enzyme (a *recognition site*) is usually four or six base pairs long but can be as long as eight. It follows from probability alone that the frequency of a four-nucleotide site appearing in a random stretch of DNA will be much greater than one of six nucleotides. An eight-nucleotide site would occur at an even lower frequency. As a result of these differences, restriction enzymes with four nucleotide sites are *frequent cutters* and those with eight nucleotide sites *rare cutters*. Since there are four base choices at each nucleotide

---

*Jacob Bronowski (1908–1974), British scientist, author. "The Creative Mind," lecture delivered Feb. 26, 1953, at the Massachusetts Institute of Technology.

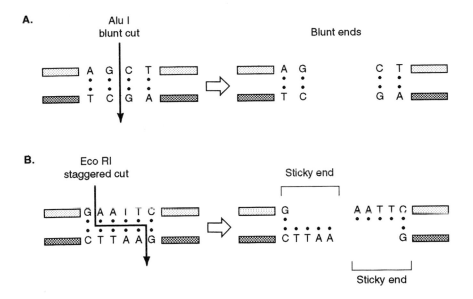

Figure 23 (A) Blunt cleavage (or cut) of DNA at a restriction site leaves two resulting ends that remain double-stranded. The restriction endonuclease Alu I (a four-cutter) shown here recognizes the sequence AGCT in double-stranded DNA. Whenever this sequence occurs in the presence of this enzyme, a blunt cut will result. (B) Staggered cuts of DNA at a restriction site leave two ends, each with a single-stranded overhang. The restriction endonuclease Eco RI (a six-cutter) recognizes the sequence GAATTC in double-stranded DNA. Whenever this sequence occurs in the presence of this enzyme, a staggered cut with a four-base overhang on the 5' end results. (Note that other enzymes can leave 3' overhangs.) Since these overhangs readily base-pair with complementary bases, they are called "sticky ends." With either blunt- or staggered-cut DNA, the ends can be covalently joined (*ligated*) by any member of a set of enzymes known as *DNA ligases*. Although one blunt end can ligate to another without specific sequence requirements, sticky ends must find partners with complementary overhanging ends for the joining reaction to occur. For example, in the ligation of an Eco RI generated end, the adjoining DNA candidate must include a single-stranded overhang with the four requisite complementary nucleotides. Such an overhang can be generated either by another Eco RI digested end or by another restriction enzyme that leaves the same complementary overhang. (Such compatible ends can still result even when two enzymes' exact recognition sequences differ, as long as they leave the identical overhangs.)

position, an eight-nucleotide site will arise 256 (4^4) times less often than a four-nucleotide site. Therefore, in general, the number of fragments generated when DNA undergoes restriction enzyme digestion depends on the class of restriction enzyme employed.

A universal characteristic of restriction endonuclease recognition sequences, regardless of length, is that the double-stranded sequence has a central axis of symmetry such that if the segment were rotated 180° about this axis the sequence and its orientation would be identical to the starting sequence. The best way to conceptualize this property is through an example. As seen in Figure 24, after rotation 180° around the axis coming out of the page, the sequence is identical when read from left to right on the top strand or from right to left (the same polarity) on the bottom strand—it is GAATTC either way.

After restriction digestion, DNA fragments can be rejoined in new combinations with the help of the enzyme ligase. Such ligations are possible as long as the ends are compatible, regardless of the origin of the DNA (even between bacterial and human DNA). Blunt to blunt ends as well as sticky overhanging ends can be ligated. These technological achievements that allow DNA segments from any source to be pieced together in new combinations lie at the core of recombinant DNA technology.

In addition to their obvious use in creating more easily managed fragment sizes from long stretches of DNA (or even chromosomes), restriction

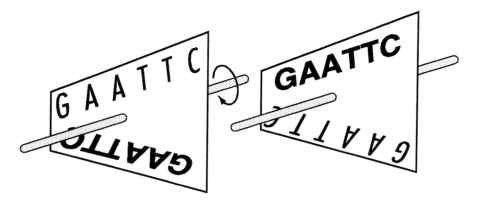

Figure 24   The restriction recognition site of the Eco RI restriction endonuclease. Restriction endonuclease sites have a central axis of symmetry; here, the hexanucleotide recognition sequence GAATTC is identical to its complementary bases when both are read in a 5′ to 3′ direction. Although the specific recognition sequences differ from one restriction endonuclease to another, they all share this type of symmetry.

enzymes are invaluable in providing rough maps of genomes ranging from viruses to humans. The unique pattern of restriction sites for a specific restriction enzyme or set of enzymes essentially generates a unique map of the DNA from which to navigate to a sequence of interest.

Gel Electrophoresis Separates Molecules by Size

DNA with its negative charge evenly distributed along its length can migrate through semisolid matrices under the influence of electrical current. DNA fragments of different lengths (different molecular mass) can be separated in such systems, in which the matrix is usually a high-water-content gel made up of polymerized plastics (polyacrylamides) or agar derivatives (agarose). In such *gels*, small fragments migrate faster through the polymer matrix than larger fragments and the distance traveled after a given time of current application is proportional to the length (size) of the DNA. The samples of dissolved DNA are placed in wells on one end of the gel (Figure 25), electrodes are attached to the ends of the gel, and the negatively charged DNA migrates toward the (positively charged) cathode placed on the opposite end. When samples of unknown size are run in parallel with samples of DNA of known lengths, the sizes of the unknown fragments can be determined. The DNA itself can be visualized with these gel systems in a number of ways (see the section below on DNA cloning), but one of the most common techniques takes advantage of the property of the dye ethidium bromide to intercalate (insert) between sequential base pairs in the double helix. When the ethidium bromide–stained DNA fragments, still in the gel, are exposed to ultraviolet light, they fluoresce as bands with an intensity proportional to the mass of DNA present. In this way, gel electrophoresis can separate DNA fragments based on size and also provide an estimate (when compared to standards) of the amount of each fragment originally in the sample. Finally, separated DNA fragments can be isolated by physically cutting the individual, visualized bands out of the gel.

Identifying Nucleic Acid Sequences with Labeled Hybridization Probes

The property of nucleic acid to form double-stranded structures through base-pairing (*hybridization*) is essential for many techniques employed in molecular biology and genetics. When heated to a critical temperature, the hydrogen bonds holding together the base pairs break and the two strands separate in a process called *unwinding*, *melting*, or *denaturation* (Figure 26). Upon cooling, the complementary strands will realign and reform their original double helix (*reannealing*). Multiple cycles of melting and reannealing of double-stranded DNA can be thought of like the opening and closing

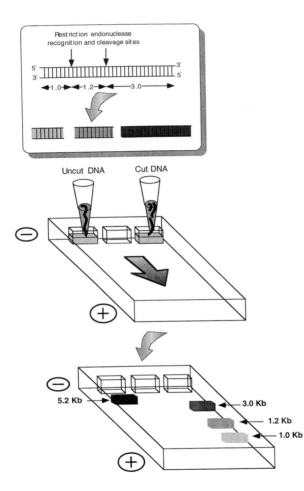

Figure 25   Separation of DNA restriction fragments by agarose gel electrophoresis. A fragment of DNA is cut (digested) by restriction endonuclease digestion (top). Dissolved samples of uncut or cut DNA are placed in either of two wells on one end of a horizontal agarose gel (center). Electrodes are then attached to the ends of the gel. The negatively charged DNA migrates toward the (positively charged) cathode placed on the opposite end. Because the agarose gel acts as a type of molecular sieve, it follows that the smallest fragments migrate the fastest and the largest fragments the slowest. In this way, a given time after the current has been applied (electrophoresis), the smallest fragments will have migrated the furthest from the well and the largest fragments the least (bottom). The distance migrated is a direct function of the DNA fragments' lengths. Therefore, when samples of unknown size are run in parallel with samples of DNA of known lengths, the sizes of the unknown fragments can be determined (for simplicity, size standards are not shown here). The DNA itself can be visualized by staining the gel with the dye ethidium bromide. Only the DNA fragments

of a zipper—each time the closed or duplexed product is faithfully regenerated. In the case of DNA, the aligned duplex represents its lowest-energy conformation (maximizing the number of base-pairing hydrogen bonds). Similarly, single-stranded RNA can hybridize to complementary segments of DNA under appropriate conditions.

If radioactive nucleotides or chemically reactive groups are incorporated into the DNA (or RNA) molecules, they can be used to detect minute amounts of unlabeled nucleic acid as vanishingly small as single genes on a chromosome. Isolated DNA fragments so labeled are often referred to as *probes* since they can be used to search, within a complex mixture, for another related piece of DNA or RNA with which it can base-pair (or hybridize). One of the more common techniques to label DNA includes the addition of trace amounts of a radioactively labeled DNA precursor (e.g., ^{32}P-dCTP)* into a DNA synthesis reaction (see below for details of in vitro and bacterial DNA synthesis techniques). In this way, a fraction of all the C residues along the newly generated DNA will be radioactive. Alternatively, specialized DNA-modifying enzymes can label the ends of linear DNA fragments with single radioactive groups. Analogous reactions with chemically active groups such as chemiluminescent compounds can also label DNA. This latter method can involve either direct coupling of such groups to the DNA fragment or indirect coupling by addition of unique chemical groups to the DNA that can then be specifically recognized by a chemiluminescent antibody. Highly labeled probes can also be made from in vitro synthesized RNA.

The Base Sequence of DNA Fragments Can Be Easily and Rapidly Determined

In 1977, Allan Maxam and Walter Gilbert devised a chemical cleavage technique to determine the nucleotide sequence of DNA. However, an enzymatic

---

*The nucleotide cytosine triphosphate (i.e., cytidine), for example, in which the phosphate group closest to the base is radioactive. The phosphate groups are designated as α, β, and γ, with α being closest to the base. Thus, ^{32}P-dCTP is actually an α-^{32}P cytidine.

---

in the gel pick up the dye. When the ethidium bromide–stained DNA fragments are exposed to ultraviolet light, they fluoresce as bands with an intensity proportional to the mass of DNA present. In this way, gel electrophoresis not only can separate DNA fragments based on size but also provides an estimate (when compared to standards) of the amount of each fragment originally in the sample. In this example, the restriction digest resulted in three bands, each migrating a distance consistent with lengths of 1, 1.2, or 3 kb. In addition, the uncut DNA migrated the least, consistent with its 5.2 kb length—the sum of the three digested fragments.

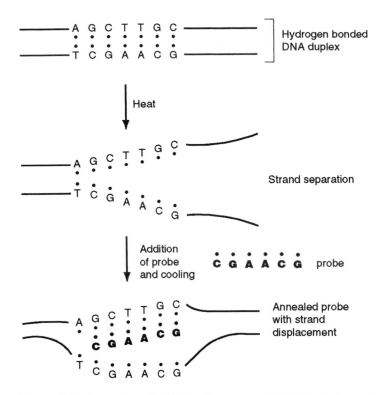

Figure 26 Annealing of a DNA probe to target DNA. DNA oligonucleotides or larger fragments can be labeled with either chemically or radioactively tagged modifications and then annealed to target DNA in order to search out regions of complementarity. In this example, the hydrogen-bonded DNA duplex (top line) is first heated to allow strand separation. Excess amount of labeled DNA probe (bold type) is then added to the strand-separated target DNA. As the reaction mixture cools, favoring reannealing of DNA strands, the probe searches out any complementary regions of the target with which it can base-pair (**bybridize**). Since the probe is added in great excess, the complementary target DNA strand will tend to hybridize with the probe out of shear probability rather than with its original complementary partner. This latter strand therefore becomes displaced as shown at the bottom of the figure. In contrast, the vast majority of the target DNA do not form good base pairs with the probe and therefore reform double strands with their original partners. After the reannealing step, the excess unhybridized probe is washed away. This technique forms the basis for nucleic acid detection in Northern and Southern analyses of cloning (see text and Figure 32). The dots between the nucleotides represent hydrogen bonding.

method devised by Frederick Sanger is currently the method of choice in most circumstances. Gilbert and Sanger shared a 1980 Nobel Prize for their work. The prize was Sanger's second.* The Sanger technique depends on the enzymatic incorporation of a small fraction of modified nucleotides along with the normal deoxynucleotides during the in vitro synthesis of DNA off its template strand. These modified precursors, called dideoxynucleotides (ddNTPs), essentially create dead ends on their 3' side. They can be incorporated into the growing DNA strand because they have a normal 5' triphosphate; however, once incorporated, no further nucleotides can be added because they lack a 3' hydoxyl group (Figure 27A). This is the reactive group that is normally required for the phosphodiester linkage with the next nucleotide. There are ddNTP analogs—ddATP, ddCTP, ddGTP, and ddTTP— corresponding to each normal dNTP.

The enzymatic sequencing involves annealing a labeled *primer* (a short, single-stranded DNA molecule complementary to the template strand) to the 3' end of the DNA fragment to be sequenced. The mixture is then divided into four equal portions. To each portion is added a mixture of the four dNTPs plus a much smaller amount of one of the four ddNTPs, along with DNA polymerase, which begins to synthesize a new DNA strand by extending the primer in a 5' to 3' direction. When by chance a ddNTP is added to the growing chain, that chain stops short. The result in each of the four reactions is a set of newly synthesized strands, varying in their length depending on where the polymers incorporated a dead-end ddNTP (Figure 27B). Optimized concentrations of the ddNTP leads to a family of fragments, each with identical, labeled 5' ends but varying in the extension of their 3' ends. In most instances, this family of fragments will represent a stop every place at which the ddNTP can be incorporated downstream from the primer. For example, in an area where the template sequence was . . . CAGCTAGCTG . . ., a reaction containing ddGTP (represented as **ddG**) would generate three newly synthesized fragments three, seven, and 10 nucleotides long on the opposite strand: CA**ddG**, CAGCTA**ddG**, and CAGCTAGCT**ddG**. The labeled fragments are then separated by their size (and therefore rate of migration) using polyacrylamide gel electrophoresis (PAGE). Similarly, three additional reactions with the identical primer but with either ddA, ddT, or ddC will each generate a distinct set of fragments. The four ddNTP reactions are run separately in adjacent lanes on the gel. Finally, when the gel is exposed to highly sensitive X-ray film, the pattern of labeled DNA fragments appears as dark bands in the *autoradiograph*. Four patterns emerge on the autoradiograph that, taken together,

---

*His first Nobel Prize was in 1958 for discovering the amino acid sequence of insulin and for his contributions to approaches to sequencing polypeptides.

generate a colinear representation of all the nucleotides downstream from the sequencing primer.

Identification of Specific DNA and RNA Segments Using Specific Probes

It is often useful to identify a particular sequence within a large collection of nucleic acid molecules, even a mixture as complex as the entire human genome. This can be accomplished by using the techniques of *Southern* and

Figure 27 Sanger sequencing of DNA using dideoxynucleotides. (A) In dideoxynucleotides (ddNTPs), a 3′ H replaces the 3′ OH group present in their deoxynucleotide counterparts (see shaded elipses). The ddNTPs can be incorporated into the growing DNA strand because they have a normal 5′ triphosphate; however, once incorporated, no further nucleotides can be added because they lack a 3′ hydroxyl group. This is the reactive group that is required for further extension of a DNA strand during synthesis. There are ddNTP analogs—ddATP, ddCTP, ddGTP, and ddTTP—corresponding to each normal dNTP. (B) Enzymatic dideoxynucleotide sequencing involves first denaturing the DNA to be sequenced and then annealing a labeled *primer* to the 3′ end of the DNA fragment to be sequenced. The mixture is then divided into four equal portions. To each portion is added a mixture of the four dNTPs plus a much smaller amount of one of the four ddNTPs, along with DNA polymerase, which begins to synthesize a new DNA strand by extending the primer in a 5′ to 3′ direction. When by chance a ddNTP is added to the growing chain, that chain stops short. The result in each of the four reactions is a set of newly synthesized strands varying in their length depending on where the polymers incorporated a dead-end ddNTP. The molar ratio of the specific ddNTP to its dNTP counterpart in the reaction determines the probability of an early chain termination. The higher this ratio, the sooner the DNA polymerase will incorporate a ddNTP into the strand and the shorter the strands will tend to be. Optimized concentrations of the ddNTP lead to a family of fragments (100 to 500 nucleotides long), each with identical, labeled 5′ ends but varying in the extension of their 3′ ends. In most instances, this family of fragments will represent a stop every place where the ddNTP can be incorporated downstream from the primer. The labeled fragments are then separated by their size (and therefore rate of migration) using polyacrylamide gel electrophoresis (PAGE). The four ddNTP reactions (ddA, ddT, ddC, and ddG) are run separately in adjacent lanes on the gel. Finally, when the gel is exposed to highly sensitive X-ray film, the labeled DNA fragments appear as dark bands in the *autoradiograph*. The four patterns that emerge on the autoradiograph, taken together, generate a colinear representation of all the nucleotides downstream from the sequencing primer. The sequence can be determined in its 5′ to 3′ direction by noting the ddNTP lane assignment of each successive fragment as one "reads" the autoradiograph bands from the shortest (bottom) to the longest (top). (B, redrawn from Alberts et al., Molecular Biology of the Cell. New York: Garland Publishing, 1994.)

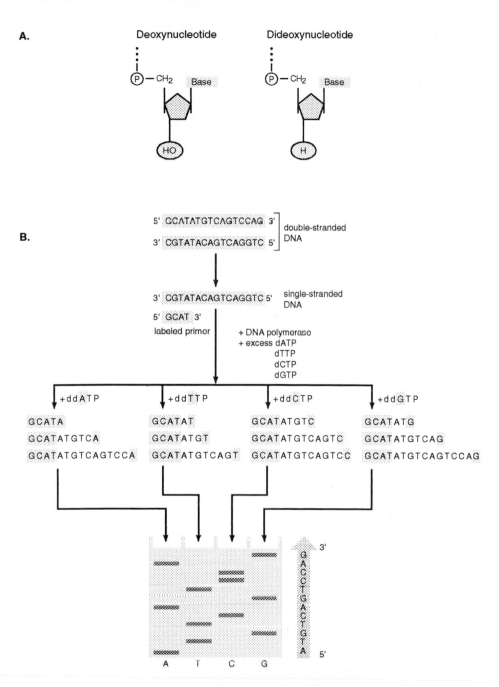

Northern analysis for DNA and RNA, respectively. Like sequencing and DNA cloning (discussed further below), the molecular basis of these techniques is once again the complementary base-pairing of nucleotides. Since the principles are nearly identical in both techniques, they will be described together. In Southern blotting (named after its inventor* and for the use of blotting paper to facilitate DNA transfer—see below), restriction enzymes are used to digest large DNA molecules, or even entire genomes, into smaller, more manageable fragments. The digested DNA is then run on agarose gels to separate the fragments by size. A thin DNA-binding membrane (previously nitrocellulose but now usually nylon) is then placed over the gel. Through a wicking process, usually using weighted blotting paper on top of the membrane, the DNA fragments migrate out of the gel at right angles to the direction of electrophoresis, and onto the membrane. In this way the fragments of DNA maintain their relative positions unchanged. Ultraviolet light then permanently bonds the nucleic acid to the membrane. After strong basic conditions permanently denature the bound DNA, the membrane is exposed to excess amounts of a strand-separated (or single-stranded) DNA probe that is labeled radioactively or chemically (an RNA probe can also be used). Free to diffuse over the membrane, the probe will hybridize selectively to the bound fragments of DNA with which it can base-pair. The better the complementarity between probe and target DNA, the stronger the hybridization. After removal of unbound probe, autoradiography (or the equivalent for a nonradioactive probe) with the membrane demonstrates whether the original DNA sample contained the sequence of interest and, if so, the size of the hybridizing restriction fragment. Similarly, in Northern analysis, a pool of RNAs can be separated by size on agarose gels, transferred to membranes, and searched for selected sequences with either denatured DNA or single-stranded DNA or RNA probes. The rest of the analysis is analogous to that employed for Southern blots. It is also worth mentioning at this juncture that if the hybridization Northern analysis conditions are relaxed to allow one or more base-pair *mismatches* between the probe and the immobilized DNA, then related sequences will also be identified on the autoradiograph. This lower stringency would be useful to search for different members of a multigene family or homologous genes in different species.

In certain cases, the size of the target DNA or RNA is either not important or already known. Here, simplified Southern and Northern analyses can be performed by applying the nucleic acid directly to the membrane rather than

---

*Many think *Southern blotting* derives from a regional discovery, like Southern-fried chicken. That is not the case. The technique was developed by Ed Southern, then at the University of Edinburgh. It was only later that variations on the technique—e.g., Western, Southwestern—were named after points of the compass.

first separating them based on size with gel electrophoresis. The samples can be applied freehand onto the membrane, but more often, in the interest of better quantitation (or just aesthetics!), investigators prefer to apply the sample to a mechanically delimited surface area. Different methods exist for doing this, but they usually employ a *slot* or *dot blot* apparatus (Figure 28). Each uses a tightly clamped rectangular or circular gasket, respectively, over the membrane, providing a well-circumscribed area on which the sample is applied. The remaining steps in these hybridization techniques are identical to those for the conventional blots described above.

The ability of DNA and RNA to hybridize to complementary nucleic acids with very high degrees of specificity also allows these probes to detect their target in fixed cells or tissue sections (*in situ hybridization*). Hence, conditions are used to denature the cellular nucleic acids while trying to maintain cellular architecture. The dissolved probe is then incubated in solution over the slide containing the fixed cells or tissue section. Upon removal of the unbound probe, the specific subcellular or chromosomal location of the target sequences can be determined by autoradiography or by direct microscopy if the probe is fluorescent [*FISH* (fluorescent in situ hybridization)].

The concept of molecular transfer from gel to membrane is also exploited in a protein technique related to Southern and Northern blotting: a *Western blot*. Rather than nucleic acids, this method first uses PAGE to size-separate

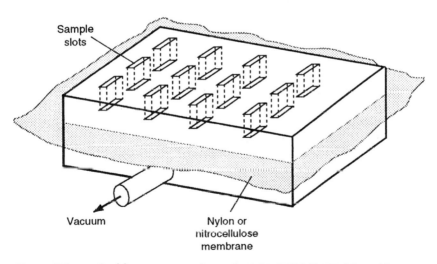

Sample slots

Vacuum

Nylon or nitrocellulose membrane

Figure 28 A slot-blot apparatus for collecting multiple nucleic acid or protein samples on retentive membranes. After drying, the membrane can be wetted with a solution containing specific labeled probes, washed, and analyzed.

proteins and then a second electric field (rather than capillary flow) at right angles to the first to transfer them to a membrane. Finally, a solution of antibodies (rather than nucleic acids) specific for the desired protein or peptide act as probes to identify candidate peptides on the membrane. Only those protein bands specifically bound by the antibody give positive signals in a colorimetric, chemiluminescent, or radioactive assay for adherent antibody.

DNA CLONING AND OTHER APPLICATIONS OF MOLECULAR BIOLOGY

*There does not exist a category of science to which one can give the name applied science. There are science and the applications of science, bound together as the fruit of the tree which bears it.**

Vectors

When DNA fragments are ligated together in novel combinations, they are called recombinant molecules, or *chimeras*, after the serpent she-goat beast of Greek mythology.[†] Furthermore, segments of DNA from any organism can be greatly amplified if inserted into certain DNA molecules (*vectors*) capable of carrying the foreign DNA into a host cell (usually bacteria) (Figure 29). To do this, a restriction fragment of the desired DNA is inserted into a vector that is digested with the same restriction enzyme or one that generates compatible ends (see the section above describing restriction enzymes). The vector is then inserted into bacteria through a process called *transformation*, which includes briefly heating the bacteria to promote their natural tendency to take up extracellular DNA. Within the bacteria (most commonly, specialized strains of *Eschericia coli*), the bacterial DNA polymerase replicates the vector into hundreds of identical copies (or *clones*) per cell. As the cells divide, they carry the replicating vectors with them such that all the cells in the culture derived from a single cell are chock-full of replicating vectors carrying the inserted DNA. The most common type of vector used to replicate and amplify a segment of DNA is a *plasmid*. Plasmids are circular DNA molecules that act like miniature chromosomes in the bacteria. They have a site, an *origin of*

---

*Louis Pasteur, in Revue Scientifique (Paris, 1871).
[†]Chimera: an imaginary monster made up of grotesquely disparate parts. In Greek mythology, the Chimera was a fire-breathing she-monster usually represented as a composite of a lion, goat, and serpent. Perhaps as an early allegorical warning to molecular geneticists, after he vanquished the monster Chimera with the aid of the winged horse Pegasus, the Greek mythological hero Bellerophon, grown proud, attempted to ride Pegasus to Mt. Olympus, but was thrown, crippled, and blinded. A luminescent fiery crevice high atop a mountain near the Turkish city of Antalya is still visible and believed by many to be the remnants of the mythological beast.

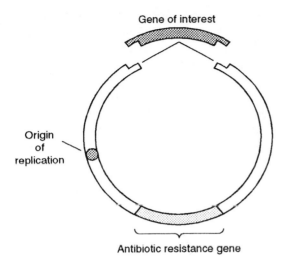

Gene of interest

Origin of replication

Antibiotic resistance gene

Figure 29 Schematic drawing of a simple bacterial plasmid cloning vector. A circular double-stranded DNA plasmid can be cleaved open at a single site by a restriction endonuclease. A DNA gene segment of interest can then be inserted into the plasmid by ligation. The newly closed plasmid with its inserted foreign DNA can be taken up by a suitable bacterium. The plasmid has an origin of DNA replication to allow its duplication by DNA polymerase and a selectable marker (an antibiotic resistance gene) to allow bacteria carrying the engineered plasmid to be selected from among those that did not take up the plasmid.

replication, that is recognized by DNA polymerase, enabling the quite small plasmid to be replicated many times in each round of bacterial replication, in turn leading to *amplication* of the plasmid and its inserted recombinant DNA. In addition, most plasmids carry one or more genes encoding the information for making a protein that provides a function to the bacterial cell in which it is expressed. For example, if the plasmid carries a gene that encodes a protein that allows the cell to become resistant to an antibiotic such as tetracycline, then by including tetracycline in the culture media, only cells carrying the recombinant plasmid will survive and amplify. Such genes are used as *selectable markers*.

Similarly, special bacteria-specific viruses called *bacteriophages* (*lambda phage* is used commonly) carrying linear DNA can act as vectors to replicate fragments of foreign DNA. The large linear DNA of these phages can be cut with a restriction enzyme into two halves (or *arms*) so that pieces of linearized foreign DNA up to approximately 15,000 bp long and having compatible sticky ends can be inserted and ligated between them. Once the DNA is packaged in vitro into empty phage protein coats, a single completed phage injects its entire nucleic acid content (in a phenomenon called *trans-duction*) into a susceptible bacterium, where it multiplies into hundreds of identical progeny (including the foreign DNA) before *lysing* (breaking open) its host. When recombinant phage are sprinkled onto a lawn of healthy recipient bacteria, individual phages infect, replicate, and kill the cell and then infect the surrounding cells. After several rounds of such phage replication, a clear spot appears in the bacterial lawn, representing the place where the dead bacteria have lysed. By sticking a sterile probe (e.g., a toothpick) into such a plaque, one comes away with a pure population of a single type of recombinant that can then be used to infect a culture of bacteria and produce large quantities of recombinant DNA.

Another specialized vector used to replicate fragments of selected DNA is a hybrid entity called a *phagemid*. These circularized, double-stranded DNA vectors are a recombinant containing both a bacterial plasmid and a single-stranded (DNA) bacteriophage called M13. The phagemids behave like normal propagating plasmids in *E. coli* but, with the addition of "helper" phage to bacteria containing the phagemid, convert into the bacteriophage form. In such cells, single-stranded phagemid DNA is synthesized, packaged, and extruded from the bacteria. The single-stranded progeny phage provide a rich source of single-stranded DNA that can be very useful in optimizing Sanger sequencing.

Another vector type that is especially helpful in replicating large pieces of selected DNA (up to several hundred thousand base pairs long) are YACs (yeast artificial chromosomes). These long linear vectors propagate in yeast in ways that are similar to the replication of their native chromosomes and, with their large capacity, are ideal for cloning entire sections of genomes.

Isolation of a DNA Sequence of Interest: The Construction of DNA Libraries

A population of DNA sequences, even an entire human genome, can be represented as a random array of fragments, each cloned into identical vectors. Such a collection is called a *DNA library*. Dividing a population of DNA molecules into smaller pieces and distributing each individually into the vector allows a systematic screening of the population of recombinants (clones) for the sequence of interest. The method of generating such libraries involves simultaneous cloning of the population of DNA fragments (each with identical restriction enzyme–generated ends) into a collection of identical vectors so that each DNA fragment molecule is inserted into its own vector. The new recombinants are then inserted into bacteria by *transformation* (if the vector is a plasmid) or by *transduction* (if it is a virus). A ratio of vectors to cells is chosen to increase the likelihood that only a single vector will invade each cell. Each bacterium in a colony contains the sequences from a single DNA fragment and, as described earlier, each of these is amplified many times in each cell. If the transformed bacteria are plated onto agar, the cells in an individual colony, formed from a single progenitor, all contain the same, amplified segment of chromosomal DNA.

The choice of starting material for a DNA library depends on the particular goals of the investigator. It can be derived from one of two major sources: genomic DNA or a DNA representation of the mRNA within a particular cell or tissue type (Figure 30). A *genomic library* allows the isolation of a gene of interest in its native chromosomal form, including its intervening sequences and some or all of its gene regulatory elements. The second type of library, representing mRNAs, is generated with the help of an enzyme called *reverse transcriptase*, which transcribes RNA into a complementary DNA copy (cDNA). The end result is a population of double-stranded cDNA copies of the original mRNA population. This cDNA population can then be used as the starting material to create a cDNA library. Since the template for the cDNA inserts is usually processed mRNA, only the coding sequences are present and the intervening sequences (introns) and gene regulatory elements are not represented. However, the advantage of cDNA libraries is that they represent the subpopulation of genes that are active in a particular cell or tissue type. In this way, a cDNA library is a subset of its genomic library counterpart.

When generating a cDNA library, the substrate population of mRNAs can be preselected to enrich for the species of interest. For example, when trying to determine the coding sequence for a particular protein, it is best to use the mRNA isolated from a cell line or tissue that produces a lot of this protein. This follows since such cells will also likely have a high proportion of the protein's corresponding mRNA among its total mRNA content.

Another way to preselect the mRNA used in generating a cDNA library is to use a technique called *subtractive hybridization* (Figure 31). This method depends on having two closely related cell types in which only one produces the protein(s) of interest. In such closely related cell types (e.g., human cardiac and skeletal myocytes), most of the mRNAs will be common to both. However, a few, perhaps the most interesting ones, will be present in only one of the two cell types. With this in mind, the mRNA from one of the two cell types is isolated and converted into single-stranded cDNA copies through the use of reverse transcriptase followed by alkali degradation of the RNA. Then a vast excess of mRNA isolated from the second cell type is hybridized to the cDNA from the first, and the resulting RNA–cDNA duplexes are removed with particular resins, such as hydroxyapatite, that selectively bind double-stranded species. The remaining unhybridized cDNAs from the first cell type reflect the differences between the two mRNA populations. As

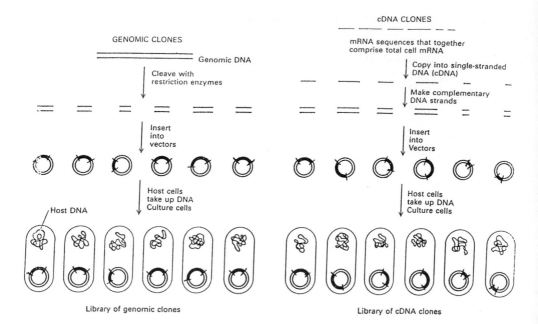

Figure 30 Genomic and cDNA libraries. To create a genomic library, the total DNA from the organism or cell line of choice is isolated and then digested with restriction enzymes into segments small enough to be inserted into plasmid vectors. Circular DNA vectors capable of replication in bacterial hosts are then digested with a compatible restriction enzyme (see legend of Figure 23). The family of genomic DNA restriction fragments are then ligated wholesale into the vectors. (Multiple *insertions* of the genomic DNA into a single vector molecule are almost always selected against by the bacterial hosts.) The recombinant plasmids carrying the inserts are then introduced into an excess number of bacteria, usually *Escherichia coli* in a process called *transformation*. The bacteria tend to allow only one plasmid per cell in addition to their own chromosomal DNA, so the isolation of a single bacterial colony is equivalent to the isolation of one recombinant plasmid. Taken as a group, the transformed bacteria represent a library of the original genomic DNA. Since the bacteria replicate the DNA as though it were their own during cell division, the library can be *amplified* by merely growing the transformed bacteria in culture. Producing a cDNA library is conceptually very similar to the process described above for genomic libraries except that the starting material is mRNA. Therefore, the end result will be a library representing not the entire genome but only those genes being expressed in the cell or tissue selected as the starting material. The first step in creating such a cDNA library involves converting the population of mRNAs to their respective cDNA forms with the use of the enzyme *reverse transcriptase* and dNTPs. This enzyme uses the mRNA as a template to make the cDNA strand. This cDNA, in turn, is then used by DNA polymerase as a template to convert the RNA-cDNA duplex into a DNA-DNA duplex (with displacement of the original mRNA). At this point, the double-stranded cDNAs all have blunt ends and can be inserted (as described above with the genomic library) into vectors similarly opened at a single site by a blunt-cutting restriction enzyme (see Figure 23). The rest of the procedure is the same as that described above for the creation of a genomic library. (Redrawn from Darnell J et al., Molecular Cell Biology. New York: Scientific American Books, 1986.)

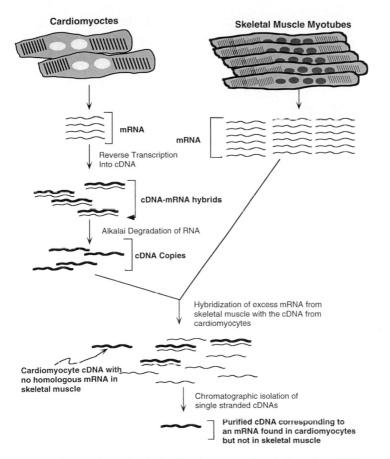

Figure 31 Subtractive hybridization is used to isolate the cDNA representation of mRNAs that are present in one cell type but absent in another. In this example, the technique is used to isolate gene products present in cardiac myocytes but absent in another muscle cell, skeletal myotubes. mRNA from cardiac myocytes is isolated and converted into single-stranded cDNA copies through the use of reverse transcriptase followed by alkali degradation of the RNA. Then a vast excess of mRNA isolated from skeletal myotubes is hybridized to the cardiac myocyte cDNAs. Since the vast majority of expressed genes in the two muscle cell types will be very similar if not identical, most of the cDNA from the cardiac myocyte will find and hybridize with their complementary mRNA from the skeletal myotubes. In this way the cDNA representing gene products not unique to the cardiac myocytes is *subtracted* by the excess skeletal myotube mRNA. The resulting RNA-cDNA duplexes are physically removed with resins such as hydroxyapatite that selectively bind double-stranded species. The remaining unhybridized cDNAs from the cardiac myocytes reflect the differences between the two mRNA populations. Thin and thick lines represent mRNA and cDNA, respectively.

a result, the cDNA library made from this subpopulation will be highly enriched for the species unique to the first cell type. To generate the analogous cDNA library for the second cell type, the sources for the single-stranded cDNA and excess mRNA are simply switched.

Cloning DNA Molecules

The identification of an individual clone in a library can be achieved by one of two major methods (Figure 32). The first involves the use of a labeled DNA or RNA probe that will selectively hybridize only to its complement in the array of clones represented in the library. With plasmid-based libraries, whether genomic or cDNA, an appropriate dilution of the bacteria results in discrete single colonies on petri dishes. Often, antibiotics in the growth media (agar) can select for bacteria with recombinant DNA plasmids when the plasmid carries an antibiotic-resistance gene in addition to the recombinant gene. A duplicate set of colonies is grown by touch transfer of the growing colonies on the master agar plate to a sterile membrane. The bacterial colonies growing on the membrane are then soaked in alkali, which lyses the adherent bacteria and denatures their DNA. At this point, the bacterial DNA (including the recombinant plasmids) is bound to the membrane and hybridized with labeled DNA or RNA probes. After the removal of excess probe, hybridized colonies on the membranes appear as exposed areas on the X-ray film (if the probe is radioactive or chemiluminescent) or pigmented areas on the membrane itself (if the probe is based on a colorimetric system). Only those bacterial colonies containing sequences complementary to the probe will give positive signals. Since the membrane represents a replica of the original master plate of bacterial colonies, the position of each positive signal correlates with a discrete colony. The investigator can therefore align the positive signal on the X-ray film, for example, with its corresponding colony on the petri dish. This allows the identification, picking, and growth of a single "clone" for the isolation of its recombinant plasmid. All the recombinant DNA molecules in the progeny of the picked colony are identical. Subsequent steps can include sequencing of the inserted fragment of DNA in the plasmid or its use as a probe to look for other related clones in the library.

A second major method of identifying a clone in cDNA libraries involves the use of antibodies (rather than DNA or RNA probes) to detect the protein product of interest. Here, the bacteria must be able not only to replicate the recombinant plasmid but also to transcribe and translate the inserted foreign DNA. In addition, the foreign protein or (as is more often the case) peptide must be sufficiently distinct from the bacteria's own proteins so that the antibody used for screening will not cross-react (i.e., bind to both the foreign peptide and a bacterial protein), giving false-positive signals. Such immunological cross-reactivity is not usually a problem for clones coding for highly specialized proteins unique to higher organisms (e.g., peptide hormones), but can be for proteins that are highly conserved throughout evolution (e.g., the enzymes involved in amino acid metabolism). The technique for screening these *expression libraries* is not unlike that described above for colony hybridization. However, rather than probing for bacterial-colony nucleic acid, this method entails screening the membrane-bound proteins with an antibody of choice. As with Western blots, the antibody is either directly attached to a fluorescent, chemiluminescent, colorimetric or radioactive group or, instead, bound in turn to a similarly labeled anti-immunoglobulin antibody (*secondary antibody*). Autoradiography of the membrane (or, if appropriate, colorimetric processing) then reveals the colonies that produced proteins reacting with the antibody. As with standard colony lifts with nucleic acid probes, the positive signals align with the bacterial colonies on the master petri dishes, enabling the amplification, isolation, and analysis of the plasmids.

Both of the above methods of library screening are adaptable to phage vectors with only minor changes in the protocols. Rather than representing an array of recombinant bacterial colonies, the membrane lifts from phage libraries reflect individual recombinant phage and their progeny along with their DNA and protein produced in and released from their lysed bacterial hosts. At the appropriate dilution, a phage library plated onto an opaque *lawn* (confluent growth) of bacteria will result in discrete areas of bacterial lysis called *plaques*, which appear as discrete clearings. Each plaque represents the result of multiple rounds of phage transduction and multiplication following an initial infection of a single bacterium by one recombinant phage. Therefore, each plaque represents an amplification of a single recombinant phage. As with the colony hybridization methods, either a nucleic acid or, if using a cDNA phage library, an antibody can serve as a probe in screening the membranes (*plaque lifts*) for the desired target.

A variety of other techniques have been used to isolate genes of interest by *functional cloning assays*. By using a cell model in which a functional attribute is missing, it is possible to select for cells that have acquired the missing phenotype following *transfection* of the cells (a process of induced DNA uptake) with a cDNA expression library. A number of cDNAs whose products can be detected by immunofluorescence in living cells have been isolated using a fluorescence-activated cell sorter (*FACS*). An alternative approach (*sib selection*) starts with isolating, for example, 10,000–20,000 individual colonies carrying members of a cDNA library, pooling them into lots of several hundred, and introducing each lot into a cell type that doesn't normally express the phenotype desired. If one of the pooled test populations transfers the desired phenotype, it is subdivided into smaller groups of about 50 clones and the process is repeated until a single clone is found that can imbue the recipient cell with the phenotype. A number of genes for ion channels have been cloned in this way by testing for the acquisition of ion fluxes in transfected cells, such as toad oöcytes, that normally do not express such channels.

Similarly, the cDNA for the myogenic determination factor MyoD was cloned by searching for a rare muscle cell that arose from nonmuscle cells transfected with a cDNA library from muscle cells. An alternative approach (often called the *two-hybrid system*) has proven to be very powerful when searching for genes whose products interact with a known molecular component (the target, or "bait"), for example, a gene whose product binds to a particular DNA segment, or a gene whose product forms heterodimers with a known protein. In the two-hybrid system, the target DNA site or cDNA sequences for the target protein are incorporated into a eukaryotic vector and cotransfected into cells with a cDNA expression library expected to include the "prey." If a cell produces the prey molecule,

---

Figure 32 Isolation of specific clones from DNA libraries. Transformed bacteria containing a genomic or cDNA plasmid library are plated out onto agar-filled petri dishes. To prevent contaminating bacterial growth, the agar usually contains a drug such as ampicillin. This selects for only those bacteria containing a recombinant plasmid that by design carries a corresponding *selectable marker*—for example, an ampicillin-resistance gene. The bacteria are diluted before plating so that single discrete bacterial colonies arise after overnight growth. (This allows easy physical isolation of one colony from another after the colony of choice is found—see below.) A thin membrane is then placed briefly over the bacterial colonies so that a portion of them adhere to the membrane when it is lifted off the plate (a *colony lift*). At this point, the bacterial colonies on the membrane (representing a duplicate set of those remaining on the plate) are lysed. This releases the bacterial DNA and protein that bind to the membrane (see text for more details on the distinct techniques in screening different library types for either plasmid DNA or the resultant protein products). Depending on the library type chosen, the membrane is probed with either labeled antibody specific for the gene protein product (bottom left) or nucleic acid complementary to the inserted gene itself (bottom right). In either case, only the clone(s) possessing the gene of interest will react with the probe, giving a positive signal on an autoradiograph when the probe is radioactive or chemiluminescent (see text for a description of other detection systems). The positive signal on the X-ray film can then be aligned with the original master bacterial plate, allowing the selection and growth of the colony that gave rise to the signal. The recombinant DNA can then be easily isolated for further analysis (see text). (Adapted from Lindahl R, Molecular medicine: A primer for clinicians—II: Recombinant DNA molecules, S Dak J Med, May 1993, and Darnell J et al., Molecular Cell Biology, New York: Scientific American Books, 1986.)

Plasmid Library

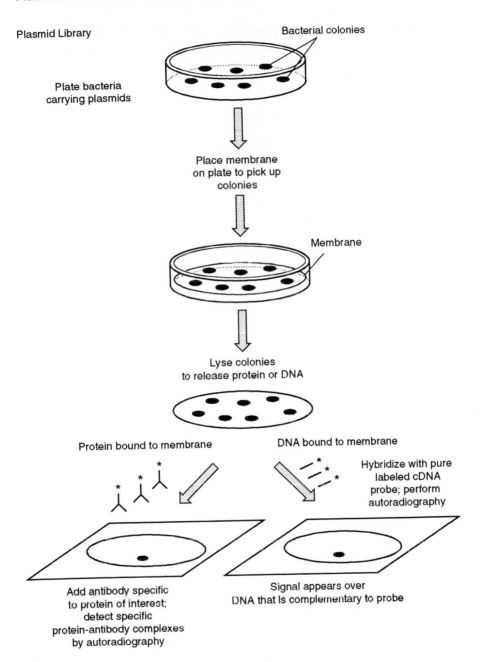

Bacterial colonies

Plate bacteria
carrying plasmids

Place membrane
on plate to pick up
colonies

Membrane

Lyse colonies
to release protein or DNA

Protein bound to membrane

DNA bound to membrane

Hybridize with pure
labeled cDNA
probe; perform
autoradiography

Add antibody specific
to protein of interest;
detect specific
protein-antibody complexes
by autoradiography

Signal appears over
DNA that is complementary to probe

it will interact with the bait in such a way as to activate the expression of a selectable marker gene. Thus, colonies of cells that grow after transfection have the prey cDNA, can be grown up, and the cDNA vector sequences recovered and analyzed.

Haute Couture of Probe Design

The design of probes for screening a library depends on both the available information about the target sequence and the goals of the project. For example, if the coding sequence for a given protein is the desired target, then the probe for an expression library could consist of either a monospecific antibody or, if at least part of the coding sequence is known, an *oligonucleotide* (a short stretch of single-stranded DNA). Likewise, if the protein of interest had been isolated and partly sequenced, an oligonucleotide probe could be designed by "recoding" the amino acid sequence back into the triplet nucleotide code. However, since this code is partly degenerate, with many amino acids being coded for by more than one triplet nucleotide set (see "Translation: 'The Other Side of a Tapestry'" above), there is usually no *a priori* way to predict the exact nucleotide sequence based on amino acid sequence alone. As a result, a *degenerate probe* can be designed that is actually a set of oligonucleotides that vary at the uncertain bases. For example, if the known peptide sequence contains an aspartate residue—which is encoded by either GAT or GAC in DNA—a degenerate probe would need to have the two complementary sequences to cover both these possibilities. Such degenerate probes can be synthesized chemically by automated machines that polymerize a polynucleotide sequentially and can generate milligram amounts of oligonucleotides up to 100 nucleotides or longer. The degeneracy in these syntheses is introduced by programming the machine to use a mixture of the appropriate nucleotides at the point in the extension of the polymerizing chain at which an uncertain base is called for (Figure 33). In the above case with aspartate, then, a mixture of two bases would be required at the variable base (A or G, the complements of T and C). In the end, the degenerate probe will actually consist of a family of related oligonucleotides, representing all the possible cDNA sequences for a given polypeptide.

Design of oligonucleotide probes can also be based on a known DNA or mRNA sequence from the homologous (equivalent) gene from another species. Here one depends on evolutionary conservation, at least in the coding sequences, since gross mutations from one species to the next would otherwise have led to profound changes in a protein's functional capabilities. However, silent mutations at the third base position accumulate randomly such that the hybridization conditions must usually be made less stringent to accommodate the mismatches between the oligonucleotide probe and the

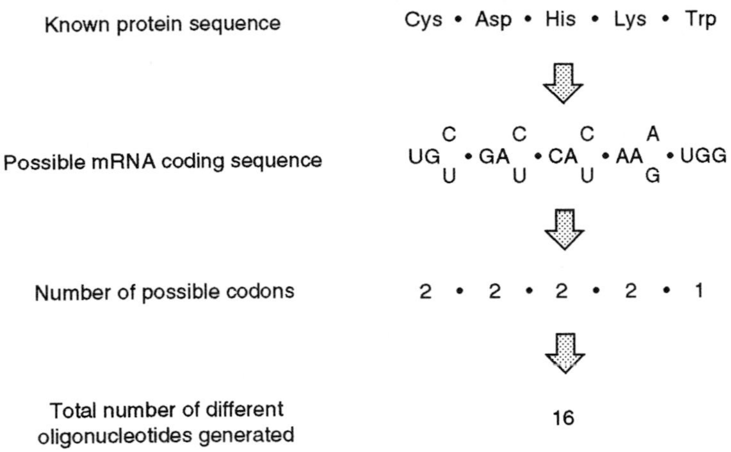

Known protein sequence   Cys • Asp • His • Lys • Trp

⇩

Possible mRNA coding sequence   UG$_U^C$ • GA$_U^C$ • CA$_U^C$ • AA$_G^A$ • UGG

⇩

Number of possible codons   2 • 2 • 2 • 2 • 1

⇩

Total number of different oligonucleotides generated   16

Figure 33 Degenerate oligonucleotide probe design. In this example, the five-amino-acid peptide is "reverse translated" employing the known degeneracy of the genetic code. Each of the first four amino acids (Cys, Asp, His, and Lys) can be coded for by two possible codons whereas the fifth (Trp) has only one possibility. The total number of potential sequences that could give rise to these same five amino acids is therefore 16 (the product of the number of possible codons at each position). As described in the text, these 16 probes can easily be generated by correctly programming an oligonucleotide synthesizer with a mixture of nucleotides at the variable positions. Under sufficiently stringent hybridization conditions, only the one oligonucleotide perfectly complementary to the actual coding sequence (i.e., 15 out of 15 base pairs) will remain hybridized in assays such as Northern, Southern, or colony hybridization.

target sequence. Similarly, degenerate oligonucleotide probes can be based on presumed sequence conservation within gene families (e.g., ion channel genes) or between different types of proteins sharing similar and evolutionary conserved functional domains (e.g., ATP binding domains).

The Polymerase Chain Reaction

The polymerase chain reaction (PCR), one of the newest and most powerful additions to the armamentarium of the biomedical scientist, allows the in vitro amplification of segments of DNA and RNA. This rapid technique (Figure 34), invented by Kerry Mullis, Norman Arnheim, Henry Erlich, and their colleagues at Cetus Corporation, is an alternative to cloning DNA and can be used to amplify and therefore to detect vanishingly small amounts of target sequence—in some cases, even a single gene from a single sperm cell. Mullis received the 1994 Nobel Prize in Physiology and Medicine for his initial observation. The method is based on repeated cycles of DNA synthesis

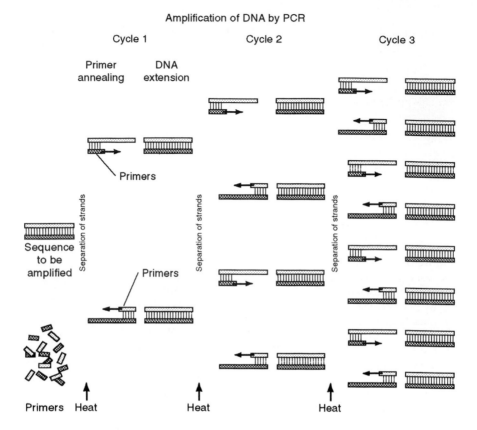

Figure 34 Polymerase chain reaction. A *thermocycler*, an automated rapid heating and cooling apparatus, carries out the ordered steps of PCR as follows: 1) the DNA sample containing the sequence to be amplified is heated to melt the strands; 2) the sample is then rapidly cooled to specifically anneal the single-stranded primers to their respective target flanking region, each on opposite strands so that the 3' end of each primer faces the target sequence; 3) the temperature is raised to an intermediate temperature (not shown) to ensure that the only hybrids remaining are perfectly matched; and 4) a heat-stable DNA polymerase extends the annealed primers past their target sequence. This effectively doubles the mass of original target DNA. This cycle of melting, annealing, and extension is then repeated until the desired level of amplification is reached. (Adapted from Rosenthal N, N Engl J Med 1994; 331:316).

followed by melting and annealing of primers (short oligonucleotides complementary to the sequences flanking the DNA segment of interest). With the primers in vast molar excess, they effectively compete with the newly synthesized strands after each cycle of DNA synthesis. In this way, since DNA is double-stranded, the number of DNA strands synthesized doubles each cycle just as it does in a single cycle of cell division. The difference in PCR, however, is that 30 or more cycles can be completed in a single test tube in just a few hours, giving a potential amplification factor of 10^8 or more!

A *thermocycler*, an automated rapid heating and cooling apparatus, carries out the ordered steps of PCR (Figure 34). The resultant DNA fragments can be analyzed directly on an ethidium bromide–stained agarose gel to determine their presence and their approximate lengths. Furthermore, if greater sensitivity is desired, a radioactive dNTP can be included in the PCR reactions and the subsequent gels exposed to X-ray film. Similarly, gels analyzing nonradioactive PCR reactions can be subjected to Southern blotting using as a probe either one of the primers or, if available, an oligonucleotide complementary to an internal sequence of the amplified fragment. Finally, the amplified fragments themselves can then be cloned, sequenced, or even used as probes.

PCR can also be used in the place of traditional gene isolation techniques such as library screening. For example, degenerate primers (see Figure 33) can be used to amplify segments of DNA when their exact sequence is not known but in which related or partial sequence data are available. Likewise, where partial protein sequence is known, degenerate primers can amplify the appropriate cDNA fragment in the same way that degenerate probes can often be used to screen cDNA libraries. This last PCR technique is often called *reverse transcription PCR (RT PCR)*, since the target is generated by using reverse transcriptase to create a population of cDNAs from a mixture of RNAs (Figure 35). In PCR reactions using degenerate primers, the annealing temperature (at least in the first few cycles) is often lowered to allow mismatches between the primer and the initial target DNA. In later cycles, the primed strands quickly become the dominant species so that with each sequential round of amplification the number of mismatches in the annealing step becomes negligible. Therefore, at later cycles, the annealing temperature can be returned to more stringent levels. Since the amount of the final PCR product is theoretically directly proportional to the number of starting molecules of RNA, it is not surprising the RT-PCR has also been used to establish quantitative assays to compare the absolute number of RNA molecules between samples.

Finally, PCR can be used to purposefully introduce mutations into stretches of DNA. This is simply accomplished, by purposefully designing and synthesizing the primers with the desired base changes and then amplifying the target DNA under moderately low-stringency annealing conditions as mentioned above. The final product, which can then be cloned into an

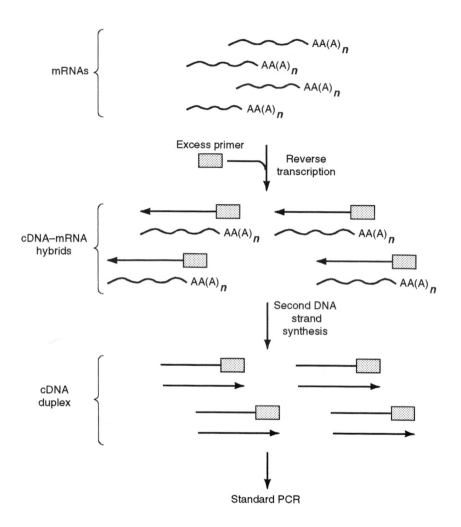

Figure 35 Reverse transcriptase polymerase chain reaction (RT PCR). Since PCR depends on a DNA target for amplification, target sequences contained within mRNA must first be converted to their cDNA form using reverse transcriptase. In the first step of this conversion, the entire population of mRNAs can be converted to cDNA using an oligonucleotide primer complementary to a region common to the 3′ ends of the mRNAs (e.g., a poly T primer would anneal to all mRNA poly A tails). Conversely, if narrower selection is desired at this first step, an RT primer complementary only to a specific mRNA sequence can be employed. Such primers are often based on known protein sequence (see text and Figure 33). Under sufficiently stringent conditions, the specific primers will anneal only to those mRNAs containing the complementary sequence. In the second step, RT extends the 3′ ends of the annealed primers forming an RNA-DNA duplex (shown here as parallel wavy and straight lines). DNA polymerase then synthesizes the second strand of DNA displacing the mRNA. The double-stranded cDNA molecules can then act as substrates for subsequent standard PCR analysis using a set of PCR primers as shown in Figure 34.

appropriate vector, will have the sequence on its ends changed to that of the primers rather than the original target sequence to which they had annealed.

FINDING OUT WHERE GENES ARE, WHAT THEY DO, AND HOW THEY ARE REGULATED

*If we do not find anything very pleasant, at least we shall find something new.**

From Phenotype to the Gene

In this introductory chapter we have continued to stress how physical alterations in genes lead to alterations in the proteins responsible for the phenotypes of cells, tissues, organs, and organisms. The clinician always deals with phenotype first, since it is by definition the presenting illness or symptom. For many diseases or syndromes with a hereditary basis or influence and a well-described phenotype, neither the protein abnormality nor the genetic target is known. As several chapters in this volume demonstrate, molecular genetic and recombinant DNA technology has made it possible to isolate genes and identify the mutations responsible for a wide variety of single-gene defects (*monogenic disorders*), such as familial hypertrophic cardiomyopathy, and methods are being developed to tackle the heritable causes of multigene diseases (*polygenic disorders*), such as essential hypertension.

The major approach being used is *genetic linkage analysis* based on many of the molecular genetic concepts described in this chapter. Given DNA samples from a few large kindreds—or a large number of smaller kindreds—with a heritable disease, geneticists can attempt to determine whether a linked DNA segment associated with a specific region of the genome (a *DNA marker*) is always present in individuals with the disease (or who are carriers of the disease gene, in the case of a recessive disorder) and absent in those without the disease (or who do not carry the recessive disease gene). DNA markers are cloned or sequenced unique stretches of the genome that exhibit allelic polymorphism in the unaffected population and that can be identified by PCR or molecular hybridization with nucleic acid probes. International efforts related to the Human Genome Project have been able to identify such polymorphic marker DNA segments spaced at useful distances along the entire genome. The markers must be as close as possible to reduce errors in linkage assignment related to the relatively high frequency of reassortment of genes (recombination frequency) that goes on normally during meiosis. If markers are far away from a gene, the chance of recombination between the marker and the gene increases and can give misleading information. Obviously, the closer the markers

---

*Voltaire, in *Candide*, 1759.

are spaced the more of them there will be, thus increasing the Herculean efforts still required to test many markers on DNA samples from many individuals.

Linkage analysis for a previously undetermined disease locus is carried out by displaying the entire genome of each individual on a blotted gel or by amplifying specific segments with PCR before using the Southern method to examine the polymorphic markers. Once a linkage has been established, the next step is to clone the DNA segments between the nearest pair of markers. An eventual goal of the genome project is to create a full set of cloned and ordered DNA fragments in YACs or even larger-capacity vectors. Even with a full segment of genome carrying the disease locus in hand, it is not uncommon to find later that the initial linkage markers are millions of base pairs (*megabases*) away from the actual site of mutation and that there are many uninvolved genes scattered throughout the segment.

Finding genes in an uncharted DNA segment is a major task. An obvious benefit of the Human Genome Project to sequence the entire genome is that most (but not all) genes in every cloned DNA segment will be known following inspection of the primary nucleotide sequence. Current strategies usually require breaking down the large segments into more manageable bits by constructing phage-based libraries. A number of strategies to identify genes in such segments have been developed, including: methods to clone exons, mapping of sequence segments conserved across evolution on the assumption that the sequence of coding segments is conserved but that noncoding information will likely have diverged, and identifying by cDNA cloning and hybridization all mRNAs transcribed from the segment. A perusal of the references in any journal article reporting the isolation of a new disease gene by such techniques should support the notion that this approach, while fundamentally rewarding, requires literally dozens of the proverbial person-years of effort. After the genomic locus for Huntington's chorea had been located by linkage markers, it still took more than 10 years of concerted hard work to identify and clone *huntingtin*, the mutated gene.

But What Does It Do?

If we make a couple of discoveries here and there we need not believe things will go on like this for ever. . . .
Just as we hit water when we dig in the earth, so we discover the incomprehensible sooner or later.*

Once a disease locus has been identified and the cDNAs for the wild-type locus are in hand, it is not always obvious what the function is of the inferred polypep-

---

*G. C. Lichtenberg, *Aphorisms*, "Notebook F," aph. 82 (written 1765–1799; translated by R. J. Hollingdale, 1990).

tide. If the polypeptide chain contains segments similar to previously described domain motifs or, by computer-aided sequence comparisons, it contains segments similar to active sites of known enzymes, some educated guesses can be made. Additional clues can be provided by Northern blotting and in situ hybridization, to determine whether the gene is expressed in a subset of tissues or organs and whether its expression is developmentally regulated. By developing antisera against potential epitopes of the inferred amino acid sequence, the subcellular location of the wild-type polypeptide can be demonstrated to provide additional information. By carrying out similar experiments on the appropriate tissues or cells isolated from an afflicted individual, it may be possible to explore further the impact of the mutation on expression of the gene. Finally, by exploiting the power of animal models, it would be possible to overexpress or knockout the expression of the gene in laboratory animals with transgenic and homologous recombination technology (for examples, see Chapters 10–16).

Only a small fraction of the 60,000 human cDNAs have been fully cloned and the function of their encoded peptides identified. However, there are many cloned cDNAs stored in freezers around the world for which no function is known. Identifying such functions remains a formidable task whose incentive, at least in the commercial pharmaceutical and biotechnology sectors, is the ability to patent sequences of known function and develop potential therapeutic recombinant compounds. An interesting question to ponder in this context: If the cDNA for a therapeutically and commercially important recombinant protein, such as tissue plasminogen activator (t-PA), were lying among a freezerful of unidentified cDNA clones, how easy would it be to determine the function of its encoded polypeptide? Numerous approaches to solving such problems have been followed, but if there are no clues as to what cell system to evaluate in vitro, or what to look for in experimental animals, the task is arduous.

Expressing Recombinant DNA in Cells

Representative cells from essentially all prokaryotic or eukaryotic sources can be solicited to take up protein-encoding recombinant DNA molecules and then to correctly transcribe them and produce the recombinant proteins. Useful amounts of recombinant peptide can be synthesized by cloning the coding sequence of interest into expression vectors designed to make otherwise rare proteins in large quantity. To express recombinant peptides in mammalian cells, precision DNA cleaving and splicing tools allow insertion of cDNA coding segments into bacterial plasmid vectors or mammalian viral vectors in such a way that the coding sequences become part of an idealized artificial eukaryotic gene. The coding sequences are inserted flanked by appropriate eukaryotic promoter/enhancer elements, polyadenylation signals, and one or more introns. Such vectors often contain other genes (*reporter genes*) whose products, if expressed, render the recipient cell identifiable by, for example, resistance to an antibiotic or expression of peptides detectable by vital dye staining or through their immunoidentity. The bacterial components of the expression vector allow the amplification of the plasmid in bacterial cells for subsequent introduction into mammalian cells. Alternatively, the cDNA segment can be inserted into a viral vector such as recombinant retrovirus or adenovirus for cell infection.

When naked DNA plasmids are used, mammalian cells can be induced to take up the DNA (transfection) in a number of ways, most of which involve endocytosis of precipitated DNA, electric fields (*electroporation*), or artificial fusion of the membranes with hydrophobic or lipophillic substances (*liposomes*) enmeshed with the DNA of interest (*lipofection*) or attached to carrier molecules known to be taken up avidly by the tissue or cells of interest.

Following insertion of the expression vector into mammalian cells as naked plasmid, the inserted DNA finds its way to the nucleus and begins to be transcribed in a matter of hours. Such *transient transfection assays* are an excellent way to test the importance of *cis* regulatory sequences, including enhancers, promoters, splice signals, and polyandenylation signals. By employing site-specific mutagenesis, the precise nucleotides participating in regulating the gene can often be identified. Such approaches have also allowed the characterization of tissue-specific transcription factors or other components of signal transduction machinery. By conducting the assays in cells lacking the required components, the addition of wild-type or mutated expression vectors carrying cDNAs coding for the missing components can test their ability to support expression of the reporter gene. Commonly used reporter genes linked to eukaryotic regulatory signals include the bacterial genes for *chloramphenicol acetyl transferase (CAT)* and β-*galactosidase* (sometimes referred to as *LacZ*) and the insect gene for firefly *luciferase*. In general, the fitness of the regulatory sequences attached to these reporter genes to produce an mRNA is directly proportional to the amount of their protein product synthesized. Thus, measurement of the CAT, β-galactosidase, or luciferase activity in the transfected cells is a direct assay of the gene expression. Substrates and techniques to assay these enzyme assays are readily available. CAT is assayed by determining whether lysates of transfected genes can transfer ^{14}[C]-acetyl groups onto unlabeled chloramphenicol as determined by thin-layer chromatographic separation of acetylated and nonacetylated forms. β-galactosidase converts its substrates in fixed tissues to an easily visible blue color and can also be used in a quantitative assay. It has proven particularly valuable in providing a light-microscopic method for analyzing the timing and location of gene expression in transgenic embryos and animals.

Luciferase production is assayed in cell lysates by measuring in a *luminometer* the light emission from its natural substrate, *luciferin*.

Following insertion of the expression vector into mammalian cells, even as naked plasmid, the DNA often integrates randomly into the host chromosome. Such integration events and the maintenance of the inserted genes (*stable transfectants*) can be selected for by inclusion in the vector of a selectable marker such as a neomycin-resistance gene (*neo*). Thus, exposure of mammalian cells to the poisonous substance g418, which is detoxified by the *neo* gene product, allows growth of g418-resistant clones following transfection or transduction of cells carrying *neo* as well as the other genes of interest. The DNA construct bearing the selectable marker does not always need to be on the same vector but can be *cotransfected*.

Genetic engineering also allows the creation and expression of novel polypeptides by fusion of the coding segments for several genes. Such chimeric fusion proteins have been used, for example, to dissect the activation and regulatory domains of various protein kinases or to localize polypeptides to subcellular compartments by immunofluorescence microscopy using antisera against a short fusion peptide.

Revealing Function by Shutting Off Genes on Purpose

Remarkable new technology has provided techniques to mutate specific genes in whole animals. By directing the location of stable integration into specific genes of interest, making use of cellular mechanisms known as *homologous recombination*, it is now possible to physically disrupt, "*knockout*," any gene. When this is done in *embryonic stem cells* (*ES cells*), whole animals can be generated using transgenic techniques in which all the cells, including germ-line cells, carry the disrupted gene. If this mutation is recessive, homozygotes can be bred. In either event, this technique allows the organismal function and consequences of almost any gene to be evaluated. Crosses between animals bearing different knockouts, for example, genes for components of the renin-angiotensin system, allow analysis of protein interactions. Such techniques are more fully described in several chapters in this volume (Chapters 10–16).

An alternative approach that allows shutting down gene expression in cells in culture or in tissues in adult animals involves delivering DNA transcription units to cells or tissues that encode RNA molecules that bind normal gene transcripts. Since RNA is a single-stranded polynucleotide, synthesis of its complementary strand in a cell would allow the formation by complementarity of a double-stranded RNA molecule. Such antisense RNAs are easily designed to interfere, once bound to the normal transcript, with nuclear processing or cytoplasmic translation initiation. The end result is the down-regulation of the endogenously encoded protein and an approach to an absent protein (null mutation). Such strategies are by their nature dominant negatives. Production of antisense molecules does not always work to completely shut off expression of the target gene, and its ultimate effect depends on the cell or tissue requirements for the gene product. Down-regulation of gene expression in vivo has proven to be a valuable tool for studying mechanisms and therapeutic approaches for smooth-muscle hyperplasia following vascular injury and the prevention of arterial restenosis after angioplasty. Antisense gene therapy strategies designed to inhibit paracrine growth factor responses are being developed for in vivo delivery (see Chapter 21).

ACKNOWLEDGMENTS

The authors acknowledge support from the Howard Hughes Medical Institute and the Borchard Foundation that helped in the preparation of this chapter.

2

A Nonmathematical Overview of Modern Gene Mapping Techniques Applied to Human Diseases

Nicholas J. Schork and Aravinda Chakravarti
*Case Western Reserve University School of Medicine
and University Hospitals of Cleveland
Cleveland, Ohio*

INTRODUCTION

Technological advances made in molecular biology in the last 10-20 years have profoundly influenced the manner in which the etiology of disease is investigated. Nowhere is this more apparent than in the study of the genetic basis of human disease. To date, genes (or mutations) influencing a number of diseases whose biochemical defects were unknown at one time have been either physically isolated or localized to a small region in the human genome (i.e., the collective genetic material possessed by each of us) by taking advantage of modern laboratory methods (1,2). Although the molecular tools that paved the way for the localization, or "mapping," of these genes were unthinkable only some 25 years ago, their development does not tell the whole story. Equally important to the development of any technology is its proper application or implementation. In this review, we describe contemporary and successful strategies for the detection and localization of genes that depend on modern molecular genetic technologies. We want to emphasize that this review focuses not on the technologies themselves, but rather on the ways in which one can exploit them to map genes influencing a disease or phenotype of interest. In addition, we do not review strategies one might use to determine whether a disease or phenotype is under genetic control (i.e., heritability or segregation assessment). With this in mind, the primary question to be addressed in this review is not "how can one determine if a disease is genetic?" but "how can one find the genes in-

fluencing a disease known to have a genetic component?" Since an understanding of a few basic concepts in genetics is required for this purpose, we
offer a brief summary of some necessary terms. We also offer tables listing
studies that make use of the strategies described. For readers interested not
only in the application of modern gene mapping methods, but also in the
details of the requisite laboratory methods, we suggest any of the contemporary textbooks describing the biological basis of modern recombinant DNA
methods (3–5).

WHY GENE MAPPING STRATEGIES?

One question that is important to answer at the outset is "why are gene
mapping strategies needed?" For most disorders the underlying dysfunctional
biochemical and physiological factors are poorly understood, if known at all.
By locating and isolating a putative disease gene, its products can be studied,
its mutant forms can be identified, and, ultimately, attempts can be made to
understand the biological pathway that impacts the disease. Thus, locating a
previously unidentified disease gene can provide medical researchers with
leads they can follow in an attempt to understand and combat that disease.
Two approaches, *candidate* gene analysis and *genome scanning*, are typically
used for disease gene mapping purposes.

Candidate gene studies involve the investigation of genes whose biological properties make them logical objects of study for a particular disease
(see, e.g., Ref. 6). For example, it is now possible with modern laboratory
methods to determine which genes "express" themselves in various tissues
throughout the body. If one can identify genes that express themselves in
tissues of relevance to a particular disease (e.g., smooth-muscle tissue in
hypertension), then one might want to ascertain if variants of those genes
can distinguish "normal" from "diseased" persons or tissues. Candidate gene
studies are plagued by numerous problems, not the least of which is that
there are literally thousands of genes that play crucial roles in human physiology, making the list of possible candidates for any disease extremely large.
Obviously, the probable success of a candidate gene study is often a function
of the prior knowledge, or understanding, of the biological properties of the
candidate.

An alternative strategy to candidate gene studies is the genome scan
strategy. Genome scans, which assume no prior knowledge of the biological
properties of a gene or genes, work by investigating the association or linkage
between variants at landmark spots along the genome (known as "marker"
loci) and loci that are hypothesized to encode putative disease-influencing
genes in related individuals or family members. Marker loci are often derived
from simple repeat sequences found with some regularity throughout the

genome (see, e.g., Ref. 7). Such loci possess unique "signatures" that can be identified with modern laboratory assays. By having a number of such marker loci at one's disposal, knowing either the physical locations on the genome or the relative order or proximity of these marker loci, and having an appropriate sample of individuals whose variants (or *genotypes*) at the marker loci have been identified and recorded, one can test the hypothesis that various points along the genome encode putative disease genes. Figure 1 offers a graphical description of the genome scan approach. Two aspects of genome scan strategies are worth emphasizing at this point. First, genome scans depend crucially on characteristics of the marker loci; for example, the more closely spaced they are, the more likely one is to not only find relevant genes, but also refine their positions on the genome. Second, since marker loci are often inert biologically, they can really only be used within the genome scan framework to provide a rough estimate of the position of relevant disease genes; i.e., once linkage to a marker locus has been found, other methods must be used to determine the actual physical chromosomal location of the disease locus and relevant mutations.

In the sections that follow, we briefly review some basic genetic concepts and distinctions, and then describe the strategies one can use to exploit candidate gene and genome scan technologies for gene mapping purposes. Problems that plague gene mapping studies are taken up in later sections.

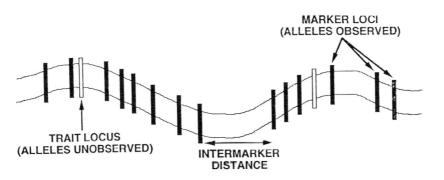

Figure 1 A graphical representation of a genomic scan. Loci exist along the genome, which can be thought of as an ordered string of 23 chromosomes. The darkened bars represent marker loci, whose variants, or alleles, can be observed in individuals. The open bars represent trait (disease) loci whose alleles are unobserved and can sometimes be predicted. The object of a genome scan is to place or position disease loci relative to the marker loci, using the information obtained from marker alleles in persons affected and unaffected by a disease. Note that the "intermarker" distance need not be (and rarely is) uniform.

SOME BASIC TERMS

Segregation

The term *segregation* refers to the genetic transmission of traits from generation to generation. These traits can be disease states, features (or "phenotypes") such as handedness, eye color, IQ, high blood pressure, etc., or genetic material unique to a specific spot on the human genome. Each human possesses 23 pairs of chromosomes. One of each pair, or homolog, is inherited from the mother and the other from the father. Each homolog contains genetic material that may have been unique to the parent from whom it was inherited. A point along a chromosome, known as a locus, may encode material (i.e., short DNA sequence) that can vary from person to person. The variants associated with a particular locus are called alleles. A person's genotype is the specific combination of (two) alleles he or she possesses at a particular locus and that were transmitted on chromosomes from his or her mother and father. Inherited genotypes (in conjunction with other factors such as the environment or genotypes at other loci) give rise to the features that uniquely identify each individual. The term *gene* is often ambiguous, in that it can be applied to actual genotypes or alleles that influence or determine a particular trait. The two most well-known segregation patterns are exhibited by alleles at a single locus. Alleles exhibiting dominance occur when only one copy of a particular (mutant) allele is necessary to produce a particular trait. Thus, dominant alleles can be inherited through either a maternal or a paternal chromosome and still cause the trait. Recessive alleles require two copies before the trait will appear. Thus, a copy of the relevant allele must be inherited from the mother and the father before the trait will appear. Figure 2 offers a graphical description of the segregation of alleles and disease. By tracing the cosegregation of a particular disease with alleles at a particular locus among related individuals, one can draw inferences about the likelihood that those alleles determine—or are near a locus with alleles that determine—a trait or disease. This is the essence of disease gene mapping.

Recombination

During meiosis—i.e., the biological process by which germ cells are produced within an individual—homologous chromosomes pair up and exchange genetic material. This exchange of material, or "recombination," results in chromosomes that differ from each of the two parental chromosomes transmitted to the relevant individual. These recombined chromosomes can then be transmitted to that individual's offspring. Figure 3 offers a simple graphical depiction of this process. The alleles at each locus along the homologous chromosomes have recombined after meiosis to form chromosomes with a

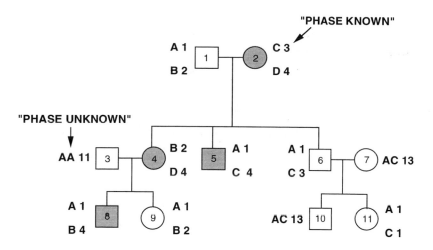

Figure 2 Graphical depiction of the segregation of a disease and marker alleles within a pedigree. Males are denoted by squares, females by circles. Affected individuals are denoted by shading. Alleles at two loci are segregating: a "letter" locus and a "number" locus. Note that the 4 allele at the number locus is responsible for the disease. Thus, the number locus is the disease locus. When the maternal and paternal origins of specific alleles (*phase*) are known, the paternal alleles are written over the maternal alleles for the two loci. When unknown, the alleles are merely written next to each other to give genotype. From the phase information, it can be seen that persons 5 and 8 are recombinant between the loci, whereas persons 4, 6, and 9 are not. Persons 10 and 11 are ambiguous with respect to recombination.

combination of maternal and paternal alleles. The points at which the homologous chromosomes hook up and exchange material are called *chiasmata*. Chiasmata occur randomly approximately every 50 centiMorgans (note: the average human chromosome is roughly 150 centiMorgans long). The closer two loci are, the less likely they are to have their allelic material recombine. This fact is at the heart of all gene mapping strategies, since one can attempt to determine chromosomal sites of recombination and trace the cosegregation (i.e., common transmission) of alleles from generation to generation. Such cosegregation, or lack of recombination between loci, is indicative of their proximity to each other. Loci that recombine with less than a 50:50 chance (random recombination) are said to be "linked" and are likely to be near each other on a particular chromosome. Unlinked loci are either on different chromosomes or far enough apart on a single chromosome to exhibit a 50:50 chance of recombining.

It is rarely possible to observe recombination events as clearly as in Figure 3, in which it was assumed that the grandmaternal and grandpaternal origins of

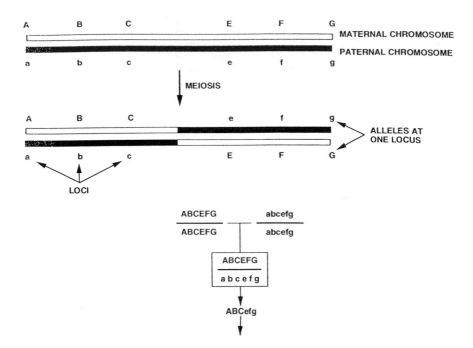

Figure 3 Graphical depiction of recombination. During meiosis, two homologous chromosomes pair up and exchange genetic material. This exchange, or recombination, results in chromosomes that are combinations of the chromosomes originally transmitted from a mother and father and allows unique chromosomes to be transmitted to offspring. The upper panel displays alleles at six loci. The mother of an individual transmitted only (hypothetical) uppercase alleles to that individual, whereas the father transmitted only lowercase alleles. During meiotic events within that individual, recombination has led to the formation of new chromosomes that contain alleles originally transmitted from his or her mother and father. These new chromosomes can then be transmitted to that individual's offspring. The lower panel depicts the origin of the chromosomes and alleles depicted in the upper panel. The boxed offspring is the individual whose maternal and paternal chromosomes recombined to form the new chromosome, which could then be transmitted to his or her offspring, as denoted by the arrows.

the alleles on the chromosome were known. When such information is available, the "phase" of the alleles is said to be known; when the phase is unknown, genetic analysis is more complicated. In addition, in Figure 3 it was also assumed that the maternal and paternal alleles were distinct (i.e., the mother passed on all uppercase alleles, whereas the father passed on all lowercase alleles). Rarely will such fully informative genotypes occur in gene mapping studies. Figure 2 also shows how recombination events can be traced. By examining the phase information of parents, it can be seen that persons 5 and 8 must have been "recombinant" between the letter and number loci, while persons 4, 6, and 9 are nonrecombinant. From the available information, it is unclear if persons 10 and 11 are recombinant or nonrecombinant. The following four mating situations may help clarify relevant concepts.

Phase Known, Fully Informative Mating

Assume that the parental genotypes at two neighboring loci are known. The father possesses genotypes AB and 12 at the two loci, and the mother possesses genotypes CD and 34. The parents produce two children with genotypes AC and 13 and AC and 14. Assume further that the origin of the parents' alleles are known. This mating can be written in the following way:

$$\left(\frac{A1}{B2}\right)_f \times \left(\frac{C3}{D4}\right)_m \rightarrow \left(\frac{A1}{C3}\right)_{o1}, \left(\frac{A1}{C4}\right)_{o2}$$

The subscripts f, m, and $o1$ and $o2$ denote father, mother, and offspring 1 and 2, respectively. The line separating the alleles denotes phase information. Thus, the father received alleles A and 1 at the two loci from his father and the alleles B and 2 from his mother. Since alleles A and 1 were transmitted together to the two children, the meiotic event leading to the chromosomal material ultimately transmitted by the father to the two children did not recombine between the two loci. However, the second child received a C allele and a 4 allele from his mother; since these alleles were on different chromosomes transmitted to the mother by her parents, a recombination event must have occurred in the meiotic event leading to the second child by the mother. The second child is thus a "recombinant" for the two loci.

Phase Unknown

If phase is unknown, the determination of recombination events is more difficult because one would not know whether alleles were inherited together on a chromosome transmitted by a parent. Thus, one would not know if the first or second child was recombinant for material transmitted from the mother.

Phase Known, One Parent Informative

If one of the parents possessed two alleles of the same type at a locus (i.e., the parent is homozygous at that locus), then recombination cannot be detected. Consider the following mating:

$$\left(\frac{A1}{A1}\right)_f \times \left(\frac{C3}{D4}\right)_m \rightarrow \left(\frac{A1}{C3}\right)_{o1}, \left(\frac{A1}{C4}\right)_{o2}$$

The children were destined to receive A and 1 alleles from the father (because that is all he had to offer), but it cannot be determined if recombination took place by examining the children, since both recombinant and nonrecombinant events would result in the transmission of A and 1 alleles. Note that this is not the case for the mother.

Uninformative mating

An uninformative mating is one in which the origin of alleles cannot be determined for either parent, making impossible the assessment of recombination events in their offspring. An example of an uninformative mating would involve the mating of parents who are both homozygous at both loci.

Note that phase information can often be gathered by tracing the origin of alleles through grandparents and other relatives. For a more complete discussion of the gene mapping information provided by different mating types, see Ott (8).

Identity by Descent vs. Identity by State

Alleles possessed by any two individuals, but particularly relatives, can be (or not be) either "identical by descent" (IBD) or "identical by state" (IBS). Alleles are IBD between two individuals if those alleles are of the same type and have been inherited from a common ancestor. Alleles are IBS if they are simply of the same type. Thus, all alleles shared IBD between two individuals are IBS, but all alleles shared IBS are not necessarily IBD. Determining IBD status from parental data is not possible if the parents are homozygous (see Figures 2 and 4). Therefore, determining IBD and IBS status is contingent on the marker or locus in question being highly polymorphic (i.e., has several variants or alleles associated with it), because then the chance of finding individuals who are homozygous is small. The utility of determining IBD and IBS status from related individuals for gene mapping purposes revolves around the following simple concept: the more similar two individuals are at relevant genomic regions or loci, the more similar they should be phenotypically. Consider some measure of phenotypic similarity, such as the squared difference between two siblings' trait values (e.g., squared difference in blood-pressure values, heights, weights, etc.). If the siblings are alike with respect to the trait, then the squared difference between their values will be small. If a particular locus or genomic region harbors alleles that influence the trait, and the siblings share alleles or genetic material in this region, one would expect the difference in their phenotype values to be small. This concept, depicted in Figure 5, forms the basis for the oft-used sib-pair test procedure developed by Haseman and Elston (9).

Linkage Disequilibrium

Linkage disequilibrium refers to the nonrandom association of alleles at linked loci in a population. Because of recombination and the size and age of populations, one would expect specific alleles at different loci to occur together in any one individual with a probability dictated merely by the frequency of those alleles. That is, one typically wouldn't expect, except under certain conditions, an allele at one locus to occur more often with a specific allele at another locus in the population at large. To formalize this, if $p(A)$ is the frequency of the A allele at one locus in the population at large, and $p(1)$ is the frequency of the 1 allele at another linked locus, then if linkage equilibrium holds, the probability of observing a person with the A and 1 alleles would be $p(A,1) = p(A)p(1)$; i.e., it would be the simple product of the individual allele frequencies. When alleles at linked loci are in fact associated, the alleles are in disequilibrium. Linkage disequilibrium can occur as a result of the recent admixture of populations with different alleles or by the introduction of a de novo mutation (i.e., a new allele) in the population, since alleles at chromosomal regions near the new mutation possessed by the relevant individual will be transmitted along with the mutation to ensuing generations. The association between alleles at the neighboring loci and the mutation will hold until enough recombination has occurred to deplete the association.

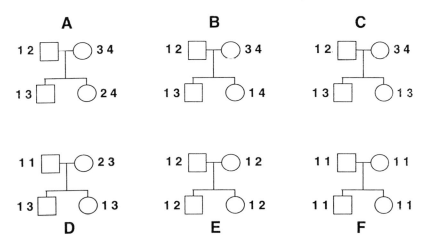

Figure 4 Graphical representation of situations in which IBD and IBS information can be computed explicitly or is ambiguous. The numbers represent alleles (genotypes) at a single locus. For situation A, the two offspring share no alleles IBS and no alleles IBD. For situation B, the two offspring share one allele IBS and the same allele IBD. For situation C, the two offspring share both alleles IBS and IBD. For situation D, the two offspring share both alleles IBS, but only one allele is unambiguously shared IBD (it is unclear if the two offspring were transmitted the same 1 allele from the father). For situations E and F, the two offspring share two alleles IBS, but it is unclear how many alleles they share IBD.

Figure 6 offers a graphical display of the concept of linkage disequilibrium. The darkened chromosomal regions observed in the last generation are what is left of the original parental chromosome bearing the mutant allele. Recombination has depleted much of the original mutant chromosome in the ensuing generations. The original mutant chromosome's alleles at loci in the regions near the actual disease mutation are in linkage disequilibrium with the mutant allele (e.g., from Figure 6, alleles from the original mutant chromosome that are at regions within the dashed lines flanking the disease mutation in the last generation will be observed in all the affected individuals, creating an association between these alleles and the disease). Thus, the A alleles on the original mutant chromosome that are at loci near the actual locus encoding the mutation are possessed by the individuals in the third generation who inherit the mutation, and an association exists between the A allele and the mutation (and ultimately the phenotype or disease it causes). Note that this holds only for the pedigree or population shown in Figure 6—it says nothing of other populations, in which different alleles may be in association with the same or different mutations.

Quantitative vs. Qualitative Phenotypes

Traits or phenotypes can be measured or observed in two intrinsic ways: as qualitative or dichotomous (yes/no) traits, such as cancer or cystic fibrosis, or

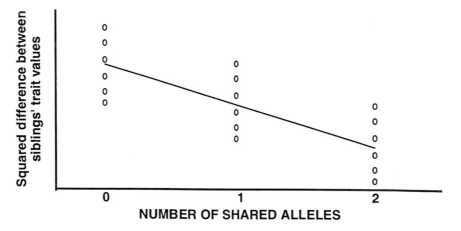

Figure 5 Graphical display of the intuition behind allele-sharing gene mapping methods. If a locus influences a quantitative trait (blood pressure, cholesterol, etc.), then persons sharing more alleles at this locus should be more alike phenotypically (i.e., the difference between their trait values should be small). Thus, a negative correlation should hold between squared trait value difference and number or fraction of alleles shared IBD or IBS.

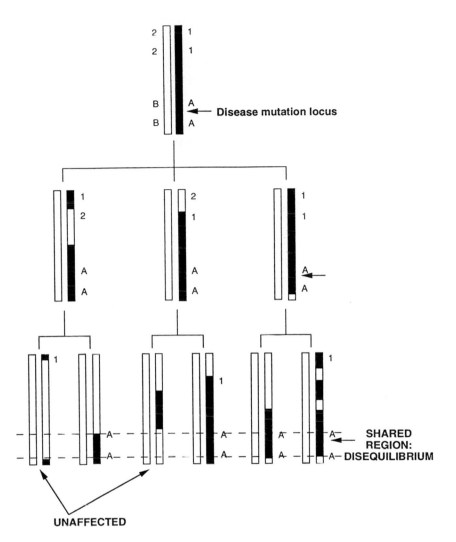

Figure 6 Graphical depiction of the transmission of chromosomes within a pedigree. If the darkened chromosome is a chromosome that bears a disease-causing allele or mutation, then recombination events will lead to transmission of only parts of this chromosome to persons in ensuing generations. However, alleles at loci near the mutation at a disease locus should be transmitted together with the mutation until enough recombination has occurred to physically separate disease and marker alleles. Persons transmitted chromosomal material in the region of the mutation not only will possess the disease (since they will have inherited the mutation), but are also likely to have been transmitted alleles that were originally associated with the mutant chromosome. Thus, the *A* alleles at loci in the immediate area of the mutation (i.e., between the dashed lines in the last generation) will have been transmitted along with the mutation. These *A* alleles will thus be in association with the mutation. This *A* allele and the mutation will thus be in linkage disequilibrium in the population at large.

as quantitative or metrical traits, such as height or blood pressure. Most quantitative traits are thought to be unlike qualitative traits in that they have numerous genetic (and possibly nongenetic) determinants that work to induce the great variation exhibited by them. However, this line of thinking is not *necessarily* true, since a qualitative trait can have multiple determinants as well. Figure 7 offers a simple graphical depiction of how genes might influence the distribution of a trait in the population at large. A good question is whether all traits are quantitative in some sense. The reason one might think so is that many of the biochemical and physiological processes underlying diseases and phenotypes are governed by quantitative variables (e.g., hormones, protein amounts, timing mechanisms). In fact, one could argue that

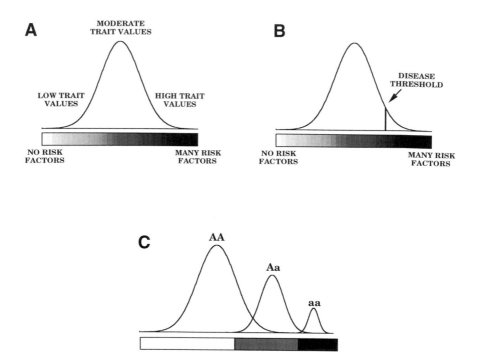

Figure 7 Graphical depiction of multiple factors (genes, environmental exposures, etc.) that can influence quantitative or qualitative phenotypes. (A) The shading underneath the trait distribution represents the number of factors possessed, with darker shading representing more factors. Thus, a person possessing more factors is likely to have a higher trait value. (B) A "liability" distribution, in which one manifests a disease only if he or she possesses enough factors to surpass an appropriate disease threshold. (C) A situation in which multiple factors work to induce quantitative variation over and above the effects of a factor with a relatively large effect on a trait distribution.

attention should be focused not on actual endpoints such as a disease state but rather on the "intermediate" quantitative phenotypes known or hypothesized to influence the relevant endpoint. Since these phenotypes are less removed from the loci being sought after in a gene mapping experiment (in terms of the biological chain of events leading from a relevant mutation to a disease phenotype), it may be easier to identify the genes influencing them. As a counterargument to the notion that all phenotypes are quantitative in nature, one could argue that the very genetic elements that influence disease or phenotype expression—namely, alleles and mutations—are discrete. Whether one treats the trait of interest as qualitative or quantitative is not a matter of convenience, because it is known that treating a quantitative trait as a qualitative trait (e.g., by splitting individuals into "low" and "high" groups based on some cutoff point of the trait distribution, as is done in the determination of "hypertensives" on the basis of blood-pressure values greater than 140/90 mm Hg) can reduce the ability to map relevant genes (Ref. 8, pages 162–164).

FOUR GENE MAPPING STRATEGIES

Linkage Analysis Within Pedigrees

Historically, the most widely used gene mapping analysis method has involved tracing the cosegregation and transmission of alleles that can be observed (e.g., at marker loci or candidate loci) with *putative* or *hypothetical* alleles that control a disease within pedigrees. Consider Figure 2, in which the "number" locus is the disease locus, with allele 4 the cause of the disease (diseased individuals are denoted by the shading—note that each possesses one copy of the 4 allele, suggesting dominance). If alleles at this locus are unobserved, as would be the case if one wanted to map the locus, then one would have to merely assume that disease alleles are segregating within the pedigree that are controlling the disease, and then attempt to trace the cosegregation of these assumed alleles with the observed marker (letter) alleles. Tests of the likelihood of recombination between the assumed disease alleles and the marker alleles would then proceed by accounting for the fact that diseased individuals are likely to have the same allele or genotype at the disease locus. If little evidence for recombination between the marker and assumed disease alleles can be extracted from the data, then the putative disease locus is taken to be near the marker locus.

Table 1 lists studies that have used standard pedigree-analysis methods for mapping genes. Since the alleles that control the disease or phenotype are not observed, assumptions about their number and action must be made. These assumptions are often summarized in two mathematical constructs: the

Table 1 Examples of Published Studies Using Linkage Analysis Techniques with Pedigree Data

Ref.	Disease or trait	Study features
Coon et al., 1993 (22)	Manic-depression	Genome scan—largely negative results
Elston et al., 1976 (23)	Hypercholesteremia	Quantitative trait
Hall et al., 1990 (24)	Breast cancer	Early-onset disease
Kravitz et al., 1979 (25)	Hemochromatosis	Quantitative trait
Shimkets et al., 1994 (26)	Hypertension	Rare form of hypertension
Lifton et al., 1992 (27)	Hypertension	Rare form of hypertension
Schellenberg et al., 1992 (28)	Alzheimer's disease	Early-onset disease
Sherrington et al., 1988 (29)	Schizophrenia	Called into question
Smith et al., 1983 (30)	Reading disability	Behavioral trait
Thein et al., 1994 (31)	Fetal hemoglobin persistance	Quantitative trait—multi-locus analysis
Tienari et al., 1994 (32)	Multiple sclerosis	Multilocus analysis
Tomforhde et al., 1994 (33)	Psoriasis	Heterogeneity detected

penetrance function, which describes the probability that an individual with a certain genotype at the hypothesized disease locus will, in fact, express the disease, and the *transmission* function, which describes the probability that a parent with certain genotypes will transmit specific alleles to his or her offspring. The transmission function is typically based on Mendel's laws and must include information about how far the putative disease locus is hypothesized to be from the marker locus so that some assessment can be made about the probability of recombination events between the observed marker locus and the unobserved disease locus. The transmission function can include information from more than one marker locus. Also, when no information about a pedigree member's parents is available, the probability that an individual possesses certain genotypes at the disease and marker locus is no longer a function of what alleles may have been transmitted to him or her, but is a function of how frequent relevant alleles are in the population at large. Such information must often be estimated prior to the analysis of the pedigree.

Pedigree analysis proceeds by considering the possibility that each person in the pedigree carries a particular genotype at the (unobserved) disease locus. Statistical inferences can then be made about the likelihood that the pedigree members possess certain underlying disease-locus genotypes given their observed marker genotypes. This kind of analysis therefore involves consideration of *all possible* disease-locus genotype configurations compatible with the structure of the pedigree. If one can find statistical evidence suggesting that little recombination between the marker-locus alleles and the putative disease-locus alleles must have occurred given the pattern of inheritance of the disease, then one must assume that the putative disease locus is in close proximity (i.e., linked) to the marker locus. By varying the position of the putative disease locus relative to a group of markers whose genomic positions are known, and reassessing cosegregation and possible recombination between the hypothetical disease alleles and relevant observed marker alleles, one can attempt to position, or map, the disease locus. The location of the marker loci producing the greatest statistical evidence for linkage to the putative disease locus can be taken to be the rough location of the disease locus.

As described below, the greatest problems with pedigree-based linkage analysis arise from the fact that one not only must assume things about the number and general effects of the alleles at the disease locus, but one must also explicitly consider the probability that each member of the pedigree possesses each of the possible genotypes at the disease locus. To give the essence of this problem, consider that there may be three possible genotypes at a marker locus (two of which, say, may cause disease) and one wants to analyze a pedigree consisting of a simple nuclear family (mother, father, offspring) with seven children; then one might have to evaluate 2703 possible disease-locus genotype configurations compatible with the pedigree. Despite the development of clever computer algorithms to analyze pedigrees, the large number of genotype configurations one must consider in this type of analysis can cause severe computational problems. This problem becomes more pronounced if one considers the action of multiple genes predisposing to a disease.

Allele Sharing Among Pairs of Relatives

An alternative to pedigree analysis is the computation of IBD and IBS information and the exploitation of the simple concept that persons sharing (relevant) genetic material should be similar phenotypically, as described in the section above on IBD vs. IBS. The analytical procedures for allele-sharing methods are fairly straightforward (although, as discussed later, some complications can arise when one uses IBS information): some measure of phenotypic similarity is computed for each pair (e.g., squared difference in trait value for quantitative traits; simple concordance for qualitative traits) and is compared to the proportion of genetic material shared IBD or IBS at a particular (marker) locus or genomic region for those pairs. If a correlation between phenotypical similarity and degree of IBD/IBS sharing exists, then it is likely that a disease locus resides within or near the genomic region for which the IBD/IBS information was computed. By testing the association between a phenotypical similarity measure and IBD/IBS information gathered from different loci, one can attempt to determine the locus most likely to be near a disease locus and thereby attempt to map the disease locus.

Table 2 lists studies that have used allele-sharing methods to map a disease locus. Three aspects of allele-sharing methods should be emphasized. First, the relationship between greater IBD/IBS sharing and phenotypical similarity will hold only if the locus (or region) for which IBD/IBS information was computed is close to the actual disease locus. The information one would normally want in an allele-sharing study would be IBD/IBS sharing information at the actual disease locus. Since this will typically not be available a priori (hence the need to map the locus!), one must compute IBD/IBS information at landmark (i.e., marker) loci and assess the relationship between this information and the measurement of phenotypical similarity. If a marker locus is not near the actual disease locus, IBD/IBS sharing at this locus will not provide a good estimate of IBD/IBS sharing at the disease locus, and no relationship between IBD/IBS sharing and phenotypical similarity will hold (due to, for example, recombination). Second, if one uses IBS information exclusively, allele-frequency information must be incorporated into the analysis. The reason for this can be best explained in the context of an example. Consider a locus with two alleles, one of which occurs with a frequency of 0.99 in the population and the other with a frequency of 0.01. Obviously, a pair of individuals will be likely to share IBS the allele with greater frequency, not because it might be associated with a disease or phenotype they share but simply because the other allele is so rare. The greater the number of equally frequent alleles at a locus, the more IBS information computed at that locus approaches the IBD information that could be computed at that locus. Third, one need not necessarily have individuals with contrasting (i.e., "dissimilar") phenotypes in a sample to carry out allele-sharing studies. One could simply compare the frequency with which alleles or regions are shared IBD or IBS among pairs of individuals both affected by a disease against what would be expected if that locus or region was *not* assumed to be near a disease locus. If the departure from this expectation is great, one could infer that a disease locus is near the region or locus tested. The use of only affected individuals to map genes in this manner is known as affected pedigree member (APM) analysis.

Case-Control Association Studies

By far the simplest, but most assumption-laden, way of trying to map disease loci is through the use of association tests testing the relationship between alleles at a putative disease locus and the disease itself. The motivation behind such studies is straightforward: if a particular allele at a locus is responsible

Table 2 Examples of Published Studies Using Allele-Sharing Methods

Ref.	Disease or trait	Study features
Caulfield et al., 1993 (34)	Hypertension	IBS/APM methods
Hamer et al., 1993 (35)	Male homosexuality	X-linkage; IBS methods
Jeunemaitre et al., 1992 (36)	Hypertension	IBS/APM methods
Marsh et al., 1994 (37)	Total serum IgE	Haseman-Elston; quantitative trait
Morrison et al., 1994 (38)	Bone density	Modified Haseman-Elston for twins
Prochazka et al., 1993 (39)	Maximal insulin action	Haseman-Elston; quantitative trait
Angrist et al., 1993 (40)	Hirschprung's disease	Multiple methods
Wilson et al., 1991 (41)	Hypertension-related traits	Haseman-Elston; quantitative trait

for a disease, then diseased individuals should possess this allele with a greater frequency than individuals not affected by the disease. To test this hypothesis, one would need a sample of affected (case) and unaffected (control) individuals and relevant genotype information from these individuals. Table 3 lists studies that have been carried out using association studies to investigate the relationship between alleles at a particular locus and a disease. The primary assumption behind association studies is that some *causal* relationship exists between the alleles at the locus being tested and the disease. Investigating a direct association between alleles at loci merely near a disease locus and a disease will rarely work since, for most populations (but see the following section), alleles at two different loci will be in linkage equilibrium; i.e., alleles at one locus will not be good "surrogates" for alleles at the other. Thus, the assumption of causality is highly problematic. In addition, there are other problems that plague the use of association studies for gene mapping purposes, one of which—the existence of a number of logical or "candidate" loci one might want to investigate for a particular disease—was mentioned earlier.

An additional problem is the choice of an appropriate control group. Obviously, if one chooses a control group from a population that possesses alleles at the locus being tested with different frequencies than the population from which the cases were selected, then one might observe a misleading association between certain alleles and the disease, i.e., one due not to a causal relationship between the disease and the alleles tested, but to the unequal allele frequencies between cases and controls (2). Since testing all possible loci on the human genome that manifest variants or alleles that could, in principle, be responsible for a disease would be too vast an undertaking, association studies can best be utilized in the very late stages of a gene mapping study in which a small genomic region has been identified that is likely to carry a disease locus and one merely wants to test if specific loci within this region actually encode alleles that determine the disease.

Population Disequilibrium Mapping

One very underutilized method for mapping genes that is gaining increased attention makes use of aspects of pedigree analysis, allele sharing, and association-study gene mapping methods. This method involves the investigation of the cosegregation of putative disease genes with marker-locus alleles among individuals living in geographically isolated, relatively young populations (10,11). The reason that one might want to use isolated populations is that the "gene pool" (i.e., the total number of variants at important, potentially disease-related, loci) is limited. The reason one might want to investigate a young population (in terms of the number of generations that have passed since the original "founding" members of the population existed) is that the

Table 3 Examples of Published Studies Using Association Studies

Ref.	Disease or trait	Locus	Study features
Ballinger et al., 1992 (42)	Diabetes/deafness syndrome	mtDNA	Mitochondrial DNA mutation
Blum et al., 1990 (43)	Alcoholism	Dopamine D_2	Equivocal further studies
Corder et al., 1994 (44)	Alzheimer's disease	APO-E	Late onset only
Dover et al., 1992 (45)	Sickle cell anemia	Xp22.2	X-linkage
Kajiwara et al., 1994 (46)	Retinitis pigmentosa	RDS; ROM	Only double heterozygotes affected
Plomin et al., 1994 (47)	IQ	60 markers	Largely negative
Schachter et al., 1994 (48)	Longevity	ACE; APO-E	Study of centenarians
Schunkert et al., 1994 (49)	Left ventricular hypertrophy	ACE	Candidate locus for hypertension
Simon et al., 1977 (50)	Idiopathic hemachromatosis	HLA	Quantitative trait
Todd et al., 1987 (51)	Insulin-dependent diabetes	HLA-DQB	Sequence analysis
Ward et al., 1993 (52)	Pre-eclampsia	Angiotensinogen	Hypertension-related

original disease mutations possessed by the founding members of the population that have been transmitted to their offspring are probably in *linkage disequilibrium* with alleles at loci neighboring the disease loci. Obviously, the older a population is, the greater the chance that recombination events will have depleted the association between alleles at neighboring loci that were on the original founding members' chromosomes (see Figure 6). Thus, with young, geographically isolated populations, one could look for regions of greater than expected IBD/IBS sharing among pairs of individuals affected by a disease. In addition, if the disequilibrium between alleles at loci is pronounced, one could simply investigate the association between alleles at various marker loci and a disease, since the alleles at the marker loci will, by virtue of disequilibrium, be good surrogates of the actual alleles at the disease locus (Figure 6). Table 4 lists studies that have used linkage-disequilibrium techniques to map a disease locus. Of the studies listed, only the study by Kerem et al. in 1989 (12) did not make use of a relatively young and isolated population for the *express* purpose of exploiting linkage disequilibrium.

COMPLEXITIES AND COMPLICATIONS

A number of biological and statistical phenomena can cause problems for disease gene mapping. Although many of the problems described may seem obvious, their solutions are far from trivial and have been the object of a great deal of debate and research. We outline some of the relevant phenomena below.

Reduced Penetrance and Phenocopies

Very often genes do not completely determine the presence of disease; rather, other factors, such as other genes, environmental factors such as diet, or completely stochastic factors, work together with a particular gene to cause

Table 4   Examples of Published Studies Using Linkage-Disequilibrium Mapping Methods

Ref.	Disease or trait	Study features
Hastbacka et al., 1992 (53)	Dwarfism	Finnish isolate
Hastbacka et al., 1994 (54)	Dwarfism	Refined analysis Hastbacka et al., 1992
Houwen et al., 1994 (55)	Intraheptic cholestasis	Isolate from the Netherlands
Kerem et al., 1989 (12)	Cystic fibrosis (CF)	Led to the identification of CF mutations
Lehesjoki et al., 1993 (56)	Myoclonus epilepsy	Finnish isolate
Puffenberger et al., 1994 (57)	Hirschprung's disease	Mennonite population

disease. If it is indeed the case that a particular gene does not determine a disease, then the mere possession of that gene is not sufficient to produce the disease. This suggests that many individuals could carry a "disease gene" and yet not manifest the disease. The phrase used to describe what occurs when it is not 100% certain that a person carrying a particular gene will manifest a disease is *reduced penetrance.* Reduced penetrance can cause problems for gene mapping studies since it will be unclear if a person does not manifest a certain disease because he or she does not have an appropriate disease gene or because of reduced penetrance associated with the disease genes. A problem similar to reduced penetrance results from *phenocopies*—individuals who display characteristics similar to those associated with the disease or trait being studied but have a nongenetic etiology. Phenocopies plague many gene mapping studies involving psychiatric illnesses because the behaviors used to define many psychiatric disease states are often exhibited by individuals who do not actually possess the underlying disease pathology.

Multiple Genes: Heterogeneity vs. Epistasis

As noted above, multiple genes can work to determine a trait. These genes can work either independently or in tandem. When alleles or genotypes at different loci (e.g., on different chromosomes or at opposite ends of a chromosome) can cause a disease independently of one another, locus *heterogeneity* is said to exist. If alleles at two or more different loci are required for a disease to be present, then *epistasis*, or epistatic interaction among the loci, is said to exist. Figure 8 offers a graphical representation of the possible biological import of the concepts of heterogeneity and epistasis. The upper panel of Figure 8 depicts a sequence or chain of biological events mediating disease or phenotype expression. If any link in the chain is upset, the entire sequence is dysfunctional and can lead to disease. This situation may characterize locus heterogeneity, since different loci may influence or control each link in the chain and different persons may have mutations that upset different links in the chain. The lower panel of Figure 8 depicts pathways influenced by different loci that mediate disease expression. If any of these pathways is upset, the normal functioning of another pathway can compensate for it and thereby prevent disease. If all the pathways are dysfunctional, disease will occur. This situation may characterize epistasis. The problems for gene mapping studies caused by heterogeneity, epistasis, and multiple locus involvement in general are enormous, because one may not be able to identify the effects of one gene without knowledge of the others. Since the goal of gene mapping studies is to identify the genes themselves, prior knowledge of the number and interdependence of genes will rarely, if ever, exist. As a result, the only strategies that can be used to combat problems induced by multiple genetic determinants for a disease would involve either the use of statistical methodologies powerful enough to detect the slight association between one of a number of genes influencing a disease and the disease itself or the use of statistical methodologies designed to simultaneously map two or more loci.

Environmental Influences

Many diseases, such as hypertension and other cardiovascular diseases, have strong environmental components, such as diet, exercise, and smoking. If such components are not accounted for properly in a gene mapping study, the power to detect relevant genetic effects will be compromised, since, as with multiple genetic influences, reduced penetrance, and phenocopies, it will be equivocal in certain instances as to whether a particular individual has (or does not have) a disease because he or she possesses a certain genotype or because he or she has been exposed to certain environmental stimuli.

Pleiotropy

One of the most underrecognized and unexplored genetic phenomena in gene mapping studies is pleiotropy, which refers to the effects of a gene on multiple phenotypical endpoints. Cystic fibrosis is a disease caused by mutations at a

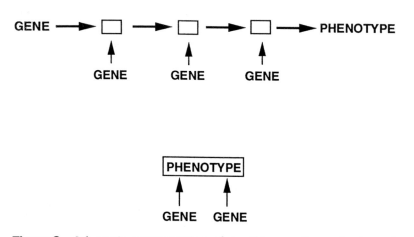

Figure 8   Schematic representation of possible genetic mechanisms that induce locus heterogeneity and epistasis. For both situations it is assumed that some biochemical or physiological pathway is influenced by genes at different loci. The upper panel suggests that if any link in the chain of events leading to the control of a phenotype is disrupted, the normal phenotype will be perturbed (e.g., disease will result). This may characterize locus heterogeneity. The lower panel suggests that disease will result only if each of two independent pathways is perturbed. This may characterize epistasis.

single genetic locus. However, its features are numerous. For example, cystic fibrosis patients exhibit problems with the pancreas and lung parenchyma. In addition, male cystic fibrosis patients are often infertile. Thus, the mutation possessed by cystic fibrosis patients influences many different physiological systems and thereby exhibits pleiotropy. Some cardiovascular diseases manifest themselves in syndromes, such as the familial dyslipidemic syndrome described by Williams et al. (13). It is an open question as to whether the features of such syndromes are due to a gene with pleiotropic effects or multiple genes, each influencing a different phenotypical feature. Obviously, making the assumption that a single gene (or a few genes) explains the multiple endpoints associated with a disease can have dramatic effects on how appropriate disease-gene mapping studies should be designed and carried out.

Development

Ignoring developmental phenomena in gene mapping studies can also have dramatic negative effects on the ability to detect relevant genes. For example, if it is the case that a gene influencing susceptibility to a particular disease cannot induce its deleterious effects until puberty is reached, then using prepubertal individuals in a sample for a gene mapping study would be inappropriate. It has been argued that hypertension may be a disorder of growth whose genetic basis has such properties (14). One simple way to avoid the problems induced by developmental phenomena is to study only those individuals who are unequivocally affected with the relevant disease. However, such strategies are limited in that they cannot investigate the *contribution* or role of development in the disease.

Implementation Problems

In addition to the biological phenomena described above, there are a number of implementation problems that plague gene mapping strategies. We briefly discuss four of these phenomena below.

Computational Problems

As pointed out in the discussion of pedigree-based linkage-analysis methods of gene mapping, consideration of all possible disease-genotype configurations compatible with a pedigree is required. The number of such configurations grows exponentially with the size of pedigree. This creates enormous computational problems for the analysis of pedigrees. In addition, if one was to try to map two or more loci simultaneously (as a way of accommodating multiple locus effects), then the number of possible disease-genotype configurations would grow even larger, causing even more extreme computational difficulties

(15,16). As a way of combating such computational problems, many research-ers have resorted to allele-sharing methods for gene mapping purposes.

Age of Onset

Many diseases do not manifest themselves until late in life. For instance, cardiovascular disease often does not manifest itself until middle age, and Alzheimer's disease usually does not appear until later in life. Correcting for the age specificity of a disease is important, since ignoring age of onset of disease can induce problems similar to those discussed in the context of developmental phenomena.

Statistical Problems

By testing a number of loci for linkage to a disease, as part of either a genome scan or a test of a number of candidate loci, the statistical problem associated with carrying out multiple-significance tests arises. In carrying out a large number of tests, one may see relevant test-statistic values that would normally indicate significance arising from chance alone (2). Methods for treating this problem have been developed (2) but need to be improved.

Data Collection

In collecting data for a disease-gene mapping study, one will almost assuredly purposely sample individuals who have the disease. The scheme used to bring in, or ascertain, the sample must then be accounted for in various aspects of the data analysis or important biases can result (see, e.g., Ref. 8, pages 217–222). Another implementation problem that plagues many disease gene mapping studies revolves around the nature of the disease itself. Most diseases are lethal. Therefore, it may be difficult to sample, for example, the parents of affected individuals because those parents may have been carriers of relevant genes and died as a result of the disease they induce.

Positional Cloning and the Biological Properties of Genes

As noted in the introductory sections, gene mapping strategies are typically used to identify the rough genomic position of influential trait or disease loci. Other methods are then used to refine the position of the locus and ultimately identify mutations responsible for the disease or trait. Methods that can be used to more precisely identify loci once an initial linkage has been found generally fall under the heading of *positional cloning* methods. Positional cloning assays can be technically difficult and time-consuming. In addition, the mere identifi-cation of a gene or mutation does not tell the whole story. Understanding how the product(s) of the relevant gene(s) function and what biological or biochem-ical pathways are influenced by these products are extremely important concerns for the correct assessment of the role of the genes in disease or phenotype expression. Thus, ultimately, gene mapping studies are merely one step (albeit a very early and usually necessary step) in a series of steps leading to the understanding of the genetic basis of a disease or trait.

THE FUTURE OF DISEASE-GENE MAPPING

New Technologies

One area relevant to gene mapping that will continue to see changes is the development of newer laboratory technologies. Already two very promising molecular genetic assays are under development that could radically change the way in which researchers will investigate the genetic basis of diseases. These two assays address two complementary questions. Representational difference analysis (RDA) takes the chromosomal material from two individuals and attempts to determine those regions that are "different" (i.e., possess different variants or alleles). Genomic mismatch scanning (GMS) seeks to identify the regions that are "alike" or inherited from a common ancestor (17). With these assays, one might be able to take, for instance, individuals affected with a disease and easily determine regions of the genome that they share (and do not share). Those regions that are common to the affected individuals may then be responsible for the disease or phenotype they also share. Many questions will have to be addressed before such assays become routinely implemented. However, they represent the same kind of technological drive that made current genome-scan and candidate-gene-analysis technologies possible.

Isolated and Admixed Populations

As noted in "Population Disequilibrium Mapping" above, the use of young, isolated populations or recently admixed populations for gene mapping purposes is advantageous because of the probable linkage disequilibrium between alleles in these populations. With this in mind, the search for and study of remote or recently admixed populations will undoubtedly gain more attention. This is not to say that anthropological and cross-culture epidemiological studies have not been pursued or given attention before, but rather that they have not been exploited for gene-mapping purposes to an appropriate degree. One problem that could emerge with the greater use of isolated populations for gene mapping purposes, however, is that the genes found in these populations may not actually be present in other populations (due to between-population locus heterogeneity or the simple unequal frequency of alleles between the isolate and other populations) and therefore will not help in isolating factors responsible for the garden-variety diseases seen in large, outbred, freely mixing populations, such as urban U.S. populations.

Better Phenotyping

One way of reducing heterogeneity and the chance of infecting a sample with phenocopies is to use stricter criteria for selecting individuals for study (1). This obvious suggestion has been given a lot of attention, but is often not utilized to great advantage. For instance, hypertension is known to be a multifactorial disease with multiple genetic determinants. It is also known that many hypertensives often possess similar distinguishing phenotypical features beyond high blood pressure (13). If one sampled hypertensive individuals who possess not only high blood-pressure values but also other features (e.g., high lipid levels, obesity, or salt sensitivity), then one might have a better chance of mapping a gene responsible for a form of hypertension characterized by those other features. Obviously, this suggestion is not without its problems, since it may be that the "other features" used to distinguish a subset of individuals sampled for the purpose of reducing possible heterogeneity are characteristic not of a unique disease syndrome with unique genetic determinants but merely occasional sequelae to a more generic form of the disease. Despite this fact, nongenetic technological advances in the medical sciences that can help better identify unique disease states and processes will undoubtedly advance the material one can draw on to carry out genetic studies.

Determining Candidate Genes

As the catalog of isolated human genes grows, so will the list of potential candidate genes for any disease (6). What is more, if all functional genes in the human genome are identified, one could in principle design simple—but more than likely costly—studies to test all of them for a certain disease. Since such studies are futuristic at best at this point in time, the determination of compelling candidate loci for study with human diseases is an important topic. Currently, the two most useful methods for determining good candidate loci involve the use of model organisms and a better appreciation of human biochemistry and physiology.

Homology Exploitation

Mapping genes that influence a relevant phenotype in a model organism such as a rat or mouse can be tremendously insightful and useful. Since model organisms can be bred and manipulated in ways that are unethical in human contexts, mapping genes through genome-scan technologies that influence model organism phenotypes is much easier than mapping genes influencing human phenotypes (2). Once a gene has been identified in a model organism, its human evolutionary homolog can be identified with the aid of modern laboratory methods (18). This human counterpart to a model organism gene can then be assessed for the disease or phenotype in question. A problem with the model organism/human homology strategy is that often genes have adopted different roles during the course of evolution, suggesting that the mere identification of a gene that influences a model organism's phenotype is no guarantee that the human homolog of that gene will influence the human counterpart of the appropriate model organism's phenotype.

Better Physiology

Obviously, one way to expedite the process of determining compelling candidate genes for a particular disease is to expand understanding of biochemical and physiological processes associated with a disease. If, for example, it can be shown that a certain system or biological phenomenon is implicated in disease pathogenesis, then studying genes known to influence, play a role in, or have their products exist in that system or phenomenon is logical. Such integrative and interdisciplinary research efforts, which are aimed at a comprehensive understanding of the disease, are often difficult to organize but ultimately seem to have no simple substitute.

Pharmacogenetics

Another strategy for understanding the genetic basis of a particular disease might involve the study or isolation of genes that modify responses to particular disease treatments or drugs. Consider the fact that many (palliative) treatments and preventive agents for certain diseases exist, and have been designed to interact with or affect particular systems or tissues known to be influenced by, or involved in, the pathogenesis of those diseases. For example, the renin-angiotensin system, which is known to be involved in blood-pressure regulation, is the target of the class of antihypertensive medications known as angiotensin-converting enzyme (ACE) inhibitors, since dysfunctions in components of the renin angiotensin system can lead to the dysregulation of blood pressure and hypertension. If mutations could be found that influence responsiveness to ACE inhibitor therapy, then not only would a hypertension-modifying gene be found, but the potential pathway that this gene works through to induce its deleterious effects (i.e., the renin-angiotensin system) would be obtained virtually by default. Such studies can be tremendously useful in providing information about the broader physiological significance of a particular gene and its relation to a disease, and would take their lead from standard pharmacogenetic analyses (19).

Better Analytic Strategies

One area that will undoubtedly receive immediate attention for the purpose of advancing disease gene mapping is the development of novel and more powerful statistical methods and analytical strategies. Although methods that can accommodate multiple gene effects, environmental influences, computational problems, multiple phenotypes, multiple comparisons, and other statistical issues are being developed, there is plenty of room for improvement. In addition, as newer technologies are developed, such as RDA and GMS, analytical devices that make appropriate use of the information they provide will also need to be developed.

CONCLUSION

The foregoing discussion has tried to make clear the strategies and issues associated with the modern disease-gene mapping. Space limitations have prevented us from doing justice to the more subtle and technical aspects behind these strategies and issues. However, a number of good books exist that discuss relevant concepts in greater detail, and we refer the interested reader to them (8,20,21). We also suggest reading papers listed in Tables 1–4 as a way of getting acquainted with the aims, scope, practical utility, and general issues surrounding disease-gene mapping. Ultimately, gene mapping and gene identification strategies in general represent some of the most exciting scientific undertakings known, and will only be surpassed by the knowledge that will be gained from their application, as the rest of this volume makes clear.

ACKNOWLEDGMENTS

We would like to thank Audrey Lynn and Steve Mockrin for critically reading and commenting on earlier versions of this chapter and Beatrix Katona for preparing manuscripts.

REFERENCES

1. McKusick VA. Mendelian Inheritance in Man: Catalogs of Autosomal Dominant, Recessive, and X-linked Phenotypes. Baltimore: Johns Hopkins University Press, 1992.
2. Lander ES, Schork NJ. Genetic dissection of complex traits. Science 1994; 265:2037–2048.
3. Thompson MW, McInnes RR, Willard HF. Genetics in Medicine. 5th ed. Philadelphia: WB Saunders, 1991.
4. Rothwell NV. Understanding Genetics. New York: Wiley-Liss, 1993.

5. Schleif RF. Genetics and Molecular Biology. Reading, MA: Addison-Wesley, 1986.
6. Sobell JL, Heston LL, Sommer SS. Delineation of genetic predisposition to multifactorial disease: A general approach on the threshold of feasibility. Genomics 1992; 12:1-6.
7. CHLC. A comprehensive human linkage map with centimorgan density. Science 1994; 265:2049-2054.
8. Ott J. Analysis of Human Genetic Linkage. 2nd ed. Baltimore: Johns Hopkins University Press, 1991.
9. Haseman JK, Elston RC. The investigation of linkage between a quantitative trait and marker locus. Behav Genet 1972; 2:3-19.
10. de la Chapelle A. Disease gene mapping in isolated human populations: the example of Finland. J Med Genet 1993; 30:857-865.
11. Jorde LB. Linkage disequilibrium as a gene-mapping tool. Am J Hum Genet 1995; 56:11-14.
12. Kerem BS, et al. Identification of the cystic fibrosis gene: Genetic analysis. Science 1989; 245:1073-1080.
13. Williams RR, et al. Familial dyslipidemic hypertension: evidence from 58 Utah families for a syndrome present in approximately 12% of patients with essential hypertension. JAMA 1988; 259:3579-3586.
14. Schork NJ, Jokelainen P, Grant EJ, Schork MA, Weder AB. Relationship between growth and blood pressure in inbred rats. Am J Physiol 1994; 266:R702-R708.
15. Schork NJ. Efficient computation of patterned covariance matrix mixed models in quantitative segregation analysis. Genet Epidemiol 1991; 8:29-46.
16. Ott J. Maximum likelihood estimation by counting methods under polygenic and mixed models in human pedigrees. Am J Hum Genet 1979; 31:165-179.
17. Aldhous P. Fast tracks to disease genes. Science 1994; 265:2008-2010.
18. Copeland NG, et al. A genetic linkage map of the mouse: Current applications and future prospects. Science 1993; 262:57-66.
19. LaDu B. Overview of pharmacogenetics. In: Kalow W, ed. Pharmacogenetics of Drug Metabolism. New York: Pergamon Press, 1992:1-12.
20. Bishop MJ. Guide to Human Genome Computing. San Diego: Academic Press, 1994.
21. Terwilliger JD, Ott J. Handbook of Human Genetic Linkage. Baltimore: Johns Hopkins University Press, 1994.
22. Coon H, et al. A genome-wide search for genes predisposing to manic-depression, assuming autosomal dominant inheritance. Am J Hum Genet 1993; 52:1234-1249.
23. Elston RC, Namboodiri KK, Go RCP, Siervogel RM, Glueck CJ. Probable linkage between essential familial hypercholesterolemia and third complement component (C3). Birth Defects 1976; 3:294-297.
24. Hall JM, et al. Linkage of early-onset familial breast cancer to chromosome 17q21. Science 1990; 250:1684-1689.
25. Kravitz K, et al. Genetic linkage between hereditary hemochromatosis and HLA. Am J Hum Genet 1979; 31:601-619.
26. Shimkets RA, et al. Liddle's syndrome: Heritable human hypertension caused by mutations in the Beta subunit of the epithelial sodium channel. Cell 1994; 79:407-414.

27. Lifton RP, et al. A chimeric 11Beta-hydroxylase aldosterone synthase gene causes gluco-corticoid-remediable aldosteronism and human hypertension. Nature 1992; 355:262-265.

28. Schellenberg GD, et al. Genetic linkage evidence for a familial Alzheimer's disease locus on chromosome 14. Science 1992; 258:668-671.

29. Sherrington R, et al. Localization of a susceptibility locus for schizophrenia on chromosome 5. Nature 1988; 336(10):164-170.

30. Smith SD, Kimberling WJ, Pennington BF, Lubs HA. Specific reading disability: Identification of an inherited form through linkage analysis. Science 1983; 219:1345-1347.

31. Thein SL, et al. Detection of a major gene for heterocellular hereditary persistence of fetal hemoglobin after accounting for genetic modifiers. Am J Hum Genet 1994; 54:214-228.

32. Tienari PJ, Terwilliger JD, Ott J, Palo J, Peltonen L. Two-locus linkage analysis in multiple sclerosis. Genomics 1994; 19:320-325.

33. Tomforhde J, et al. Gene for familial psoriasis susceptibility mapped to the distal end of human chromosome 17q. Science 1994; 264:1141-1145.

34. Caulfield M, et al. Linkage of the angiotensinogen gene to essential hypertension. N Engl J Med 1994; 330:1629-1633.

35. Hamer DH, Hu S, Magnuson VL, Hu N, Pattalucci AM. A linkage between DNA markers on the X chromosome and male sexual orientation. Science 1993; 261:321-327.

36. Jeunemaitre X, et al. Molecular basis of human hypertension: role of angiotensinogen. Cell 1992; 71:169-180.

37. Marsh DG, et al. Linkage analysis of IL4 and other chromosome 5q31.1 markers and toal serum immunolobulin E concentrations. Science 1994; 264:1152-1155.

38. Morrison NA, et al. Prediction of bone density from vitamin D receptor alleles. Nature 1994; 367:284-287.

39. Prochazka M, et al. Linkage of chromosomal markers on 4q with a putative gene determining maximal insulin action in Pima indians. Diabetes 1993; 42:514-519.

40. Angrist M, et al. A gene for Hirschprung disease (megacolon) in the pericentromeric region of human chromosome 10. Nature Genet 1993; 4:351-356.

41. Wilson AF, Elston RC, Tran LD, Siervogel RM. Use of robust sib pair methods to screen for single locus, multiple loci, and pleiotropic effects: application to traits related to hypertension. Am J Hum Genet 1991; 48:862-871.

42. Ballinger SW, et al. Maternally transmitted diabetes and deafness associated with a 10.4 kb mitochondrial DNA deletion. Nature Genet 1992; 1:11.

43. Blum K, et al. Allelic association of human dopamine D2 receptor gene in alcoholism. JAMA 1990; 263:2055-2060.

44. Corder EH, et al. Protective effect of apolipoprotein E type 2 allele for late onset Alzheimer disease. Nature Genet 1994; 7:180-183.

45. Dover G, et al. Fetal hemoglobin levels in sickle cell disease and normal individuals are partially controlled by an X-linked gene located at Xp22.2. Blood 1992; 80(3):816-824.

46. Kajiwara K, Berson EL, Dryja TP. Digenic retinitis pigmentosa due to mutations

at the unlinked peripherin/RDS and ROM 1 loci. Science 1994; 264(10):1604–1608.

47. Plomin R, et al. DNA markers associated with high versus low IQ: the IQ Quantitative Trait Loci (QTL) Project. Behav Genet 1994; 24:107–118.

48. Schachter F, et al. Genetic association with human longevity at the APOE and ACE loci. Nature Genet 1994; 6:29–32.

49. Schunkert H, et al. Association between a deletion polymorphism of the angiotensin-converting-enzyme gene and left ventricular hypertrophy. N Engl J Med 1994; 330:1634–1638.

50. Simon M, Bourel M, Genetet B, Fauchet R, Edan G, Brissot P. Idiopathic hemochromatosis and iron overload in alcoholic liver disease: differentiation by HLA phenotype. Gastroenterology 1977; 73:655–658.

51. Todd JA, Bell JI, McDevitt HO. HLA-DQ gene contributes to susceptibility and resistance to insulin-dependent diabetes mellitus. Nature 1987; 329(15):599–604.

52. Ward K, et al. A molecular variant of angiotensinogen associated with preeclampsia. Nature Genet 1993; 4:59–61.

53. Hastbacka J, delaChapelle A, Kaitila I, Sistonen P, Weaver A, Lander ES. Linkage disequilibrium mapping in isolated founder populations: diastrophic dysplasia in Finland. Nature Genet 1992; 2:204–211.

54. Hastbacka J, et al. The diastrophic dysplasia gene encodes a novel sulfate transporter: Positional cloning by fine-structure linkage disequilibrium mapping. Cell 1994; 78:1073–1087.

55. Houwen RHJ, et al. Genome screening by searching for shared segments: mapping a gene for benign recurrent intrahepatic cholestasis. Nature Genet 1994; 8:380–386.

56. Lehesjoki AE, et al. Localization of the EPM1 gene for progressive myoclonus epilepsy on chromosome 21: linkage disequilibrium allows high resolution mapping. Hum Molec Genet 1993; 2:1229–1234.

57. Puffenberger EG, et al. Identity-by-descent and association mapping of a recessive gene for Hirschprung disease on human chromosome 13q22. Hum Molec Genet 1994.

3

Molecular Genetics of Human Hypertension

Richard P. Lifton
Howard Hughes Medical Institute
Boyer Center for Molecular Medicine
Yale University School of Medicine
New Haven, Connecticut

HYPERTENSION AS A GENETIC TRAIT

Hypertension is a common trait affecting 50 million Americans and contributing to over 200,000 deaths annually from myocardial infarction, stroke, and end-stage renal disease. Despite intensive physiological investigation, the primary determinants of hypertension remain unknown in the overwhelming majority of affected individuals. Major reasons for this include recognition that hypertension is not a single disease with a single cause in all affected subjects. Moreover, the complex interplay of different physiological systems regulating blood pressure has made it difficult to determine whether physiological abnormalities found in hypertensive patients are primary contributors to the hypertensive process, or mere secondary consequences of the true, and elusive, primary causes. A consequence of this ignorance is that our therapeutic approach to this disease is necessarily empiric, and not directed toward underlying primary abnormalities.

A complementary approach to physiological analysis is the use of genetic approaches to identify mutations that contribute to the development of hypertension. Identification of such mutations would clearly establish where primary genetic abnormalities lie, and thus would provide a clear starting point from which to determine how normal physiology is deranged to result in hypertension, permitting use of physiologically more homogeneous patient groups, as well as study of the molecular physiology of relevant variants. Identification of these elusive primary defects has the potential to explain disease pathophysiology, permit early identification of at-risk individuals, permit improved

111

therapy by tailoring treatment to primary abnormalities, and perhaps even prevent development of hypertension by preclinical intervention.

Such an approach presumes that interindividual variation in blood pressure is at least in part genetically determined. Concordant evidence from many lines of investigation indicates that this is indeed the case. Twin studies document the greater concordance of blood pressures of monozygotic compared with dizygotic twins (1); large epidemiological studies demonstrate familial aggregation of blood-pressure levels with much stronger sibling–sibling or parent–offspring correlations than spouse correlations (2). Further evidence that this familial resemblance is due to shared genes rather than shared environment comes from adoption studies that demonstrate greater concordance of blood pressures of biological siblings than of adoptive siblings living in the same household (3). Finally, the fact that single genetic determinants can indeed have strong influences on blood pressure is documented by the rare Mendelian hypertensive syndromes (4).

Estimates of the proportion of the population variance in blood pressure attributable to genetic factors are presently thought to be roughly 30–40%, indicating that identification of these genetic determinants would have substantial impact on our understanding of the pathogenesis of human hypertension (5).

The major caveat to identification of relevant genetic variants remains the multifactorial determination and etiological heterogeneity of hypertension. These features are well documented from a variety of studies, ranging from differences in clinical responses to various pharmacological agents among hypertensive subjects to differences in physiological measurements such as hemodynamic parameters or responses to saline loading, to genetic studies indicating that essential hypertension does not segregate as a single-gene trait even within hypertensive families (5). These genetic observations are further supported by experimental animal models of hypertension in which hypertension is again determined by multigenic inheritance (6). A consequence of this complexity is that hypertension does not typically segregate within families in a fashion consistent with determination by inheritance at a single locus. As a consequence of this complex determination, the enormous power of traditional genetic studies comparing the segregation of traits to the segregation of genetic markers is blunted. Genetic approaches to hypertension must permit detection of genetic effects despite these apparent complexities.

APPROACHES TO IDENTIFICATION OF GENES CONTRIBUTING TO HYPERTENSION

In the setting of multifactorial determination, a number of potentially informative genetic approaches may be considered, and it is crucial to contrast the relative merits of these in designing informative genetic studies. Among potential approaches, case-control or association studies using polymorphic markers in candidate genes, analysis of genetic linkage using either candidate genes or anonymous markers, or direct/indirect search for mutations in candidate genes followed by case-control or linkage studies may ultimately prove successful. Each approach has strengths and weaknesses that are dependent on a number of factors, including 1) the prevalence of mutations in the relevant genes and the magnitude of the effect imparted, 2) the true and unknown model of inheritance of the trait and the degree of etiological heterogeneity, and 3) the rate at which new mutations in relevant loci are introduced into the population.

Association Studies

Case-control studies contrasting frequencies of alleles at polymorphic loci in, for example, hypertensive and normotensive subjects, critically rely on the presence of linkage disequilibrium between the predisposing alleles at the trait locus and particular alleles at the marker locus (linkage disequilibrium refers to a situation in which particular alleles at one locus are found to be linked in cis with particular alleles at another locus more often than expected by chance alone). Over long periods of time, true linkage disequilibrium between a functional variant and marker alleles will typically occur only when 1) the marker loci are so close that meiotic recombination rarely if ever occurs in the interval between them and 2) the functional variants at the trait locus have been introduced into the population only one or a few times such that particular marker alleles are preferentially associated with functional variants. An exception to this characterization is seen when a variant has been introduced recently enough into a population that only a small number of meioses have occurred, permitting linkage disequilibrium to extend over larger genetic distances. From such considerations, it can be seen that some classes of mutation will be unlikely to be detectable by disequilibrium studies of this sort. By example, for mutations that significantly impair reproductive fitness, a high proportion of independent cases will have new or recently introduced mutations, reducing the likelihood of finding a significant proportion of disease alleles descendant from a common ancestral allele.

As a consequence of such considerations, association studies are generally appropriate only with markers tightly linked to the presumed trait locus, as for the case of a candidate gene study in which the recombination fraction between the test marker and the (presumed) trait locus is postulated to be zero. An exception to this rule is in the setting of populations recently descendant from a small number of founders, in which case more extended disequilibrium may been seen. Moreover, because of the assumptions that underlie disequilibrium, association studies of this sort may miss even extremely common variants contributing to hypertension if causal mutations have been introduced into the population on multiple occasions. In addition, studies of this design are commonly plagued by unrecognized stratification between cases and controls such that the cases are more common in individuals of a particular racial or ethnic or geographic origin. Because this may not be recognized or accounted for in study design, false-positive association studies are very common. Nonetheless, in the setting of pure polygenic inheritance in which inheritance at any one locus imparts only a very small phenotypical effect, association studies may have the best chance of success.

A recent modification of traditional association studies has the potential to reduce the propensity for false-positive tests of association. Tests of transmission disequilibrium are based on the principle that if inheritance of a functional variant at a locus influences a trait, parents heterozygous for such variants will transmit the allele bearing the functional variant to an affected offspring more often than the 50% frequency expected by Mendelian segregation ratios under the null hypothesis. By analyzing association in the context of segregation from heterozygous parents, this approach should reduce or eliminate false-positive studies arising from differing allele frequencies in contrasting populations, and represents an approach with great promise in the setting of complex traits such as hypertension (8).

Direct Search for Molecular Variants in Candidate Genes

A limitation of case-control studies is the requirement of linkage disequilibrium between underlying functional variants and the marker loci used in the analysis. It is readily apparent that the power of such tests could be increased if one could identify these functional variants directly. In recent years, powerful methodology for circumventing this requirement for linkage disequilibrium has been developed with the advent of methods for direct/indirect search for potentially functional variants in candidate genes. In this approach, molecular variants in candidate genes are sought by any of a number of methods that permit screening of samples from many individuals. At present, several methods with high sensitivity and high throughput are available, including single-strand conformational polymorphism (SSCP) (9) and denaturing gradient gel electrophoresis (DGGE) (10).

Direct detection of heterozygous variants by DNA sequence analysis also shows considerable promise in application to this problem. The altered DNA sequence on variants identified by any of these methods can be rapidly determined, and the potential functional significance of variants so identified can be assessed in a number of ways:

1. *Association studies*—similar to studies outlined above, with the exception that association is directly sought between putative functional variants and endpoints such as hypertension.
2. *Linkage studies*—in this case, pedigrees are constructed post hoc around branches of the pedigree segregating the relevant variants, permitting assessment of the impact of the variant on the trait of interest.
3. *Functional expression*—expression of identified variants in appropriate cells or development of transgenic/knockout animals may permit direct examination of the physiological consequences of particular variants, providing important evidence for functional consequences of identified variants.

The power of direct search approaches is that 1) they have the ability to directly identify functional genetic variants; 2) they can detect uncommon or rare variants that might be difficult or impossible to find within a large population by linkage analysis; 3) they can detect variants that impart only small effects on the phenotype that might also be difficult to detect by linkage analysis. A drawback to this approach is that negative results with direct searches can rarely exclude the possibility that undetected variants either in regions studied or in unstudied regions in cis could have substantial effects on the trait of interest.

Analysis of Genetic Linkage

In contrast to association studies, analysis of genetic linkage is not directly affected by assumptions of linkage disequilibrium, and false-positive studies are less likely to be produced by admixture of patients of different genetic backgrounds. For common and multifactorial traits, however, the power of traditional linkage strategies utilizing large extended pedigrees may be greatly diminished. Different genetic factors may contribute to development of the trait in different individuals, even within the same pedigree. Moreover, owing to delayed and/or incomplete penetrance, individuals inheriting predisposing alleles may be normotensive, further complicating analysis.

In the setting of relatively high degrees of etiological heterogeneity, delayed or incomplete penetrance, and unknown model of inheritance, an alternative approach to linkage in large pedigrees is the use of affected sibling pairs (11-13). This approach has considerable merit, since the power to detect linkage is relatively well preserved. The primary determinants of the power of affected sibling pair linkage tests are 1) the true (and unknown) number of trait-influencing loci and the magnitude of the effects imparted by each trait locus, 2) the number of affected relative pairs available for study and the ability to determine identity by descent at marker loci, 3) the informativeness of the genetic markers used for study, and 4) the recombination fraction between the trait and marker loci. From these considerations, a number of strategies to maximize the power of linkage analysis can be employed.

Reduction of the etiological heterogeneity of the study population is one avenue by which power of analysis can be increased. This can be pursued in a number of fashions. One approach is to identify intermediate phenotypes that define homogeneous subgroups of the hypertensive population. Many such intermediate phenotypes have been proposed for human hypertension; however, few of these have been shown to have Mendelian or major gene inheritance (14). For example, salt sensitivity, long purported to define a subgroup of hypertensives, shows a continuous distribution in the population, reducing the likelihood that this trait shows high heritability (15). Two traits for which evidence of major gene inheritance has been found are elevated erythrocyte sodium–lithium countertransport (16–18) and urinary kallikrein levels (19), raising the possibility of using these traits as surrogates for hypertension in linkage studies. Identification of informative kindreds segregating for these traits nonetheless remains a difficult issue.

The most extreme versions of major gene determination are the Mendelian forms of hypertension in which mutation in a single gene may be sufficient to produce elevated blood pressure; many of the currently recognized syndromes are associated with phenotypes that can distinguish affected from unaffected subjects on a basis other than blood pressure alone. Examples of these include:

1. Glucocorticoid-remediable aldosteronism, in which the ectopic production of aldosterone by the adrenal fasciculata is recognized by secretion of high levels of the signature steroid 18-oxocortisol or by complete suppression of aldosterone secretion by exogenous glucocorticoids (4)
2. Liddle's syndrome, in which affected patients have suppressed PRA and low aldosterone levels (20)
3. Apparent mineralocorticoid excess, in which type I patients have high ratios of metabolites of cortisol:cortisone in urine (21)
4. Gordon's syndrome, which is characterized by hypertension and hyperkalemia in the setting of normal glomerular filtration rate (22)

In families segregating for these traits, linkage analysis may be performed with the same power as for other monogenic traits.

Despite the dearth of solid intermediate phenotypes for use in linkage studies of human hypertension, the power of linkage studies may potentially be increased by employing measures to increase the homogeneity of the group being analyzed by stratifying the population being studied for various features such as race, severity of hypertension, age of onset, or body mass index. If, for example, the higher prevalence of salt-sensitive, low-renin hypertension seen in African Americans reflects an underlying genetic effect that is more common in this group than in others, the separate analysis of this group will increase the power to detect relevant underlying loci.

The power of linkage analysis using affected relative pairs is also highly dependent on the informativeness of the genetic markers employed. Simple restriction fragment length polymorphisms (RFLPs) generally have only two alleles and typically are heterozygous in only 40–50% of subjects. Such markers are generally of unacceptably low informativeness for such studies. An alternative type of genetic marker that can have many different alleles, and consequently high heterozygosity, is based on a variable number of repeats of simple sequence motifs such as $(GT)_n$. Such markers are common in the human genome, and are frequently highly polymorphic, with heterozygosities in excess of 70%. These markers are conveniently typed by gel electrophoresis after amplification of target segments by polymerase chain reaction, permitting rapid typing and high throughput. Dense maps of such markers spanning the human genome are now available (23,24).

The power of affected sib pair linkage tests drops off quickly as the recombination fraction between the trait and marker loci increases (25). The recombination fraction between marker locus and (putative) trait locus can be minimized by employing a candidate gene strategy, in which case the recombination fraction is assumed to be zero, again maximizing the power of linkage analysis. Nonetheless, the development of complete genetic maps of *Homo sapiens* with highly informative genetic markers now permits application of general linkage strategies to complex traits. The feasibility of such an approach has been recently demonstrated by linkage studies of type I diabetes mellitus.

A major limitation on the power of linkage analysis for complex traits is the possibility that trait loci may impart such a small incremental effect on the phenotype that the effects of any single locus may be too small to detect. In this sense an approach combining direct search for variants with linkage analysis is complementary: direct search can detect rare variants or variants imparting small effect; linkage analysis can detect effects on the trait imparted by variants anywhere in the candidate locus, and, moreover, negative results can place upper limits on the effects of inheritance at the locus on the trait.

When nothing is known about the physiological determinants of a trait, linkage with anonymous markers may prove to be the most direct route to identification of trait loci; even in the best of circumstances, however, proceeding from location of the trait locus to identification of the correct gene remains an arduous task. Moreover, for complex diseases in which recombination events between trait and marker loci cannot readily be discerned, this positional cloning will likely prove even more difficult.

Alternatively, if the physiological determinants of a trait have been well studied, a number of candidate genes may be suggested. In this case, linkage with the candidate can test the specific hypothesis that inheritance at this locus influences the trait. Hypertension is perhaps the ideal complex trait for study by a candidate approach. The leading candidate genes in hypertension are not implicated merely because they might plausibly be related to this trait, but rather because the extensive physiological study of blood-pressure regulation has demonstrated that these physiological systems are directly involved in the tonic regulation of blood pressure. These systems have been exploited in the development of therapeutic agents that have utility in treating hypertension, proving their role in blood-pressure regulation. The genes that have to date been implicated in the pathogenesis of human hypertension have been identified in large part because of this advanced understanding of relevant physiology.

Finally, another approach to identifying genetic determinants of human hypertension is to exploit experimental animal models of hypertension in which ideal genetic crosses can be constructed to permit maximal informativeness for genetic linkage studies. It is certain that chromosomal locations of genes relevant to rat hypertension can be mapped in this fashion, although proceeding from map location to mutant gene remains a daunting challenge except when linkage implicates a previously known candidate gene. Importantly, the question of the potential relevance of any loci implicated in animal models to human hypertension remains open; it clearly will be necessary to have established populations and approaches to investigate the relevance of genes found in animal models to human hypertension.

FINDING GENES THAT CAUSE HUMAN HYPERTENSION

Linkage Studies in Essential Hypertension

Over the last several years, a number of genetic studies of candidate genes in human essential hypertension have been performed. The general approach of linkage studies has been to identify highly informative genetic markers in candidate genes of interest, and genotype these markers in collections of hypertensive sibling pairs. The results are analyzed by comparing the expected number of alleles shared among sib pairs under the null hypothesis of no linkage to the experimentally observed number of alleles shared. The significance of observed results is calculated by use of a single-tailed t-test. A number of case-control studies have also been performed, in some cases in conjunction with linkage studies and in others as standalone studies. As indicated above, these latter case-control studies are often difficult to interpret because of conflicting results and incomplete documentation of the sources for ascertainment of cases and controls.

Linkage studies using affected sibling or relative pairs have to date been performed with a number of candidate genes, including the sodium-hydrogen exchanger NHE-1 (27), the renin gene (28), the angiotensin-converting enzyme gene (29), angiotensin II receptor AT1 (30), and the S_A gene (31). Linkage studies of each of these genes have been negative, with no evidence of increased allele-sharing among hypertensive sibling pairs (Table 1). It is important to note that all of these studies have been performed in two patient populations—one a population-based sample from Salt Lake City, Utah, and the second a referral population from Paris, France. These two populations are virtually exclusively Caucasian subjects of northern European descent, and consequently little inference can be made regarding the possibility that inheritance at these loci might play a role in some other populations. Similarly, these studies will have low power to exclude effects of inheritance of either common variants imparting small effects on blood pressure or rare variants with either small or even large effects.

In contrast to these negative studies, in the last few years the feasibility of using genetic approaches to identify molecular variants relevant to the pathogenesis of human hypertension has been documented by the identification of three such genes; two of these encode uncommon variants that impart large effects on blood pressure in affected subjects, while the other encodes a very common variant that may impart a modest effect on blood pressure in millions of individuals.

Table 1 Linkage Studies of Candidate Genes in Hypertensive Sibling Pairs

| Locus | No. HT sib pairs | Alleles shared | | | | |
		Observed	Expected	% excess	P value	Ref.
NHE-1	93	88	95	0	NS	27
Renin	98	140	132	5.8	NS	28
	258	328	330	0	NS	a
ACE	237	255	255	0	NS	29
AT1	267	353	362	0	NS	30
S_A	224	293	311	0	NS	31
Angiotensinogen						
Total	379	487	466	5%	0.02	32
More severe	110	159	135	17%	<0.001	32

NHE-1: sodium-hydrogen exchanger, isoform 1; ACE: angiotensin-converting enzyme; AT1: angiotensin II receptor type 1; HT: hypertensive.
[a]Jeunemaitre, Lifton, and Lalouel, unpublished data.

Role of Angiotensinogen Variants in Hypertension

A collaborative study of Salt Lake City and Paris sib pairs reported by Jeune-maitre et al. (32) has implicated molecular variants in the angiotensinogen gene in Caucasian essential hypertension. The strengths of this study were the relatively large sample, the use of independent patient sets to permit independent replication of findings, and the multiple lines of inference supporting an effect.

Evidence potentially supporting a potential role of angiotensinogen in hypertension has been present for many years, although little appreciated (33,34). Generation of angiotensin I from angiotensinogen via cleavage by renin is the rate-limiting step in the renin-angiotensin cascade (Figure 1). In

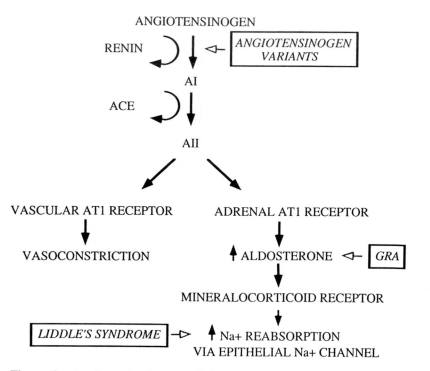

Figure 1 A schematic diagram of the renin-angiotensin system and location of inherited variants affecting blood pressure in humans. The locations at which inherited variants have been shown to affect blood pressure in humans are indicated. AI, angiotensin I; AII, angiotensin II; AT1, type I angiotensin II receptor; GRA, glucocorticoid-remediable aldosteronism.

humans, plasma levels of angiotensinogen are poised near the Km for cleavage by renin, raising the possibility that variation in angiotensinogen levels or susceptibility to cleavage by renin could alter activity of the renin-angiotensin system.

A highly informative genetic marker in the angiotensinogen gene locus was used to genotype hypertensive sibling pairs. Analysis of the results demonstrated significantly greater sharing of alleles among the combined set of sib pairs from Salt Lake City and Paris than would have been expected by chance alone (17% excess allele-sharing in more severely affected sib pairs, $p < 0.001$; Table 1) (32). These findings were significant in both populations independently when younger hypertensive sib pairs or more severely hypertensive sib pairs were considered, suggesting that molecular variants in or near the angiotensinogen gene locus affected blood pressure in these study populations. This finding was all the more striking because of the very different means of ascertainment of each of these patient samples.

This finding motivated the search for unique molecular variants in the angiotensinogen gene. Fifteen such variants were identified, and the prevalence of each was contrasted in hypertensive and control subjects. One of the identified variants, in which the amino acid threonine is substituted for methionine at codon 235 of the angiotensinogen gene (T235), was found to be significantly more prevalent in hypertensive than normotensive subjects in both the Salt Lake City and Paris patient populations [prevalence of 36% in controls, 47% in all index cases ($p < 0.001$), and 51% in more severely affected index cases ($p < 0.001$).

Because of the potential effect of varying levels of angiotensinogen on activity of the renin–angiotensin system, plasma angiotensinogen levels were contrasted in individuals of differing genotype at codon 235. Individuals homozygous for T235 had significantly higher mean levels of angiotensinogen (1582 ± 459) than either MT heterozygotes (1425 ± 344) or MM homozygotes (1313 ± 283) ($p < 0.0001$). Again, these findings were replicated independently in the Salt Lake City and Paris populations.

These three lines of evidence, which replicate in independent patient samples, implicate molecular variants of the angiotensinogen gene in the pathogenesis of human hypertension. At present, the mechanism by which these variants impart effect on blood pressure is uncertain. One possibility is that elevation of angiotensinogen levels results in modestly increased activity of the renin–angiotensin system by increasing the rate of formation of angiotensin I; in this case, one might expect such patients to reduce renin secretion by feedback inhibition, in which case such patients might constitute a subset of low-renin hypertension. The observed elevation in angiotensinogen levels could arise either by prolongation of the plasma half-life of the protein or by increasing its rate of synthesis/secretion. Alternatively, the higher angiotensinogen levels could reflect reduced susceptibility to cleavage by renin; in this case, the hypertension would result from activation of other pathways to maintain homeostasis. Further experimentation will clearly be required to distinguish among these possibilities.

Similarly, at present it is uncertain whether the variant T235 is itself the functional variant, or whether it is merely in linkage disequilibrium with the truly functional variant; at present no such subset of T235 alleles has been reported. For these reasons, it is uncertain whether all T235 alleles confer a small effect on blood pressure, or whether a subset of such alleles impart a larger effect. Further study of diverse patient populations as well as further molecular investigation will help clarify these issues.

Allele T235 comprises about 36% of all angiotensinogen alleles in the Caucasian populations studied, with the consequence that about 12% of such populations are homozygous for this allele (32). These findings suggest that angiotensinogen variants affect blood pressure in literally millions of Caucasian individuals. These observations naturally raise the question of whether this variant plays a role in other populations in which the prevalence of hypertension is higher than in Caucasians. This has been addressed by the study of a group of African Americans from Birmingham, Alabama (35). Allele T235 shows a prevalence of 85% in this population, such that 70% of subjects are homozygous for T235. Haplotype analysis as well as the nature of the mutation indicate that the T235 variant is actually ancestral to the M235 variant. Studies in small samples of African Americans have not shown association of T235 with hypertension (36). T235 is similarly highly relevant in Japanese populations, where this allele has also shown association with hypertension (37).

Recently, Caulfield et al. (38) reported finding very strong linkage disequilibrium of alleles of the highly polymorphic GT repeat at the angiotensinogen locus with hypertension, a finding not observed in the Paris or Utah populations. This report also found strong evidence of linkage of hypertension with this marker in affected relative pairs; however, the use of control allele frequencies for the expected identity by state calculations in this analysis, given the strong linkage disequilibrium, leaves open the possibility that the observed linkage could be artifactual.

In sum, the present evidence implicating angiotensinogen variants in human hypertension is quite strong, and indeed is exactly what one would expect of a common variant imparting modest effect on blood pressure. Further studies will be required to determine how variants at this locus alter normal physiology to contribute to hypertension, as will genetic epidemiological studies to determine the quantitative effect of inheritance of such variants on blood pressure. The success with this candidate approach provides considerable optimism that other loci imparting effects on blood pressure will be found.

Molecular Genetics of Mendelian Forms of Human Hypertension

The simplest forms of hypertension to analyze genetically will likely prove to be the mendelian forms of human hypertension, in which mutation in a single gene is sufficient to produce large effects on blood pressure. These studies may provide new insight into pathophysiology, and also will provide early opportunities to examine issues such as gene–gene and gene–environment interaction by studying the observed phenotypical variation in subjects who have inherited the same mutations.

The Molecular Basis of Glucocorticoid-Remediable Aldosteronism

Glucocorticoid-remediable aldosteronism (GRA) is a mendelian form of human hypertension in which inheritance of a single genetic abnormality is sufficient to result in elevated blood pressure. First described by Sutherland, Ruse, and Laidlaw in 1966 (39), and independently by New and Peterson in 1967 (40), GRA is an autosomal dominant trait characterized by early onset of hypertension mediated via the mineralocorticoid receptor. GRA is associated with suppressed PRA, and secretion of high levels of the abnormal adrenal steroids 18-hydroxycortisol and 18-oxocortisol (41,42). The sine qua non of GRA is the aberrant control of aldosterone (as well as the abnormal steroids) by ACTH rather than its usual secretagogue angiotensin II, with the consequence that aldosterone secretion can be suppressed to subnormal levels by administration of exogenous glucocorticoids (43).

While GRA has been thought to be a rare disorder, difficulty in establishing the diagnosis has likely resulted in underdiagnosis. Clues to the diagnosis include early onset of hypertension; signs of aldosteronism, although spontaneous hypokalemia is by no means necessary or even common in GRA; suppressed PRA; and strong family histories of early hypertension or cerebral hemorrhage.

Genetic studies of an extended GRA kindred indicated strong linkage of the intermediate biochemical phenotype of elevated urinary 18-oxocortisol metabolites to a genetic marker derived from the candidate gene aldosterone synthase, strongly motivating further study of this gene in GRA patients. Investigation of this gene ultimately revealed a surprising finding: The GRA allele carried an additional copy of an aldosterone synthase-like gene (44).

How might such a duplication have occurred? The aldosterone synthase gene is closely related to another gene involved in steroid biosynthesis—the steroid 11-beta-hydroxylase gene (45,46). These two genes are 95% identical in DNA sequence, have identical intron–exon boundaries (47), and were both known to be present on human chromosome 8 (48). If these two genes were tightly linked and arranged in tandem, an aberrant recombination event between them—called unequal crossing over—could occur (Figure 2). One of the chromosomal products of this event would retain normal copies of the 11-hydroxylase and aldosterone synthase genes, but would in addition carry an extra gene; this extra gene would be a chimera, fusing portions of the 11-hydroxylase and aldosterone synthase genes to one another.

Molecular analysis of the gene duplication in this GRA kindred confirmed that it had indeed arisen via unequal crossing over, demonstrating that the 5′ end of the 11-hydroxylase gene had been fused to more distal sequences of the aldosterone synthase gene (44).

Is this chimeric gene the cause of GRA? The 11-hydroxylase gene is highly expressed in adrenal fasciculata, and that its expression in this tissue is positively regulated by ACTH (Figure 3). Consequently, by virtue of the 5′ regulatory sequences from the 11-hydroxylase gene, we anticipate that the chimeric gene will be expressed in fasciculata under ACTH control. Owing to the coding sequences of aldosterone synthase, this gene might have aldosterone synthase enzymatic activity. The consequence of this would be ectopic expression of this activity in fasciculata; as a result, the enzyme could use cortisol as a substrate to synthesize 18-hydroxycortisol and 18-oxocortisol, the two signature steroids of GRA. Moreover, due to high levels of corticosterone in the adrenal fasciculata, aldosterone will also be produced in this adrenal

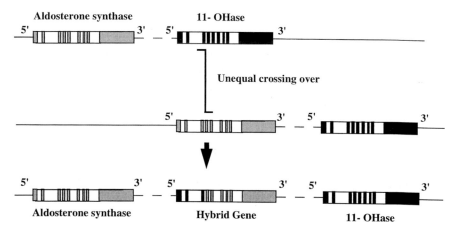

Figure 2 Generation of a chimeric gene duplication by unequal crossing over between aldosterone synthase and steroid 11-beta-hydroxylase (11-OHase) genes. One of the chromosomes resulting from unequal crossing over between these genes will carry three genes instead of two; the duplicated gene will fuse sequences of the two genes at the site of recombination. These chimeric gene duplications are found in all patients with glucocorticoid-remediable aldosteronism. (From Ref. 44.)

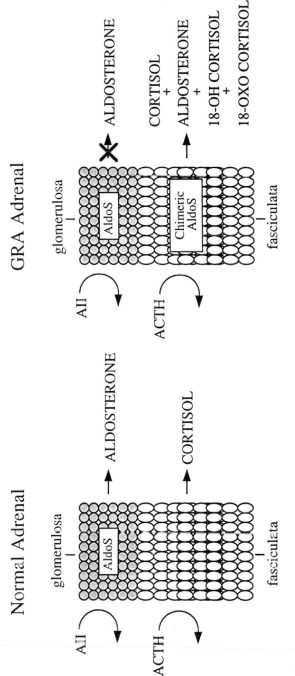

Figure 3 Model of the adrenal gland in GRA. The chimeric gene duplication shown in Figure 2 is expressed in adrenal fasciculata due to ACTH-responsive regulatory sequences of the 11-hydroxylase gene; the encoded product has aldosterone synthase (AldoS) enzymatic activity. Ectopic expression of aldosterone synthase enzymatic activity in adrenal fasciculata results in synthesis of 18-hydroxycortisol and 18-oxocortisol using cortisol as a substrate; aldosterone is also produced in fasciculata from corticosterone. The result is constitutive mineralocorticoid secretion leading to hypertension. (From Ref. 49.)

zone; this can account for the regulation of aldosterone secretion by ACTH in GRA, and the suppression of aldosterone secretion by glucocorticoids. This mutation would be a gain-of-function variant, which would be expected to display dominant inheritance, as is seen in GRA. Thus the structure of this mutation can explain the known physiology and genetics of GRA.

By similar methods, 17 additional GRA patients from 12 additional kindreds were studied (49). All 17 patients showed chimeric gene duplications like that seen in the first pedigree, while such duplications were absent in over 400 unaffected subjects. Analysis of the site of fusion of 11-hydroxylase and aldosterone synthase genes indicates that most of the duplications in different pedigrees arose independently; these findings provide conclusive proof that these duplications are the cause of GRA. Strong support for the model of the pathogenesis of GRA comes from Pascoe et al. (50), who have demonstrated that the products of chimeric genes constructed in vitro do indeed retain aldosterone synthase enzymatic activity.

These findings suggest the use of a direct genetic test for GRA—the presence of characteristic chimeric gene duplications. These can be readily screened for by analysis of DNA prepared from a small sample of blood. The simplicity of this test circumvents the problems of previous diagnostic tests, which required either pharmacological intervention or use of mass spectrometry to measure steroid levels in urine.

This test will likely be applicable in a number of clinical settings: patients presenting with unexplained pediatric hypertension, patients with unexplained aldosteronism, or patients with severe or refractory familial hypertension. In addition, one parent and, on average, 50% of the siblings and offspring of GRA patients will have GRA. The possibility of finding many cases through sequential screening of at-risk relatives is apparent.

Making the diagnosis of GRA is important for two principal reasons. First, patients with GRA are often refractory to therapy with conventional antihypertensive medications; however, they may respond very well to therapy directed toward the underlying abnormality. The importance of therapy for GRA is seen in the high prevalence of morbid outcomes, such as early stroke by cerebral hemorrhage, in GRA kindreds. Potential specific therapies for GRA include inhibition of secretion of aldosterone and 18-oxocortisol by exogenous glucocorticoids, inhibition of mineralocorticoid action by competition at the receptor using spironolactone, and inhibition of the mineralocorticoid-responsive distal renal epithelial sodium channel by amiloride. Which of these agents alone or in combination with other agents is most effective has not to date been studied systematically. Second, in general, many cases can be found for each index case, providing the opportunity to diagnose many cases at early ages without reliance on symptomatic presentation; this should permit earlier diagnosis and therapeutic intervention.

In addition, further investigation of GRA kindreds will be of interest to establish the phenotypical variation seen in patients who have inherited the same mutation by descent from a common ancestor. This will provide a unique opportunity to begin to determine the role of gene–environment and gene–gene interactions in the pathogenesis of hypertension and its clinical sequelae.

Molecular Genetics of Liddle's Syndrome

Liddle's syndrome represents another mendelian form of human hypertension. This disorder is characterized by early onset of hypertension in association with low plasma renin activity and, in contrast to GRA, suppressed aldosterone secretion (20). Variable hypokalemia is also found, again suggesting that the hypertension might be mediated via increased sodium reabsorption by the epithelial sodium channel of the distal nephron, which indirectly exchanges potassium excretion for sodium reabsorption. Support for this contention came from the recognition that the hypertension and hypokalemia of this disorder could be ameliorated by triamterine or amiloride, specific inhibitors of this channel, but not by spironolactone, an inhibitor of the mineralocorticoid receptor (20). Moreover, renal transplantation apparently has corrected the defect in one affected subject (51).

Collection of a large Liddle's syndrome kindred permitted the search for the chromosomal position of the presumed single gene underlying this trait. In this case, early onset of hypertension was used as the trait for linkage analysis, and eventually strong linkage was established between early hypertension in this kindred and a segment of chromosome 16 (52). In parallel, the gene encoding the beta subunit of the epithelial sodium channel was shown to be linked to precisely the same location (52), motivating the study of this gene as a candidate gene for Liddle's syndrome. The renal amiloride-sensitive epithelial sodium channel has been shown to be composed of at least three subunits of similar structure: each has two transmembrane domains with intracytoplasmic amino and carboxy termini with a large extracellular loop (53,54). The alpha subunit will support sodium conductance when expressed alone in Xenopus oocytes; in contrast, while the beta and gamma subunits will not support sodium conductance on their own, they greatly augment the current induced when expressed in conjunction with the alpha subunit (54). Molecular analysis of the beta subunit gene in Liddle's original kindred revealed a point mutation alterating an arginine codon just distal to the second transmembrane domain to a stop codon, thereby truncating the encoded protein by 75 amino acids but leaving intact the transmembrane domains believed to constitute the functional pore of the channel (52) (Figure 4). Investigation of affected members from four additional Liddle's kindreds revealed the presence of premature termination codons or frameshift mutations in all four; in each case these occurred in a 95 bp segment of the gene

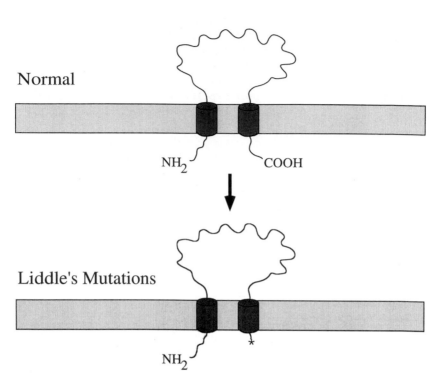

Figure 4 The molecular basis of Liddle's syndrome. The mineralocorticoid regulated epithelial sodium channel of the kidney is a hetero-oligomer composed of three subunits, each of which has the topology shown. Mutations that truncate the cytoplasmic carboxy terminus of the beta subunit of the epithelial sodium channel lead to constitutive activation of channel activity, resulting in increased renal sodium reabsorption, expanded plasma volume, and salt-sensitive hypertension (52).

encoding the cytoplasmic carboxy terminus. These mutations were not found in a large number of control subjects, demonstrating the specificity of these mutations for Liddle's syndrome (52).

The phenotype of subjects with Liddle's syndrome is compatible with constitutive activation of the epithelial sodium channel, suggesting that the mutations observed in Liddle's patients in some fashion result in an inability to maintain the channel in a properly closed state. The finding of truncated beta subunits of the epithelial sodium channel in Liddle's patients is consistent with this hypothesis. This possibility has been tested by expressing truncated channel subunits and measuring their effects in Xenopus oocytes (L. Schild et al., Proc Natl Acad Sci USA, in press). When the truncated beta subunit is expressed in Xenopus oocytes in conjunction with normal alpha and gamma subunits, a highly significant threefold increase in whole-cell sodium conductance is observed, confirming that the observed mutations result in constitutive activation of the channel complex. These findings indicate that the normal carboxy terminus of the beta subunit must be involved, either directly or indirectly, in the normal regulation of channel activity and that loss of this segment results in increased activity of the channel. This confirmation of the inference from clinical and genetic studies demonstrates the molecular basis for this form of hypertension, and opens a new area of investigation in order to determine how this channel is normally regulated.

In addition, it will be of interest to determine whether molecular variants in the genes encoding subunits of the epithelial sodium channel contribute to blood pressure variation in other hypertensive subjects. In this regard, these channels are particularly attractive candidate genes in subjects with salt-sensitive hypertension characterized by low plasma renin activity and low aldosterone levels, a situation commonly encountered in hypertensive African American subjects (55).

Finally, it can now be appreciated that in humans subunits of the epithelial sodium channel map to the same location as a locus that has shown linkage to blood-pressure variation in a number of experimental animal strains. The locus of the SA gene has shown linkage to blood-pressure variation in the spontaneously hypertensive rat, the stroke-prone spontaneously hypertensive rat, and the Dahl salt-sensitive hypertensive rat (56–58). Linkage studies in humans have demonstrated that the SA gene and the beta subunit of the epithelial sodium channel are both tightly linked to locus D16S412 (31,52). This observation suggests that these loci are also tightly linked in rat, and raise the possibility that the observed linkage of blood pressure and the SA locus in rat results from mutation in a subunit of the epithelial sodium channel.

FUTURE DIRECTIONS

Investigation of the molecular genetics of human hypertension remains in its infancy. A small number of candidate genes and mendelian forms of the disease have been exploited. It is of interest that all three of the genes in which functional variants have been identified or inferred are involved in the regulation of intravascular volume in the renin-angiotensin system (Figure 1). Similarly, most of the other known sufficient causes of hypertension, such as renal artery stenosis and aldosterone-producing adenoma, also act to increase net renal sodium reabsorption. These findings support the notion of a common pathway to the development of hypertension in many settings, and it may consequently be surprising if further functional variants are not found in this pathway. Nonetheless, it is important to point out that few meaningful studies have been performed in other major pathways known to be involved in the tonic regulation of blood pressure, leaving a wide territory open for investigation.

Clearly, major advances in the field lie in the future. Approaches that will likely have impact include those involving genome-wide screens for loci imparting effects on blood pressure, screens employing isolated populations or populations with recent admixture in which linkage disequilibrium may be fully exploited, as well as examination of candidate genes and investigation of mendelian forms of hypertension. The power of these studies is dependent on many factors, including the unknown number of loci affecting blood pressure, the magnitude of the effect imparted by each locus, the prevalence of each allele in the population, and the number of independent mutations affecting blood pressure at each trait locus. Five years ago the power of genetic studies of this sort was relatively weak due to absence of complete and highly informative genetic maps. As a consequence, studies focused largely on candidate gene studies with or without use of intermediate phenotypes. The available genetic tools have now improved to such an extent that identification of trait loci by linkage may not require extensive investigation of individual subjects to define intermediate phenotypes at the outset. This should streamline study design and patient recruitment. In addition, parallel studies of clinical endpoints of hypertension can be contemplated and executed. Success in the identification of underlying trait loci and mutations will doubtless renew the need for physiological studies of individuals newly recognized to harbor particular genetic variants. This should permit characterization of the underlying physiological derangement in individually affected patients, as well as assessment of issues such as gene–gene and gene–environment interactions. There is reason for considerable optimism that use of these approaches will continue to lead to new insights into the pathogenesis and treatment of human hypertension and its consequences.

ACKNOWLEDGMENTS

The author would like to thank his many colleagues and collaborators whose work contributed to this review. RPL is an investigator of the Howard Hughes Medical Institute.

REFERENCES

1.   Christian JC. Twin studies of blood pressure. In: Filer LJ, Lauer RM, eds. Children's Blood Pressure. Columbus, OH: Ross Laboratories, 1985:51-55.
2.   Longini IM, Higgins MW, Hinton PC, Moll PC, Keller JB. Environmental and genetic sources of familial aggregation of blood pressure in Tecumseh, Michigan. Am J Epidemiol 1984; 120:131-44.

3.   Biron P, Mongeau JG, Bertrand D. Familial aggregation of blood pressure in 558 adopted children. Can Med Assoc J 1976; 115:773–774.
4.   Rich GM, Ulick S, Cook S, Wang JZ, Lifton RP, Dluhy RG. Glucocorticoid-remediable aldosteronism in a large kindred: Clinical spectrum and diagnosis using a characteristics biochemical phenotype. Ann Intern Med 1992; 116:813–820.
5.   Ward R. Familial aggregation and genetic epidemiology of blood pressure. In: Laragh JH, Brenner BM, eds. Hypertension: Pathophysiology, Diagnosis and Management. New York: Raven Press, 1990:81–100.
6.   Rapp JP. Dissecting the primary causes of genetic hypertension in rats. Hypertension 1991; 18(suppl I):I-18–I-28.
7.   Risch N, et al. Genetic analysis of idiopathic torsion dystonia in Ashkenazi Jews and their recent descent from a small founder population. Nature Genet 1995; 9:152–159.
8.   Spielman RS, McGinnis RE, Ewens RJ. Transmission test for linkage disequilibrium: the insulin gene region and insulin dependent diabetes mellitus. Am J Hum Genet 1993; 52:506–516.
9.   Orita M, Suzuki Y, Sekiya T, Hayashi K. Rapid and sensitive detection of point mutations and DNA polymorphisms using the polymerase chain reaction. Genomics 1989; 5:874–879.
10.   Sheffield V, Cox DR, Lerman LS, Myers RM. Attachment of a 40-base-pair G+C-rich sequence (GC-clamp) to genomic DNA fragments by the polymerase chain reaction results in improved detection of single base changes. Proc Natl Acad Sci USA 1989; 86:232–236.
11.   Blackwelder WC, Elston RC. A comparison of sib-pair linkage tests for disease susceptibility loci. Genet Epidemiol 1985; 2:85–97.
12.   Cudworth AG, Woodrow JC. Evidence for HLA linked genes in "juvenile" diabetes mellitus. Br Med J 1975; III:133–135.
13.   Lange K. A test statistic for the affected sib set method. Ann Hum Genet 1986; 50:283–290.
14.   Williams RR, et al. Multigenic human hypertension: Evidence for subtypes and hope for haplotypes. J Hypertension 1990; 8(suppl 7):S39–46.
15.   Weinberger ML, Miller JZ, Luft FC, Grim CE, Fineberg NS. Definitions and characteristics of sodium sensitivity and blood pressure resistance. Hypertension 1986; 8:II127–134.
16.   Hasstedt SJ, et al. Hypertension and sodium-lithium countertransport in Utah pedigrees: evidence for major locus inheritance. Am J Hum Genet 1988; 43:14–22.
17.   Boerwinkle E, et al. The role of the genetics of sodium lithium countertransport in the determination of blood pressure variability in the population at large. In: Brewer GJ, ed. The Red Cell: Sixth Ann Arbor Conference. New York: Alan R Liss, 1984:479–507.
18.   Motulsky AG, et al. Hypertension and the genetic of red cell membrane abnormalities. In: Bock G, Collins G, eds. Molecular Approaches to Human Polygenic Disease. Ciba Symposium 130. Chichester: Wiley, 1987:150–166.
19.   Berry TD, et al. A gene for high urinary kallikrein may protect against hypertension in Utah kindreds. Hypertension 1989; 13:3–8.

20.   Liddle GW, Bledsoe T, Coppage WS. A familial renal disorder simulating primary
       aldosteronism but with negligible aldosterone secretion. Trans Am Assoc Phys
       1963; 76:199–213.
21.   Lifton RP, Dluhy RG. Inherited forms of mineralocorticoid hypertension:
       Glucocorticoid-remediable aldosteronism and the syndrome of apparent
       mineralocorticoid excess. In: Laragh J, Brenner BM, eds. Hypertension: Patho-
       physiology, Diagnosis and Management. New York: Raven Press, 1995:2163–
       2176.
22.   Gordon RD, et al. Gordon's syndrome: A sodium-volume-dependent form of
       hypertension with a genetic basis. In: Laragh J, Brenner BM, eds. Hypertension:
       Pathophysiology, Diagnosis and Management. New York: Raven Press,
       1995:2111–2123.
23.   Gyapay G, et al. The 1993–94 Genethon human genetic linkage map. Nature
       Genet 1994; 7:246–339.
24.   Cooperative Human Linkage Center. A comprehensive human linkage map with
       centimorgan density. Science 1994; 265:2049–2070.
25.   Risch N. Linkage strategies for genetically complex traits. II. The power of
       affected relative pairs. Am J Hum Genet 1990; 46:242–253.
26.   Davies JL, et al. A genome-wide search for human type I diabetes susceptibility
       genes. Nature 1994; 371:130–136.
27.   Lifton RP, Hunt SC, Williams RR, Pouyssegur J, Lalouel JM. Exclusion of the
       Na+/H+ antiporter as a candidate gene in human hypertension. Hypertension
       1991; 17:8–14.
28.   Jeunemaitre X, et al. Sib pair linkage analysis of renin gene haplotypes in human
       essential hypertension. Hum Genet 1992; 88:301–306.
29.   Jeunemaitre X, Lifton RP, Hunt SC, Williams RR, Lalouel JM. Absence of linkage
       between the angiotensin converting enzyme and human essential hypertension.
       Nature Genet 1992; 1:72–75.
30.   Bonnardeaux A, et al. Angiotensin II receptor gene polymorphisms in human
       essential hypertension. Hypertension 1994; 24:63–69.
31.   Nabika T, et al. Evaluation of the SA locus in human hypertension. Hypertension
       1995; 25:6–13.
32.   Jeunemaitre X, et al. Molecular basis of human hypertension: Role of an-
       giotensinogen. Cell 1992; 71:169–180.
33.   Gould AB, Green B. Kinetics of the human renin and human renin substrate
       reaction. Cardiovasc Res 1971; 5:86–89.
34.   Watt GCM, et al. Abnormalities of glucocorticoid metabolism and the renin
       angiotensin system: a four-corners approach to the identification of genetic
       determinants of blood pressure. J Hypertens 1992; 10:473–482.
35.   Lifton RP, Warnock D, Acton RT, Harmon L, Lalouel JM. High prevalence of
       hypertension-associated allele T235 in African Americans. Clin Res 1993; 41:260A.
36.   Cooper R, Ward R. Angiotensinogen gene in human hypertension: Lack of an
       association of the 235T allele among African Americans. Hypertension 1994;
       24:591–594.
37.   Ward K, et al. A molecular variant of angiotensinogen associated with preeclamp-
       sia. Nature Genet 1993; 4:59–61.

38. Caulfield M, et al. Linkage of the angiotensinogen gene to essential hypertension. N Engl J Med 1994. 330:1629–1633.
39. Sutherland DJ, Ruse JL, Laidlaw JC. Hypertension, increased aldosterone secretion and low plasma renin activity relieved by dexamethasone. Can Med Assoc J 1966; 95:1109–1119.
40. New MI, Peterson RE. A new form of congenital adrenal hyperplasia. J Clin Endocrinol Metab 1967; 27:300–305.
41. Ulick S, Chu MD, Land MJ. Biosynthesis of 18-oxocortisol by aldosterone producing adrenal tissue. J Biol Chem 1983; 258:5498–5502.
42. Gomez-Sanchez CE, et al. Elevated urinary excretion of 18-oxocortisol in glucocorticoid-suppressible aldosteronism. J Clin Endocrinol Metab 1984; 59:1022–1024.
43. Ulick S, et al. Defective fasciculata zone function as the mechanism of glucocorticoid-remediable aldosteronism. J Clin Endocrinol Metab 1990; 71:1151–1157.
44. Lifton RP, et al. A chimaeric 11β-hydroxylase/aldosterone synthase gene causes glucocorticoid-remediable aldosteronism and human hypertension. Nature 1992; 355:262–265.
45. Kawamoto T, Mitsuuchi Y, Ohnishi T, et al. Cloning and expression of a cDNA for human cytochrome P-450$_{aldo}$ as related to primary aldosteronism. Biochem Biophys Res Commun 1990; 173:309–316.
46. Ogishima T, Shibata H, Shimada H, et al. Aldosterone synthase cytochrome P-450 expressed in the adrenals of patients with primary aldosteronism. J Biol Chem 1991; 266:10731–10734.
47. Mornet E, Dupont B, Vitek A, White PC. Characterization of two genes encoding human steroid 11β-hydroxylase (P-450$_{11β}$) J Biol Chem 1989; 264:20961–20967.
48. Chua SC, Szabo P, Vitek A, Grzeschik K-H, John M, White PC. Cloning of cDNA encoding steroid 11β-hydroxylase (P450c11). Proc Natl Acad Sci USA 1987; 84:7193–7197.
49. Lifton RP, et al. Hereditary hypertension caused by chimeric gene duplications and ectopic expression of aldosterone synthase. Nature Genetics 1992; 2:66–74.
50. Pascoe L, et al. Glucocorticoid-suppressible hyperaldosteronism results from hybrid genes created by unequal crossovers between CYP11B1 and CYP11B2. Proc Natl Acad Sci USA 1992; 89:8327–8331.
51. Botero-Velez M, Curtis JJ, Warnock DG. Brief report: Liddle's syndrome revisited. N Engl J Med 1994; 330:178–181.
52. Shimkets RA, et al. Liddle's syndrome: Heritable human hypertension caused by mutations in the beta subunit of the epithelial sodium channel. Cell 1994; 79:407–414.
53. Cannessa CM, Horisberger JD, Rossier BC. Epithelial sodium channel related to proteins involved in neurodegeneration. Nature 1993; 361:467–470.
54. Canessa C, et al. Amiloride-sensitive epithelial Na+ channel is made of three homologous subunits. Nature 1994; 367:463–467.
55. Pratt JH, Jones JJ, Miller JZ, Wagner MA, Fineberg NS. Racial differences in aldosterone excretion and plasma aldosterone concentrations in children. N Engl J Med 1989; 321:1152–1157.

56.   Samani NJ, et al. A gene differentially expressed in the kidney of the spontane-
      ously hypertensive rat cosegregates with increased blood pressure. J Clin Invest
      1993; 92:1099–1103.
57.   Lindpaintner K, et al. Molecular genetics of the SA-gene: cosegregation with
      hypertension and mapping to rat chromosome 1. J Hypertens 1993; 11:19–23.
58.   Harris EL, Dene H, Rapp JP. SA gene and blood pressure cosegregation using Dahl
      salt-sensitive rats. J Hypertens 1993; 6:330–334.

4

Arrhythmias and Vascular Disease

Mark T. Keating
Howard Hughes Medical Institute
Eccles Institute of Human Genetics
University of Utah
Salt Lake City, Utah

CARDIAC ARRHYTHMIAS

Introduction

Although cardiac arrhythmias are a major source of morbidity and mortality, our pathogenic understanding of these disorders is poor. This is particularly true for the most serious cardiac arrhythmias, ventricular arrhythmias. Stratification and treatment of these most severe cardiac arrhythmias have also been challenging. One of the mainstays of diagnosis and therapy has been the use of invasive electrophysiological testing coupled with pharmacological intervention. These tests, however, are time-consuming and expensive, and have met with only modest therapeutic success. These procedures have also done relatively little to improve our understanding of the disease and, most importantly, are postsymptomatic. As a result, treatment has been moving toward nonspecific, invasive, and expensive implantable automatic defibrillators. The strength of this approach is that it works in most circumstances; the weakness is that it is merely palliative.

To improve presymptomatic diagnosis, prevention, and treatment of cardiac arrhythmias, we need to define these disorders at the molecular, cellular, and organ levels. Simplistically, cardiac arrhythmias, like all disorders, result from the interplay of genetic and environmental factors over time. Although we have some clues about the environmental factors that play a role in cardiac arrhythmias, we know almost nothing about the genetic factors. Before we can understand the complex interplay between these factors at the level of the organism, it is essential to understand the elements.

Family studies have provided clear evidence that heritable factors can be important in cardiac arrhythmias. Genetic abnormalities have been identified at all levels of the cardiac electrical system, causing atrial arrhythmias, ventricular arrhythmias, sinus node dysfunction, or AV block. Specific arrhythmias believed to have a genetic component are listed in Table 1. With the exception of long QT syndrome (LQT), little is known about the genetic abnormalities causing these disorders.

Cardiac arrhythmia can also be an important component of inherited disorders that cause structural heart disease. The most important of these are ischemia heart disease and dilated cardiomyopathy, both of which can cause a spectrum of cardiac arrhythmias, including ventricular arrhythmias. Atherosclerosis, cardiomyopathy, and structural heart disease are discussed elsewhere in this book.

Long QT Syndrome

In the long QT syndrome, individuals suffer from syncope and sudden death due to ventricular arrhythmias, specifically torsade de pointes and ventricular fibrillation (Figure 1) (24–28). Many of these individuals also have prolongation of the QT interval on surface electrocardiograms. LQT can be broadly classified as acquired or inherited, although most cases are probably caused by some combination of environmental and genetic factors. The most common acquired form of LQT is drug-induced. Quinidine therapy is probably the most common drug-induced LQT, but other antiarrhythmics, antidepressants, and general anesthetics have also been implicated. Metabolic abnormalities, particularly electrolyte disturbances such as hypokalemia, hypocalcemia, and hypomagnesemia, can cause LQT. Neurological abnormalities, notably subarachnoid bleeding and CNS surgery, can induce LQT. LQT has also been reported in the setting of severe bradycardia and ischemia.

LQT can be caused by the apparent inheritance of a single gene. Two inherited forms of LQT are clearly defined. One is inherited as an autosomal recessive trait and is associated with congenital neural deafness (37). This form of LQT is rate. The second, more common, form of LQT is inherited as an autosomal dominant trait; these individuals have normal hearing and no other obvious phenotypic abnormalities (29,30).

The mechanism of QT prolongation is acquired and inherited LQT is unknown, but it probably involves prolonged action potential duration in individual cardiac myocytes and dispersion of cardiac repolarization. The former could be caused by prolonged cellular repolarizing currents (sodium and calcium) or delayed repolarizing currents (potassium and chloride) (46). In theory, anything that affects these currents could cause LQT. In hypokalemia, for example, reduced extracellular potassium decreases myocyte potassium conductance through potassium-sensitive potassium channels, thereby

Table 1 Inherited Disorders

Disease	Rhythm abnormality	Inheritance	Chromosomal location	Refs.
Supraventricular arrhythmias				
Familial total atrial standstill	SND, AF	AD		1–5
Familial absence of sinus rhythm	SND, AF	AD		6–9
Wolff-Parkinson-White Syndrome	AVRT, AF, VF	AD		10–12
Familial PJRT	AVRT	AD		13
Conduction abnormalities				
Familial AV block	AVB, AF, SND, VT, SD	AD		14–19
Familial bundle branch block	RBBB	?		10–23
Ventricular arrhythmias				
LQT Romano-Ward syndrome	TdP, VF	AD	11; at least one other unknown	24–36
LQT-Jervell Lange-Nielsen syndrome	Tdp, VF	AR		37,38
Familial VT	VT	AD		39,44–45
Familial bidirectional VT	VT	AD		

AF = atrial fibrillation; AVB = atrioventricular block; AVRT = atrioventricular reciprocating tachycardia; RBBB = right bundle branch block; SD = sudden death; SND = sinus node dysfunction; TdP = torsade de pointes; VT = ventricular tachycardia; VF = ventricular fibrillation; AD = autosomal dominant; AR = autosomal recessive.

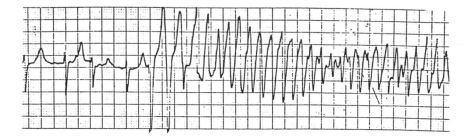

Figure 1 Torsade de pointes ventricular arrhythmia in long QT syndrome (LQT).

delaying repolarization. Quinidine is a potassium channel antagonist and has a similar effect. Inherited LQT could be caused by a mutation in a cardiac ion channel gene like a sodium channel. If such a mutation caused a delay in sodium channel inactivation, for example, continued depolarization would result and could cause prolonged action potential duration and LQT.

How does QT prolongation increase the risk of cardiac arrhythmias in these patients? The precise mechanism is not yet certain, but presumably involves the generation of secondary depolarization or after-depolarizations. Most evidence seems to support a role for after-depolarizations that occur during phase 2 or 3 of the action potential, so-called "early after-depolarizations," or EADs. EADs are thought to be mediated by the reactivation of L-type calcium channels. It is also possible, however, that secondary repolarizations that occur in phase 4, called "delayed after-depolarizations," or DADs, play a role.

The environmental factors that trigger arrhythmias in individuals with LQT differ depending on the underlying mechanism. These distinctions are important as therapy can be quite different. Arrhythmias in acquired LQT frequently result from bradycardia, so agents that increase heart rate, for example, adrenergic agonists, reduce the risk of arrhythmia until the underlying problem can be corrected. In inherited LQT, on the other hand, stress and exercise frequently increase the risk of arrhythmia and therapy generally includes beta-adrenergic blockade. This picture is complicated by the likelihood that several different genetic and environmental factors play a role in both acquired and inherited LQT and that the genetic factors are poorly defined.

The diagnosis of LQT is based on a history of syncope, sudden death, congenital deafness, a family history of LQT, and electrocardiographic findings. The last include prolongation of the QT interval corrected for heart rate (the QT interval divided by the square root of the RR interval, or the QTc) (47). In acquired LQT, it is helpful to see a change in the QT interval that can be attributed to some environmental factor. When previous ECGs are unavailable, QT prolongation can be difficult to define. In the past, a QTc of more than 0.44 seconds was used to define QT prolongation. As will be discussed later, however, this is probably not an adequate diagnostic determination. T-wave abnormalities, including T-wave alternans, have also been described in association with LQT, but these abnormalities are not a constant feature of LQT and are difficult to quantify. Stress electrocardiography is becoming increasingly important in the diagnosis of LQT. Preliminary data suggest that the QT interval fails to shorten appropriately in response to increased heart rate in LQT patients. As will be described later, molecular genetics also offers promise for improved presymptomatic diagnosis of LQT.

Initial goals for acquired LQT are to identify and eliminate etiological factors. As mentioned above, arrhythmias and acquired LQT are frequently rate-dependent, and arrhythmia management often involves overdrive pacing or adrenergic agonists. By contrast, inherited LQT is often treated with beta-adrenergic blockade. For therapeutic decision-making, individuals with inherited LQT can be subdivided into two groups, asymptomatic and symptomatic. It is not yet clear if asymptomatic LQT patients should be treated. As these individuals are at increased risk for sudden death if untreated, it seems reasonable to treat them with beta-blockers until a therapeutic trial is organized. Symptomatic LQT patients can again be divided into two groups, those suffering from syncope and those who have survived sudden death. The first line of treatment in the latter group should probably be implantation of an automatic defibrillator. If symptoms have recurred frequently, beta-blockade may also be included. However, if the patient has had only one episode of aborted sudden death, beta-blocker therapy may be unnecessary. Patients with inherited LQT and a history of syncope but without a history of aborted sudden death should be treated with one or more of the following: beta-blockers, left cervicothoracic sympathectomy, overdrive pacing, or implantation of an automatic defibrillator. Anecdotal data suggest that beta-blocker therapy reduces the incidence of syncope in these patients and most clinicians choose this initial therapy; however, these individuals are not completely protected from sudden death.

We have been studying the autosomal dominant form of inherited LQT using the techniques of molecular genetics. The goal of this work is to improve presymptomatic diagnosis of LQT, to elucidate the molecular mechanism of the disorder, and to improve treatment. In 1991, we described tight linkage between LQT and the Harvey ras-1 gene on chromosome 11p15.5 (Figure 2) (31). The maximum lod score of 16 was identified at a recombination fraction of 0, making Harvey ras-1 a candidate for LQT. Direct sequencing of the gene, however, showed no evidence for disease associated mutation, indicating that the Harvey ras-1 gene is not the LQT gene.

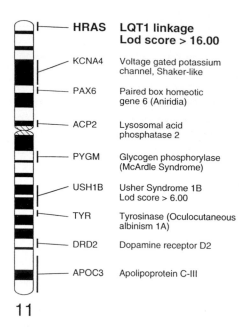

	HRAS	LQT1 linkage Lod score > 16.00
	KCNA4	Voltage gated potassium channel, Shaker-like
	PAX6	Paired box homeotic gene 6 (Aniridia)
	ACP2	Lysosomal acid phosphatase 2
	PYGM	Glycogen phosphorylase (McArdle Syndrome)
	USH1B	Usher Syndrome 1B Lod score > 6.00
	TYR	Tyrosinase (Oculocutaneous albinism 1A)
	DRD2	Dopamine receptor D2
	APOC3	Apolipoprotein C-III

11

Figure 2 A gene for LQT is located on chromosome 11p15.5, near the Harvey *ras*-1 gene. Idiogram at left shows approximate chromosomal location of a LQT gene. Other genes and loci of interest are also shown.

The discovery of genetic markers for LQT made presymptomatic diagnosis in one family feasible. To determine if these markers were generally useful, we characterized six additional autosomal dominant LQT families. In these families, the LQT gene was also linked to markers on chromosome 11p15.5, indicating that presymptomatic diagnosis was possible (Figure 3) (32). Recently, however, we and others (33,35,36) identified locus heterogeneity for LQT, demonstrating that at least two distinct genes can cause this phenotype. We have initiated new linkage studies to identify additional LQT loci in these families.

We have used molecular diagnosis of LQT in families to help define the clinical spectrum of the disorder (34). In the families we have studied, 63% of LQT gene carriers have had at least one syncopal episode; 37% have never had syncope, 7% of non–gene carriers in the family have a history of syncope. Only 5% of the gene carriers had a history of aborted sudden death. This number underestimates the risk of sudden death in this population, as we were not able to determine the gene-carrier status of individuals who died before the study. Even if we assume that everyone who died suddenly in this family was a gene carrier, however, the incidence of sudden death was less than 1% per

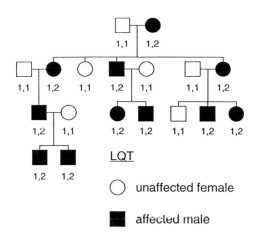

LQT

⬤ unaffected female

■ affected male

Figure 3 Pedigree structure and Harvey *ras*-1 genotypes for a family with LQT. Individuals affected by LQT are indicated by filled circles (females) or squares (males). Unaffected individuals are indicated by empty circles or squares. Note that the disease phenotype segregates with the 2 allele.

year. The age of onset of symptoms in LQT gene carriers was a mean of 8 years in males and 14 in females. The sexes were evenly distributed in the symptomatic group. These findings are quite different from previous reports; this may be due to differences in the populations studied, but also may reflect more accurate diagnosis of gene carriers in our studies.

We have also examined the electrocardiographic findings of LQT gene carriers (Figure 4) (34). As expected, the electrocardiogram was neither completely sensitive nor specific for diagnosis of LQT. While the mean QTc for gene carriers at 0.49 seconds was longer than that for noncarriers at 0.42 seconds, 63% of the population had overlapping QTc's of 0.41–0.47 seconds. Diagnosis of LQT using a cutoff of 0.44 seconds, therefore, led to misclassifications. In our study, 5% of LQT gene carriers were falsely classified as normal while 15% of noncarriers were misclassified as affected. Among gene carriers, the range of static QTc's was similar to that for noncarriers and was not useful for predicting risk of symptoms. It is clear, therefore, that other clinical or genetic tools must be used for presymptomatic diagnosis of LQT. Nevertheless, in our studies the ECG was helpful at the extremes, and a QTc of more than 0.47 seconds was completely predictive for gene carriers while a QTc of less than 0.41 seconds was completely predictive for noncarriers.

The focus of our work is to identify the 11p15.5-linked LQT gene. We have developed a refined linkage map of this locus, confining the disease gene to flanking markers separated by approximately 700 kb. Since additional recombination events remain to be defined, the LQT gene will soon be

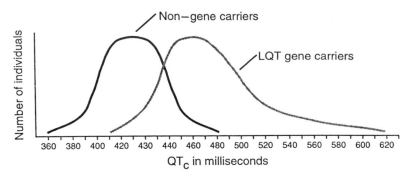

Figure 4   The spectrum of corrected QT intervals in LQT gene carriers. The corrected QT intervals (QTc) for a population of LQT gene carriers and controls and indicated. Note the considerable overlap between these two populations.

localized to the region of 200–300 kb. At that point, we will identify candidate genes from the region. The LQT gene will be identified from among these candidates by mutational analysis.

We are taking a second approach to LQT. Since the physiological features of this disorder could be caused by abnormal cardiac ion channels, we identified and characterized a human potassium channel gene (48). Several techniques were used to map this gene to the short arm of chromosome 12, indicating that it is not the chromosome 11-linked LQT gene, but remains a candidate for other nonlinked families. We have mapped additional candidates, including two potassium channel genes on chromosome 11 (one on chromosome 11p15.4). By using a combination of molecular genetic, candidate gene, and cellular electrophysiological approaches, we will soon understand the mechanismns underlying LQT. This, in turn, should help predict, prevent and treat this and other arrhythmagenic disorders.

OBSTRUCTIVE VASCULAR DISEASE

Vascular disease is the major cause of morbidity and mortality in the industrialized world, with more than 500,000 new cases per year in the United States alone. Environmental risk factors for obstructive vascular disease include cigarette and drug abuse and a diet high in saturated fat and calories. Vascular risk factors that have a genetic component include diabetes, hypertension, male sex, and hypercholesterolemia. The latter has been the subject of intense research, resulting in improved prediction, prevention, and treatment. Although a contributing factor in many cases, hypercholesterolemia does not account for most vascular pathology. Instead, other, as of yet undefined, risk factors must account for the bulk of vascular disease.

Supravalvular aortic stenosis (SVAS) is an inherited vascular disease that causes obstruction of large arteries such as the aorta and pulmonary arteries (Figure 5) (14,15,25,49–51). This disorder frequently requires surgical repair early in life and can lead to heart failure, cerebral ischemia, myocardial infarction, and death. Recently, we demonstrated that SVAS is genetically linked to the elastin locus on chromosome 7q11, suggesting that abnormalities in elastin may cause this disorder (Figure 6) (17,18,52,53). To test this hypothesis we screened DNA from SVAS patients for mutations in the elastin gene. In one family, a balanced translocation disrupted the elastin gene, generating a new stop codon at exon 28 (human elastin has 34 exons) (54,55). In a second family, a deletion beginning at elastin exon 29 and extending through the remaining exons of the gene was identified (Figure 6) (56). These findings, coupled with existing knowledge of vascular histology and physiology, suggest that mutations in the elastin gene cause this vascular disorder.

Williams syndrome (WS) is a developmental disorder affecting the vascular, connective tissue, and central nervous systems (57–59). Features of

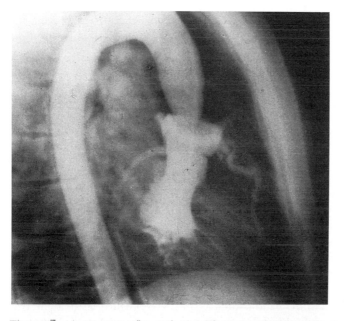

Figure 5 Aortogram of a patient with supravalvular aortic stenosis (SVAS). SVAS causes obstruction of large arteries, particularly the aorta and pulmonary arteries.

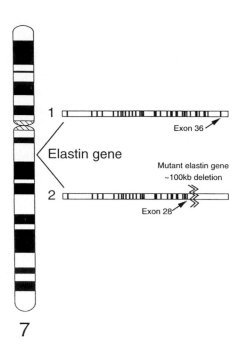

Figure 6 Ideogram of chromosome 7 showing location of the elastin locus and one elastin gene that is responsible for SVAS. A 100 kb deletion (allele 2) that disrupts the elastin gene at exon 28 and extending 3′ cosegregates with the disease in one SVAS family, suggesting that this mutation is responsible for the disease in this family.

this disorder include dysmorphic facial appearance, joint laxity early in life followed by joint contractures, premature aging of skin, vascular disease, hypertension, infantile hypercalcemia, unusually gregarious personality, and mental retardation with a mean IQ of 57. Specific cognitive deficits include attention deficit disorder and poor visual–motor integration. As a result, individuals with WS can see only part of a picture, not the whole. Language skills, by contrast, are relatively spared in WS—some elements of language are actually enhanced, particularly the quality and quantity of vocabulary, auditory memory, and social use of language. Many WS patients sing or play musical instruments with considerable skill, and they rarely forget a name.

Since SVAS is a common feature of WS, we hypothesized that elastin was involved in the pathogenesis of this disorder. In mutational analyses of DNA from familial and sporadic cases, we discovered that WS is caused by submicroscopic deletions within chromosome 7q11.23 (60). One elastin allele was completely deleted in all subjects, indicating that hemizygosity at this locus is a mechanism of vascular and connective-tissue pathology in WS. These findings suggest that a developmental reduction in elastin can lead to vascular obstruction, hypertension, premature aging of skin, and other connective-tissue abnormalities.

It is not yet clear why the connective-tissue pathology of WS is more prominent than that found in autosomal dominant SVAS. Additional connective-tissue abnormalities have been identified in some SVAS patients (hernias, mild dysmorphic facial features, fifth-finger clinodactyly, hoarseness), but SVAS is primarily a vascular disease. The only SVAS mutations defined to date involve the 3' end of the elastin gene, whereas all known WS-associated deletions affect the entire gene. The specific elastin mutations, therefore, may be responsible for some of the phenotypic differences between SVAS and WS.

The neurobehavioral features of WS are not easily explained by hemizygosity at the elastin locus and our data suggest an alternative mechanism. Our findings indicate that WS-associated deletions extend beyond the elastin locus, spanning at least 150 kb. Additional genes, therefore, may be deleted, and one or more of these genes could be involved in the disease pathogenesis. It seems likely for example, that the severity of mental retardation in WS is related to the size of the deletion, an mechanism that could involve several genes. Refined linkage and physical mapping of the chromosomal subunit of 7q11.23, identification of genes adjacent to the elastin locus, and continued deletional analysis will determine if WS is a contiguous gene disorder.

Our work has made genetic diagnosis of WS possible and has supported the hypothesis that abnormalities in elastin can cause vascular disease. Several additional lines of evidence support a role for elastin in SVAS. Histological studies indicate that elastic fibers are an important component of the media of arteries, particularly large arteries, e.g., the aorta and pulmonary arteries. Pathological studies have identified abnormalities in elastic fibers in SVAS patients, consistent with a primary abnormality of elastin. Other pathological features, including smooth-muscle hypertrophy and increased collagen, presumably represent secondary phenomena.

Our understanding of vascular physiology is also consistent with the hypothesis that SVAS is a disorder of elastin metabolism. The primary function of the ascending aorta is to absorb hemodynamic energy during cardiac systole and release energy in the form of sustained blood pressure during cardiac relaxation. This function is mediated by elastic fibers in the blood vessel, the main component of which is elastin (61,62). One would expect, therefore, that a primary defect in elastin would have an impact on the structure and function of the ascending aorta. Finally, another disorder of the elastic fiber, Marfan syndrome, has been associated with transient SVAS. Fibrillin, the protein implicated in the pathogenesis of Marfan syndrome, is an important component of the microfibular scaffold and appears to be critical for elastogenesis (63,64).

The mechanism of SVAS may involve quantitative or qualitative defects in elastin. If is not yet known if the mutant elastin alleles described here are expressed, but the resultant proteins would lack two consensus sites for desmosine cross-linking and two conserved cysteine residues near the carboxyl terminus (Figure 7). These residues are thought to be important for interaction with the cysteine-rich protein fibrillin in arrays of microfibrils (Figure 8). A truncated protein lacking domains critical for intermolecular interaction might have a dominant-negative effect on elastin encoded by the normal allele, disrupting posttranslational processing and development of complex elastic fibers. Alternatively, the mechanism of SVAS may involve a quantitative loss of normal elastin resulting from reduced production or stability of messenger RNA or protein. Analysis of DNA, RNA, and protein from SVAS patients will distinguish between these possibilities.

The mechanism of vascular obstructions in SVAS likely involves increased hemodynamic damage to the endothelium of inelastic arteries, causing intimal proliferation of smooth muscles and fibrosis (Figure 9). This hypothesis is supported by clinical improvement of pulmonary artery stenosis in SVAS patients after postnatal reduction of pulmonary artery pressure. Obstruction of the aorta, by contrast, progresses over time, coincident with sustained increases in systemic blood pressure after birth. In future studies, this mechanism will be tested by examining the temporal course of the disease in animal models.

Our work on SVAS may have implications for treatment of common vascular disease. Given or data, it is not surprising that venous grafts, which contain no elastin, are prone to vascular disease and obstruction after bypass

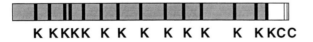

K KKKK K K   K   K K K   K K KCC

K = Lysine   important for intra– and inter–
molecular crosslinking

C = Cysteine

▨ = hydrophobic domain

Figure 7 The elastin protein, consisting of alternating hydrophobic domains and desmosine cross-linking domains. The hydrophobic domains are responsible for the protein's resilience; the cross-linking domains maintain the protein in a globular structure. The carboxyl terminus contains the only cystine domains. These cystines are thought to form an intramolecular disulfide bridge, resulting in a structure that is important for intermolecular interaction during elastogenesis.

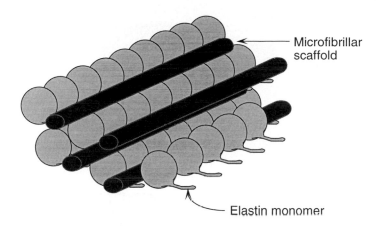

Figure 8 Elastin and elastogenesis. Elastogenesis is incompletely understood, but inelastic microfibrillar proteins are initially laid down in the extracellular space. These microfibrillar proteins are thought to form a scaffold for organization of elastin monomers. The carboxyl terminus of each elastin monomer, which is deleted in some patients with SVAS, is thought to be important for interaction between elastin and the microfibrillar scaffold. Aberrant elastogenesis is the likely molecular mechanism for SVAS.

surgery. The concept of using rigid vascular stents to hold back intimal proliferation of vascular cells appears contrary to what is actually needed in damaged arteries: integrity and resilience of the vascular wall. Future studies aimed at understanding the developmental regulation of elastin gene expression may lead to new ideas for preventing and treating vascular disease.

CONCLUSION

We are working to identify genetic factors that are important in cardiovascular disease. We have identified one genetic factor, the elastin gene, that appears to be important in the pathogenesis of supravalvular aortic stenosis and Williams syndrome (Table 2). We have also identified the chromosomal location of a long QT syndrome gene (Table 3). This work has helped define the pathogenesis of these relatively rare cardiovascular disorders. The mechanistic knowledge that we gain from these studies, however, will be applicable to more common cardiovascular diseases. Presymptomatic diagnosis of Williams syndrome is now straightforward using fluorescent in situ hybridization (FISH) with DNA markers from the elastin locus. Presymptomatic diagnosis for SVAS and long QT syndrome is more complicated and currently requires linkage analysis. Technological advances in mutational analyses and DNA

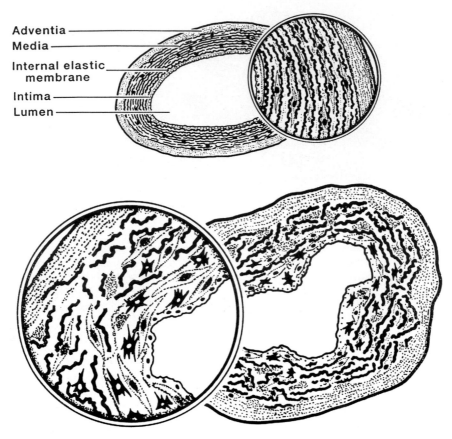

Figure 9 The mechanism of vascular obstruction in SVAS probably results from primary defect of elastin during vascular development, leaving an inelastic artery that is prone to recurrent injury and repair (top). Recurrent damage to the endothelium induces intimal proliferation of vascular cells and fibrosis, finally culminating in vascular obstruction (bottom).

sequencing, however, and identification of specific mutations that cause these disorders, will make presymptomatic diagnosis possible. Improved treatment and prevention of these disorders—the ultimate goals of this work—are likely long-term outcomes.

ACKNOWLEDGMENT

The major contributors to the work described here include Colleen Morris (University of Nevada), Amanda Ewart, Mark Curran, Donald Atkinson, Weish-

Table 2 Molecular Genetics of SVAS and WS

Disease	Inheritance	Chromosomal location	Gene
Supravalvular aortic stenosis	Autosomal dominant	7q11.23	Elastin
Williams syndrome	Autosomal dominant, sporadic	7q11.23	Elastin, +?

Table 3 Molecular Genetics of LQT

Disease	Inheritance	Chromosomal location	Gene	Ref.
Long QT syndrome	Autosomal recessive	?	?	37
Long QT syndrome	Autosomal dominant	11p15.5	?	31,32
Long QT syndrome	Autosomal dominant	?	?	33–36

an Jin, Mark Leppert, Katherine Timothy, and G. M. Vincent (University of Utah).

REFERENCES

1. Disetori M, Guarnerio M, Vergara G, et al. Familial endemic persistent atrial standstill in a small mountain community: review of eight cases. Eur Heart J 1983; 4:354-361.
2. Shah MK, Subramanyan R, Tharakan J, et al. Familial total atrial standstill. Am Heart J 1992; 123:1379-1382.
3. Ward DE, Ho Sy, Shinebourne EA. Familial atrial standstill and inexcitability in childhood. Am J Card 1984; 53:965-967.
4. Pierard LA, Henrard, L, Demoulin JC. Persistent atrial standstill in familial Ebstein's anomaly. Br Heart J 1985; 53:594-597.
5. Maeda S, Tanaka T, Hayashi T. Familial atrial standstill caused by amyloidosis. Br Heart J 1988; 59:498-500.
6. Surawicz B, Hariman RJ. Follow-up of the family with congenital absence of sinus rhythm. Am J Cardiol 1988; 61:467-469.
7. Spellberg RD. Familial sinus node disease. Chest 1971; 60:246-251.
8. Caralis DG, Varghese PJ. Familial sinoatrial node dysfunction: increased vagal tone a possible aetiology. Br Heart J 1976; 38:951-956.
9. Bharati S, Surawicz B, Vidaillet HJ, et al. Familial congenital sinus rhythm anomalies: clinical and pathological correlations. PACE; 15:1720-1729.
10. Vidaillet HJ, Pressley JC, Henke E, et al. Familial occurrence of accessory atrioventricular pathways (Preexcitation Syndrome). N Engl J Med 1987; 317:65-69.

11. Chung KY, Walsh TJ, Massie E. Wolff-Parkinson-White syndrome. Am Heart J 1965; 691:116.
12. Durakovic Z, Durakovic A, Kastelan A. The preexcitation syndrome: epidemiological and genetic study. Int J Card 1992; 35:181–186.
13. Perticone F, Marsico SA. Familial case of permanent form of junctional reciprocating tachycardia:possible role of the HLA system. Clin Cardiol 1988; 11:345–348.
14. Waxman MB, Catching JD, Felderhof CH, et al. Familial atrioventricular heart block. Circulation 1975; 51.
15. Sarachek NS, Leonard JL. Familial heart block and sinus bradycardia: classification and natural history. Am J Cardiol 1972; 29:451.
16. Segall HN. Congenital arrhythmias and conduction abnormalities in a father and four children. Can Med Assoc J 1961; 84:1283
17. Gazed PC, Culter RM, Taber E, et al. Congenital familial cardiac conduction defects. Circulation 1965; 32:32.
18. Connor AC, McFadden JF, Houston BJ, et al. Familial congenital complete heart block: case report and review of the literature. Am J Obstet Gynecol 1959; 78:75.
19. Veracochea O, Zerpa F, Morales J, et al. Pacemaker implantation in familial congenital AV block complicated by Adams Stokes attacks. Br Heart J 1967; 29:810.
20. Combrink JM, Davis WH, Snyman HW. Familial bundle branch block. Am Heart J 1962; 64:397.
21. Simonsen EE, Madsen EG. Four cases of right sided bundle branch block in three generations of a family. Br Heart J 1970; 32:501
22. Lorber A, Maisuls E, Naschitz J. Hereditary right axis deviation: electrocardiographic pattern of pseudo left posterior hemiblock and incomplete right bundle branch block. Int J Cardiol 1988; 20:399–402.
23. Schall SF, Seidenstricker J, Goodman R, Wooley CF. Familial right bundle branch block, left axis deviation, complete heart block and early death. A heritable disorder of cardiac conduction. Ann Intern Med 1973; 79:63.
24. Schwartz PJ, Periti M, Malliani A. The long QT syndrome. Am Heart J 1975; 89:378–90.
25. Schwartz PJ. Idiopathic long QT syndrome: progress and questions. Am Heart J 1985; 109:399–411.
26. Moss AJ, Schwartz PJ, Crampton RS, et al. The long QT syndrome: a prospective international study. Circulation 1985; 71:17–21.
27. Moss AJ. Prolonged QT-interval syndrome. JAMA 1986; 256:2985–2987.
28. Moss AJ, Schwartz PJ, Crampton RS, et al. The long QT syndrome. Prospective longitudinal study of 328 families. Circulation 1991; 84:1136–1144.
29. Romano C, Gemme G, Pongiglione R. Aritmie cardiache rare dell'eta pediatrica. Clin Pediatr 1963; 45:658–83.
30. Ward OC. New familial cardiac syndrome in children. J Irish Med Assoc 1964; 54:103–106.
31. Keating M, Dunn C, Atkinson D, et al. Linkage of a cardiac arrhythmia, the long QT syndrome, and the Harvey ras-1 gene. Science 1991; 252:704–706.
32. Keating M. Linkage analysis of the long QT syndrome, using genetics to study cardiovascular disease. Circulation 1992; 85:1973–1986.

33. Curran M, Atkinson D, Timothy K, Vincent GM, Moss A, Leppert M, Keating M. Locus heterogeneity of autosomal dominant long QT syndrome. J Clin Invest 1993; 92:799–803.

34. Vincent GM, Timothy KW, Leppert M, et al. The spectrum of symptoms and QT intervals in carriers of the gene for the long QT syndrome. N Engl J Med 1992; 327:846–852.

35. Towbin JAL, Pagotto B, et al. Romano-Ward long QT syndrome (RWLQTS): evidence of genetic heterogeneity. Pediatr Res 1992; 31:125A.

36. Kerem B, Benhorin J, et al. Evidence for genetic heterogeneity in the long QT syndrome. Am J Hum Genet 1992; 51:192A.

37. Jervell A, Lange-Nielsen F. Congenital deaf mutism, functional heart disease with prolongation of the QT interval, and sudden death. Am Heart J 1957; 54:59–78.

38. Jeffery S, Jamieson R, et al. Long QT and Harvey-ras. Letter to the Editor. Lancet 1991; 339:225.

39. Wren C, Rowland E, Burn J. Familial ventricular tachycardia: a report of four familes. Br Heart J 1990; 63:169–74.

40. Rubin DA, O'Keefe A, Kay RH, et al. Autosomal dominant inherited ventricular tachycardia. Am Heart J 1982; 123:1082–1084.

41. Schwensen C. Ventricular tachycardia as the result of the administration of digitalis. Heart 192; 9:199–203.

42. Cohen TJ, Liem BL, Hancock W. Association of bidirectional ventricular tachycardia with familial sudden death syndrome. Am J Cardiol 1989; 64:1078–1079.

43. Glikson M, Constantini N, Grafstein Y, et al. Familial bidirectional ventricular tachycardia. Eur Heart J 1991; 12:741–745.

44. Gault JH, Cantwell J, Lev M, et al. Fatal familial cardiac arrhythmias: histologic observations on the cardiac conduction system. Am J Cardiol 1972; 29:548–553.

45. Rosenbaum MB, Elizari MV, Lazzari JO. The mechanism of bidirectional tachycardia. Am Heart J 1969; 78:4–12.

46. Katz AM. Cardiac ion channels. N Engl J Med 1993; 328:1244–1251.

47. Bazett HC. An analysis of the time-relations of electrocardiograms. Heart 1920; 7:353–369.

48. Curran M, Landes G, Keating M. Molecular cloning, characterization and chromosomal localization of a human potassium channel gene. Genomics 1992; 12:729–737.

49. Beuren AJ. Supravalvular aortic stenosis: a complex syndrome with and without mental retardation. Birth Defects 1972; 8:45–46.

50. Eisenberg R, Young D, Jacobsen, B. Voito A. Familial supravalvular aortic stenosis. Am J Dis Child 1964; 108:341–347.

51. O'Connor W. Supravalvular aortic stenosis: clinical and pathologic observations in six patients. Arch Pathol Lab Med 1985; 109:179–185.

52. Ewart A, Morris C, Ensing G, Loker J, Moore C, Leppert M, Keating M. A human vascular disease, supravalvular aortic stenosis, maps to chromosome 7. Proc Natl Acad Sci USA 1993; 90:3226–3230.

53. Olson TM, Michels VV, Lindor NM, et al. Autosomal dominant supravalvular aortic stenosis: localization to chromosome 7. Hum Mol Genet 1993; 2:869–873.

54. Curran ME, Atkinson D, Ewart A, Morris C, Leppert M, Keating MT. The elastin

gene is disrupted by a translocation associated with supravalvular aortic stenosis. Cell 1993; 73:159–168.

55.   Morris CA, Loker J, Ensing G, Stock AD. Supravalvular aortic stenosis cosegregates with a familial 6;7 translocation which disrupts the elastin gene. Am J Med Genet 1993; 46:737–744.

56.   Ewart AK, Jin W, Atkinson D, Morris CA, Keating MT. Supravalvular aortic stenosis associated with a deletion disrupting the elastin gene. J Clin Invest 1993; 93:1071–1077.

57.   Bellugi U, Bihrle A, Jernigan T, Trauner D, Doherty S. Neurophysical, neurological, and neuroanatomical profile of Williams syndrome. Am J Genet 1990; 6:115–125.

58.   Dilts C, Morris C, Leonard C. Hypothesis for development of a behavioral phenotype in Williams syndrome. Am J Med Genet 1990; 6:126–131.

59.   Morris CA, Demsey SA, Leonard CO, Dilts C, Blackburn BL. Natural history of Williams syndrome. J Pediatr 1988; 113:318–326.

60.   Ewart AK, Morris CA, Atkinson D, Jin W, Sternes K, Spallone P, Stock AD, Leppert M, Keating MT. Hemizygosity at the elastin locus in a developmental disorder, Williams syndrome. Nature Genet 1993; 5:11–16.

61.   Uitto J, Christiano A, Kahari VM, Bashir M, Rosenbloom J. Molecular biology and pathology of human elastin. Biochem Soc Trans 1991; 19:824–829.

62.   Davidson JM. Elastin: structure and biology. In: Uitto J, Perejda A, eds. Connective Tissue Disease: Molecular Pathology of the Extracellular Matrix. New York: Marcel Dekker, 1987:29–54.

63.   Sakai L, Keene D, Engvall E. Fibrillin, a new 350-kD glycoprotein, is a component of extracellular microfibrils. J Cell Biol 1986; 103:2499–2509.

64.   Dietz H, Cutting G, Pyeritz R, Maslen C, Sakai L, Corson G, Puffenberger E, Hamosh A, Nanthakumar E, Curristin S, Stetten G, Meyers D, Francomano C. Marfan syndrome caused by a recurrent de novo missense mutation in the fibrillin gene. Nature 1991; 352:337–339.

5

Molecular Genetic Studies of Inherited Cardiomyopathies

Christine E. Seidman
Howard Hughes Medical Institute
Brigham and Women's Hospital
and Harvard Medical School
Boston, Massachusetts

Jonathan G. Seidman
Howard Hughes Medical Institute
Harvard Medical School
Boston, Massachusetts

Cardiomyopathies are a heterogeneous group of primary myocardial disorders characterized by impaired ventricular function (1). These are traditionally classified as dilated, hypertrophic, or restrictive based on anatomical and hemodynamic features. Dilated cardiomyopathies are the most common form of cardiomyopathies and occur with an estimated prevalence of 36.5 per 100,000 individuals in the population (2). Diagnosis of this disorder is based on findings of cardiac chamber dilation and depressed ventricular contractility. Hypertrophic cardiomyopathies are defined by an increase in myocardial mass with myocyte and myofibrillar disarray. This pathology perturbs both diastolic and systolic function. Hypertrophic cardiomyopathy is estimated to be present in 19.7 per 100,000 individuals (2). Restrictive cardiomyopathies are the least common of these disorders. These are characterized by normal cardiac structure and abnormal ventricular diastolic function with preserved systolic function.

Clinical and research efforts have demonstrated that both dilated and hypertrophic cardiomyopathies are often familial. In contrast, restrictive cardiomyopathies are rarely inherited. The reported inheritance patterns of familial cardiomyopathies have included autosomal (dominant or recessive), X-linked, and matrilinear. These observations, combined with recent advances in human molecular genetics, have led to significant progress toward identifying the genetic etiology of inherited cardiomyopathies. Such endeavors will have a substantial impact on clinical medicine. Most immediately, identification of a disease gene (or initially the chromosome location of the disease gene) provides the opportunity for accurate and even preclinical diagnosis. Gene-based diagnosis also provides a precise framework for stratifying analyses of clinical outcomes and therapies. A more distant goal of genetic studies is the development of new targets for therapeutics. Although therapies are currently available to alleviate symptoms associated with cardiomyopathies, none has been shown to retard or reverse the progressive nature of these disorders. Hence, for many affected individuals, cardiac transplantation is the only hope for long-term survival. In addition, definition of the genetic etiologies of these rare and inherited cardiomyopathies may provide insights into the cellular pathways that are perturbed in common and nonfamilial heart diseases.

This chapter reviews recent molecular genetic studies of familial dilated and hypertrophic cardiomyopathies. The clinical relevance of patterns of inheritance, chromosomal localization, and gene identification are updated for each disorder.

GENERAL PRINCIPLES OF HUMAN MOLECULAR GENETICS

Modes of Inheritance of Human Disorders

The technological improvements and widespread use of noninvasive cardiac imaging have both fostered earlier diagnosis of dilated and hypertrophic cardiomyopathies and contributed to the recognition that these disorders are often familial. Familial cardiomyopathies can be classified by their patterns of inheritance (Figure 1) as autosomal dominant, autosomal recessive, X-linked, or matrilinear. For the clinician, understanding the mode of inheritance is essential for identifying individuals at risk of inheriting these disorder. For the researcher, patterns of transmission can direct strategies for identification of the responsible gene defect.

Genetic information is transmitted through genes contained in nuclear chromosomes (autosomes 1–22 and sex chromosomes X and Y) as well as on a single mitochondrial chromosome (reviewed in Ref. 3). By defining the clinical inheritance pattern of a cardiomyopathy, one can discern which type of chromosome carries the disease gene. Hence, matrilinear transmission of a cardiomyopathy suggests a mitochondrial gene defect; a familial cardiomyopathy that occurs only in male family members suggests an X-linked disorder;

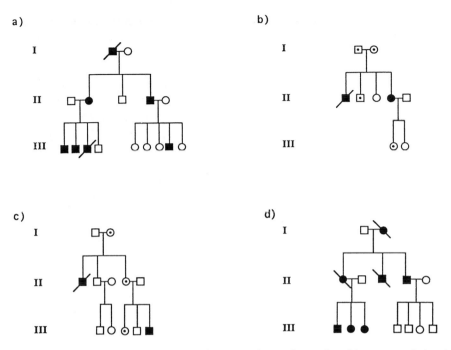

Figure 1 Pedigrees of familial cardiomyopathies inherited as (a) autosomal dominant, (b) autosomal recessive, (c) X-linked, or (d) mitochondrial traits. Squares denote
men; circles denote women. Deceased individuals are slashed. Affection status is
shown: solid symbols, affected; clear symbols, unaffected; central dot, carrier.

a cardiomyopathy that equally affects male and female members of a family is
likely to be autosomal.

Strategies for Identifying Disease-Causing Genes

Significant progress in the field of human genetics has led to the identification
of the gene defects that are responsible for many inherited disorders. The first
step in these studies is to clinically study a large kindred affected by the
disorder and to define the mode of inheritance. For gene defects transmitted
via mitochondrial DNA, the small size of this chromosome permits direct
mutation analyses. In contrast, when the inheritance pattern suggests a
nuclear gene mutation, then genetic linkage studies are appropriate to define
the chromosome location of the defect. Linkage studies involve analyzing DNA
loci from defined regions of the genome to find one that is coinherited with
the disease. (A locus is a specific position or location on a chromosome; a
locus is said to be polymorphic if there is variation either in the nucleotide

sequence or in the copy number of a repetitive sequence at a given location.) If the condition is X-linked, only loci on that chromosome will be analyzed. If the condition is autosomal, then loci from each of the 22 autosomes will be analyzed.

Genetic mapping studies have become increasingly more feasible because of the substantial number of polymorphic loci that have recently been identified on nuclear chromosomes. These polymorphic loci are usually di-, tri-, or tetranucleotide sequences that vary in length, based on the number of times the sequence is repeated in an individual. Short tandem repeat sequences can be readily analyzed from DNA (usually derived from a peripheral blood sample) using the polymerase chain reaction (PCR). The efficiency of PCR amplification combined with the abundance of polymorphic loci has enabled mapping studies to be successfully performed in less time and using smaller families than was possible with older techniques.

By tracing the inheritance of a polymorphism *and* a clinical disease through several generations of a family, one can ascertain whether these are coinherited (genetically linked). Genetic linkage occurs when two traits that are close together in the same chromosomal region are coinherited. If there is substantial genetic distance between two traits, physical exchange of genetic material will likely occur between paired chromosomes during meiosis and traits will segregate. Geneticists validate the coinheritance of a disease and a particular marker by calculating a lod (logarithm of the odds) score (4). A lod score less than –2 is generally accepted as evidence of nonlinkage between the disorder and a given locus. A lod score greater than +3 indicates that coinheritance of a disorder and a locus are 1000-fold more likely to occur because these are genetically linked. This often provides the first clue regarding the chromosome location of a disease gene.

Once the chromosome location is known, two strategies can be used to identify a gene defect. Genes that have been previously mapped to the same chromosome location, and that are expressed in the disease organ, are analyzed for potential mutations. Such genes are generally called candidate genes. If candidate genes are not found to have mutations, then molecular techniques are used to refine the map location and to identify novel genes that are subsequently analyzed for mutations.

These research strategies have led to the identification of the chromosome location of several inherited dilated and hypertrophic cardiomyopathies. Further, disease genes and precise mutations have been identified for some disorders. The molecular genetic studies reviewed below provide the current status of chromosome mapping and gene identification. It is worth noting, however, that the pace of research in this area will rapidly expand the current compendium of inherited cardiomyopathies for which a genetic etiology has been defined.

AUTOSOMAL DOMINANT CARDIOMYOPATHIES

Familial Hypertrophic Cardiomyopathy

Clinical Features

Familial hypertrophic cardiomyopathy (FHC) is an autosomal dominant disorder, characterized by increased ventricular mass (Figure 2, left), hyperkinetic systolic function, and impaired diastolic relaxation. The disease is characterized by a wide spectrum of possible symptoms, including exertional dypsnea, angina pectoris, and sudden death, which can occur even in asymptomatic individuals (reveiwed in Ref. 1). The mechanism for sudden death is often unknown, and both primary arrhythmias and hemodynamic factors have been proposed. While several studies of the natural history of FHC have demonstrated an annual mortality rate of 2–4% due to sudden death (5), reports on ambulatory FHC patients suggest a more benign prognosis (6).

Physical findings (particularly a systolic murmur that increases with maneuvers that decrease preload) or electrocardiographic abnormalities (such as inferior Q-waves and left ventricular hypertrophy) may suggest FHC. However, diagnosis in the adult rests on the demonstration by two-dimensional echocardiography of unexplained ventricular hypertrophy. In the young, diagnosis is often complicated because hypertrophy may not develop until after adolescent growth has been completed (7).

Genetic Studies

Linkage analyses in kindreds with FHC have demonstrated that hypertrophic genes are located on chromosomes 1q31 (8), 11p13–q13 (9), 14q1 (10), and 15q2 (11). Hence, FHC is a genetically heterogeneous disorder; a mutation in one of several genes can produce a clinically indistinguishable phenotype. Mapping a gene responsible for FHC to chromosome 14q1 led to the analysis of the myosin heavy-chain genes for mutations. Several myosin genes are encoded in the human genome. However, only the α and β cardiac myosin heavy-chain (MHC) genes are abundantly expressed in the heart and each maps to chromosome 14q1 (12), suggesting that these are excellent candidates for disease-causing mutations.

Extensive analyses of FHC patients have led to the identification of more than 20 mutations in the β cardiac MHC gene (reviewed in Refs. 13 and 14). In contrast, no mutations have been identified in the α cardiac MHC. The gene defects that cause FHC are typically family-specific. That is, unique mutations are found in different families (termed allelic or intragenic heterogeneity). These are missense mutations: the nucleotide change results in the replacement of the normally encoded amino acid with a different residue. Frequently the FHC mutation also results in a change of charge in the encoded amino acid.

Studies that correlate the clinical features of FHC in individuals with

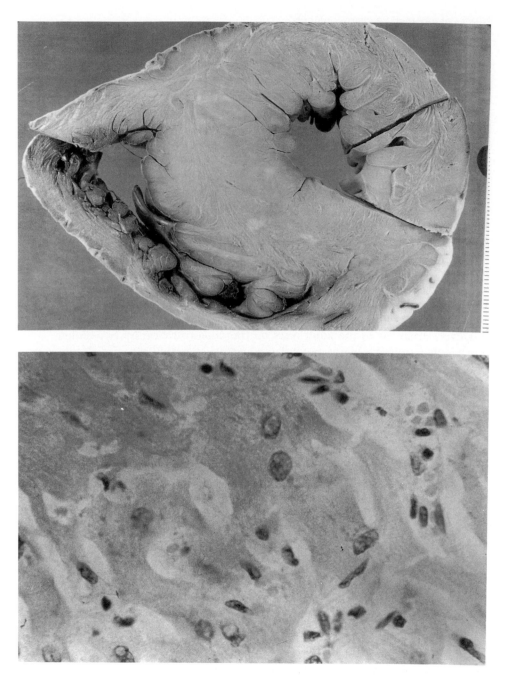

Figure 2   Gross anatomical and histological features of familial hypertrophic and dilated cardiomyopathy. (Above) Autopsy specimen from an individual with an FHC inherited as an autosomal dominant trait, showing marked hypertrophy of both the interventricular septum and left ventricular free wall. Pale areas within the left ventricle are due to interstitial fibrosis. Histological section demonstrates hypertrophied cells

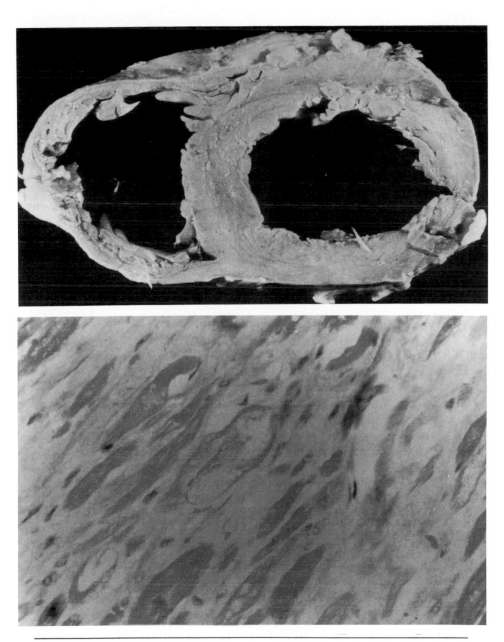

with myocytes and myofibrillar disarray (hematoxylin and eosin stain). (Above) Autopsy specimen from an individual with a progressive conduction system disease and dilated cardiomyopathy inherited as an autosomal dominant trait. There is cardiac hypertrophy (total weight − 540 grams) and biventricular dilatation. Pale areas within the left ventricle are due to interstitial fibrosis. Histology of atrial myocytes demonstrates myocyte loss with severe interstitial fibrosis; residual myocytes are hypertrophied and show prominent degeneration with cytoplasmic vacuolization.

distinct genetic etiologies are an area of active research. Previous investigations have demonstrated that there can be considerable variation in the extent
and distribution of hypertrophy even within a given family, all of whom will
have the same genetic basis of FHC (15). Recent analyses (16) of the anatomical features of FHC categorized according to different genetic etiologies have
confirmed earlier work, thereby suggesting that other factors (genetic and/or
environmental) are important in modulating the hypertrophic phenotype. In
contrast, different β cardiac MHC gene mutations appear to correlate with
survival in patients with FHC (13,14,17). Several mutations have been identified that are associated with a significant risk of sudden and/or premature
death (malignant mutations), whereas other mutations correlate with long-
term survival (benign mutations).

Figure 3 compares Kaplan-Meier curves for survival with FHC due to two

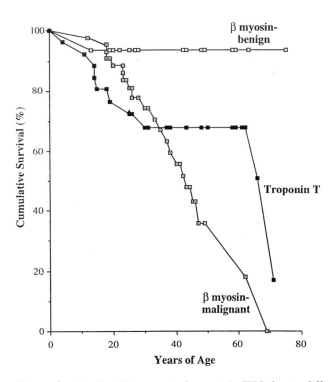

Figure 3   Kaplan-Meier survival curves in FHC due to different mutations. Survival
of individuals with FHC caused by some mutations in the β myosin heavy-chain gene
is normal (benign), whereas other mutations in this gene are associated with significant
premature mortality (malignant). Survival of individuals with mutations in cardiac
troponin T is comparable to that observed with malignant myosin mutations.

β cardiac MHC mutations and one cardiac troponin T mutation (see below). Near-normal survival is found in individuals with the Val606Met (benign) mutation. In contrast, approximately half of individuals with the Arg403Gln (malignant) mutation are dead by age 50. Survival data of individuals with different mutations within the β cardiac MHC gene continue to be accumulated. Preliminary analyses suggest that the location of mutations within the β cardiac MHC gene does not appear to predict those with a benign versus a malignant clinical phenotype. However, most malignant missense mutations alter the charge of the encoded amino acid residue, whereas benign missense mutations are generally not associated with a change in amino acid charge (13). Many other factors undoubtedly contribute to the high incidence of premature death in FHC. It is hoped that assessing those factors in light of the genetic background against which they occur will improve risk stratification in patients.

The demonstration that β cardiac MHC gene mutations caused FHC, combined with the evidence of genetic heterogeneity (18), led to the hypothesis that FHC gene defects located on other chromosomes might encode proteins of similar function. Genetic mapping of an FHC gene to chromosome 1q3 (8) led to the analyses of several candidate genes encoding proteins that could interact with myosin heavy chains (19). These included four genes (skeletal actin, non-muscle tropomyosin, slow-twitch troponin I, and myosin-binding protein H) that had previously been mapped to chromosome 1q. Molecular analyses failed to implicate any of these genes as causing FHC.

Another FHC gene was identified on chromosome 15q2 by linkage analyses (11). While there were initially no candidate genes present on this chromosome region in humans, the murine α tropomyosin gene was recently mapped in mice (20) to a region that is syntenic to human chromosome 15. (Synteny indicates conservation of the organization of genes on chromosomes of different species.) Based on this finding, the human α tropomyosin gene would be expected to reside on human chromosome 15. Because this gene encodes an important component of thin filaments that interact with myosin in the thick filaments, α tropomyosin was considered an excellent candidate for FHC-causing mutations.

Mapping studies confirmed that human α tropomyosin resided on chromosome 15. Mutational analyses were then performed and demonstrated in exon 5 of α tropomyosin two gene defects that caused FHC in two families (19). Both mutations are missense, and alter the amino acid encoded in the normal peptide. Analyses of other families are expected to demonstrate other disease-causing mutations in the α tropomyosin gene. However, based on preliminary genetic linkage studies, it is unlikely that a significant fraction of disease can be accounted for by mutations in this gene.

Because mutations in two sarcomere genes—α tropomyosin and β cardiac MHC—caused FHC, other components of the thin and thick filaments became candidates for causing hypertrophic cardiomyopathy that mapped to other chromosomes. Components of the troponin complex (troponins C, T, and I) are known to bind α tropomyosin. Further, the two missense mutations in α tropomyosin occur in a putative binding domain for cardiac troponin T. Because the chromosome location of genes encoding the troponin complex was unknown, initial studies addressed whether any of these genes mapped to chromosome 1q. Cardiac troponin T was localized to this region, and analyses for FHC-causing mutations were then performed.

Cardiac troponin T gene mutations were initially demonstrated to cause FHC in three families (19). Affected members in two families have missense mutations in exons encoding a highly conserved region of the molecule. Affected members in another family have a point mutation in the splice donor sequence of intron 15, predicting that the encoded cardiac troponin T would be truncated. Recent studies (21) have demonstrated that a wide range of mutations in cardiac troponin T can cause FHC, as was noted in the β cardiac MHC gene. The hypertrophic phenotype associated with these mutations is often subtle. That is, affected individuals appear to have modest increases in the maximum left ventricular hypertrophy (detected by echocardiography). Despite this, survival of individuals with FHC caused by cardiac troponin T mutations (Figure 3) is comparable to that seen with malignant β cardiac MHC gene mutations. These findings suggest that genetic diagnosis may be particularly important in families with cardiac troponin T mutations.

Collectively, the molecular studies on FHC have demonstrated that this is a disease of the sarcomere. Mutations in β cardiac MHC, α tropomyosin, or cardiac troponin T can cause this pathology. Given these data, it is likely that the FHC gene that maps to chromosome 11 will encode a sarcomere (or sarcomere-related) protein. Current research is centered on identification of this and other FHC genes. In addition, strategies for understanding how sarcomere mutations result in cardiac hypertrophy are an avenue of active investigation. Definition of the intracellular signals and pathways that respond to perturbation in sarcomere structure and/or function should provide the framework for rational design of therapeutic agents.

More immediately, the definition of three distinct genetic etiologies for FHC permits gene-based diagnosis in some families. As noted above, FHC is an autosomal dominant disorder that is equally inherited by men and women (Figure 1a). Since only one mutated gene is sufficient to cause autosomal dominant diseases, 50% of the offspring of an affected individual will inherit the disease-causing mutation. In individuals with a mutation in β cardiac MHC, α tropomyosin, or cardiac troponin T, gene-based diagnosis of all family members is now possible. This targets longitudinal evaluation to children who are genotype-positive (22), and prevents unnecessary restrictions on those children who are at no risk of developing FHC.

Familial Dilated Cardiomyopathies

Clinical Features

Dilated cardiomyopathy, a disease characterized by cardiac dilation and impaired ventricular function, occurs secondary to toxic, metabolic, or infectious agents, or as a primary disorder of the myocardium (reviewed in Ref. 1). Dilated cardiomyopathy describes a clinically disparate group of disorders. Even within familial dilated cardiomyopathies, there is evidence for significant heterogeneity. While signs and symptoms of congestive heart failure are commonly reported, other features—including pericardial effusion, conduction disturbances, and unexplained sudden death or syncopal episodes—are frequently noted (1,3). Pathological findings similarly suggest that familial dilated cardiomyopathy is heterogeneous. Myocyte death with and without inflammation, fibrosis, and conduction system disease have all been noted (1,23). These data suggest that a multiplicity of gene defects can cause myocardial dysfunction that results in chamber dilation and dysfunction. While elucidation of the initiating process that results in idiopathic dilated cardiomyopathy in a single affected individual may be virtually impossible, identification of families with a defined set of clinical features permits application of molecular genetic approaches to identify gene defects that cause this important form of heart failure.

Wooley and colleagues (24) have identified and clinically characterized a large American kindred with an inherited cardiomyopathy that presents initially as atrial arrhythmia (stage I, second to third decade), progresses to atrioventricular block (stages II–III, third–fifth decade), and is accompanied by progressive cardiac dilation (Figure 2, bottom) and congestive heart failure (stage IV, fifth–sixth decade). Serial endomyocardial biopsies have documented progressive fibrosis with myofibril loss throughout these stages (Figure 2, right). These findings may precede hemodynamic abnormalities, suggesting a primary myocyte defect. Therapy has been directed largely to treatment of arrhythmias; however, antiarrhythmics and electronic pacing do not prevent the development of heart failure.

Genetic Studies

Inherited or familial dilated cardiomyopathy can be transmitted as X-linked (3,25) or dominant traits (3,24). At least some X-linked dilated cardiomyopathies appear to be due to mutations in the dystrophin gene (see below) and can occur without clinical signs of skeletal myopathy (26). Dominant inheritance of dilated cardiomyopathy has often been considered a rare disorder; however, recent studies of the relatives of affected probands have demonstrated that 20% (27,28) of dilated cardiomyopathy is familial. Further, some studies (28) reported a significantly accelerated progression of disease in familial cases. These studies have important clinical implications for affected individuals and family members. Moreover, they suggest that molecular genetic approaches to the study of dilated cardiomyopathies are feasible.

The family characterized by Wooley and colleagues (24) has been followed for more than 20 years, and autosomal dominant transmission has been documented (Figure 1a). Using genetic linkage analyses (detailed above), the disease gene responsible for the progressive conduction system defect and dilated cardiomyopathy in this kindred has recently been mapped near the centromere of chromosome 1 (29). While genetic mapping of other affected families will be required to assess whether this condition is genetically heterogeneous, the report of a similar phenotype in several studies of inherited cardiomyopathies (30,31) suggests that this disease locus may be a significant cause of unexplained heart failure. The unique clinical features of disease in this family, combined with genetic mapping studies, suggest that the gap junction protein Cx 40 may be a candidate for causing this condition. The Cx gene family encodes proteins that form cell-to-cell channels (gap junctions) and allow intercellular communication by passage of small ions and molecules (32). Cx40 is present in heart tissues (33) and specialized cardiac conduction cells (34). Mutations in Cx40 might account for both the conduction system disease in this family and the progressive heart failure.

Several family studies have demonstrated atrioventricular conduction defects that precede the development of heart failure (24,30,31). Atrioventricular conduction abnormalities are also frequently noted in individuals with dilated cardiomyopathies and in part contribute to the high incidence of sudden cardiac death associated with this disorder. Ultimately, gene identification of the defect responsible for the disorder detailed above and other inherited dilated cardiomyopathies will permit accurate preclinical diagnosis. This should contribute to our understanding of the natural history of these complex disorders.

AUTOSOMAL RECESSIVE CARDIOMYOPATHIES

Genetic Studies

As with all autosomal recessive disorders, individuals with a recessive cardiomyopathy are homozygous at the disease locus and inherit one defective copy of the disease gene from each parent (Figure 2, lower left). Hence, autosomal recessive diseases are more frequent when there is a high prevalence of carriers within the population (such as for sickle cell disease or cystic fibrosis) or within a family because of parental consanguinity. Marriage between relatives increases the probability that two parents will both have inherited a mutated gene found originally in a common ancestor.

Inherited disorders of energy metabolism can result in a hypertrophic or dilated cardiomyopathies that are transmitted as an autosomal recessive trait (1,35). The responsible mutations for these disorders occur in nuclear genes that encode transport proteins or enzymes involved in cardiac fatty acid β-oxidation. These mutations occur commonly in the population, with an estimated incidence of 1 in 10,000 to 15,000 individuals (36). Molecular genetic analyses have demonstrated that mutations in the medium-chain acyl-CoA dehydrogenase (MCAD) gene on chromosome 1p (37) and the long-chain acyl-CoA dehydrogenase gene on chromosome 7 (38) can cause recessive cardiomyopathies. Unlike with dominant cardiomyopathies, genetic studies of MCAD deficiency have demonstrated that approximately 90% of affected individuals share a common mutation at amino acid 304 (39). This enables rapid genetic testing with patients suspected of having cardiomyopathies due to MCAD deficiency.

In addition to these enzymatic defects, mutations in genes involved in carnitine transport and metabolism can also cause recessive cardiomyopathies. Carnitine is required for entry of long-chain fatty acids into mitochondria; carnitine deficiency prevents metabolism of long-chain fatty acids. Gene defects have been identified in the transporter of carnitine into cells (40); in the translocase, which shuttles carnitine into the mitochondrial (41); and in carnitine palmitoyltransferase II (42), which catalyzes carnitine derivatives into acyl-CoA.

Defects in fatty acid β-oxidation can damage the myocardium by preventing adequate energy by intracellular accumulation of the intermediary metabolites. In addition, intermediary metabolites (particularly long-chain acylcarnitines) may be arrhythmogenic, thereby contributing to the high incidence of sudden death associated with these disorders.

Clinical Findings

The clinical phenotypes resulting from these gene defects and associated sequelae vary considerably (37–39, 43). Dilated and hypertrophic cardiomyopathies have been described that range from mild and stable to severe and progressive. These cardiomyopathies can cause heart failure, pulmonary edema, arrhythmias, and sudden death. Symptoms are usually precipitated by fasting states, in which fatty acid oxidation becomes a critical source of cardiac energy (44). The anatomical and hemodynamic features of these recessive dilated cardiomyopathies can be clinically indistinguishable from other causes of idiopathic dilated cardiomyopathy. However, an endocardial biopsy that demonstrates cytoplasmic inclusions of lipid may suggest the correct diagnosis. In contrast, the hypertrophic phenotype associated with defects in fatty acid oxidation typically reveals impaired systolic function, while hyperdynamic systolic function is characteristic of FHC that is transmitted as a dominant trait.

Associated findings in other organ systems, the natural history, and patterns of inheritance further distinguish the recessive and dominant cardiomyopathies. Inherited defects in fatty acid oxidation typically perturb the physiology of other organ systems that, like the heart, have high energy requirements. Hence, skeletal myopathies, neurological findings, and metabolic abnormalities are often present in affected individuals. Recessive cardiomyopathies are typically diagnosed during childhood, with the most severe manifestations usually occurring by the second decade. Autosomal dominant cardiomyopathies are often subclinical before the second or third decade. Family studies readily discriminate between the different genetic etiologies of inherited cardiomyopathies. Because defects in fatty acid oxidation are recessive, associated cardiomyopathies are found only in homozygous individuals; heterozygous carriers (i.e., parents of affected children) will have no clinical manifestations of disease (35 and Figure 1b). In contrast, one parent of an affected child and 50% of first-degree relatives will also be affected in families in which cardiomyopathies are transmitted as dominant traits.

The identification of precise gene defects in fatty acid oxidation that cause recessive cardiomyopathies has improved clinical management of families with these disorders. Gene-based diagnoses can accurately predict the risk in siblings of being affected or carriers. Genetic diagnosis also helps to define therapy. Disorders of primary carnitine deficiency can be treated with pharmacological supplementation, which can substantially improve the cardiomyopathy and other neuromuscular symptoms (43). Severe episodic cardiac dysfunction due to gene defects in other pathways of fatty acid metabolism are managed by avoidance of fasting and, if necessary, intravenous dextrose (35).

X-LINKED CARDIOMYOPATHIES

Dilated cardiomyopathies can be inherited as an X-linked trait (3,25,26). X-linked disorders typically result in affected males (Figure 1c); women who are heterozygotes for these mutations typically will be asymptomatic, but, on careful physical examination, subclinical findings may be present. Pedigree analysis is notable for the absence of male-to-male transmission.

Genetic linkage studies have demonstrated that X-linked dilated cardiomyopathy maps to the short arm, close to the dystrophin gene (25,26). The dystrophin gene encodes a cytoskeletal protein that is abundantly expressed in skeletal as well as cardiac muscle. Deletions in this gene cause Duchenne's and Becker's muscular dystrophy (3). While the predominant symptom of these disorders is skeletal muscle weakness, an associated dilated cardiomyopathy is not uncommon. These data prompted analyses of the dystrophin gene in individuals with X-linked dilated cardiomyopathies. Studies to date have identified only one mutation (26), a deletion of promoter sequences (which regulated gene expression) and first exon (encoding a portion of the dystrophin protein). The consequences of such a mutation might be expected to reduce the amounts of dystrophin. A recent report (45) examined whether such mutations might account for cases of idiopathic dilated cardiomyopathy. Analyses of promoter sequences of the dystrophin gene from 27 male patients failed to demonstrate any mutations, suggesting that this is not a common genetic etiology of idiopathic dilated cardiomyopathy.

The clinical value of defining an X-linked dilated cardiomyopathy is restricted to defining risk to family members. The male offspring of a known female carrier have a 50% chance of inheriting the gene defect. Clinical evidence of disease in these individuals can usually be demonstrated by age 20. Further, longitudinal studies of one family identified women who developed late-onset disease, suggesting that heterozygous women can also be affected (26). Such data suggest that it may be difficult to distinguish between X-linked and autosomal dominant dilated cardiomyopathies. The absence of disease in the offspring of affected parents and the variable age of onset between affected men and women can help to distinguish between these modes of inheritance. Therapy of dystrophin-associated cardiomyopathies is at present limited to supportive measures.

MITOCHONDRIAL CARDIOMYOPATHIES

Mutations in mitochondrial genes produce complex disorders affecting a variety of tissues, including the skeletal muscles, myocardium, and the central nervous system. Because these diverse organ systems each have high energy requirements, defects in proteins involved in oxidative phosphorylation that are encoded by mitochondrial genes may be particularly detrimental to these organ systems.

The transmission pattern of mitochondrial cardiomyopathies is matrilinear (i.e., mother to offspring; Figure 1d). This occurs because virtually all mitochondria are derived from the cytoplasm of the oocyte; the spermatocyte contributes almost no mitochondria to the zygote (3,46). The absence of paternal transmission of these disorders (as well as involvement of other organ systems) can often be an important clue that the cardiomyopathy is due to a mitochondrial mutation. There is also substantial phenotypic variation in the expressivity of mitochondrial mutations. That is, while all offspring of affected females will inherit a mitochondrial gene mutation, both the severity of symptoms and the range of organ-system involvement vary considerably between offspring. This variability in phenotype has been attributed to heteroplasmy, or variation in the fraction of abnormal mitochondria in cells of different organs.

Several recent studies (47,48) have demonstrated that mitochondrial disorders have an associated cardiomyopathy, which can be clinically important. These cardiomyopathies are characterized by either hypertrophy or dilation and are frequently associated with conduction-system disease. The diagnosis of cardiac involvement in mitochondrial disorders relies on standard clinical diagnostics (electrocardiogram, two-dimensional echocardiography) as well as ultrastructural findings of endomyocardial biopsy specimens and assays for oxidative phosphorylation. Molecular characterization of mitochondrial mutations are eminently feasible. This is because mitochondrial genes are all encoded on a single circular chromosome, containing approximately 16,500 nucleotides. Because the entire nucleotide sequence of the normal mitochondria is known, polymerase chain amplification of sequences to detect deletions (which are common in mitochondrial disorders) or point mutations can be rapidly completed. However, as noted above, because mitochondrial heteroplasmy may cause organ-specific findings, analyses must be conducted on the target tissue. Hence, to demonstrate cardiomyopathy due to mitochondrial gene mutations, endomyocardial biopsy specimens must be genetically analyzed.

There appears to be an age-related increase in deletions in mitochondrial DNA derived from human hearts (49). This finding has led to the hypothesis that the accumulation of deletions throughout life may contribute to the aging process in general and to the occurrence of presbycardia in the elderly. This observation has importance with regard to the potential role of mitochondrial mutations in cardiomyopathies. A recent study of idiopathic dilated cardiomyopathy demonstrated a similar frequency of mitochondrial deletions in control and cardiomyopathic hearts (50). Hence, the demonstration of cardiac mitochondrial gene mutations in patients with dilated or hypertrophic cardiomyopathies must be assessed in light of other findings. If the affected individual has clinical evidence of other organ involvement or evidence of matrilinear transmission of the cardiomyopathy, or if biochemical studies demonstrate altered oxidative phosphorylation in myocardial tissues, then a causal relationship probably exists between a mitochondrial mutation and the cardiomyopathy. While there are no specific therapies currently available for these gene defects, assessment of progressive conduction-system disease combined with management of heart failure is essential. In addition, assessment of other organ systems and genetic counseling are appropriate.

SUMMARY

Molecular genetic studies of inherited cardiomyopathies have begun to define the defects responsible for several disorders. Chromosome linkage studies have targeted research efforts to identify genes and mutations within nuclear genes that cause both hypertrophic and dilated cardiomyopathies. Definition of the precise genetic etiologies of these complex disorders will greatly improve diagnosis and may provide insights that will also impact on therapy and prevention.

ACKNOWLEDGMENTS

This work was supported by grants (to C.E.S and J.G.S) from the Howard Hughes Medical Foundation, the National Institutes of Health, and Bristol-Myers Squibb company.

REFERENCES

1. Wynne J, Braunwald E. The cardiomyopathies and myocardities: Toxic, chemical, and physical damage to the heart. In: Braunwald E, ed. Heart Disease: A Textbook of Cardiovascular Medicine. 4th ed. Philadelphia: WB Saunders, 1992:1394–1450.
2. Codd MB, Sugrue DD, Gersh BJ, Melton LJ III. Epidemiology of idiopathic dilated and hypertrophic cardiomyopathy: a population-based study in Olmsted County, Minnesota, 1975–1984. Circulation 1989; 80:564–572.
3. McKusick VA. Mendelain Inheritance in Man. 10th ed. Baltimore: Johns Hopkins University Press, 1992.
4. Ott J. A computer program for linkage analysis of general human pedigrees. Am J Hum Genet 1967; 28:528–529.
5. McKenna W, Deanfield J, Faruqui A, England D, Oakley C, Goodwin J. Prognosis in hypertrophic cardiomyopathy: Role of age and clinical, electrocardiographic and hemodynamic features. Am J Cardiol 1981; 47:532–538.
6. Spirito P, Chiarella F, Carratino L, Berisso MZ, Bellotti P, Vecchio C. Clinical course and prognosis of hypertrophic cardiomyopathy in an outpatient population. N Engl J Med 1989; 320:749–755.
7. Maron BJ, Spirito P, Wesley Y, Arce J. Development and progression of left ventricular hypertrophy in children with hypertrophic cardiomyopathy. N Engl J Med 1986; 315:610–614.
8. Watkins H, MacRae CA, Thierfelder L, Chou Y-H, Frenneaux M, McKenna WJ, Seidman JG, Seidman CE. A disease locus for familial hypertrophic cardiomyopathy maps to chromosome 1q. Nature Genetics 1993; 3:333–337.
9. Carrier L, Hengstenberg C, Beckmann JS, Guicheny P, Dufour C, Bercovici J, Daussc E, Berebbi-Bertrand I, Wisnewsky C, Pulvenis D, Fetler L, Vignal A, Weissenbach J, Hillaire D, Fcingold J, Bouhour J-B, Hagege A, Desnos M, Isnard R, Duborg O, Komajda M, Schwartz K. Mapping of a novel gene for familial

hypertrophic cardiomyopathy to chromosome 11. Nature Genet 1993; 4:311–313.

10. Jarcho J A, McKenna W, Pare JAP, Solomon SD, Holcombe RF, Dickie S, Levi T, Donis-Keller H, Seidman JG, Seidman CE. Mapping a gene for familial hypertrophic cardiomyopathy to chromosome 14q1. N Eng J Med 1989; 321:1372–1378.

11. Thierfelder L, MacRae C, Watkins H, Tomfohrde J, Williams M, McKenna W, Bohm K, Noeske G, Schlepper M, Bowcock A, Vosberg HP, Seidman JG, Seidman CE. A familial hypertrophic cardiomyopathy locus maps to chromosome 15q2. Proc Natl Acad Sci USA 1993; 90:6270–6274.

12. Saez LJ, Gianola KM, McNally EM, Feghali R, Eddy R, Shows TB, Leinwand LA. Human cardiac myosin heavy chain genes and their linkage in the genome. Nucl Acids Res 1987; 15:5443–5459.

13. Anan R, Greve G, Thierfelder L, Watkins H, McKenna WJ, Solomon S, Vecchio C, Shono H, Nakao S, Tanaka H, Mares A, Towbin JA, Spirito P, Roberts R, Seidman JG, Seidman CE. Prognostic implications of novel β myosin heavy chain gene mutations that cause familial hypertrophic cardiomyopathy. J Clin Invest 1994; 93:280–285.

14. Fananapazir L, Epstein ND. Genotype-phenotype correlations in hypertrophic cardiomyopathy. Circulation 1994; 89:22–32.

15. Maron BJ, Gottdiener JS, Epstein SE. Patterns and significance of distribution of left ventricular hypertrophy in hypertrophic cardiomyopathy: A wide angle, two dimensional echocardiographic study of 125 patients. Am J Cardiol 1981; 48:418–428.

16. Solomon SD, Wolff S, Watkins H, Ridker PM, Come P, Seidman CE, McKenna WJ, Lee RT. Left ventricular hypertrophy and morphology in familial hypertrophic cardiomyopathy associated with mutations in the β myosin heavy chain gene. J Am Coll Cardiol 1993; 22:498–505.

17. Watkins H, Rosenzweig A, Hwang D-S, Levi T, McKenna WJ, Seidman CE, Seidman JG. Distribution and prognostic significance of myosin missense mutations in familial hypertrophic cardiomyopathy. N Engl J Med 1992; 326:1108–1114.

18. Solomon SD, Jarcho JA, McKenna W, Geisterfer-Lowrance AAT, Germain R, Salerni R, Seidman JG, Seidman CE. Familial hypertrophic cardiomyopathy is a genetically heterogeneous disease. J Clin Invest 1990; 86:993–999.

19. Thierfelder T, Watkins H, MacRae C, McKenna W, Lamas R, Vosberg H-P, Seidman JG, Seidman CE. α Tropomyosin and cardiac troponin T mutations cause familial hypertrophic cardiomyopathy, a disease of the sarcomere. Cell 1994; 77:701–712.

20. Schleef M, Werner K, Satzger U, Kaupmann K, Jokusch H. Chromosomal location and genomic cloning of the mouse α-tropomyosin gene Tpm-1. Genomics 1993; 17:519–521.

21. Watkins H, McKenna W, Thierfelder L, Suk HJ, Anan R, O'Donoghue A, Spirito P, Matsumori A, Maravec CS, Seidman JG, Seidman CE. The role of cardiac troponin T and α-tropomyosin mutations in hypertrophic cardiomyopathy. N Engl J Med 1995; 332:1058–1064.

22. Rosenzweig A, Watkins H, Hwang D-S, Miri M, McKenna W, Traill TA, Seidman JG, Seidman CE. Preclinical diagnosis of familial hypertrophic cardiomyopathy by genetic analysis of blood lymphocytes. N Engl J Med 1991; 325:1753–1760.

23. Bharati S, Surawicz, Vidaillet HJ, Lev M. Familial congenital sinus rhythm anomalies: Clinical and pathological correlations. PACE 1992; 15:1720-1729.

24. Graber H, Unverferth DV, Baker PB, Ryan JM, Baba N, Wooley CV. Evolution of hereditary cardiac conduction and muscle disorder: a study involving a family with six generations affected. Circulation 1986; 74:21-35.

25. Berko BA, Swift M. X-linked dilated cardiomyopathy. N Engl J Med 1987; 316:1186-1191.

26. Muntoni F, Cau M, Banau A, Congiu R, Arvedi G, Mateddu A, Marrosu MG, Cianchetti C, Realdi G, Cao A, Melis MA. Deletion of the dystrophin muscle-promoter region associated with X-linked dilated cardiomyopathy. N Engl J Med 1993; 329:921-925.

27. Michels V, Moll PP, Miller FA, Tajik J, Chu JS, Driscoll DJ, Burnett JC, Rodeheffer RJ, Chesebro JH, Tazelaar HD. The frequency of familial dilated cardiomyopathy in a series of patients with idiopathic dilated cardiomyopathy. N Engl J Med 1992; 326:77-82.

28. Miklos C. Familiaris dilatativ cardiomyopathia. Orvosi Hetilap 1990; 134:50711.

29. Kass S, MacRae C, Graber IIL, Sparks EA, McNamara D, Boudoulas H, Basson CT, Baker PB III, Cody RJ, Fishman MC, Cox N, Kong A, Wooley CF, Seidman JG, Seidman CE. A gene defect that causes conduction system disease and dilated cardiomyopathy maps to chromosome 1p1-1q1. Nature Genet 1994; 7:546-552.

30. Voss EG, Reddy CVR, Detrano R, Virmani R, Zabriskie JB, Fotino M. Familial dilated cardiomyopathy. Am J Cardiol 1984; 54:456-457.

31. Lynch HT, Mohiuddin S, Sketch MH, Krush AJ, Carter S, Runco V. Hereditary progressive atrioventricular conduction defect: A new syndrome? JAMA 1973; 225:1465-70.

32. Beyer EC, Paul DL, Goodenough DA. Connexin family of gap junction proteins. J Membr Biol 1990; 116:187-194.

33. Kanter HL, Saffitz JE, Beyer EC. Cardiac myocytes express multiple gap junction proteins. Circ Res 1992; 70:438-444.

34. Kanter HL, Laing JG, Beau SL, Beyer EC, Saffitz JE. Distinct patterns of connexin expression in canine purkinje fibers and ventricular muscle. Circ Res 1993; 72:1124-1131.

35. Kelly DP, Strauss AW. Inherited cardiomyopathies. N Engl J Med 1994; 330:913-919.

36. Coates PM. Historical perspective of medium-chain acyl-CoA dehydrogenase deficiency: a decade of discovery. In: Coates PM, Tanaka K, eds. New Developments of Fatty Acid Oxidation. New York: Wiley-Liss 1992:409-423.

37. Kelly DP, Whelan AJ, Ogden ML, Alpers R, Zhang ZF, Bellus G, Gregersen N, Dorland L, Strauss AW. Molecular characterization of medium-chain acyl-CoA dehydrogenase deficiency. Proc Natl Acad Sci USA 1990; 87:9236-9240.

38. Rocchiccioli F, Wanders RJ, Aubourg P, Vianey-Liaud C, Ijlst L, Fabre M, Cartier N, Bougneres PF. Deficiency of long-chain 3-hydroxylacyl-CoA dehydrogenase: a cause of lethal myopathy and cardiomyopathy in early childhood. Pediatr Res 1990; 28:657-662.

39. Yokota I, Indo Y, Coates PM, Tanaka K. Molecular basis of medium chain acyl-coenzyme A dehydrogenase deficiency: an A to G transition at position 985

that causes a lysine-304 to glutamate substitution in the mature protein is the single prevalent mutation. J Clin Invest 1990; 86:1000–1003.

40. Tein I, De Vivo DC, Bierman F, Pulver P, DeMeirleir LJ, Cvitanovic-Sojat L, Pagon RA, Bertini E, Dionisi-Vici C, Servidei S. Impaired skin fibroblast carnitine uptake in primary systemic carnitine deficiency manifested by childhood carnitine-responsive cardiomyopathy. Pediatr Res 1990; 28:247–255.

41. Stanley CA, Hale DE, Berry GT, Deleeuw S, Boxer J, Bonnefont J-P. A deficiency of carnitine-acylcarnitine translocase in the inner mitochondrial membrane. N Engl J Med 1992; 327:19–23.

42. Taroni F, Verderio E, Fiorucci S, Cavadini P, Finocchiaro G, Uziel G, Lamantea E, Gellera C, DiDonato S, Molecular characterization of inherited carnitine palmitoyltransferase II deficiency. Proc Natl Acad Sci USA 1992; 89:8429–8433.

43. Waber LJ, Valle D, Neill C, DiMauro S, Shug A. Carnitine deficiency presenting as familial cardiomyopathy: A treatable defect in carnitine transport. J Pediatr 1982; 101:700–705.

44. Kelly DP, Hale DE, Rutledge SL, Ogden ML, Whelan AJ, Zhang Z, Strauss AW. Molecular basis of inherited medium-chain acyl-CoA dehydrogenase deficiency causing sudden child death. J Inherit Metab Dis 1992; 15:171–180.

45. Michels VV, Pastores GM, Moll PP, Driscoll DJ, Miller FA, Burnett JC, Rodeheffer RJ, Tajik JA, Beggs AH, Kunkel LM, Thibodeau SN. Dystrophin analysis in idiopathic dilated cardiomyopathy. J Med Genet 1993; 30:955–957.

46. Clarke A. Mitochondrial genome: Defects, disease, and evolution. J Med Genet 1990; 27:451–456.

47. Ozawa T, Tanaka M, Sugiyama S, Hattori K, Ito T, Ohno K, Takahashi A, Sato W, Takada G, Bayumi B, Yamamoto K, Adachi K, Koga Y, Toshima H. Multiple mitochondrial DNA deletions exist in cardiomyocytes of patients with hypertrophic or dilated cardiomyopathy. Biochem Biophys Res Commun 1990; 170:830–836.

48. Anan R, Nakagawa M, Miyata M, Higuchi I, Nakao S, Suehara M, Osame M, Tanaka H. Cardiac involvement in mitochondrial diseases. Circulation 1995; 91:955–961.

49. Sugiyama S, Hattori K, Hayakawa M, Ozawa T. Quantitative analysis of age-associated accumulation of mitochondrial DNA with deletion in human hearts. Biochem Biophys Res Commun 1991; 180:894–899.

50. Remes AM, Hassinen IE, Ikaheimo MJ, Herva R, Hirvonen J, Peuhkurinen KJ. Mitochondrial DNA deletions in dilated cardiomyopathy: A clinical study employing endomyocardial sampling JACC 1994; 23:935–942.

6

Hyperlipidemia: Molecular Defects of Apolipoproteins B and E Responsible for Elevated Blood Lipids

Robert W. Mahley and Stephen G. Young
Gladstone Institute of Cardiovascular Disease
University of California
San Francisco, California

Researchers have made great strides in the past few years in determining the cellular and molecular events controlling lipoprotein metabolism. Detailed structure–function correlates have highlighted the unique roles of specific apolipoproteins in lipid transport, and the recent advances in molecular biology have extended these insights to new levels. The discussion to follow focuses on our understanding of the roles of apolipoproteins B and E in normal metabolism and on the impact of specific mutations of these proteins in causing dyslipidemia.

APOLIPOPROTEINS B AND E: STRUCTURE AND FUNCTION IN LIPOPROTEIN METABOLISM

Apolipoproteins (apo) B and E play critical roles in cholesterol homeostasis and the regulation of plasma lipoprotein concentrations (for review, see Refs. 1–4). Their principal function is to mediate lipoprotein binding to lipoprotein receptors, especially the low-density lipoprotein (LDL) receptor, and to initiate the uptake of these lipoproteins by cells where they are degraded and their components utilized. In addition, they serve as structural components of the lipoproteins and are involved in biosynthesis and organization of the lipid components of the lipoprotein particles. The biochemical properties and physiological roles of these apolipoproteins in lipid metabolism are summarized in Table 1.

Apolipoproteins B and E are crucial in the exogenous (chylomicron pathway) and endogenous [very-low-density lipoprotein (VLDL)-LDL path-

Table 1 Biochemical Properties and Physiological Functions of Human Apolipoproteins B and E

	Apolipoprotein B100	Apolipoprotein B48[a]	Apolipoprotein E
Chromosomal location	Chromosome 2	Chromosome 2	Chromosome 19
Molecular weight	513,000	246,000	34,000
Lipoprotein distribution	VLDL	Chylomicrons	Chylomicron remnants
	IDL	Chylomicron remnants	VLDL
	LDL		IDL
			HDL with apo E
Sites of synthesis	Liver	Intestine	Liver (hepatocytes), brain (astrocytes), skin, adrenals, ovaries, macrophages
Major functions	Structural (lipid binding)	Structural (lipid binding)	Structural (lipid binding)
	Ligand for LDL receptor		Ligand for LDL receptor and LDL receptor–related protein
			Cofactor for hepatic lipase

[a]Apolipoprotein B48 represents the amino-terminal 48% of the full-length apo B100.

way] lipid-transport systems (for review, see Refs. 1–5). Apolipoprotein B (specifically apo B48) is a structural component of chylomicrons and is important in the biosynthesis of these lipoproteins by the intestinal epithelial cells. In the circulation, chylomicron remnants are generated by lipolytic processing, and these cholesterol-enriched lipoproteins acquire apo E. The apo E of the chylomicron remnants serves as the critical ligand for plasma clearance and receptor-mediated endocytosis by the liver. The apo B48 lacks the portion of the apo B100 molecule that binds to LDL receptors.

Likewise, apo B (specifically apo B100) is important in the biosynthesis of VLDL by hepatocytes. It is envisioned that the newly synthesized apo B100 is incorporated into the inner leaflet of the endoplasmic reticulum where, by a poorly understood mechanism, lipid (especially triglyceride) is acquired and forms the core of the nascent VLDL particle. Evidence in support of this VLDL biosynthetic pathway comes in part from recent data demonstrating that abetalipoproteinemia is caused by an absent or defective microsomal triglyceride transfer protein (6). The lack of triglyceride transfer activity interferes with both VLDL synthesis in the liver and chylomicron synthesis in the intestine; thus, apo B–containing lipoproteins are absent from the plasma.

The fate of VLDL in the circulation can take two paths. A fraction of the particles is cleared by the liver via apo E–mediated receptor clearance. The remainder of the particles is lipolyzed to form small VLDL or intermediate-density lipoproteins (IDL), which collectively represent VLDL remnants. Some of these particles are cleared from the plasma by the liver, primarily through apo E–mediated LDL receptor binding and uptake. Approximately 50% of VLDL are ultimately converted to the end product of the lipolytic pathway (LDL). The cholesterol-rich LDL containing only apo B100 are ultimately catabolized primarily by the liver and, to a much lesser extent, by peripheral cells via the LDL receptor.

MOLECULAR DEFECTS IN APOLIPOPROTEINS B AND E

Determination of the structures of apo B and apo E and identification of the functional domains within these molecules have provided insights not only into normal metabolism but also into the genetic disorders causing hyperlipoproteinemia (4). In the following sections, we review the discoveries that have unraveled the molecular causes of the disorders familial defective apolipoprotein B100 (FDB) and type III hyperlipoproteinemia (dysbetalipoproteinemia).

Apolipoprotein B: Familial Defective
Apolipoprotein B100

Introduction

The landmark investigations of Michael Brown and Joseph Goldstein during the 1970s and early 1980s (7) established that defects in the LDL receptor can interfere with the clearance of LDL from the plasma and cause familial hypercholesterolemia (FH), a syndrome characterized by markedly elevated levels of LDL cholesterol, tendon xanthomas, and premature coronary artery disease. The overall prevalence of the heterozygous form of FH has been estimated to be about 1 in 500. To date, more than 100 mutations in the LDL receptor gene that cause FH have been described (8). The characterization of different classes of LDL receptor mutations has yielded important insights into the structural features of the LDL receptor protein (7).

For many years, it had been postulated that retarded clearance of LDL from the plasma of hypercholesterolemic patients could also be caused by a genetic defect in apo B100. In the mid-1980s, clinical investigations of human subjects with moderate hypercholesterolemia established that some hypercholesterolemic subjects have LDL that are cleared slowly from the plasma and are defective in binding to cultured cells expressing a normal complement of LDL receptors (9,10). This genetic defect, familial defective apolipoprotein B100 (FDB), may be as prevalent as 1 in 500, at least in certain human populations. Only one mutation, which causes an arginine-to-glutamine substitution at apo B100 amino acid residue 3500, has been clearly and unequivocally associated with FDB (11).

Discovery of Human Subjects with FDB

In 1985, Vega and Grundy (9) reported the results of metabolic studies designed to investigate alterations in lipoprotein metabolism in human subjects with moderate hypercholesterolemia (total cholesterol levels of 250–300 mg/dl). In double-label LDL turnover studies in human subjects, they directly compared the clearance rate of autologous LDL isolated from a hypercholesterolemic human subject with the clearance rate of homologous LDL isolated from a normolipidemic control subject. In several hypercholesterolemic subjects, the clearance of the hypercholesterolemic subject's own LDL was significantly slower than that of the LDL from the normolipidemic control subject. Vega and Grundy speculated that these patients had abnormal LDL that were cleared slowly because they could not bind normally to the LDL receptor. Later, in collaborative studies, Innerarity and coworkers (10) tested the binding of the LDL from these patients to the LDL receptors of cultured human fibroblasts. The LDL from one of the hypercholesterolemic patients bound to the LDL receptor with only 32% of the affinity of normal LDL. An identical binding defect was observed with the LDL that were isolated from the plasma of several other hypercholesterolemic family members. Innerarity and coworkers proposed that these individuals were heterozygous for a genetic defect in apo B100, which they designated familial defective apolipoprotein B100.

The discovery of FDB using lipoprotein turnover techniques and fibroblast binding studies occurred nearly simultaneously with several other important developments in the apo B field. First, the amino acid sequence of apo B100 had been determined from the apo B cDNA sequence (12), and the structure of the apo B gene had also been determined (13). An analysis of the amino acid sequence of the carboxyl-terminal portion of apo B100 revealed two short stretches of amino acids containing multiple positively charged amino acid residues (14). One of these sequences, spanning apo B100 amino acids 3359–3367, was homologous to the receptor-binding domain of apo E. Knott et al. (14) suggested that one or both of these two positively charged carboxyl-terminal sequences might be important in the binding of LDL to the LDL receptor.

Second, immunochemical studies had established that several monoclonal antibodies recognizing epitopes located between apo B100 amino acids 3000 and 3600 were capable of blocking the binding of LDL to the LDL receptor (14). With these developments in hand, Soria et al. (15) cloned and sequenced a large portion of the carboxyl-terminal region of the apo B gene (spanning the region coding for amino acids 2488–3901) from an FDB heterozygote. No mutations were identified within the short stretches of positively charged amino acids that were identified in the cDNA sequencing studies; however, Soria and coworkers did identify a single unique mutation (a G→A transition at apo B cDNA nucleotide 10708) that results in an Arg→Gln substitution at apo B100 amino acid residue 3500. The amino acid 3500 mutation was identified in the apo B gene of all hypercholesterolemic family members with defective LDL and, within months, the same mutation was identified in many unrelated hypercholesterolemic human subjects with defective LDL (11).

Does the Substitution at Residue 3500 Result in LDL That Are Defective in Binding to the LDL Receptor?

It is important to point out that no one has sequenced the entire apo B gene of an affected subject and proven that the amino acid substitution at residue 3500 is the *only* amino acid substitution in the mutant allele. And no one has introduced the 3500 mutation into an apo B expression vector, expressed the mutant apo B in cultured cells, and directly shown that the mutant apo B produced by the cells has a binding defect. In the absence of these types of studies, one cannot be absolutely certain that the amino acid substitution at residue 3500 causes the defective binding of LDL to the LDL receptor. It would be conceivable, for example, that another, as yet unidentified, mutation on the same allele might either account for the defective binding or interact with the 3500 mutation to cause the defective binding. However, we believe for several reasons that it is extremely likely that the amino acid substitution at residue 3500 causes the binding defect. First, the association between the amino acid 3500 substitution and the phenotype is perfect. To date, every human subject with the 3500 mutation has LDL that are defective in binding to the LDL receptor. Second, the mutation is located in a region of the apo B molecule known to be involved in the binding of LDL to the LDL receptor. Monoclonal antibodies recognizing epitopes located between amino acids 3000 and 3600 completely block the binding of LDL to the LDL receptor (16). One of these receptor-blocking monoclonal antibodies, MB47 (17), can be used in immunoassays to identify subjects with FDB because it binds to the LDL of FDB heterozygotes more avidly than to normal LDL. Studies with β-galactosidase–apo B fusion proteins have shown that antibody MB47 binds to a discontinuous epitope that flanks residue 3500; the epitope includes apo B100 amino acids 3429–3453 and amino acids 3507–3523 (18). It seems likely that the substitution at residue 3500 alters the conformation of the epitope for antibody MB47, and at the same time alters the conformation of the receptor-binding domain of apo B so that it cannot bind normally to the LDL receptor.

A third reason to believe that the 3500 substitution disrupts the receptor-binding region of apo B comes from ^{13}C nuclear magnetic resonance studies on LDL from FDB subjects. Lund-Katz and coworkers (19) examined the ^{13}C nuclear magnetic resonance spectra of control LDL and FDB LDL containing $(^{13}CH_3)_2Lys$ residues. Lysine residues are known to be involved in the binding of LDL to the LDL receptor (20). With control LDL preparations, they found that the chemically modified lysine residues existed in two distinct microenvironments: either "normal," with pK 10.5, or "active," with pK 8.9. There is evidence that the subset of lysines with pK 8.9 is particularly important in the binding of LDL (21). With the control LDL samples, Lund-Katz et al. found about 50 lysine residues with pK 8.9 and 170 lysines with pK 10.5. In the mutant LDL from five different FDB patients, there were consistently fewer pK 8.9 lysine residues; at least seven lysines were redistributed from the pK 8.9 pool to the pK 10.5 pool. Although the study by Lund-Katz et al. did not address whether residue 3500 was directly involved in binding to the LDL receptor, the authors suggested that it was unlikely that the substitution of glutamine for the positively charged arginine at residue 3500 was *directly* responsible for interfering with the binding of apo B to the negatively charged ligand-binding domain of the LDL receptor. Rather, based on the nuclear magnetic resonance data, Lund-Katz et al. suggested that the 3500 substitution might be *indirectly* responsible for the binding defect, by disrupting the local conformation of the receptor-binding region in such a way that critical lysine residues could no longer interact with the ligand-binding domains of the LDL receptor. Of note, the 3500 substitution did not cause gross alterations in the secondary structure of apo B100, as the circular dichroism spectra of the control LDL and the mutant LDL were virtually identical.

Perhaps the most persuasive argument in favor of the amino acid 3500 substitution causing the binding defect is that the 3500 mutation has been documented on several different apo B haplotypes. In a study of eight unrelated FDB patients and their families, Ludwig and McCarthy (22) reported that the 3500 mutation was found on a single apo B haplotype. The apo B haplotype that is usually associated with the 3500 mutation was carefully defined using eight different diallelic restriction-site polymorphisms and two different hypervariable markers flanking the apo B gene. Other investigators have confirmed the finding that FDB generally occurs on a single apo B haplotype (23–26), leading to the conclusion that *most* of the cases of FDB in the general population are due to the inheritance of a single ancestral mutant allele. Recently, Bersot and coworkers (27) described a Chinese man with defective LDL in whom the 3500 mutation occurred on the background of a different apo B haplotype. Similarly, Rauh et al. (25) identified a German family in which several affected individuals had the 3500 mutation on the background of a unique apo B haplotype. In these two cases, it seems very likely that the 3500 mutation occurred independently on different apo B haplotypes, although it is not possible to exclude definitively the alternative hypothesis that the distinct haplotypes occurred by recombination events. In either case, finding the amino acid 3500 substitution on the background of different apo B haplotypes greatly strengthens the argument that this mutation *causes* the defective binding.

What Is the Extent of the Binding Defect with the 3500 Mutation?

The LDL of FDB heterozygotes (which consist of defective LDL from the mutant allele and normal LDL from the normal allele) bind to the LDL receptor about 30% as well as normal LDL (10). Because the defective LDL in the plasma of FDB heterozygotes are cleared from the plasma at a slower rate than normal LDL, the LDL of these subjects are enriched in defective LDL. One can directly document the enrichment of the defective LDL in the total LDL of FDB heterozygotes by using radioimmunoassays with the apo B-specific monoclonal antibody MB19 (28). Antibody MB19 detects a two-allele genetic polymorphism in apo B [which is almost certainly caused by a single amino acid substitution at apo B100 amino acid residue 71 (29)] and binds to the two different apo B allotypes, $MB19_1$ and $MB19_2$, with high and low affinity, respectively. It is possible to use antibody MB19 immunoassays to judge the allotype composition of human LDL samples (30). In normal individuals who are heterozygous for the MB19 polymorphism, the plasma LDL contain equal concentrations of the apo B allotypes, $MB19_1$ and $MB19_2$ (28,30). In contrast, in FDB subjects that are heterozygous for the MB19 polymorphism, the plasma LDL consist of about 75% allotype $MB19_2$ (the defective apo B) and 25% allotype $MB19_1$ (the normal apo B) (31). This type of analysis strongly suggests that the 3500 mutation results in the accumulation of one apo B allotype (allotype $MB19_2$); however, this information alone does not permit one to calculate the extent of the binding defect. Recently, Arnold et al. (31) used antibody MB19 immunoaffinity chromatography (28) to purify the $MB19_2$ LDL (the defective LDL) from the total LDL of several FDB heterozygotes, and then tested the defective LDL for binding to the LDL receptor. The defective LDL bound to the LDL receptor only about 9% as well as the normal LDL (Figure 1). Because the purified preparations of defective LDL were probably contaminated by small amounts of normal LDL, the actual level of binding of the defective LDL was probably somewhat less than 9%.

Recently, März et al. (32) examined the receptor-binding properties of the LDL isolated from an FDB *homozygote* (33). Interestingly, they found that the LDL from the homozygote bound to the LDL receptor about 20% as well as control LDL. Most of the binding of the FDB LDL was due to relatively high levels of binding by the more buoyant LDL subfractions; virtually all of this binding was due to the contamination of these buoyant subfractions with apo

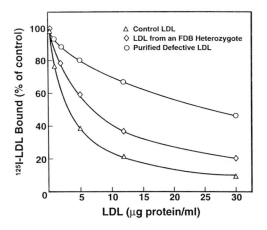

Figure 1   Ability of LDL samples to compete with a fixed concentration of normal ^{125}I-LDL for binding to LDL receptors on normal human fibroblasts. O, Defective LDL that was purified from the LDL of an FDB subject by monoclonal antibody MB19 immunoaffinity chromatography. The LDL from the FDB heterozygote competed 30% as well as normal LDL; the purified defective LDL competed about 9% as well. (Modified from Ref. 31.)

E. When the apo E was removed by immunoaffinity chromatography, the buoyant subfractions no longer bound to the LDL receptor. The dense LDL subfractions from the FDB homozygote had no capacity to bind to the LDL receptor. Interestingly, the intermediate LDL subfractions retained minimal ability to bind to the LDL receptor, even when the apo E was removed. The latter observations suggested that the extent of the binding defect associated with the 3500 mutation could be modulated by the size and density of the LDL particle.

Prevalence of FDB

The estimated prevalence of FDB obviously depends on the population that is screened. Most of the reported cases of FDB have been identified in populations of hyperlipidemic patients, and many of the estimates of the incidence of FDB have been calculated from the incidence of FDB in hypercholesterolemic populations. Innerarity et al. (11) identified 11 FDB heterozygotes from a total of 1100 hypercholesterolemic individuals from Montreal, San Francisco, Dallas, and Salzburg. By adjusting for the effects of age and gender and for the fact that the population that was screened had a high incidence of hypercholesterolemia, they estimated the prevalence of FDB in the general population to be about 1 in 500. Tybjærg-Hansen et al. (34) identified nine cases of FDB among 374 hyperlipidemic patients from the United Kingdom; they estimated that the prevalence of FDB in the general population was about 1 in 600. Schuster et al. (35) identified eight FDB heterozygotes among 243 hypercholesterolemic subjects and estimated the prevalence of FDB in the general population in Germany to be about 1 in 700. Although the population that was examined was small, the highest prevalence of FDB has been reported in Switzerland, where Miserez et al. (36) identified three cases of FDB from a group of 728 military recruits, for a prevalence of about 1 in 240 in the general Swiss population. They also identified seven cases from a study of 142 index cases of hypercholesterolemic kindreds; from these data, by adjusting for the prevalence of hypercholesterolemia in Switzerland, they calculated the prevalence of FDB to be about 1 in 190 in the general Swiss population.

There have been few large surveys to determine the prevalence of FDB in the general population. Bersot et al. (27) screened 5160 employees of a bank in California and identified four subjects with FDB, for an overall prevalence of FDB of 0.08% (90% confidence interval of 0.01–0.14%). Three of the four subjects were Caucasian and born in North America, and one was a native of China. Rust et al. (37) surveyed subjects from the general population in Germany and found 10 patients among 7069 subjects, for an overall prevalence of about 1 in 700.

The 3500 mutation has been identified in hypercholesterolemic patients in the United States, the United Kingdom, Canada, The Netherlands, Italy, Austria, Switzerland, Australia, and Denmark. Although the surveys of hyper-cholesterolemic patients are subject to considerable selection bias, it would appear that the mutation is most prevalent in Germany and Switzerland. It is noteworthy that the mutation was *not* identified within a population of 552 hypercholesterolemic patients from western and southern Finland (38), 625 Israelis (100 index hypercholesterolemic patients and 525 offspring) (39), and 250 Russians, including 120 myocardial infarction patients (40).

As noted above, the vast majority of FDB patients have an identical apo B haplotype. This finding implies that most of the cases that have been identified to date are due to the inheritance of a mutation that likely occurred more than 10,000 years ago (36) and has subsequently been widely dispersed in European and North American populations.

Lipid Levels in FDB

The vast majority of FDB heterozygotes have significant elevations in their plasma levels of LDL cholesterol. From the study of the families of 11 FDB heterozygotes (nine of which were from the United States or Canada), Innerarity et al. (11) identified a total of 41 FDB heterozygotes. The average total and LDL cholesterol levels in the FDB heterozygotes were 269 mg/dl and 199 mg/dl, respectively. These levels are 81 mg/dl and 71 mg/dl higher, respectively, than the 50th percentile for age- and sex-matched controls in the Lipid Research Clinics study (41). In other surveys, both the total and LDL cholesterol levels in FDB heterozygotes have tended to be higher. In 22 FDB heterozygotes identified in the lipid clinic of Roger Illingworth in Oregon, the mean total and LDL cholesterol levels were 329 mg/dl and 247 mg/dl, respectively (26). In 54 FDB heterozygotes identified in Germany, the mean total and LDL cholesterol levels were 308 mg/dl and 242 mg/dl, respectively (42). In this study, the LDL cholesterol levels exceeded the 50th percentile value of the Lipid Research Clinics study controls by 120 mg/dl. Triglyceride levels and high-density-lipoprotein (HDL) cholesterol levels were similar in FDB heterozygotes and unaffected family members. Table 2 lists the mean plasma lipid levels in several series of FDB heterozygotes.

It is noteworthy that there is heterogeneity in the plasma lipid levels in heterozygous FDB (23,24,43), just as there is heterogeneity in lipid levels in FH (44). Several FDB heterozygotes have been identified that have completely normal lipid levels, with total cholesterol levels around 200 mg/dl. In one such case, investigated by Friedl et al. (23), it appeared that the normal cholesterol levels in a young man with heterozygous FDB may have been due to the fact that his other apo B allele was a mutant associated with low levels of apo B in the plasma. Even though some individuals with heterozygous FDB may have relatively normal cholesterol levels, it is important to emphasize that the vast

Table 2 Mean Lipid Levels in FDB Heterozygotes

Series	No. of patients	TC (mg/dl)	TG (mg/dl)	LDL-C (mg/dl)	HDL-C (mg/dl)
Innerarity et al. (11)	41	269	NR	199	NR
Rauh et al. (42)	54	308	111	242	45
Defesche et al. (47)	18	353	116	271	56
Schuster et al. (35)	18	353	113	251	49

TC, total cholesterol; TG, triglycerides; LDL-C, low-density-lipoprotein cholesterol; HDL-C, high-density-lipoprotein cholesterol; NR, not reported.

majority of FDB heterozygotes have elevated cholesterol levels. Ninety percent of FDB heterozygotes that have been identified in the large series have total cholesterol levels at or above the 95th percentile for age- and sex-matched controls (26).

Only two FDB homozygotes have been reported to date, and both were identified in Germany (33,45). Both had only moderate hypercholesterolemia. A 54-year-old male homozygote with xanthelasmas and arcus lipoides had total and LDL cholesterol levels of 331 mg/dl and 265 mg/dl, respectively (32,33). Four weeks later, the total and LDL cholesterol levels were 365 mg/dl and 292 mg/dl, respectively. A 31-year-old female homozygote with arcus lipoides had total cholesterol levels that ranged between 299 mg/dl and 331 mg/dl (45).

A 35-year-old male FDB heterozygote who was also a heterozygote for FH has been reported (46). This individual had coronary and carotid atherosclerosis, tendon xanthomas, and total and LDL cholesterol levels of 387 mg/dl and 289 mg/dl, respectively. The lipid levels and clinical signs in this FH/FDB heterozygote were similar to those in other family members who had only FDB or only FH.

Clinical Manifestations of FDB

Most of the published information on the clinical characteristics of heterozygous FDB has come from European studies in which the affected subjects tended to have relatively high cholesterol levels. Many of these affected subjects had arcus lipoides, tendon xanthomas, and coronary artery disease. In a study of 54 German FDB heterozygotes of all ages, 26% had tendon xanthomas and 22% had arcus lipoides (42). The prevalence of these findings increased significantly with age. Of the 15 patients between ages 51 and 73, 47% had tendon xanthomas and 40% had arcus lipoides. Similar findings were reported in series of Dutch patients and patients from the United Kingdom and Denmark (26).

Several large collections of FDB patients leave little doubt that this disorder is associated with atherosclerosis. In the series of Rauh et al. (42),

22% of all the FDB patients and 60% of the FDB patients older than age 51 had coronary artery disease; 73% of the older patients had plaques in their carotid arteries. Tybjærg-Hansen and Humphries (26) have reported that the cumulative frequency of coronary artery disease in FDB patients increases with age, with about one-half of FDB males having coronary artery disease by the age of 50. Of 18 Dutch patients with FDB, eight had documented coronary artery disease or angina (47). In that series, the cholesterol levels in FDB patients with coronary artery disease were much higher than the levels in FDB patients with no signs of coronary artery disease. The latter finding would help to explain the low percentage of FDB patients with coronary disease (5%) in the series of Innerarity et al. (11); in that series, the cholesterol levels were lower than in the FDB patients in the European series.

Rust et al. (37) have reported that the FDB mutation is overrepresented in coronary artery disease patients. They found the FDB mutation in 10 of 7069 patients in the general population (about 1 in 700) but in 11 of 2450 patients (about 1 in 223) with documented coronary artery disease ($p < 0.01$).

Comparison of the Phenotypes of FDB and Familial Hypercholesterolemia

To date, no metabolic studies that directly compare lipoprotein metabolism in patients with FH and FDB have been published. However, from our current understanding of lipoprotein metabolism (1), one could persuasively argue that FH would affect LDL metabolism more profoundly than FDB. In heterozygous FH, the defective LDL receptor is "double trouble," as it affects the metabolism of both LDL and remnant lipoproteins. Not only are LDL removed slowly from the circulation, but the precursors of LDL (IDL and VLDL) are also removed slowly from the circulation. The decreased hepatic uptake of IDL and VLDL through apo E–mediated clearance is associated with an increased production of LDL. In contrast, in heterozygous FDB, LDL clearance is retarded, but one would expect the apo E–mediated clearance of IDL and VLDL to be normal. Consequently, it would be reasonable to predict that the LDL cholesterol levels in FDB heterozygotes would be lower than those in FH heterozygotes. In several of the series of FDB patients, it appears that the phenotype of FDB might indeed be milder than that observed in FH. In the series of 41 FDB heterozygotes identified by Innerarity et al. (11), the mean plasma cholesterol level of 269 mg/dl was significantly lower than levels observed in large groups of FH heterozygotes (~360 mg/dl). Similarly, Schuster et al. (35) found that the total and LDL cholesterol levels in 18 FDB heterozygotes were 50–60 mg/dl lower than levels observed in 49 well-defined FH heterozygotes. However, in other series of FDB patients, the lipid levels and clinical manifestations observed in FDB heterozygotes were virtually identical to those observed in FH heterozygotes. For example, Defesche et al. (47)

found that the lipid, lipoprotein, and apo B levels in 18 Dutch FDB patients were virtually identical to levels observed in two comparably sized groups of Dutch patients with specific LDL receptor mutations. Tybjærg-Hansen and Humphries (26) reviewed the existing literature and suggested that the cumulative incidence of coronary atherosclerosis was similar in those with FH and FDB. However, because all of the series of FDB heterozygotes involve obvious selection biases, the current data do not permit unequivocal conclusions regarding whether patients with FDB have a milder phenotype than patients with FH. However, it is quite clear that the clinical presentations of patients with the two disorders overlap considerably; it has been repeatedly demonstrated that FDB can cause relatively severe hypercholesterolemia and can be associated with all of the clinical manifestations of classic FH.

Although only two patients with homozygous FDB have been described (32,45), it seems quite clear that the phenotype of homozygous FDB is considerably milder than that of homozygous FH. Both FDB homozygotes had total cholesterol levels <350 mg/dl, whereas FH homozygotes typically have total cholesterol levels >600 mg/dl. It seems very likely that the apo E-mediated clearance of LDL precursors protects the FDB homozygotes from the extremely severe hypercholesterolemia observed in FH homozygotes. Of interest, drug treatment of one of the FDB homozygotes with a β-hydroxy-β-methylglutaryl–coenzyme A (HMG-CoA) reductase inhibitor resulted in a pronounced lowering of LDL cholesterol level (45). Reductase inhibitors lower cholesterol levels by stimulating LDL receptor-mediated uptake of lipoproteins. Presumably, the favorable effect of this drug in the FDB homozygote reflected an increased uptake of apo E–containing lipoproteins by the LDL receptor. Typically, FH homozygotes have a poor response to HMG-CoA reductase inhibitors because of the complete absence of LDL receptors.

Diagnosis of FDB

It is possible to identify patients with FDB by testing the ability of their LDL to bind to cultured fibroblasts (10) or by testing their ability to bind to the apo B–specific monoclonal antibody MB47 (20). However, these techniques are cumbersome and time-consuming. On a practical level, the diagnosis of the 3500 mutation is now made exclusively by analyzing a fragment of the apo B gene that is enzymatically amplified from genomic DNA prepared from peripheral blood leukocytes. At the Gladstone Institute of Cardiovascular Disease, we initially tested the amplified DNA fragment by slot-blot analysis using ^{32}P-labeled allele-specific oligonucleotide probes (15). This approach yielded definitive results, but it was labor-intensive and required the use of radioactivity. More recently, we have adopted a polymerase chain reaction (PCR) methodology, which permits the identification of the mutation by restriction digestion with *Msp*I (40,48). The 3500 mutation does not change any restriction endonuclease sites. However, one can modify one of the oligonucleotides used in the PCR reaction to create a single base pair mismatch at the 3′ penultimate base of the oligonucleotide primer. This single base substitution does not interfere with the PCR reaction. When this modified oligonucleotide is used to amplify a normal apo B allele, the mismatch in the oligonucleotide results in the creation of an *MspI* site in the amplified DNA product. However, when this oligonucleotide is used to amplify a mutant allele, the *MspI* site is not formed. Thus, it is possible to subject the amplified DNA to restriction digestion with *MspI* and then resolve the products of the restriction digestion on a polyacrylamide or agarose gel. Because the mutant apo B gene fragment is not cleaved with *MspI*, the presence of a mutant allele can readily be detected on an ethidium bromide–stained gel. Newer methods, using the Amplification Refractory Mutation System, have been devised for large-scale screening of pooled blood samples (49,50).

Efficacy of Drug Treatment

In an early report, Corsini et al. (51) treated two patients with FDB and found that they responded poorly to treatment with simvastatin, an HMG-CoA reductase inhibitor, and suggested that their poor response to this drug might be characteristic of patients with FDB. Subsequent studies, however, have established that FDB patients respond very favorably to HMG-CoA reductase inhibitors. The most comprehensive studies on the treatment of FDB have been performed in the lipid clinic of Roger Illingworth at the University of Oregon. Illingworth et al. (52) treated FDB patients with either 20 or 40 mg of lovastatin per day. The mean LDL cholesterol levels decreased by 22% on 20 mg lovastatin daily and by 32% on 40 mg lovastatin daily. Illingworth and coworkers did observe heterogeneity in the responses of individual patients. However, the extent of cholesterol-lowering in the FDB patients was virtually identical to that observed in patients with FH, familial combined hyperlipidemia, or moderate hypercholesterolemia.

Patients with FDB also respond very favorably to treatment with bile acid sequestrants and nicotinic acid. Illingworth's group found that FDB patients treated with either 20 g of colestipol or 16 g of cholestyramine had a 32% reduction in LDL cholesterol levels (53). Treatment of four FDB patients with nicotinic acid (3 g/day) resulted in a 24% reduction in LDL cholesterol (54). In both of these studies, the therapeutic response in the FDB patients was at least as good as, and possibly somewhat better than, the response observed in other studies in patients with FH.

Could Other Mutations, Aside from the 3500 Mutation, Cause FDB?

As outlined above, monoclonal antibodies recognizing epitopes in the region of apo B100 amino acid residues 3000–3600 block the binding of LDL to the

LDL receptor (16). And, within this region, there is one cluster of positively charged residues (apo B100 amino acids 3359–3367) that is homologous to the receptor-binding region of apo E (14). We believe that it is likely that other apo B mutations will be found that disrupt the binding of LDL to the LDL receptor, either by disrupting the conformation of the receptor-binding domain or by eliminating positively charged residues that are directly involved in receptor binding. It seems quite possible, for example, that mutations within amino acids 3359–3367 could prevent LDL binding to the LDL receptor and cause clinical manifestations identical to those found with the 3500 mutation.

Several laboratories have screened hypercholesterolemic populations for new mutations that could cause FDB, either by directly sequencing the receptor-binding domain or by testing LDL samples for defective receptor binding. To date, no mutation that yields a defect as profound as the 3500 mutation has been identified. However, Pullinger and coworkers (55) have published an abstract in which they reported the identification of two unrelated patients with atherosclerotic disease who have an Arg→Cys mutation at apo B100 residue 3531. This mutation can be easily identified because it creates a new *Nsi*I site in the apo B gene. Family studies on one of these patients resulted in the identification of eight subjects that were heterozygous for the defect. The mean LDL cholesterol level in the affected subjects was 190 ± 13 mg/dl versus 150 ± 12 mg/dl in unaffected subjects. Pullinger et al. found that the LDL of affected individuals had an affinity for the LDL receptor that was ~70% that of control LDL. The defect associated with the amino acid substitution at residue 3531 did not appear to be as severe as the defect with the 3500 mutation because, in their control experiments, Pullinger et al. found that the LDL of heterozygotes with the 3500 mutation had an affinity for the LDL receptor that was 39% that of control LDL. An additional piece of data suggesting that the defect at amino acid 3531 was not as severe as the 3500 mutation came from immunochemical studies with antibody MB19 (C. Pullinger, S. G. Young, unpublished data). In antibody MB19 immunoassays, we studied the relative levels of the two MB19 allotypes in 1) individuals who were heterozygous for the 3531 mutation and heterozygous for the MB19 polymorphism and 2) an individual who was heterozygous for the 3500 mutation and heterozygous for the MB19 polymorphism. The antibody MB19 immunoassays were not capable of detecting an increased amount of either of the MB19 allotypes in the plasma in the 3531 heterozygotes, suggesting that there was no marked accumulation of either of the two MB19 allotypes in the 3531 heterozygotes. In control experiments, the 3500 heterozygote had a markedly increased amount of allotype $MB19_2$ in the plasma (compared to the amount of allotype $MB19_1$).

We anticipate that additional apo B mutations causing hypercholesterolemia will be reported over the next few years.

Developing an Animal Model of FDB

Several investigators have worked toward developing animal models for FDB. Within the past year, our laboratory has used a full-length apo B gene clone to develop a transgenic mouse that expresses human apo B100 (56). Several of the transgenic lines express plasma levels of human apo B100 and LDL cholesterol that are comparable to those observed in normolipidemic human subjects, even when the mice are fed a chow diet containing low levels of fats. On a high-fat diet, the transgenic animals develop even higher levels of LDL cholesterol and develop severe atherosclerotic lesions in the proximal aorta (D. Purcell-Huynh, R. V. Farese, Jr., D. A. Sanan, and S. G. Young, unpublished data). Cell culture studies as well as turnover studies in mice have indicated that human apo B is taken up poorly by the murine LDL receptor. Thus, even without mutating the receptor-binding domain of the human apo B gene construct, the human apo B100 transgenic animals probably represent an animal model of defective apo B clearance. In the future, it should be possible to introduce mutations into the full-length human apo B gene construct and develop transgenic mice that express human apo B100 molecules with a variety of subtle mutations. This approach should allow investigators to understand more thoroughly the important structural features of the receptor-binding region of human apo B.

In addition to transgenic approaches, researchers in the laboratory of Nobuyo Maeda are currently using gene targeting in mouse embryonic stem cells to introduce mutations into the receptor-binding region of mouse apo B100 (57). The gene-targeting approach will undoubtedly be useful for generating a mouse model of defective apo B clearance and for adding to our knowledge of the important structural features of the receptor-binding region of apo B.

Apolipoprotein E: Type III Hyperlipoproteinemia (Dysbetalipoproteinemia)

Introduction

The dyslipidemia now referred to as type III hyperlipoproteinemia was initially characterized by analytical ultracentrifugation as an increase in small VLDL and IDL (58,59). In addition, these patients possessed distinctive xanthomas on the palms of their hands (58,60–62). Fredrickson et al. (62) described this disorder as type III hyperlipoproteinemia and demonstrated the appearance of $d < 1.006$ g/ml floating, broad-β-migrating lipoproteins, now referred to as β-VLDL. The term dysbetalipoproteinemia is also used to denote the lipoprotein abnormality, describing the chemical and physical properties of the VLDL. However, this term is probably best restricted to normolipidemic patients possessing β-VLDL in their plasma (as discussed below).

The β-VLDL characteristic of these patients represent two distinct lipoprotein classes—chylomicron remnants of intestinal origin (apo B48–apo E particles) and VLDL remnants of hepatic origin (apo B100–apo E particles) (63,64). These lipoproteins are cholesteryl ester-enriched and resemble the remnant lipoproteins that accumulate in the plasma of cholesterol-fed animals (65).

Apolipoprotein E was shown to be increased in the plasma of type III hyperlipoproteinemia patients (66), and the important observations of Utermann and associates (67–71) linked type III hyperlipoproteinemia to a unique form of apo E. Determination of the amino acid sequence of apo E provided the key data unraveling the structural differences among the common isoforms of this protein (72–74) and established the role of mutant apo E in the development of type III hyperlipoproteinemia (for review, see Refs. 75 and 76). The finding that mutant forms of apo E associated with type III hyperlipoproteinemia (apo E2 being the most common mutant) were defective in LDL receptor binding provided the insight necessary to link the mutant apo E to defective clearance of remnant (β-VLDL) lipoproteins (3,4).

It is now appreciated that the most common form of type III hyperlipoproteinemia is associated with the occurrence of apo E2(Arg$_{158}$→Cys), that homozygosity for the defective allele is required, and that the hyperlipoproteinemia is modulated by a variety of genetic and environmental factors (apo E2 homozygosity plus a second factor equals overt hyperlipidemia) (for review, see Refs. 75 and 76). On the other hand, the dominant expression of type III hyperlipoproteinemia has now been described in association with rare mutations of apo E that directly affect the remnant clearance pathway and do not require secondary factors to cause the hyperlipidemia (75,76).

Discovery of Human Subjects with Type III Hyperlipoproteinemia

In 1975, Utermann and associates (68,69) established the familial nature of type III hyperlipoproteinemia and linked the disorder to an abnormality in the pattern of the apo E isoforms. The apo E polymorphism and nomenclature for the isoforms were established in 1982 (77). The three genetically determined isoforms, referred to as apo E2, apo E3, and apo E4, were separated by isoelectric focusing and displayed six common phenotypes: apo E2/2, apo E3/2, apo E3/3, apo E4/3, apo E4/2, and apo E4/4. The corresponding alleles were designated as ε2, ε3, and ε4. It was established that apo E2 homozygosity was most commonly associated with type III hyperlipoproteinemia.

In 1981 and 1982, the molecular basis of the apo E polymorphism was established. It was determined that apo E2, apo E3, and apo E4 differed by a single amino acid substitution occurring at residue 112 or 158 in the molecule (72–74). Apolipoprotein E3, the most common allelic form of apo E, possessed cysteine at residue 112 and arginine at residue 158, whereas apo E4 had arginine at residue 112. On the other hand, the only difference between the sequences of apo E3 and apo E2 was the occurrence of cysteine, rather than arginine, at residue 158 in apo E2.

The next crucial discovery establishing the molecular basis of type III hyperlipoproteinemia was that apo E2 was defective in binding to the LDL receptor (78). In contrast, both apo E3 and apo E4 bound very well (78,79). In addition, it was shown that other variant forms of apo E associated with type III hyperlipoproteinemia also did not bind normally to LDL receptors (for review, see Refs. 75 and 76). The defective binding of apo E variants from patients with type III hyperlipoproteinemia was demonstrated by in vitro binding studies of apo E–phospholipid recombinants with cultured fibroblasts. The in vivo demonstration of defective clearance of apo E2 was shown by comparing the disappearance of radiolabeled apo E3 and apo E2 in normal subjects and subjects with type III hyperlipoproteinemia (80). Apolipoprotein E2 clearance was significantly retarded.

These studies clearly implicated apo E2 in the accumulation of chylomicron remnants and VLDL remnants (β-VLDL) in the plasma of subjects with type III hyperlipoproteinemia. However, it was also appreciated that, whereas apo E2 (or defective apo E) was necessary for the development of type III hyperlipoproteinemia, it was not always sufficient to cause the marked hyperlipidemia. Other genetic and environmental factors were required to precipitate the remnant lipoprotein accumulation (for review, see Refs. 75 and 76). In addition, it was puzzling that subjects with type III hyperlipoproteinemia often had very low levels of LDL. However, Ehnholm et al. (81) presented data indicating that apo E2 did not support lipolytic conversion of hepatic β-VLDL to LDL. More recently it was shown that apo E does play a role in the lipolytic processing of VLDL and that apo E3 activates hepatic lipase more avidly than does apo E2 (82,83).

Other rare mutants of apo E causing type III hyperlipoproteinemia have now been described. In addition, an absence or deficiency of apo E has been shown to cause remnant lipoprotein accumulation and type III hyperlipoproteinemia (for review, see Refs. 75 and 76).

Correlation of the Structure and Function of Apolipoprotein E with Type III Hyperlipoproteinemia

Detailed structural studies of apo E have provided important insights into how apo E mediates binding of apo E–containing lipoproteins to the LDL receptor and to a postulated chylomicron remnant or apo E receptor, which now appears to be the LDL receptor–related protein (LRP) (for review, see Refs. 5 and 84–86). The amino-terminal two-thirds of apo E (residues ~1–165) contains the receptor-binding and heparin-binding domains, whereas the carboxyl-terminal one-third (residues ~200–299) contains the lipid-binding domain (87). It has now been established that the basic residues within the 136 to 150 region of apo E represent the LDL receptor-binding and heparin-binding domains. The identification of key arginine and lysine residues has been established using a variety of complementary approaches (for review, see Refs. 3, 75, 76, 87, and 88): selective chemical modification of specific amino acid residues, identification of natural mutants of apo E causing defective receptor binding, generation of fragments of apo E, mapping of the binding site of an apo E monoclonal antibody that selectively blocks receptor binding of apo E–containing lipoproteins, production of site-specific mutations that identify critical residues for receptor binding, and, finally, determination of the three-dimensional structure of the amino-terminal domain of apo E (89–91).

The X-ray crystal structure of apo-E3 revealed that the clustered basic residues (Arg_{136}, Arg_{142}, Arg_{145}, Lys_{146}, Arg_{147}, and Arg_{150}) were all on the surface of the fourth helical bundle of the apo E structure and were oriented away from the protein backbone, toward the solvent plane (89). Thus, these residues might be expected to be available to interact directly with the negatively charged ligand-binding domains of the LDL receptor. Site-directed mutagenesis and the identification of several natural mutants had already implicated these residues in receptor binding (75,76).

Residue 158, the site for the common mutation in apo E2 (i.e., $Arg_{158} \rightarrow Cys$), was postulated to cause defective binding by indirectly altering the conformation of the receptor-binding domain (92). In fact, the three-dimensional structure of apo E2 demonstrates the mechanism for the effect. In apo E3, arginine 158 forms salt bridges with aspartic acid 154 and glutamic acid 96 (89). However, the substitution of cysteine at residue 158 within apo E2 rearranges these salt bridges and causes aspartic acid 154 to form a new salt bridge with arginine 150 (a critical residue envisioned to be within the receptor-binding domain) (90,91). The side chain of arginine 150 swings into a new plane and apparently alters the conformation of the basic residues in the 136 to 150 region. Therefore, it appears that direct receptor interaction is mediated by residues in the vicinity of residues 136 to 150 and that residue 158 impacts receptor binding secondarily by altering the orientation of residues 136 to 150. Mutations involving residues 136 to 150 appear to cause the dominant form of type III hyperlipoproteinemia, whereas the substitution of residue 158 (the most common mutant form of defective apo E) causes the recessive form of the disorder (to be discussed later).

Apolipoprotein E plays a critical role in the clearance of remnant lipoproteins from the plasma by the liver. In the case of chylomicron remnants, it appears that the first step in the pathway for their clearance involves their interaction with heparan sulfate proteoglycans on the surface of hepatocytes and within the space of Disse (93,94). This step has been referred to as sequestration. The second step appears to involve further processing of the remnant lipoproteins by lipoprotein lipase and/or hepatic lipase. Lipoprotein lipase (95) and hepatic lipase (96) have both been implicated in the clearance of remnant lipoproteins. The third step involves the interaction of the remnants with lipoprotein receptors for internalization; the LDL receptor (97) and the LRP (98–101) have both been shown to be involved. The defective-binding forms of apo E [e.g., apo E2(Arg$_{158}$→Cys) or other variants] have substitutions at residues 136, 142, 145, or 146 that can directly or indirectly alter the sequestration, lipolytic processing, and/or uptake of the remnant lipoproteins. The various mutations have been implicated at different steps in the pathway (102).

Mode of Inheritance of Type III Hyperlipoproteinemia

Recently it was shown that the aforementioned specific mutations of apo E correlate with the mode of inheritance of type III hyperlipoproteinemia (for review, see Refs. 75, 76, and 88). The most common form of type III hyperlipoproteinemia is recessive and is associated with apo E2(Arg$_{158}$→Cys). Rare forms of type III hyperlipoproteinemia involving several other variant forms of apo E have been shown to have a dominant mode of inheritance (Table 3).

Recessive Type III Hyperlipoproteinemia

Homozygosity for the apo E2(Arg$_{158}$→Cys) is required for the development of the overt hyperlipidemia (Table 3). Homozygosity for ε2 occurs in about 1% of most populations in the world; however, only about 5–10% of these individuals actually develop hyperlipidemia. Most apo E2/2 homozygotes actually have low levels of plasma cholesterol. Development of elevated levels of triglyceride and cholesterol requires the defective apo E2, but also requires a second factor to cause the overt type III hyperlipoproteinemia. These secondary factors may include: 1) obesity and excessive calorie consumption, which may overwhelm the normal clearance pathways by stimulating overproduction of remnant lipoproteins, 2) hypothyroidism, which appears to lower the expression of LDL receptors, and 3) low estrogen levels (menopause), which also may lower the expression of LDL receptors (for review, see Refs. 75 and 76).

The defective binding of the apo E2(Arg$_{158}$→Cys) appears to be modulated—from very defective to near-normal binding—by a variety of parameters. For example, the binding activity can be altered by the lipid composition of the lipoprotein particles. This effect may be explained by the fact that the defective binding results from a secondary effect of the Cys$_{158}$ on the conformation of the receptor-binding domain, and that this indirect effect can be modulated by other factors that alter the three-dimensional structure of the apo E molecule (for review, see Refs. 75 and 76).

Table 3 Apolipoprotein E Variants Associated with Type III Hyperlipoproteinemia and Parameters Modulating Expression of Hyperlipidemia

Mutation[a]	Mode of inheritance	β-VLDL present	LDL receptor binding[b]	Heparin binding defect	VLDL preference	Ratio mutant:normal apo E	Lipase processing defect	Ref.
Arg$_{158}$→Cys	Recessive	Yes	2%	No	No	~1:1	Yes (HL)	79,81
Arg$_{136}$→Ser	Unknown	Yes	40%	ND	ND	4:1	ND	131
Arg$_{142}$→Cys, Cys$_{112}$→Arg	Dominant	Yes	20%	Yes	Yes	3:1	ND	132–134
Arg$_{145}$→Cys	Dominant	Yes	45%	Yes	ND	ND	ND	73,135,136
Lys$_{146}$→Gln	Dominant	Yes	40%	ND	ND	ND	Yes (LPL)	137,138
Lys$_{146}$→Glu	Dominant	Yes	<5%	Yes	ND	ND	ND	139,140
7-aa insertion,[c] Cys$_{112}$→Arg	Dominant	Yes	25%	Yes	Yes	>4:1	No	141,142

HL, hepatic lipase; LPL, lipoprotein lipase; ND, not yet determined.
[a] Lists changes compared to apo E3 structure (e.g., Arg$_{158}$→Cys, arginine at residue 158 changed to cysteine at that site).
[b] Presented as a percentage of apo E3 binding.
[c] 7-amino acid insertion, tandem repeat of residues 121–127.
Source: From Ref. 75.

The reason most individuals with apo E2/2 are hypocholesterolemic is poorly understood. It is speculated that the defective clearance of apo E2–containing lipoproteins up-regulates the LDL receptor and that the increase in LDL receptors mediates an enhanced clearance of LDL. In addition, it has been shown that apo E2 does not support lipolytic conversion of VLDL to LDL as effectively as does apo E3. Therefore, the impaired conversion of VLDL and IDL to LDL could contribute to low LDL cholesterol levels. Accompanying the impaired conversion to LDL, there would have to be effective clearance of the VLDL remnants and IDL for hypocholesterolemia to develop (for review, see Refs. 75 and 76). Clearly, a greater understanding of the impact of apo E2 on lipoprotein catabolism and processing is required.

Dominant Type III Hyperlipoproteinemia

The rare variants of apo E involving the substitution of basic residues in the 136 to 150 region for neutral or acidic amino acids are associated with the dominant mode of inheritance (for review, see Refs. 75 and 76). These substitutions appear to impact receptor-binding activity directly by interfering with the interaction of the apo E receptor-binding domain with the receptor. The presence of a single copy of the mutant allele is sufficient to disrupt remnant clearance. In addition, the variant form of apo E that has a tandem repeat of residues 121–127 (called apo E_{Leiden}) also results in the dominant mode of inheritance (for review, see Refs. 75 and 76). Insertion occurs at the junction of helix 3 and helix 4 of the four-helix bundle of the apo E molecule and is likely to disrupt markedly the orientation of the receptor-binding domain of apo E (89). However, this hypothesis remains to be proven.

As suggested, dominant type III hyperlipoproteinemia may reflect the defective receptor-binding activity of these variants, but these variants may also alter remnant catabolism by affecting lipoprotein metabolism at other steps (Table 3). It is known that apo E variants with arginine at residue 112 (the mutation occurring on the $\varepsilon 4$ allele) preferentially associate with triglyceride-rich lipoproteins (including remnants), as do apo E_{Leiden} and apo $E(Arg_{142} \rightarrow Cys)$. So-called "normal" apo E4 also has a preference for triglyceride-rich lipoproteins. An enrichment of the remnants with defective apo E at the expense of normal apo E may cause the dominant pattern of hyperlipidemia (for review, see Refs. 75 and 76).

A second means by which the variants associated with the dominant mode of inheritance may impact remnant metabolism is at the level of binding of apo E to heparan sulfate proteoglycans or heparin. In fact, it has been shown that these variants display defective interactions with heparan sulfate proteoglycans, which could disrupt the sequestration step of remnant clearance (102).

In addition, the apo E variants may alter the interaction of the remnants with lipases and thus alter the lipolytic processing of these particles. Such defects have already been demonstrated (for review, see Refs. 75 and 76).

Lipid Levels in Type III Hyperlipoproteinemia

Type III hyperlipoproteinemia is associated with hypertriglyceridemia and hypercholesterolemia (for review, see Refs. 75 and 76). Typically, the triglyceride and cholesterol levels are approximately equal (300–500 mg/dl). Ultracentrifugation reveals a cholesterol-enriched lipoprotein that has a prominent broad-β band upon electrophoresis of the $d < 1.006$ g/ml fraction. In addition, the lipoproteins in the $d = 1.006-1.019$ g/ml fraction (IDL) are typically increased. Those cholesterol-enriched lipoproteins floating at $d = 1.006$ g/ml and $d = 1.006-1.019$ g/ml are the chylomicron remnants and VLDL remnants (β-VLDL). Apolipoprotein E2 homozygosity is the diagnostic hallmark of this disorder, since the other defective variants of apo E are extremely rare. Determination of apo E phenotype requires isoelectric focusing gel electrophoresis and is primarily available in research laboratories.

It is important to note that apo E2/2 individuals, even those with hypolipidemia, appear to have increased remnant lipoproteins in their plasma. Thus, the term dysbetalipoproteinemia applies to all apo E2 homozygous individuals, but type III hyperlipoproteinemia applies only to those with overt hyperlipidemia.

Clinical Manifestations of Type III Hyperlipoproteinemia

Type III hyperlipoproteinemia is more prevalent in men than in women, and women usually do not develop the disorder until after menopause. The onset of the recessive form of hyperlipidemia is very rare in childhood, usually occurring in adulthood. These parameters apply to the most common form of type III hyperlipoproteinemia [the recessive apo E2($Arg_{158} \rightarrow Cys$) form]. It is known that in the dominant form of type III hyperlipoproteinemia the hyperlipidemia first manifests itself in childhood (103,104).

One of the most striking findings in patients with type III hyperlipoproteinemia is the occurrence of the pathognomonic xanthoma striata palmaris (yellow lipid deposits in the creases of the palms of the hands). Approximately 50% of patients with this disorder have this type of xanthoma. Tuberous and tuberoeruptive xanthomas, which are also present in other types of hyperlipidemia, also occur in type III hyperlipoproteinemia (58,60–62,103).

Accelerated atherosclerosis involving the peripheral arteries of the lower extremities and the coronary arteries occurs in 33–50% of patients with type III hyperlipoproteinemia (62,103–105). Peripheral atherosclerosis may be more common than coronary artery disease. It appears that the remnant lipoproteins (β-VLDL) are atherogenic. The remnant lipoproteins from humans with type III hyperlipoproteinemia and from animals fed diets high in fat and cholesterol are capable of causing massive cholesterol accumulation in cultured macrophages (106–108). It has been speculated that the delivery of β-VLDL cholesterol to macrophages results in foam cell formation in the artery wall and causes premature atherosclerosis in type III patients and in saturated fat- and cholesterol-fed animals (65,106).

As discussed, a variety of conditions can precipitate the development of type III hyperlipoproteinemia in apo E2 homozygous individuals. These conditions include obesity, hypothyroidism, and menopause. Hyperuricemia occurs in about 50% of patients, and glucose intolerance is common.

Diagnosis of Type III Hyperlipoproteinemia

The diagnosis is suspected when the plasma triglyceride and cholesterol levels are elevated and approximately equal (300–500 mg/dl). The $d = 1.006$ g/ml cholesterol:plasma triglyceride is usually ≥0.3, reflecting the presence of cholesterol-enriched remnant lipoproteins. Upon electrophoresis the $d < 1.006$ g/ml lipoproteins migrate as a broad-β band. The most prominent diagnostic features are the unique palmar xanthomas (62,103,104).

Although not yet readily available, determination of the apo E phenotype is the best biochemical diagnostic feature. As has been stated, most patients with type III hyperlipoproteinemia have the apo E2/2 phenotype. Other rare mutations cannot usually be diagnosed by phenotyping with isoelectric focusing (109). Recently, apo E genotyping using restriction enzyme digestion of amplified genomic DNA has been used to distinguish between arginine and cysteine at residues 112 and 158 (110). A comparison of the results from phenotyping and genotyping can identify mismatches that may well turn out to be rare mutants. However, this methodology is available only in research laboratories and usually requires DNA sequencing to identify the specific mutation involved.

Treatment of Type III Hyperlipoproteinemia

Diet therapy alone is usually effective in treating type III hyperlipoproteinemia (for review, see Refs. 111–113). Normalization of body weight (if necessary) plus a low-saturated-fat, low-cholesterol diet usually results in rather prompt reduction in plasma triglyceride and cholesterol levels. Obviously, treatment of secondary disorders—e.g., hypothyroidism, obesity, uncontrolled diabetes, and excessive alcohol consumption—is essential.

Drugs may be required in some cases (114–122). Gemfibrozil (600 mg twice a day) may be the drug of choice. Nicotinic acid is also effective, as are the HMG-CoA reductase inhibitors.

Treatment is effective not only in reducing plasma lipids, but also in reducing the size of xanthomas. Furthermore, studies have demonstrated that treatment reduces signs and symptoms of preipheral and coronary vascular disease (123,124).

Animal Models for Type III Hyperlipoproteinemia

Using gene-targeting techniques, Plump et al. (125) and Zhang et al. (126) have knocked out the apo E gene in mice and created apo E–null mice. These mice develop a profound hypercholesterolemia and an increased suscepti-bility to atherosclerosis (127,128). Fazio et al. (129) have created a mouse model of type III hyperlipoproteinemia by overexpressing a mutant apo E($Arg_{142} \rightarrow Cys$) associated with dominant inherited type III hyperlipoproteine-mia. These mice develop a spontaneous hypertriglyceridemia and hyperchol-esterolemia due to the accumulation of β-VLDL (129), and have an increased susceptibility to atherosclerosis (130).

CONCLUSIONS

In this chapter, we have reviewed two disorders of lipoprotein metabolism, familial defective apolipoprotein B100 and type III hyperlipoproteinemia. Advances in molecular genetics have proven to be instrumental in understand-ing both disorders. The tools of biochemistry and molecular biology permitted lipoprotein investigators to identify the mutations responsible for these disor-ders, and DNA-based diagnostic tests can now be used to identify these disorders in family and population studies. The techniques of molecular gen-etics have also enabled investigators to create animal models for both disorders; the study of these animals will undoubtedly provide new insights into lipoprotein metabolism and the atherogenic potential of various classes of lipoproteins.

The disorders of human lipoprotein metabolism reviewed in this chapter have also yielded pivotal insights into the important functional domains of apo B100 and apo E, particularly the regions of these molecules that bind to the LDL receptor. The structural information provided by mutations has yielded many new scientific hypotheses regarding apolipoprotein structure/function. With these hypotheses in hand, investigators have been able to use the tools of molecular biology to produce new mutant proteins in order to refine our understanding of the structure of these molecules. Even though it is possible to generate new mutant proteins in the laboratory, we nevertheless believe that we will continue to learn from human patients with lipoprotein disorders. We fully expect that newly identified apo B and apo E mutations in human patients will provide important new insights into the structure and function of these proteins.

REFERENCES

1. Young SG. Recent progress in understanding apolipoprotein B. Circulation 1990; 82:1574-1594.
2. Young SG, Linton MF. Genetic abnormalities in apolipoprotein B. Trends Cardiovasc Med 1991; 1:59-65.
3. Mahley RW. Apolipoprotein E: Cholesterol transport protein with expanding role in cell biology. Science 1988; 240:622-630.
4. Mahley RW, Weisgraber KH, Innerarity TL, Rall SC Jr. Genetic defects in lipoprotein metabolism: Elevation of atherogenic lipoproteins caused by impaired catabolism. J Am Med Assoc 1991; 265:78-83.
5. Mahley RW, Hussain MM. Chylomicron and chylomicron remnant catabolism. Curr Opin Lipidol 1991; 2:170-176.
6. Wetterau JR, Aggerbeck LP, Bouma M-E, Eisenberg C, Munck A, Hermier M. Schmitz J, Gay G, Rader DJ, Gregg RE. Absence of microsomal triglyceride transfer protein in individuals with abetalipoproteinemia. Science 1992; 258:999-1001.
7. Brown MS, Goldstein JL. A receptor-mediated pathway for cholesterol homeostasis. Science 1986; 232:34-47.
8. Hobbs HH, Brown MS, Goldstein JL. Molecular genetics of the LDL receptor gene in familial hypercholesterolemia. Hum Mutat 1992; 1:445-466.
9. Vega GL, Grundy SM. *In vivo* evidence for reduced binding of low density lipoproteins to receptors as a cause of primary moderate hypercholesterolemia. J Clin Invest 1986; 78:1410-1414.
10. Innerarity TL, Weisgraber KH, Arnold KS, Mahley RW, Krauss RM, Vega GL, Grundy SM. Familial defective apolipoprotein B-100: Low density lipoproteins with abnormal receptor binding. Proc Natl Acad Sci USA 1987; 84:6919-6923.
11. Innerarity TL, Mahley RW, Weisgraber KH, Bersot TP, Krauss RM, Vega GL, Grundy SM, Friedl W, Davignon J, McCarthy BJ. Familial defective apolipoprotein B100: A mutation of apolipoprotein B that causes hypercholesterolemia. J Lipid Res 1990; 31:1337-1349.
12. Knott TJ, Pease RJ, Powell LM, Wallis SC, Rall SC Jr, Innerarity TL, Blackhart B, Taylor WH, Marcel Y, Milne R, Johnson D, Fuller M, Lusis AJ, McCarthy BJ, Mahley RW, Levy-Wilson B, Scott J. Complete protein sequence and identification of structural domains of human apolipoprotein B. Nature 1986; 323:734-738.
13. Blackhart BD, Ludwig EM, Pierotti VR, Caiati L, Onasch MA, Wallis SC, Powell L, Pease R, Knott TJ, Chu M-L, Mahley RW, Scott J, McCarthy BJ, Levy-Wilson B. Structure of the human apolipoprotein B gene. J Biol Chem 1986; 261:15364-15367.
14. Knott TJ, Rall SC Jr., Innerarity TL, Jacobson SF, Urdea MS, Levy-Wilson B, Powell LM, Pease RJ, Eddy R, Nakai H, Byers M, Priestley LM, Robertson E, Rall LB, Betsholtz C, Shows TB, Mahley RW, Scott J. Human apolipoprotein B: Structure of carboxyl-terminal domains, sites of gene expression, and chromosomal localization. Science 1985; 230:37-43.
15. Soria LF, Ludwig EH, Clarke HRG, Vega GL, Grundy SM, McCarthy BJ. Association between a specific apolipoprotein B mutation and familial defective apolipoprotein B-100. Proc Natl Acad Sci USA 1989; 86:587-591.
16. Milne R, Théolis R Jr., Maurice R, Pease RJ, Weech PK, Rassart E, Fruchart J-C,

Scott J, Marcel YL. The use of monoclonal antibodies to localize the low density lipoprotein receptor-binding domain of apolipoprotein B. J Biol Chem 1989; 264:19754–19760.

17. Young SG, Witztum JL, Casal DC, Curtiss LK, Bernstein S. Conservation of the low density lipoprotein receptor-binding domain of apoprotein B: Demonstration by a new monoclonal antibody, MB47. Arteriosclerosis 1986; 6:178–188.

18. Young SG, Koduri RK, Austin RK, Bonnet DJ, Smith RS, Curtiss LK. Definition of a nonlinear conformational epitope for the apolipoprotein B100-specific monoclonal antibody, MB47. J Lipid Res 1994; 35:399–407.

19. Lund-Katz S, Innerarity TL, Arnold KS, Curtiss LK, Phillips MC. ^{13}C NMR evidence that substitution of glutamine for arginine 3500 in familial defective apolipoprotein B-100 disrupts the conformation of the receptor-binding domain. J Biol Chem 1991; 266:2701–2704.

20. Weisgraber KH, Innerarity TL, Newhouse YM, Young SG, Arnold KS, Krauss RM, Vega GL, Grundy SM, Mahley RW. Familial defective apolipoprotein B-100: Enhanced binding of monoclonal antibody MB47 to abnormal low density lipoproteins. Proc Natl Acad Sci USA 1988; 85:9758–9762.

21. Lund-Katz S, Ibdah JA, Letizia JY, Thomas MT, Phillips MC. A ^{13}C NMR characterization of lysine residues in apolipoprotein B and their role in binding to the low density lipoprotein receptor. J Biol Chem 1988; 263:13831–13838.

22. Ludwig EH, McCarthy BJ. Haplotype analysis of the human apolipoprotein B mutation associated with familial defective apolipoprotein B100. Am J Hum Genet 1990; 47:712–720.

23. Friedl W, Ludwig EH, Balestra ME, Arnold KS, Paulweber B, Sandhofer F, McCarthy BJ, Innerarity TL. Apolipoprotein B gene mutations in Austrian subjects with heart disease and their kindred. Arterioscler Thromb 1991; 11:371–378.

24. Myant NB, Gallagher JJ, Knight BL, McCarthy SN, Frostegard J, Nilsson J, Hamsten A, Talmud P, Humphries SE. Clinical signs of familial hypercholesterolemia in patients with familial defective apolipoprotein B-100 and normal low density lipoprotein receptor function. Arterioscler Thromb 1991; 11:691–703.

25. Rauh G, Schuster H, Schewe CK, Stratmann G, Keller C, Wolfram G, Zöllner N. Independent mutation of arginine$_{(3500)}$→glutamine associated with familial defective apolipoprotein B-100. J Lipid Res 1993; 34:799–805.

26. Tybjærg-Hansen A, Humphries SE. Familial defective apolipoprotein B-100: A single mutation that causes hypercholesterolemia and premature coronary artery disease. Atherosclerosis 1992; 96:91–107.

27. Bersot TP, Russell SJ, Thatcher SR, Pomernacki NK, Mahley RW, Weisgraber KH, Innerarity TL, Fox CS. A unique haplotype of the apolipoprotein B-100 allele associated with familial defective apolipoprotein B-100 in a Chinese man discovered during a study of the prevalence of this disorder. J Lipid Res 1993; 34:1149–1154.

28. Young SG, Bertics SJ, Scott TM, Dubois BW, Curtiss LK, Witztum JL. Parallel expression of the MB19 genetic polymorphism in apoprotein B-100 and apoprotein B 48: Evidence that both apoproteins are products of the same gene. J Biol Chem 1986; 261:2995–2998.

29. Young SG, Hubl ST. An *Apa*LI restriction site polymorphism is associated with the MB19 polymorphism in apolipoprotein B. J Lipid Res 1989; 30:443–449.

30. Young SG, Bertics SJ, Curtiss LK, Dubois BW, Witztum JL. Genetic analysis of a kindred with familial hypobetalipoproteinemia. Evidence for two separate gene defects: One associated with an abnormal apolipoprotein B species, apolipoprotein B-37; and a second associated with low plasma concentrations of apolipoprotein B-100. J Clin Invest 1987; 79:1842–1851.

31. Arnold KS, Balestra ME, Krauss RM, Curtiss LK, Young SG, Innerarity TL. Isolation of allele-specific, receptor-binding-defective low density lipoproteins from familial defective apolipoprotein B-100 subjects. J Lipid Res 1994; 35:1469–1476.

32. März W, Baumstark MW, Scharnagl H, Ruzicka V, Buxbaum S, Herwig J, Pohl T, Russ A, Schaaf L, Berg A, Böhles H-J, Usadel KH, Groβ W. Accumulation of "small dense" low density lipoproteins (LDL) in a homozygous patient with familial defective apolipoprotein B-100 results from heterogenous interaction of LDL subfractions with the LDL receptor. J Clin Invest 1993; 92:2922–2933.

33. März W, Ruzicka C, Pohl T, Usadel KH, Gross W. Familial defective apolipoprotein B-100: Mild hypercholesterolaemia without atherosclerosis in a homozygous patient. Lancet 1992; 340:1362.

34. Tybjærg-Hansen A, Gallagher J, Vincent J, Houlston R, Talmud P, Dunning AM, Seed M, Hamsten A, Humphries SE, Myant NB. Familial defective apolipoprotein B-100: Detection in the United Kingdom and Scandinavia, and clinical characteristics of ten cases. Atherosclerosis 1990; 80:235–242.

35. Schuster H, Rauh G, Kormann B, Hepp T, Humphries S, Keller C, Wolfram G, Zöllner N. Familial defective apolipoprotein B-100: Comparison with familial hypercholesterolemia in 18 cases detected in Munich. Arteriosclerosis 1990; 10:577–581.

36. Miserez AR, Laager R, Chiodetti N, Keller U. High prevalence of familial defective apolipoprotein B-100 in Switzerland. J Lipid Res 1994; 35:574–583.

37. Rust S, Funke H, Assmann G. Analysis of pooled samples from nearly 10000 individuals with mutagenically separated PCR (MS-PCR) shows a significant overrepresentation of familial defective apo B-100 in coronary artery disease patients (abstr). Circulation 1992; 86:I-420.

38. Hämäläinen T, Palotie A, Aalto-Setälä K, Kontula K, Tikkanen MJ. Absence of familial defective apolipoprotein B-100 in Finnish patients with elevated serum cholesterol. Atherosclerosis 1990; 82:177–183.

39. Friedlander Y, Dann EJ, Leitersdorf E. Absence of familial defective apolipoprotein B-100 in Israeli patients with dominantly inherited hypercholesterolemia and in offspring with parental history of myocardial infarction. Hum Genet 1993; 91:299–300.

40. Schwartz EI, Shevtsov SP, Kuchinski AP, Kovalev YP, Plutalov OV, Berlin YA. Approach to identification of a point mutation in apo B100 gene by means of a PCR-mediated site-directed mutagenesis. Nucleic Acids Res 1991; 19:3752.

41. Lipid Research Clinics Program. Population Studies Data Book. The Prevalence Study. Vol 1. Washington, DC: US Department of Health and Human Services, Public Health Service, National Institutes of Health (NIH Publication no. 80-1527), 1980.

42. Rauh G, Keller C, Kormann B, Spengel F, Schuster H, Wolfram G, Zöllner N. Familial defective apolipoprotein B_{100}: Clinical characteristics of 54 cases. Atherosclerosis 1992; 92:233–241.

43. Gallagher JJ, Myant NB. Variable expression of the mutation in familial defective apolipoprotein B-100. Arterioscler Thromb 1993; 13:973–976.

44. Hobbs HH, Leitersdorf E, Leffert CC, Cryer DR, Brown MS, Goldstein JL. Evidence for a dominant gene that suppresses hypercholesterolemia in a family with defective low density lipoprotein receptors. J Clin Invest 1989; 84:656–664.

45. Funke H, Rust S, Seedorf U, Brennhausen B, Chirazi A, Motti C, Assmann G. Homozygosity for familial defective apolipoprotein B-100 (FDB) is associated with lower plasma cholesterol concentrations than homozygosity for familial hypercholesterolemia (FH) (abstr). Circulation 1992; 86:I-691.

46. Rodgers GM, Shuman MA. Characterization of the interaction between Factor X_a and bovine aortic endothelial cells. Biochim Biophys Acta 1985; 844:320–329.

47. Defesche JC, Pricker KL, Hayden MR, van der Ende BE, Kastelein JJP. Familial defective apolipoprotein B-100 is clinically indistinguishable from familial hypercholesterolemia. Arch Intern Med 1993; 153:2349–2356.

48. Hansen PS, Rüdiger N, Tybjærg-Hansen A, Faergeman O, Gregersen N. Detection of the apoB-3500 mutation (glutamine for arginine) by gene amplification and cleavage with MspI. J Lipid Res 1991; 32:1229–1233.

49. Ruzicka V, März W, Russ A, Gross W. Apolipoprotein B($Arg_{3500}\rightarrow Gln$) allele specific polymerase chain reaction: Large-scale screening of pooled blood samples. J Lipid Res 1992; 33:1563–1567.

50. Wenham PR, Newton CR, Houlston RS, Price WH. Rapid diagnosis of familial defective apolipoprotein B-100 by Amplification Refractory Mutation System. Clin Chem 1991; 37:1983–1987.

51. Corsini A, Mazzotti M, Fumagalli R, Catapano AL, Romano L, Romano C. Poor response to simvastatin in familial defective apo-B-100 (letter to the editor). Lancet 1991; 337:305.

52. Illingworth DR, Vakar F, Mahley RW, Weisgraber KH. Hypocholesterolaemic effects of lovastatin in familial defective apolipoprotein B-100. Lancet 1992; 339:598–600.

53. Schmidt EB, Illingworth DR, Bacon S, Mahley RW, Weisgraber KH. Hypocholesterolemic effects of cholestyramine and colestipol in patients with familial defective apolipoprotein B-100. Atherosclerosis 1993; 98:213–217.

54. Schmidt EB, Illingworth DR, Bacon S, Russel SJ, Thatcher SR, Mahley RW, Weisgraber KH. Hypolipidemic effects of nicotinic acid in patients with familial defective apolipoprotein B-100. Metabolism 1993; 42:137–139.

55. Pullinger CR, Hennessy LK, Love JA, Frost PH, Mendel CM, Liu W, Malloy MJ, Kane JP. Familial ligand-defective apolipoprotein B: Identification of a new mutation that decreases LDL receptor binding affinity (abstr). Circulation 1993; 88:I-322.

56. Linton MF, Farese RV Jr., Chiesa G, Grass DS, Chin P, Hammer RE, Hobbs HH, Young SG. Transgenic mice expressing high plasma concentrations of human apolipoprotein B100 and lipoprotein(a). J Clin Invest 1993; 92:3029–3037.

57. Homanics GE, Smith TJ, Zhang SH, Lee D, Young SG, Maeda N. Targeted mod-

ification of the apolipoprotein B gene results in hypobetalipoproteinemia and developmental abnormalities in mice. Proc Natl Acad Sci USA 1993; 90:2389–2393.

58. McGinley J, Jones H, Gofman J. Lipoproteins and xanthomatous diseases. J Invest Dermatol 1952; 19:71–82.

59. Gofman JW, deLalla O, Glazier F, Freeman NK, Lindgren FT, Nichols AV, Strisower B, Tamplin AR. The serum lipoprotein transport system in health, metabolic disorders, atherosclerosis and coronary heart disease. Plasma 1954; 2:413–484.

60. Haber C, Kwiterovich PO Jr. Dyslipoproteinemia and xanthomatosis. Pediatr Dermatol 1984; 1:261–280.

61. Polano MK. Xanthomatosis and hyperlipoproteinemia: A review. Dermatologica 1974; 149:1–9.

62. Fredrickson DS, Levy RI, Lees RS. Fat transport in lipoproteins—an integrated approach to mechanisms and disorders. N Engl J Med 1967; 276:34–44, 94–103, 148–156, 215–225, 273–281.

63. Fainaru M, Mahley RW, Hamilton RL, Innerarity TL. Structural and metabolic heterogeneity of β-very low density lipoproteins from cholesterol-fed dogs and from humans with type III hyperlipoproteinemia. J Lipid Res 1982; 23:702–714.

64. Kane JP, Chen GC, Hamilton RL, Hardman DA, Malloy MJ, Havel RJ. Remnants of lipoproteins of intestinal and hepatic origin in familial dysbetalipoproteinemia. Arteriosclerosis 1983; 3:47–56.

65. Mahley RW. Atherogenic lipoproteins and coronary artery disease: Concepts derived from recent advances in cellular and molecular biology. Circulation 1985; 72:943–948.

66. Havel RJ, Kane JP. Primary dysbetalipoproteinemia: Predominance of a specific apoprotein species in triglyceride-rich lipoproteins. Proc Natl Acad Sci USA 1973; 70:2015–2019.

67. Utermann G, Pruin N, Steinmetz A. Polymorphism of apolipoprotein E. III. Effect of a single polymorphic gene locus on plasma lipid levels in man. Clin Genet 1979; 15:63–72.

68. Utermann G, Hees M, Steinmetz A. Polymorphism of apolipoprotein E and occurrence of dysbetalipoproteinaemia in man. Nature 1977; 269:604–607.

69. Utermann G, Jaeschke M, Menzel J. Familial hyperlipoproteinemia type III: Deficiency of a specific apolipoprotein (apo E-III) in the very-low-density lipoproteins. FEBS Lett 1975; 56:352–355.

70. Utermann G, Vogelberg KH, Steinmetz A, Schoenborn W, Pruin N, Jaeschke M, Hees M, Canzler H. Polymorphism of apolipoprotein E. II. Genetics of hyperlipoproteinemia type III. Clin Genet 1979; 15:37–62.

71. Utermann G, Langenbeck U, Beisiegel U, Weber W, Genetics of the apolipoprotein E system in man. Am J Hum Genet 1980; 32:339–347.

72. Weisgraber KH, Rall SC Jr, Mahley RW. Human E apoprotein heterogeneity. Cysteine-arginine interchanges in the amino acid sequence of the apo-E isoforms. J Biol Chem 1981; 256:9077–9083.

73. Rall SC Jr, Weisgraber KH, Innerarity TL, Mahley RW, Structural basis for receptor binding heterogeneity of apolipoprotein E from type III hyperlipoproteinemic subjects. Proc Natl Acad Sci USA 1982; 79:4696–4700.

74. Rall SC Jr., Weisgraber KH, Mahley RW. Human apolipoprotein E. The complete amino acid sequence. J Biol Chem 1982; 257:4171–4178.

75. Mahley RW, Rall SC Jr. Type III hyperlipoproteinemia (dysbetalipoproteinemia): The role of apolipoprotein E in normal and abnormal lipoprotein metabolism. In: Scriver CR, Beaudet AL, Sly WS, Valle D, eds. The Metabolic and Molecular Bases of Inherited Disease. 7th ed. New York: McGraw-Hill, 1995:1953–1980.

76. Rall SC Jr, Mahley RW. The role of apolipoprotein E genetic variants in lipoprotein disorders. J Intern Med 1992; 231:653–659.

77. Zannis VI, Breslow JL, Utermann G, Mahley RW, Weisgraber KH, Havel RJ, Goldstein JL, Brown MS, Schonfeld G, Hazzard WR, Blum C. Proposed nomenclature of apoE isoproteins, apoE genotypes, and phenotypes. J Lipid Res 1982; 23:911–914.

78. Schneider WJ, Kovanen PT, Brown MS, Goldstein JL, Utermann G, Weber W, Havel RJ, Kotite L, Kane JP, Innerarity TL, Mahley RW. Familial dysbetalipoproteinemia. Abnormal binding of mutant apoprotein E to low density lipoprotein receptors of human fibroblasts and membranes from liver and adrenal of rats, rabbits, and cows. J Clin Invest 1981; 68:1075–1085.

79. Weisgraber KH, Innerarity TL, Mahley RW. Abnormal lipoprotein receptor-binding activity of the human E apoprotein due to cysteine-arginine interchange at a single site. J Biol Chem 1982; 257:2518–2521.

80. Gregg RE, Zech LA, Schaefer EJ, Brewer HB Jr. Type III hyperlipoproteinemia: Defective metabolism of an abnormal apolipoprotein E. Science 1981; 211:584–586.

81. Ehnholm C, Mahley RW, Chappell DA, Weisgraber KH, Ludwig E, Witztum JL. Role of apolipoprotein E in the lipolytic conversion of β-very low density lipoproteins to low density lipoproteins in type III hyperlipoproteinemia. Proc Natl Acad Sci USA 1984; 81:5566–5570.

82. Thuren T, Weisgraber KH, Sisson P, Waite M. Role of apolipoprotein E in hepatic lipase catalyzed hydrolysis of phospholipid in high-density lipoproteins. Biochemistry 1992; 31:2332–2338.

83. Thuren T, Wilcox RW, Sisson P, Waite M. Hepatic lipase hydrolysis of lipid monolayers: Regulation by apolipoproteins. J Biol Chem 1991; 266:4853–4861.

84. Mahley RW, Innerarity TL. Lipoprotein receptors and cholesterol homeostasis. Biochim Biophys Acta 1983; 737:197–222.

85. Brown MS, Herz J, Kowal RC, Goldstein JL. The low-density lipoprotein receptor-related protein: Double agent or decoy? Curr Opin Lipidol 1991; 2:65–72.

86. Herz J. The LDL-receptor-related protein—portrait of a multifunctional receptor. Curr Opin Lipidol 1993; 4:107–113.

87. Weisgraber KH. Apolipoprotein E: Structure–function relationships. Adv Protein Chem 1994; 45:249–302.

88. Mahley RW, Innerarity TL, Rall SC Jr, Weisgraber KH, Taylor JM. Apolipoprotein E: Genetic variants provide insights into its structure and function. Curr Opin Lipidol 1990; 1:87–95.

89. Wilson C, Wardell MR, Weisgraber KH, Mahley RW, Agard DA. Three-dimensional structure of the LDL receptor-binding domain of human apolipoprotein E. Science 1991; 252:1817–1822.

90. Wilson C, Mau T, Weisgraber KH, Wardell MR, Mahley RW, Agard DA. Salt bridge relay triggers defective LDL receptor binding by a mutant apolipoprotein. Structure 1994; 2:713–718.

91. Wardell MR, Wilson C, Agard DA, Mahley RW, Weisgraber KH. Crystal structures of the common apolipoprotein E variants: Insights into functional mechanisms. In: Sirtori CR, Franceschini G, Brewer BH Jr, eds. NATO Advanced Research Workshop. Human Apolipoprotein Mutants III. Apolipoproteins in the Diagnosis and Treatment of Disease. Berlin: Springer-Verlag, 1993:81–96.

92. Innerarity TL, Weisgraber KH, Arnold KS, Rall SC Jr, Mahley RW. Normalization of receptor binding of apolipoprotein E2: Evidence for modulation of the binding site conformation. J Biol Chem 1984; 259:7261–7267.

93. Ji Z-S, Brecht WJ, Miranda RD, Hussain MM, Innerarity TL, Mahley RW. Role of heparan sulfate proteoglycans in the binding and uptake of apolipoprotein E-enriched remnant lipoproteins by cultured cells. J Biol Chem 1993; 268: 10160–10167.

94. Ji Z-S, Fazio S, Lee Y-L, Mahley RW. Secretion-capture role for apolipoprotein E in remnant lipoprotein metabolism involving cell surface heparan sulfate proteoglycans. J Biol Chem 1994; 269:2764–2772.

95. Beisiegel U, Weber W, Bengtsson-Olivecrona G. Lipoprotein lipase enhances the binding of chylomicrons to low density lipoprotein receptor-related protein. Proc Natl Acad Sci USA 1991; 88:8342–8346.

96. Ji Z-S, Lauer SJ, Fazio S, Bensadoun A, Taylor JM, Mahley RW. Enhanced binding and uptake of remnant lipoproteins by hepatic lipase-secreting hepatoma cells in culture. J Biol Chem 1994; 269:13429–13436.

97. Choi SY, Fong LG, Kirven MJ, Cooper AD. Use of an anti–low density lipoprotein receptor antibody to quantify the role of the LDL receptor in the removal of chylomicron remnants in the mouse in vivo. J Clin Invest 1991; 88:1173–1181.

98. Kowal RC, Herz J, Goldstein JL, Esser V, Brown MS. Low density lipoprotein receptor-related protein mediates uptake of cholesteryl esters derived from apoprotein E-enriched lipoproteins. Proc Natl Acad Sci USA 1989; 86:5810–5814.

99. Kowal RC, Herz J, Weisgraber KH, Mahley RW, Brown MS, Goldstein JL. Opposing effects of apolipoproteins E and C on lipoprotein binding to low density lipoprotein receptor-related protein. J Biol Chem 1990; 265:10771–10779.

100. Hussain MM, Maxfield FR, Más-Oliva J, Tabas I, Ji Z-S, Innerarity TL, Mahley RW. Clearance of chylomicron remnants by the low density lipoprotein receptor-related protein/α_2-macroglobulin receptor. J Biol Chem 1991; 266:13936–13940.

101. Willnow TE, Sheng Z, Ishibashi S, Herz J. Inhibition of hepatic chylomicron remnant uptake by gene transfer of a receptor antagonist. Science 1994; 264: 1471–1474.

102. Ji Z-S, Fazio S, Mahley RW. Variable heparan sulfate proteoglycan binding of apolipoprotein E variants may modulate the expression of type III hyperlipoproteinemia. J Biol Chem 1994; 269:13421–13428.

103. Morganroth J, Levy RI, Fredrickson DS. The biochemical, clinical, and genetic features of type III hyperlipoproteinemia. Ann Intern Med 1975; 82:158–174.

104. Hazzard WR, Primary type III hyperlipoproteinemia. In: Rifkind BM, Levy RI, eds.

Hyperlipidemia: Diagnosis and Therapy. New York: Grune & Stratton, 1977:137–175.

105. Stuyt PMJ, Van 't Laar A. Clinical features of type III hyperlipoproteinaemia. Neth J Med 1983; 26:104–111.

106. Mahley RW. Development of accelerated atherosclerosis: Concepts derived from cell biology and animal model studies. Arch Pathol Lab Med 1983; 107:393–398.

107. Goldstein JL, Ho YK, Brown MS, Innerarity TL, Mahley RW. Cholesteryl ester accumulation in macrophages resulting from receptor-mediated uptake and degradation of hypercholesterolemic canine β–very low density lipoproteins. J Biol Chem 1980; 255:1839–1848.

108. Mahley RW, Innerarity TL, Brown MS, Ho YK, Goldstein JL. Cholesteryl ester synthesis in macrophages: Stimulation by β-very low density lipoproteins from cholesterol-fed animals of several species. J Lipid Res 1980; 21:970–980.

109. Menzel H-J, Utermann G. Apolipoprotein E phenotyping from serum by Western blotting. Electrophoresis 1986; 7:492–495.

110. Hixson JE, Vernier DT. Restriction isotyping of human apolipoprotein E by gene amplification and cleavage with HhaI. J Lipid Res 1990; 31:545–548.

111. Gotto AM Jr., Jones PH, Scott LW. The diagnosis and management of hyperlipidemia. Dis Mon 1986; 32:245–311.

112. Schaefer EJ, Levy RI. Pathogenesis and management of lipoprotein disorders. N Engl J Med 1985; 312:1300–1310.

113. Connor WE, Connor SL. The dietary treatment of hyperlipidemia. Rationale, technique and efficacy. Med Clin North Am 1982; 66:485–518.

114. Choice of cholesterol-lowering drugs. Med Lett Drugs Ther 1993; 35:19–22.

115. Prihoda JS, Illingworth DR. Drug therapy of hyperlipidemia. Curr Probl Cardiol 1992; 17:545–605.

116. Illingworth DR. Management of hyperlipidemia: Goals for the prevention of atherosclerosis. Clin Invest Med 1990; 13:211–218.

117. Hoogwerf BJ, Bantle JP, Kuba K, Frantz ID Jr, Hunninghake DB. Treatment of type III hyperlipoproteinemia with four different treatment regimens. Atherosclerosis 1984; 51:251–259.

118. Vega GL, East C, Grundy SM. Lovastatin therapy in familial dysbetalipoproteinemia: Effects on kinetics of apolipoprotein B. Atherosclerosis 1988; 70:131–143.

119. Illingworth DR, O'Malley JP. The hypolipidemic effects of lovastatin and clofibrate alone and in combination in patients with type III hyperlipoproteinemia. Metabolism 1990; 39:403–409.

120. Stuyt PMJ, Mol MJTM, Stalenhoef AFH. Long-term effects of simvastatin in familial dysbetalipoproteinaemia. J Intern Med 1991; 230:151–155.

121. Feussner G, Eichinger M, Ziegler R. The influence of simvastatin alone or in combination with gemfibrozil on plasma lipids and lipoproteins in patients with type III hyperlipoproteinemia. Clin Invest 1992; 70:1027–1035.

122. Wiklund O, Angelin B, Bergman M, Berglund L, Bondjers G, Carlsson A, Lindén T, Miettinen T, Ödman B, Olofsson S-O, Saarinen I, Sipilä R, Sjöström P, Kron B, Vanhanen H, Wright I. Pravastatin and gemfibrozil alone and in combination for the treatment of hypercholesterolemia. Am J Med 1993; 94:13–20.

123. Zelis R, Mason DT, Braunwald E, Levy RI. Effects of hyperlipoproteinemias and their treatment on the peripheral circulation. J Clin Invest 1970; 49:1007.
124. Kuo PT, Wilson AC, Kostis JB, Moreyra AB, Dodge HT. Treatment of type III hyperlipoproteinemia with gemfibrozil to retard progression of coronary artery disease. Am Heart J 1988; 116:85–90.
125. Plump AS, Smith JD, Hayek T, Aalto-Setälä K, Walsh A, Verstuyft JG, Rubin EM, Breslow JL. Severe hypercholesterolemia and atherosclerosis in apolipoprotein E-deficient mice created by homologous recombination in ES cells. Cell 1992; 71:343–353.
126. Zhang SH, Reddick RL, Piedrahita JA, Maeda N. Spontaneous hypercholesterolemia and arterial lesions in mice lacking apolipoprotein E. Science 1992; 258:468–471.
127. Nakashima Y, Plump AS, Raines EW, Breslow JL, Ross R. ApoE-deficient mice develop lesions of all phases of atherosclerosis throughout the arterial tree. Arterioscler Thromb 1994; 14:133–140.
128. Reddick RL, Zhang SH, Maeda N. Atherosclerosis in mice lacking apo E. Evaluation of lesional development and progression. Aterioscler Thromb 1994; 14:141–147.
129. Fazio S, Lee Y-L, Ji Z-S, Rall SC Jr. Type III hyperlipoproteinemic phenotype in transgenic mice expressing dysfunctional apolipoprotein E. J Clin Invest 1993; 92:1497–1503.
130. Fazio S, Sanan DA, Lee Y-L, Ji Z-S, Mahley RW, Rall SC Jr. Susceptibility to diet-induced atherosclerosis in transgenic mice expressing a dysfunctional human apolipoprotein E(Arg112, Cys142). Arterioscler Thromb 1995; 14:1873–1879.
131. Wardell MR, Brennan SO, Janus ED, Fraser R, Carrell RW. Apolipoprotein E2-Christchurch (136 Arg → Ser). New variant of human apolipoprotein E in a patient with type III hyperlipoproteinemia. J Clin Invest 1987; 80:483–490.
132. Havel RJ, Kotite L, Kane JP, Tun P, Bersot T. Atypical familial dysbetalipoproteinemia associated with apolipoprotein phenotype E3/3. J Clin Invest 1983; 72:379–387.
133. Horie Y, Fazio S, Westerlund JR, Weisgraber KH, Rall SC Jr. The functional characteristics of a human apolipoprotein E variant (cysteine at residue 142) may explain its association with dominant expression of type III hyperlipoproteinemia. J Biol Chem 1992; 267:1962–1968.
134. Rall SC Jr., Newhouse YM, Clarke HRG, Weisgraber KH, McCarthy BJ, Mahley RW, Bersot TP. Type III hyperlipoproteinemia associated with apolipoprotein E phenotype E3/3: Structure and genetics of an apolipoprotein E3 variant. J Clin Invest 1989; 83:1095–1101.
135. Emi M, Wu LL, Robertson MA, Myers RL, Hegele RA, Williams RR, White R, Lalouel J-M. Genotyping and sequence analysis of apolipoprotein E isoforms. Genomics 1988; 3:373–379.
136. Lohse P, Mann WA, Stein EA, Brewer HB Jr. Apolipoprotein E-4Philadelphia (Glu[13] → Lys, Arg[145] → Cys). Homozygosity for two rare point mutations in the apolipoprotein E gene combined with severe type III hyperlipoproteinemia. J Biol Chem 1991; 266:10479–10484.

137. Rall SC Jr., Weisgraber KH, Innerarity TL, Bersot TP, Mahley RW, Blum CB. Identification of a new structural variant of human apolipoprotein E, E2(Lys$_{146}$→Gln), in a type III hyperlipoproteinemic subject with the E3/2 phenotype. J Clin Invest 1983; 72:1288–1297.

138. Smit M, de Knijff P, van der Kooij-Meijs E, Groenendijk C, van den Maagdenberg AMJM, Gevers Leuven JA, Stalenhoef AFH, Stuyt PMJ, Frants RR, Havekes LM. Genetic heterogeneity in familial dysbetalipoproteinemia. The E2(lys 146→gln) variant results in a dominant mode of inheritance. J Lipid Res 1990; 31:45–53.

139. Mann WA, Gregg RE, Sprecher DL, Brewer HB Jr. Apolipoprotein E-1 Harrisburg: a new variant of apolipoprotein E dominantly associated with type III hyperlipoproteinemia. Biochim Biophys Acta 1989; 1005:239–244.

140. Moriyama K, Sasaki J, Matsunaga A, Arakawa F, Takada Y, Araki K, Kaneko S, Arakawa K. Apolipoprotein E1 Lys-146 → Glu with type III hyperlipoproteinemia. Biochim Biophys Acta 1992; 1128:58–64.

141. de Knijff P, van den Maagdenberg AMJM, Stalenhoef AFH, Leuven JAG, Demacker PNM, Kuyt LP, Frants RR, Havekes LM. Familial dysbetalipoproteinemia associated with apolipoprotein E3-Leiden in an extended multigeneration pedigree. J Clin Invest 1991; 88:643–655.

142. Wardell MR, Weisgraber KH, Havekes LM, Rall SC Jr. Apolipoprotein E3-Leiden contains a seven-amino acid insertion that is a tandem repeat of residues 121 to 127. J Biol Chem 1989; 264:21205–21210.

7

Genetic Causes of Aortic Aneurysms

Helena Kuivaniemi,* Gerard Tromp,* and Darwin J. Prockop
Jefferson Medical College
Thomas Jefferson University
Philadelphia, Pennsylvania

INCIDENCE

Aortic aneurysms are a pathological condition characterized by dilatation of the aorta and involve the expansion and thinning of all layers of the arterial wall (1,2; see Figure 1). Aortic aneurysms are the thirteenth leading cause of death in the United States (3), which means that about 15,000 individuals die every year from the rupture of an aortic aneurysm. In the United States in 1984, 1.2% of all deaths in men and 0.6% of all deaths in women over the age of 65 were due to aortic aneurysms (4). Abdominal aortic aneurysms (AAAs) are responsible for about 1.3% of the deaths among men aged over 50 years in both the United Kingdom and Sweden, and the proportion has been increasing (5,6). A study of 22,765 autopsies in Finland during the years 1959 to 1979 revealed an abdominal aortic aneurysm in 193 individuals, corresponding to a prevalence of 0.85% (7). Most AAAs do not produce symptoms until they suddenly rupture. Aortic aneurysms most commonly occur in the infrarenal segment of the abdominal aorta, and most patients with an AAA do not have aneurysms in the other parts of aorta (8).

CAUSES

There is no consensus as to the cause of aortic aneurysms (9-11). Hypertension exists in about half of patients and is obviously an aggravating condition. Tertiary syphilis was once an important cause of aneurysms, particularly of the ascending thoracic aorta, but is a less common cause now. Another cause

---

Current affiliation: Wayne State University School of Medicine, Detroit, Michigan.

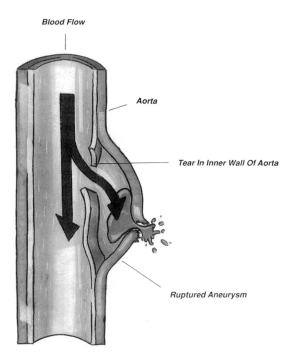

Blood Flow

Aorta

Tear In Inner Wall Of Aorta

Ruptured Aneurysm

Figure 1   A schematic drawing of a ruptured aortic aneurysm. Aneurysm is a balloon-like swelling of the aorta. Rupture of an aneurysm leads to massive bleeding, and often to sudden death.

frequently cited is arteriosclerosis. Still another cause cited, particularly when cystic medial necrosis is prominent, is activation of enzymes such as collagenases and elastases that degrade the extracellular matrix.

Recently, however, a number of reports have focused on the possibility that the primary cause of many aortic aneurysms may be genetic defects (11–26). One recent report indicated that 29% of asymptomatic brothers of patients with AAA had dilatations of the aorta detectable by ultrasound examination (20). Another recent report indicated that 15.1% of 542 patients undergoing surgery for AAA had a first-degree relative with aneurysms (19). The results, therefore, indicated that a large proportion of AAAs are familial and, therefore, may be genetic.

EVIDENCE FOR THE CLUSTERING OF AORTIC ANEURYSMS IN FAMILIES

A family in which three brothers had AAAs that were diagnosed between the ages of 60 and 70 years led to the suggestion by Clifton (12) that there may be a heritable trait that caused aortic aneurysms. Tilson and Seashore (13) reported on 16 families with AAAs in first-degree relatives. From the pedigree data, the authors concluded that the mode of inheritance was either X-linked or autosomal dominant, with the X-linked mode being more common. Subsequently, Tilson and Seashore (14) reported on a collection of 50 families and favored an autosomal dominant mode of inheritance. These reports established that AAAs, at least in some cases, did occur in family clusters.

To establish what fraction of AAAs were familial, Norrgård et al. (15) surveyed 200 consecutive patients who underwent surgery for repair of AAAs over a period of 16 years in Umeå, Sweden. Of the 200 patients, 89 were alive and 87 responded to the questionnaire. Abdominal aortic aneurysms occurred in the families of 16. The familial incidence was, therefore, 18.4%. Johansen and Koepsell (16) interviewed 250 patients with AAAs as well as 250 control subjects and found that 48 (19.2%) of the patients, as compared to six (2.4%) of the controls, had a first-degree relative with an AAA. The highest familial incidence of AAAs was reported by Powell and Greenhalgh (17), who interviewed 60 patients and were able to collect family data on 56. They reported a positive family history in 20 patients (35.7%). The lowest familial incidence was reported by Johnston and Scobie (27) in a Canadian study based on the histories of 666 patients that were collected by 72 surgeons. A positive family history was recorded in only 6.1% of the cases. In a second Canadian study, based on telephone interviews of 305 patients who underwent surgery for repair of AAAs, Cole et al. (18) reported that 34 patients had a positive family history (11.1%). Darling et al. (19) reported that 82 of 542 patients (15.1%) had a positive history for AAAs in their families. Most recently, Majumder et al. (24) reported that 13 of 91 patients (14.3%) had a positive history. Summing the number of patients in each of the studies and the number of patients with a positive history in their family, the cumulative value for the familial incidence of AAAs is about 12.7%.

In addition to the above studies, all of which relied on questionnaires and interviews to determine the familial incidence of AAAs, there are four studies that used ultrasonography to determine the incidence of undiagnosed aneurysms.

The first and second studies were reported from Sweden (20) and the United Kingdom (21) in 1989, the third study from the United States in 1991 (22), and the fourth study from the United Kingdom in 1992 (23). In the Swedish study 29% (10/35) of the brothers and 6% (3/52) of the sisters were found to have AAAs (20). The individuals screened were between 39 and 82 years of age. In the second study, 16 brothers and 15 sisters between the ages of 50 and 79 were screened with ultrasound and four brothers (25%) were found to have AAAs (21). In addition, one brother declined ultrasound examination because he was soon to have an AAA resected in his local hospital.

Thus, of 17 brothers examined, altogether five (29%), who were related to four probands, had an AAA. In the third study, only 6.7% (7/103) of all first-degree relatives screened who were 40 years or older had AAAs, but 25% (5/20) of brothers who were 55 years or older had AAAs (22). The fourth study included 25 brothers and 28 sisters from 28 families (23). Twenty percent (5/25) of the brothers and 11% (3/28) of the sisters had AAAs (23). Eleven additional sibs were found to have a wide aorta (2.5 to 3 cm) (23).

Ultrasound screening to detect AAAs has been used in several studies of patients with hypertension (28–30), peripheral vascular disease (31–34), and coronary artery disease (35). Efforts have been made to carry out wide population screening studies to detect AAAs before they rupture (6,32,35–43) because of the relatively high incidence of AAAs in the general population, the dramatic nature of the disease in that it often leads to sudden death of the patient, and a survival rate of greater than 90% after elective surgery for AAA. All the ultrasonography screening studies of large populations have been reported in the United Kingdom. A total of 7773 men and 5346 women have been examined by ultrasonography in the above-mentioned studies (6,28–43). About 8% (599) of the men and 1.3% (70) of the women were found to have AAAs (Table 1). In contrast, the prevalences were 3.4- and 4.8-fold greater when brothers and sisters of patients with AAA were screened.

Table 1 Results of Studies Using Ultrasonography to Detect Abdominal Aortic Aneurysms

	No. screened	Aneurysms found
Population screening[a]		
Unrelated men	7,733	599 (7.7%)
Unrelated women	5,346	70 (1.3%)
Unspecified	826	26 (3.1%)
Total	13,905	695 (5.0%)
Screening of relatives of AAA patients[b]		
Brothers	77	20 (26%)
Sisters	95	6 (6.3%)
Unspecified	103	7 (6.8%)
Total	275	33 (12%)

[a]Data from Refs. 28–43.
[b]Data from Refs. 20–23.

IDENTIFICATION OF FAMILIAL AORTIC ANEURYSMS USING DNA TECHNIQUES

DNA Linkage Analysis

There are two general approaches using DNA techniques that can be used to identify the gene harboring the mutations causing genetic diseases. The first approach is to carry out linkage studies using markers that have been mapped to a particular locus on the genome and are used to test whether any of the markers is coinherited with aortic aneurysms in families. Linkage can be established by analyzing large families. There are, however, several problems in the approach when studying a late-onset disease such as aortic aneurysms. It is rare to find large index families that have more than two generations of living, affected members so that pertinent samples can be obtained. In addition, diagnosis in young individuals is a problem, since the data on the incidence of aneurysms in the population indicate that few individuals develop aneurysms before the age of 50 and that the incidence increases with age.

Figure 2 shows a typical family with aortic aneurysms. The father (in generation I) died of a ruptured aortic aneurysm when he was about 60 years old. Three of his sons have subsequently been operated on for AAAs; two other sons died suddenly, but no autopsy was carried out and the possibility exists that these two sons also had aneurysms. Each of the three brothers has several children. None of them shows any signs of aortic aneurysms, but they are still too young (30 to 40 years old) to be definitively diagnosed as unaffected individuals.

To date, no linkage study that would involve a genome wide search in families with aortic aneurysms has been reported.

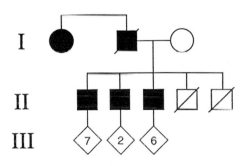

Figure 2   A family with aortic aneurysms. The black squares indicate males who were diagnosed with abdominal aortic aneurysms. The black circle indicates a female who has an abdominal aortic aneurysm.

Candidate Gene Approach

The second approach, the candidate gene approach, involves the analysis of a candidate gene for absence or presence of mutations. The modern techniques of molecular biology have made the approach feasible (see Ref. 44). For example, detailed DNA sequencing analysis of the gene for type III procollagen of 55 individuals from 51 unrelated families revealed a mutation in obligatory glycine (see Ref. 45) in two unrelated patients (46,47). The substitution of arginine for a glycine at amino acid positions 136 and 619 in type III procollagen is likely to disrupt the triple-helical structure of the protein and make the protein less stable (see Ref. 45). The results indicated that mutations in the triple-helical domain of type III procollagen are the cause of about 4% (2/51) of aortic aneurysms. In addition, the results suggest that mutations in the promoter region or other control regions of the gene for type III procollagen are not a common cause for aneurysms, since at least 40 of the 50 patients studied had mRNA that was derived from both alleles of the gene.

From a theoretical perspective, DNA sequencing provides a complete record of all of the bases that comprise a gene or its coding sequence. Therefore, DNA sequencing will lead to the detection of all mutations in a target region. Furthermore, it makes it possible to exclude definitively the region from those that possibly contain mutations. Our results demonstrate that sequencing has become a feasible approach not only to establish rapidly and definitively whether a candidate gene harbors the mutations causing a disease phenotype, but also to determine what fraction of affected individuals have a mutation in the particular candidate gene. This approach could, therefore, be used to establish whether other genes that have been suggested as candidate genes, such as the genes for fibrillin (48), elastin (49), collagenase (50), elastase (51), and tissue inhibitor of metalloproteinase (52), harbor mutations causing aortic aneurysms.

Benefits of DNA Analysis in the Future

Although we do not yet know the location of a gene or genes harboring the genetic defect responsible for the majority of aortic aneurysms, it is evident from the family studies mentioned above that aortic aneurysms are a genetic disorder. Aortic aneurysms frequently go undetected until rupture and are therefore a significant cause of mortality. The mortality from aneurysms that are repaired electively is low (1 to 7%, depending on the study cited), whereas the mortality from ruptured aneurysms is as high as 90%. Once a DNA test becomes available, it would be possible to use a small blood sample or even buccal smear or saliva to test whether a person has the gene defect (46). If he had the defect, he would be at increased risk to develop aortic aneurysms, but he could be offered an annual ultrasonography to monitor the aorta. If dilatation was detected, an elective surgery could be planned to repair the aneurysms. If he did not carry the gene defect, no ultrasonography would be necessary. We firmly believe that such "preventive DNA testing" would have enormous impact on the quality of life of our patients, and many lives could be saved this way.

ACKNOWLEDGMENTS

The studies on familial aortic aneurysms in our laboratory are supported by National Institutes of Health grant HL45996.

REFERENCES

1. Johnston KW, Rutherford RB, Tilson MD, Shah DM, Hollier L, Stanley JC. Suggested standards for reporting on arterial aneurysms. J Vasc Surg 1991; 13:444–450.
2. Crawford ES, Hess KR. Abdominal aortic aneurysm. N Engl J Med 1989; 321: 1040–1042.
3. Silverberg E, Lubera J. Cancer Statistics. New York: American Cancer Society, 1983.
4. National Center for Health Statistics. Vital Statistics of the US, 1984. Vol 2. Mortality, Part A. Hyattsville, MD: US Department of Health and Human Services, 1987.
5. Bergqvist D, Bengtsson H. Ökat antal patienter dör av bukaorta-aneurysm. Ökad diagnostisk skärpa krävs. Läkartidn 1986; 83:3010–3012.
6. Collin J. Screening for abdominal aortic aneurysms. Br J Surg 1985; 72:851–852.
7. Rantakokko V, Havia T, Inberg MV, Vänttinen E. Abdominal aortic aneurysms: A clinical and autopsy study of 408 patients. Acta Chir Scand 1983; 149:151–155.
8. Roberts WC. Pathology of arterial aneurysms. In: Bergan JJ, Yao JST, eds. Aneurysms: Diagnosis and Treatment. New York: Grune & Stratton, 1982:17–42.
9. Dzau VJ, Creager MA. Diseases of the aorta. In: Wilson JD, Braunwald E, Isselbacher KJ, Petersdorf RG, Martin JB, Fauci AS, Root RK, eds. Harrison's Principles of Internal Medicine. 12th ed. New York: McGraw-Hill, 1991:1015–1018.
10. DePalma RG, Sidawy AN, Giordano JM. Associated aetiological and atherosclerotic risk factors in abdominal aneurysms. In: Greenhalgh RM, Mannick JA, Powell JT, eds. The Cause and Management of Aneurysms. London: WB Saunders, 1990:37–46.
11. Reilly JM, Tilson MD. Incidence and etiology of abdominal aortic aneurysms. Surg Clin North Am 1989; 69:705–711.
12. Clifton MA. Familial abdominal aortic aneurysms. Br J Surg 1977; 64:765–766.
13. Tilson MD, Seashore MR. Human genetics of the abdominal aortic aneurysm. Surg Gynecol Obstet 1984; 158:129–132.
14. Tilson MD, Seashore MR. Fifty families with abdominal aortic aneurysms in two or more first-order relatives. Am J Surg 1984; 147:551–553.

15. Norrgård Ö, Rais O, Ängquist KA. Familial occurrence of abdominal aortic aneurysms. Surgery 1984; 95:650-656.

16. Johansen K, Koepsell T. Familial tendency for abdominal aortic aneurysms. JAMA 1986; 256:1934-1936.

17. Powell JT, Greenhalgh RM. Multifactorial inheritance of abdominal aortic aneurysm. Eur J Vasc Surg 1987; 1:29-31.

18. Cole CW, Barber GG, Bouchard AG, et al. Abdominal aortic aneurysm: consequences of a positive family history. Can J Surg 1989; 32:117-120.

19. Darling RC III, Brewster DC, Darling RC, LaMuraglia GM, Moncure AC, Cambria RP, Abbott WM. Are familial abdominal aortic aneurysms different? J Vasc Surg 1989; 39-43.

20. Bengtsson H, Norrgärd Ö, Ängquist KA, Ekberg O, Öberg L, Bergqvist D. Ultrasonographic screening of the abdominal aorta among siblings of patients with abdominal aortic aneurysms. Br J Surg 1989; 76:589-591.

21. Collin J, Walton J. Is abdominal aortic aneurysm familial? Br Med J 1989; 299: 493.

22. Webster MW, Ferrell RE, St. Jean PL, Majumder PP, Fogel SR, Steed DL. Ultrasound screening of first-degree relatives of patients with an abdominal aortic aneurysm. J Vasc Surg 1991; 13:9-14.

23. Adamson J, Powell JT, Greenhalgh RM. Selection for screening for familial aortic aneurysms. Br J Surg 1992; 79:897-898.

24. Majumder PP, St Jean PL, Ferrell RE, Webster MW. On the inheritance of abdominal aortic aneurysm. Am J Hum Genet 1991; 48:164-170.

25. Kuivaniemi H, Tromp G, Prockop DJ. Genetic causes of aortic aneurysms: Unlearning at least part of what the textbooks say. J Clin Invest 1991; 88:1441-1444.

26. Norrgård Ö. Looking for the familial connection in aortic aneurysm. In: Greenhalgh RM, Mannick JA, Powell JT, eds. The cause and management of aneurysms. London: WB Saunders, 1990:29-36.

27. Johnston KW, Scobie TK. Multicenter prospective study of nonruptured abdominal aortic aneurysms. I. Population and operative management. J Vasc Surg 1988; 7:69-81.

28. Twomey A, Twomey EM, Wilkins RA, Lewis JD. Unrecognized aneurysmal disease in male hypertensive patients. Br J Surg 1984; 71:307-308.

29. Lindholm L, Ejlerstsson G, Forsberg L, Norgren L. Low prevalence of abdominal aortic aneurysm in hypertensive patients: A population-based study. Acta Med Scand 1985; 218:305-310.

30. Allen PIM, Gourevitch D, McKinley J, Tudway D, Goldman M. Population screening for aortic aneurysms. Lancet 1987; ii:736-737.

31. Cabellon S, Moncrief CL, Pierre DR, Cavanaugh DG. Incidence of abdominal aortic aneurysms in patients with atheromatous arterial disease. Am J Surg 1983; 146:575-576.

32. Allardice JT, Allwright GJ, Wafula JMC, Wyatt AP. High prevalence of abdominal aortic aneurysm in men with peripheral vascular disease: screening by ultrasonography. Br J Surg 1988; 75:240-242.

33. Bengtsson H, Ekberg O, Aspelin P, Källerö S, Bergqvist D. Ultrasound screening

of the abdominal aorta in patients with intermittent claudication. Eur J Vasc Surg 1989; 3:497–502.

34. Berridge DC, Griffith CDM, Amar SS, Hopkinson BR, Makin GS. Screening for clinically unsuspected abdominal aortic aneurysms in patients with peripheral vascular disease. Eur J Vasc Surg 1989; 3:421–422.

35. Lederle FA, Walker JM, Reinke DB. Selective screening for abdominal aortic aneurysms with physical examination and ultrasound. Arch Intern Med 1988; 148:1753–1756.

36. Scott RAP, Ashton H, Sutton GLJ. Ultrasound screening of a general practice population for aortic aneurysm. Br J Surg 1986; 73:318.

37. Thurmond AS, Semler HJ. Abdominal aortic aneurysm: incidence in a population at risk. J Cardiovasc Surg 1986; 27:457–460.

38. Collin J, Araujo L, Lindsell D. Screening for abdominal aortic aneurysm. Lancet 1987; ii:736–737.

39. Collin J, Araujo L, Walton J, Lindsell D. Oxford screening programme for abdominal aortic aneurysm in men aged 65 to 74 years. Lancet 1988; ii:613–615.

40. O'Kelly TJ, Heather BP. General practice-based population screening for abdominal aortic aneurysms: a pilot study. Br J Surg 1989; 76:479–480.

41. Collin J, Araujo L, Walton J. A community detection program for abdominal aortic aneurysm. Angiology 1990; 41:53–58.

42. Bengtsson H, Bergqvist D, Ekberg O, Janzon L. A population based screening of abdominal aortic aneurysms (AAA). Eur J Vasc Surg 1991; 5:53–57.

43. Scott RAP, Wilson NM, Ashton HA, Kay DN. Is surgery necessary for abdominal aortic aneurysm less than 6 cm in diameter? Lancet 1993; 342:1395–1396.

44. Tromp G, Kuivaniemi H, Identification of familial aortic aneurysms using DNA technique. In: Yao JST, Pearce WH, eds. Technologies in Vascular Surgery. Philadelphia: WB Saunders, 1992:27–39.

45. Kuivaniemi H, Tromp G, Prockop DJ. Mutations in collagen genes: causes of rare and some common diseases in humans. FASEB J 1991; 5:2052–2060.

46. Kontusaari S, Tromp G, Kuivaniemi H, Romanic AM, Prockop D. A mutation in the gene for type III procollagen (COL3A1) in a family with aortic aneurysms. J Clin Invest 1990; 86:1465–1473.

47. Tromp G, Wu Y, Prockop DJ, Madhatheri SL, Kleinert C, Earley JJ, Zhuang J, Norrgård Ö, Darling RC, Abbott WM, Cole CW, Jaakkola P, Ryynänen M, Pearce WH, Yao JST, Majamaa K, Smullens SN, Gatalica Z, Ferrell RE, Jimenez SA, Jackson CE, Michels VV, Kaye M, Kuivaniemi H. Sequencing of cDNA from 50 unrelated patients reveals that mutations in the triple-helical domain of type III procollagen are an infrequent cause of aortic aneurysms. J Clin Invest 1993; 91:2539–2545.

48. McKusick VA. The defect in Marfan syndrome. Nature (Lond) 1991; 352:279–281.

49. Menashi S, Campa JS, Greenhalgh RM, Powell JT. Collagen in abdominal aortic aneurysms: typing, content and degradation. J Vasc Surg 1987; 6:578–582.

50. Busuttil RW, Abou-Zamzam AM, Machleder HI. Collagenase activity of the human aorta: a comparison of patients with and without abdominal aortic aneurysms. Arch Surg 1980; 115:1373–1378.

51. Cannon D, Read R. Blood elastolytic activity in patients with aortic aneurysms. Ann Thorac Surg 1982; 34:10-15.
52. Brophy CM, Sumpio B, Reilly JM, Tilson MD. Decreased tissue inhibitor of metalloproteinases (TIMP) in abdominal aortic aneurysmal tissue: a preliminary report. J Surg Res 1991; 50:653-657.

8

Marfan Syndrome

Harry C. Dietz
The Johns Hopkins University School of Medicine
Baltimore, Maryland

INTRODUCTION

Marfan syndrome is a systemic disorder of connective tissue with autosomal dominant inheritance and a prevalence of approximately 1 in 10,000 individuals. It is estimated that between 15 and 25% of cases occur in the absence of a family history, presumed new mutations due to parental germ line defects (1). There are no documented examples of skipped generations (high penetrance), but marked interfamilial—and, to a lesser extent, intrafamilial—variability in the distribution and severity of tissue involvement is the rule. The lack of a sensitive or specific genetic or biochemical test for the disorder, compounded by a high rate of new mutation and marked clinical variability, has frustrated the accurate diagnosis of equivocal cases and may have perpetuated an underestimation of disease prevalence.

The classic manifestations in Marfan syndrome are found in the skeletal, ocular, and cardiovascular systems (1). Long-bone overgrowth is the hallmark of musculoskeletal involvement (2). While the mechanism remains unclear, abnormally lax connective-tissue components of the periosteum might allow excessive linear bone growth. Manifestations include tall stature, mainly due to excessive leg length; arachnodactyly; anterior chest deformity; scoliosis; highly arched and narrow palate with tooth-crowding; joint laxity; and variable forms of foot and leg deformity. Sequelae may include joint dislocation, chronic pain, difficulty with ambulation, and cardiopulmonary impairment.

The eye pathology is characterized by early (preadolescent) and severe myopia and dislocation of the ocular lenses (3). The ciliary zonules—the suspensory apparatus of the lens—are thinned, stretched, and often frag-

219

mented, leading to the characteristic upward dislocation of the lens in up to 80% of patients. While high myopia is often seen in isolation, ectopia lentis can predispose to secondary refractive defects, glaucoma, and retinal detachment.

Retrospective study in the early 1970s demonstrated that the majority of patients with Marfan syndrome died by the third decade of life due to cardiovascular complications (4). An early age of onset of heart disease was clearly correlated with a less favorable outcome. The majority of children and nearly 80% of adults with Marfan syndrome show dilatation of the aorta. Aortic disease can be seen anywhere along the course of the vessel but classically begins at the sinuses of valsalva and extends to the proximal ascending arch (5). Aortic regurgitation is a secondary event that manifests stretching of the commissures at the sinotubular junction. Dissection of the adult aortic root rarely occurs before the maximal dimension at the sinuses has reached 55 to 60 mm. In childhood, such factors as the aortic root dimension in relation to body surface area, family history of early dissection, and the rate of dilatation over time must be considered. Aortic regurgitation can lead to myocardial dysfunction and can accelerate aortic dilatation by increasing stroke volume and hence wall stress. Dysplasia and prolapse of the mitral valve, with or without regurgitation, is as common as aortic dilatation, occurring in approximately three-quarters of patients. Sequelae include myocardial dysfunction, dysrhythmia, thromboembolic events, and bacterial endocarditis. Less common cardiovascular features include tricuspid valve prolapse, pulmonary artery dilatation, and primary cardiomyopathy.

MANAGEMENT AND DIAGNOSIS

Effective management of the cardiovascular manifestations of Marfan syndrome relies on frequent clinical and echocardiographic assessment, institution of medical therapy to slow the rate of aortic dilatation, and prophylactic surgical intervention to replace the diseased aorta prior to dissection. Current recommendations call for yearly examination for adults with aortic root dimensions of less than 50 mm or for children with slowly progressive disease. Individuals with more advanced or aggressive disease should be followed at more frequent intervals. Prophylactic surgery is indicated once the aortic root exceeds 55 mm in adults; precise guidelines have not been established for children, but rapid progression, measurements that exceed twice the expected size for body size, or ventricular dysfunction should precipitate early intervention. Surgery for aortic disease generally involves replacement of the aortic root with a valved synthetic conduit; a homograft is commonly used for children in order to avoid the complications associated with anticoagulation (6,7). Advanced mitral valve pathology with consequent left ventricular dilatation and dysfunction is the most common indication for surgery in childhood (8).

It has now been demonstrated that the use of β-adrenergic blockade to slow the rate of aortic dilatation and delay the need for surgery has the desired effect in at least a subpopulation of patients (9). Another mainstay of medical management is the prescription of antibiotic prophylaxis for dental work or any other procedure expected to contaminate the bloodstream with bacteria.

Other less common manifestations in the Marfan syndrome include spontaneous pneumothorax, striae distensae, dural ectasia (with rare nerve impingement), protrusio acetabulae, and mild learning disability with or without hyperactivity. These associated features can be quite useful for clinical diagnosis when classic features in other organ systems are mild or lacking. In the absence of a family history, current diagnostic guidelines require involvement of the skeleton and at least two other organ systems with at least one "major" manifestation (aortic dilatation, ectopia lentis, or dural ectasia) (10). In the presence of an unequivocally affected first-degree relative, features must only involve two organ systems; a "major" manifestation is preferred but not required.

GENETICS

The search for the site of genetic defect in the Marfan syndrome was long and arduous. The pattern of tissue involvement suggested that the disease manifests a primary defect in the elastic fiber system, a hypothesis that was supported by early histological studies demonstrating elastic fiber abnormalities in tissues from Marfan syndrome patients (11). Although inconclusive and later refuted, preliminary evidence was also presented for a primary abnormality in β-glucuronidase, decorin, collagen, and hyaluronic acid metabolism (12-18). Upon successful cloning and mapping of the tropoelastin gene, this locus was excluded as the site of primary defect.

Lacking additional biochemical data for further advancement of the "classical" approach for disease gene identification, efforts focused on a positional cloning strategy to localize the Marfan gene. This method makes use of localized sites of normal sequence variation, termed polymorphic markers, that are scattered throughout the human genome. Such landmarks can be used to track the inheritance pattern of specific regions of DNA through families displaying a phenotype of interest. If the disease phenotype is consistently coinherited with a specific copy of a region of DNA, then it is likely that the disease gene is close by. Linkage analysis relies on the propensity for recombination to occur between two fixed points in the genome. The greater the physical distance between loci, the greater the frequency of recombination and the greater the chance that the two markers will be separated during meiosis and therefore will not appear in the same offspring.

Using large and well-characterized families, it was reasoned that the identification of a polymorphic marker allele that showed cosegregation with the Marfan phenotype would allow precise localization of the defective gene. Studies using polymorphic markers that mapped to either candidate genes or anonymous DNA segments excluded approximately 75% of the human genome as the site for the Marfan gene (19). In 1990, Peltonen and colleagues (20) mapped the disease locus to the long arm of chromosome 15 in a small group of Finnish families. Additional studies confirmed the chromosome 15 assignment in a large number of families with diverse ethnic origins, and established a localization of the disease gene to 15q15-q21.3 near anonymous marker D15S1 (21). Subsequent analysis revealed that virtually all families segregating the classic Marfan phenotype demonstrated linkage to chromosome 15 markers, suggesting that defects in a gene in close physical proximity to marker D15S1 accounted for most, if not all, cases of Marfan syndrome (20–24). With a location in hand, two approaches could be taken to identify the disease gene. Either genes that had already been localized to the region could be scrutinized (candidate gene approach) or new genes with coincident map positions could be found (positional cloning).

By this time, more information was becoming available about a 350 kD glycoprotein called fibrillin (25). First identified by Sakai and colleagues, fibrillin was found to be a major constitutive element of the 10 nanometer extracellular microfibril. The microfibril was known to be an abundant component of all elastic tissues, and localization of microfibrils to the maturing ends of amorphous elastic fibers suggested that they play a regulatory role in organizing the deposition of tropoelastin molecules (26). Microfibrils were also known to be the predominant component of the ciliary zonules. This information suggested that fibrillin or other yet uncharacterized microfibrillar-associated proteins might be defective in patients with the Marfan syndrome. To test this hypothesis, Hollister and colleagues (27) examined fibrillin metabolism in cell culture and tissue samples using fluorescent-labeled monoclonal antibodies. Quantitative and qualitative abnormalities of extracellular fibrillin were detected in the majority of patient samples, and these defects segregated with the disease phenotype within families (28). This observation was not entirely specific to patients with Marfan syndrome, however, as evidenced by the erroneous assignment of "affected" status to selected patients with other connective tissue disorders, including pseudoxanthoma elasticum, homocystinuria, ectodermal dysplasia, and a variant of epidermolysis bullosa.

Next, McGookey-Milewicz and colleagues (29) examined fibrillin processing in Marfan fibroblast cell lines using quantitative pulse-chase methods.

These experiments allowed determination of the level of fibrillin synthesis and observation of the fate of radiolabeled fibrillin monomers. Pro-fibrillin (350 kD) is normally secreted within 4 hours of synthesis, is processed to a smaller form (320 kD), and ultimately aggregates with itself and perhaps with other microfibril-associated proteins during incorporation into the extracellular matrix as microfibrils. In contrast, 7 of 26 samples (27%) from Marfan syndrome patients demonstrated approximately half-normal synthesis of monomeric fibrillin, and 7 of 26 (27%) showed normal synthesis but impaired intracellular trafficking of the monomers as evidenced by delayed secretion of the protein. The majority of these cell lines also demonstrated a significant reduction in the amount of fibrillin that was incorporated into the extracellular matrix. Fibroblasts from a third subset of patients (8 of 26; 31%) showed normal synthesis and secretion of fibrillin, but an isolated defect in matrix utilization. Finally, in samples from 4 of 26 patients (15%), there was no evident abnormality in fibrillin synthesis or processing.

Although these data suggested that fibrillin metabolism was altered in Marfan syndrome patients, it remained to be determined whether the fibrillin gene was the site of primary defect. Alternatively, defects in another gene and protein could have a secondary effect on fibrillin synthesis and processing. Interestingly, whatever "class" of defect that was detected in a proband was also evident in all other affected family members. This Mendelian segregation of the same cellular phenotype within families, with variation between families, suggested that multiple defects would be found as the basis of Marfan syndrome—probably different mutations in the same gene (allelic variants)—and that that mutant genotype has a strong correlation with cellular phenotype.

Characterization of a large portion of fibrillin coding sequence revealed a remarkably redundant organization (30–33). The gene encodes stretches of tandem domains with homology to a domain first described in epidermal growth factor (EGF-like domains). These long stretches of EGF-like domains are occasionally interrupted by a domain with homology to transforming growth factor-β-1 binding protein. Both of these domains have a high cysteine content (six and eight residues, respectively), an observation in keeping with the evidence that fibrillin monomers fold and interact via disulfide linkage (34).

Identification of intragenic fibrillin polymorphisms allowed characterization of the linkage relationship between fibrillin, markers at the previously established Marfan locus, and the disease phenotype. Linkage analysis revealed no recombination between fibrillin and the Marfan disease gene (30,35). Simultaneous work by Magenis and colleagues (36) mapped the fibrillin gene to 15q21.1 using in situ hybridization methods (36). In addition, during identification of fibrillin clones, other fibrillin-like sequences were characterized and mapped to chromosome 5 (30). The gene on chromosome 15, now termed FBN1, was linked to the Marfan phenotype, while the related gene on chromosome 5 (FBN2) was genetically linked to congenital contractural arachnodactyly, a disorder with significant clinical overlap with Marfan syndrome, including all typical skeletal features but no ocular or aortic involvement (30).

Final demonstration that FBN1 gene defects caused Marfan syndrome awaited the identification of disease-producing mutations in affected individuals. Short overlapping regions of fibrillin coding sequence from patients and control individuals were amplified using the polymerase chain reaction (PCR). These short fragments of DNA were then screened for a sequence alteration using a technique called single-strand conformation polymorphism analysis (Figure 1). This method makes use of the facts that the sequence of a fragment of single-stranded DNA will influence how the DNA folds, that even a single base change can alter the folding pattern, and that the conformation of folded

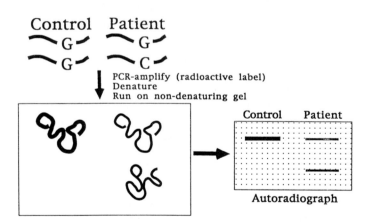

Figure 1   Single-strand conformation polymorphism (SSCP) analysis for the detection of sequence variants within short fragments of DNA that have been amplified and radioactively labeled using the polymerase chain reaction (PCR). This method makes use of the facts that even a single base substitution can alter the folding pattern of single-stranded DNA and that the conformation of a fragment of DNA will influence its rate of migration through a gel when exposed to an electrical current. The patient shown here carries a G-to-C substitution in one copy of the fibrillin gene. DNA amplified from the mutant copy folds abnormally, and therefore migrates faster through the gel. When the gel is exposed to X-ray film, an abnormally migrating band can be detected when compared to the pattern resulting from analysis of a control individual who is homozygous for G at the point of mutation.

DNA determines how quickly it will migrate through a gel when exposed to an electrical current. Once an abnormally migrating fragment is identified, this region of DNA can be sequenced to determine the basis for abnormal configuration. In this manner a guanine-to-cytosine change was identified in one copy of the fibrillin gene from a patient with neonatal presentation of severe and rapidly progressive disease (35). This patient had no family history of Marfan syndrome and the single base change was not carried by either parent. In addition, the identical alteration was found as a new mutation in an unrelated patient with severe and sporadic disease that presented at birth. This mutation causes the substitution of proline (P) for arginine (R) in one of the FBN1 EGF-like domains at codon 1137 (R1137P). It is known that proline residues introduce bends in proteins and it was therefore predicted that this mutation would alter the secondary structure of the domain, perhaps with consequence for the folding of a larger portion of the monomer and hence the function of the protein. Thus, a combination of the characterization of the clinical disorder complemented by biochemical inquiry, positional cloning, and mutation detection culminated in the knowledge that FBN1 gene defects are the predominant, if not the sole cause of Marfan syndrome.

A firm understanding of etiology immediately allowed investigation of several puzzling clinical issues. For example, the molecular basis for marked intrafamilial clinical variability remained to be determined. Some light was shed on this situation with the identification and characterization of a second mutation (37). The substitution of serine for cysteine at codon 2307 (numbering reflects cloning of the entire FBN1 coding sequence) in an EGF-like domain was found to segregate with disease in a large kindred with classic Marfan syndrome. EGF-like domain cysteine residues are known to interact via intramolecular disulfide linkage such that the first bonds to the third, the second to the fourth, and the fifth to the sixth to form an antiparallel β-pleated sheet conformation (38). Replacement of one cysteine would have an obligate effect on confirmation of the domain. Interestingly, the phenotype observed in family members varied widely with respect to age of onset, tissue distribution, and severity of manifestations (37). The only two individuals who shared the identical phenotype were monozygotic twins. These data suggested the presence of genetic modifiers that influenced the clinical expression of a FBN1 gene defect. Such modifiers would vary with genetic background, explaining divergent phenotypes in and between families.

Recent experience offers a second molecular explanation for intrafamilial clinical variability. A panel of four polymorphisms within the FBN1 gene was typed in families. Observation of combinations of FBN1 polymorphic variants that were inherited as a unit (haplotypes) allowed identification of distinct copies of the fibrillin gene (FBN1 alleles) and observation of their inheritance pattern in families. The goal was to discriminate between the many normal copies and the single (disease-producing) abnormal copy in given families. One would expect that the same copy of the fibrillin gene would be inherited by all affected individuals in a family, and that none of the unaffected family members would carry this allele.

In Figure 2 the different haplotypes that define distinct copies of the fibrillin gene are assigned individual numbers. The proband (III-3) first presented with spontaneous pneumothorax. Further evaluation revealed myopia, dural ectasia, foot deformity, and mild aortic dilatation, features shared with her father, uncle, and cousin. The family requested presymptomatic diagnosis for the brother of the proband (III-2). Haplotype segregation analysis revealed that the disease was coinherited with the number 2 allele. III-2 did not inherit this copy of the fibrillin gene from his affected father; he is therefore predicted to be unaffected.

The two families illustrated in Figure 3 contained selected individuals with classic Marfan syndrome and other members with a related but apparently milder connective-tissue phenotype that included typical skeletal abnormality, mitral valve prolapse, early myopia, joint laxity, and/or striae distensae but no "major" manifestations. Interestingly, because of the presence of an unequivocally affected first-degree relative, all the members with the milder disorder satisfied the diagnostic criteria for Marfan syndrome and all were counseled and treated accordingly. The proband (II-6) in family A with classic Marfan syndrome passed the same copy of FBN1 (allele 4) to each of her classically affected daughters. This allele was not carried by family members with the

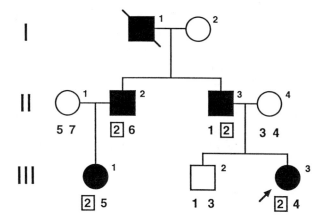

Figure 2 Haplotype segregation analysis in a family segregating classic Marfan syndrome (blackened symbols). Haplotypes were generated after typing of each individual for the four intragenic FBN1 polymorphisms. Disease in this family is coinherited with the number 2 allele of the fibrillin gene (boxed).

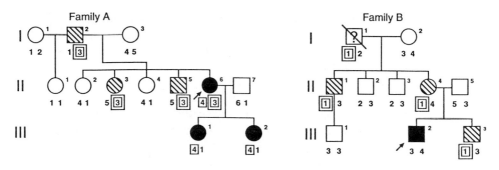

Figure 3 Haplotype segregation analysis in two families segregating both Marfan syndrome (blackened symbols) and a related but milder connective-tissue phenotype (hatched symbols). In family A, Marfan syndrome is inherited on the number 4 allele (boxed), while the milder disorder cosegregates with the number 3 allele (double box). In that multiple unaffected individuals carry the number 4 allele, it appears that a new mutation has occurred on this copy of the fibrillin gene in the proband (arrow). In family B, a mild connective-tissue disorder cosegregates with the number 1 allele (double box), a copy of the fibrillin gene that is not carried by the proband with more severe disease involving the aorta. SSCP analysis allowed the detection of a new mutation in III-2. (Adapted from Ref. 39.)

milder phenotype; rather, they all shared a different copy (allele 3). Allele 4 originated from the unaffected mother and is also carried by two of the proband's unaffected sisters. These data suggest that the proband carries a new mutation on allele 4 that causes Marfan syndrome, and that she passed this allele and disease to her affected daughters. This occurred upon the genetic background for a milder, albeit related, connective-tissue phenotype that is not associated with aortic disease or ectopia lentis. The data are consistent with, but not conclusive for, the possibility that the milder phenotype is also caused by a FBN1 gene defect. In family B, members with a milder phenotype all carry FBN1 allele 1. This allele is not carried by the proband (III-2), who has a distinctly more severe phenotype involving the aorta. A new mutation in FBN1 was identified in III-2 that is not carried by any other family members (40).

These results demonstrate that the occurrence of phenotypically related but etiologically distinct disorders in the same kindred can be the basis for phenotypic variability in selected families. Molecular analysis should allow for stratification of cardiovascular risk within families. Implications include the need for individualized counseling and management, and for judicious caution when interpreting linkage, natural history, or therapeutic outcome studies. It appears that the diagnostic criteria for Marfan syndrome should be modified to require a "major" manifestation even in the presence of an unequivocal family history.

To date, a total of 34 mutations causing Marfan syndrome have been reported in the literature (35,37,40–47). At this early stage in the correlation of mutant genotype to clinical phenotype, certain patterns are emerging. First, with the exception of the first mutation, all have been specific to single families. This observation probably reflects a relatively high rate of new mutation and impaired reproductive fitness in Marfan syndrome due to early morbidity and mortality. Second, mutations have been identified with approximately equal frequency along the length of the gene. There is no definitive correlation between the position of a mutation and the severity of the resultant phenotype. One possible caveat to this statement stems from the observed clustering of mutations associated with neonatal presentation of severe disease at the beginning of a central region of the cDNA sequence that encodes a long stretch of EGF-like domains (47). It must be considered that the number of cases is small and that other mutations in this region have been associated with nonneonatal presentation. A definitive correlation awaits further experience.

An early and continuing observation is that the majority of identified mutations causing Marfan syndrome substitute cysteine residues in EGF-like domains (37,42,46,47). The study of other proteins has demonstrated that EGF-like domains can bind calcium, an event that may be necessary to stabilize secondary structure, to resist the activity of proteases, and to promote protein-to-protein interactions (48). As shown in Figure 4, the requirements for calcium binding include an antiparallel β-sheet conformation, as dictated by the six predictably spaced cysteine residues, and the presence of aspartic acid (D) and asparagine (N) residues at critical locations (49–50). Forty-three of FBN1 EGF-like domains satisfy the calcium binding consensus. The many mutations that directly substitute cysteine residues would cause an obligatory alteration in conformation and hence calcium binding. Interestingly, all four naturally occurring mutations in EGF-like domains that do not substitute cysteines replace the D or N residues that are believed essential for calcium binding (40,44,47).

While these data suggest that calcium binding to EGF-like domains is critical to the normal function of fibrillin and that a majority of missense mutations causing Marfan syndrome may act by perturbation of this process, the precise role of disrupted domain conformation and altered calcium binding in the pathogenesis of this disorder remains a mystery. Recent quantitative pulse-chase studies examining fibrillin metabolism in cell lines with known mutant genotypes has begun to shed light on this matter (51). While all cell lines showed normal levels of fibrillin synthesis (relative to control lines), nearly all lines carrying mutations that substitute cysteine residues in EGF-like domains showed a dramatic delay in the secretion of fibrillin and reduced fibrillin deposition into the extracellular matrix. In

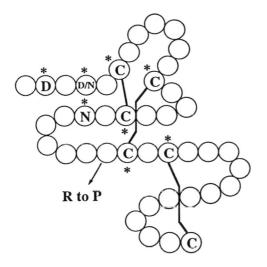

Figure 4 Structure of fibrillin EGF-like domain as modeled after the solution structure of native EGF. Disulfide linkages (bold lines) between the six predictably spaced cysteine (C) residues create an antiparallel β-pleated sheet conformation. The positions of the aspartic acid (D) and asparagine (N) residues that are believed critical for calcium binding are indicated. *Sites of naturally occurring mutations. The position of the only mutation that does not substitute a highly conserved residue—arginine (R) to proline (P)—is shown by the arrow.

contrast, mutations predicted to have an isolated effect on calcium binding do not delay secretion but also impair extracellular incorporation; this cellular phenotype was reproducible when control lines were deprived of calcium. These data suggest that proper folding of EGF-like domains is critical to normal intracellular trafficking of fibrillin. Delayed secretion could manifest failure of folding or the formation of abnormal intracellular protein–protein complexes. Both abnormal conformation and altered calcium binding lead to a failure of incorporation of fibrillin into the extracellular matrix. The mechanism remains unclear. Possibilities include impaired intermolecular interactions and increased susceptibility of abnormal monomers or multimers to the activity of proteases.

A broad second class of mutations are those predicted to result in the formation of shortened fibrillin monomers. This could result from deletion of a portion of the gene or from mutations that create a premature signal for the ribosome to stop making protein. The stop signals are encoded by the triplet base pairs TAG, TAA, or TGA. Frameshift mutations occur when insertion or deletion of nucleotides shifts the 3 base reading code that the ribosome machinery uses to determine which amino acid to add to an elongating peptide. Shifting of the reading frame generally leads to the creation of a premature termination codon shortly downstream. In contrast, nonsense mutations create a premature termination signal by substituting a single nucleotide with another (e.g., TAT to TAG). All types of defects can be associated with classic Marfan syndrome (40,41,43,45–47). In that microfibrillogenesis is believed to involve an orderly aggregation of fibrillin monomers, perhaps in association with other proteins, it could be postulated that the presence of shortened fibrillin polypeptides could interfere with intermolecular interactions. The abnormal structure of these aggregates could directly impair their functional integrity or they could be predisposed to untimely degradation. In one circumstance a genomic deletion causes removal of three EGF-like repeats (41). Since the deleted region is an equal multiple of three nucleotides in length, this mutation does not create a frameshift. A stable truncated peptide, in an amount apparently equal to that expressed from the normal allele, was detected from a fibroblast cell line carrying this mutation.

Additional insight into the pathogenesis of Marfan syndrome has come from the study of mutant alleles that encode premature termination codons (PTCs). It has been demonstrated that RNA molecules that contain premature stop signals are prone to early degradation (52,53). The first such mutation identified in FBN1 involved a four-base-pair insertion (40). This frameshift mutation leads to the formation of a PTC and is associated with a mutant RNA level that is only 6% of that observed from the normal allele. Interestingly, unlike all patients with missense mutations and normal RNA levels, this patient had an extremely mild phenotype that failed to meet the established diagnostic criteria for Marfan syndrome. These data suggested a "dominant negative" pathogenesis for the Marfan syndrome. Such a model presumes that the presence of an abnormal protein that impairs processing of the normal product, rather than a relative deficiency of normal peptide, is central to expression of the disease phenotype (54).

The mild phenotype observed in this patient can be reconciled in either of two ways (40). First, the predicted truncated peptide from the mutant allele may be so dissimilar to the normal monomer that it fails to participate in multimer formation. Alternatively, the severely reduced mutant transcript level predicts very low levels of mutant peptide. Either situation predicts a preponderance of normal multimer and hence an abbreviated disease phenotype (Figure 5).

Additional information stems from the study of two additional reduced-RNA mutants and from quantitative biosynthetic study of cell lines carrying these mutant alleles. One patient with a nonsense mutation and another with a second frameshift mutation showed mutant RNA levels of 25% and 16%, respectively (40,43). Interestingly, both of these patients have a classic and severe form of Marfan syndrome. These data suggest that a relatively small

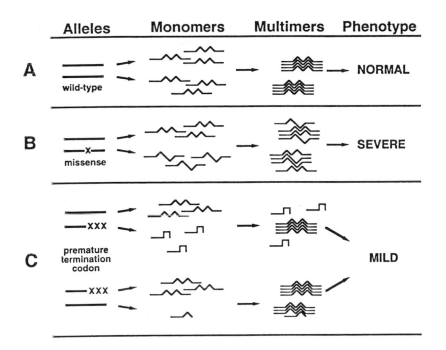

Figure 5 A dominant negative model for Marfan syndrome. (A) Two normal copies of the fibrillin gene leads to homogeneous populations of normal monomer and multimer, and hence a normal phenotype. (B) Missense mutations that substitute one amino acid for another lead to equal populations of normal and mutant monomer, a homogeneous population of abnormal multimer, and hence severe disease. (C) The mild phenotype observed in the patient with a premature termination codon and only 6% of the expected amount of mutant RNA can be reconciled in two ways. The truncated protein could be so dissimilar to the wild-type product that the two can no longer interact. Alternatively, the reduced RNA level predicts a very low amount of mutant monomer. Either situation predicts a preponderance of normal multimer and hence mild disease.

amount of mutant peptide can cause disease. Specifically, a critical threshold, between 6 and 16% of wild type levels, may be necessary to cause sufficient disruption of microfibrillar assembly in order to produce the classic phenotype. This apparent correlation between the amount of abnormal protein and disease severity is analogous to what has been observed with collagen gene defects, another multimeric protein system. Quantitative pulse-chase analysis of these cell lines provides data that are absolutely consistent with this dominant negative hypothesis (55). All three cell lines synthesize approximately half of the amount of fibrillin made by control lines. The mutant line with 6% mutant transcript, associated with a mild phenotype, deposits all synthesized fibrillin into the matrix. The cell lines carrying the other two reduced-transcript mutants, associated with a classic phenotype, deposit only a fraction of synthesized fibrillin, identical to the situation observed with missense mutations.

To further test the dominant negative hypothesis, the mutant allele associated with 16% mutant RNA levels was subcloned into a eukaryotic expression vector and was stably introduced into normal human and mouse fibroblasts. The cellular phenotype of cultured cells was then assayed by immunofluorescence using a monoclonal antibody to fibrillin. The isolated expression of the extreme amino terminus of human FBN1, upon a normal genetic background, was sufficient to recreate a Marfan-like cellular phenotype, including a marked decrease in the amount of fibrillin that was deposited into the extracellular matrix and disorganized and fragmented microfibrillar bundles (56).

CONCLUSION

These data suggest that the amino-terminal end of the molecule plays a critical role in the polymerization of fibrillin aggregates, in keeping with the previously predicted head-to-tail model of microfibrillar assembly (34). In addition, the fact that expression of a truncated human peptide in cells from another species is sufficient to disrupt the cellular phenotype has clear implication for the creation of an animal model of Marfan syndrome. It is plausible that introduction of a mutant allele into fertilized mouse pronuclei would be sufficient to recreate the clinical phenotype.

The medical significance of fibrillin may not be limited to its role in the pathogenesis of MFS. In fact, the FBN1 gene has been genetically linked to the isolated ectopia lentis phenotype with no recombination (57), and a mutation has been identified in a patient with the MASS phenotype, which includes mitral valve prolapse and borderline aortic dilatation without progression to dissection (40). It is reasonable to speculate that fibrillin gene defects may play an etiological role in patients with other isolated or atypical features of the Marfan phenotype, for example, mitral valve prolapse syndrome, idiopathic scoliosis, or annuloaortic ectasia.

To date, molecular studies of the Marfan syndrome and the FBN1 gene have resulted in a firm understanding of the molecular origins of the disorder, have begun to shed some light on the origins of clinical variability, and have contributed to a greater understanding of pathogenesis. In addition, direct clinical application has ensued with successful efforts at presymptomatic and prenatal diagnosis. The greatest challenges and rewards await attempts at a full understanding of pathogenesis and efforts to relieve the clinical burden of this disorder using novel therapeutic strategies. While correlation of mutant genotype to phenotype in patients has provided some insight, this approach is limited by a number of factors, including the low number of patients carrying the same mutation, the great diversity in genetic background (and of potential genetic modifiers) in and between affected families, the inability to study protein expression and interactions at different stages of early fetal development, a restricted ability to follow the natural history of Marfan syndrome serially using invasive diagnostic techniques, and the hardships associated with execution and interpretation of experimental therapeutic trials in a limited number of human subjects. In this light, the creation of a transgenic animal model of MFS will prove invaluable to future studies.

To gain greater knowledge of the biology of normal and mutant fibrillin, the protein and its interactions will need to be studied in vivo. The application of transgenic technology will provide insight into the biochemical and developmental significance of the normal protein and the consequence of expressing an altered gene in a system that mimics the physiological complexity of the human system. Such a model will provide a means to evaluate conventional medical therapies such as the optimal preparation, time of initiation, dosage, or route of administration of pharmacological agents that slow the rate of aortic root dilatation. Animal research will also be necessary to test novel gene-therapy strategies. For example, if the dominant negative hypothesis is confirmed, selective inhibition of expression from the mutant allele should help to alleviate the clinical burden carried by patients with Marfan syndrome.

REFERENCES

1. Pyeritz RE, McKusick VA. The Marfan syndrome: diagnosis and management. N Engl J Med 1979; 300:772-777.
2. Magid D, Pyeritz RE, Fishman EK. Musculoskeletal manifestations of the Marfan syndrome: radiologic features. Am J Roentgenol 1990; 155:99-104.
3. Maumenee IH. The eye in the Marfan syndrome. Trans Am Ophthalmol Soc 1981; 79:684-733.
4. Murdoch JL, Walker BA, Halpern BI, Kuzma JW, McKusick VA. Life expectancy and causes of death in the Marfan syndrome. N Engl J Med 1972; 286:804-808.
5. Roberts WC, Honig HS. The spectrum of cardiovascular disease in the Marfan syndrome: A clinico-morphologic study of 18 necropsy patients and comparison to 151 previously reported necropsy patients. Am Heart J 1982; 104:115-135.
6. Bentall HH, DeBono AA. A technique for complete replacement of the ascending aorta. Thorax 1987; 23:338-339.
7. Gott VL, Pyeritz RE, Cameron DE, Green PS, McKusick VA. Composite graft repair of Marfan aneurysm of the ascending aorta: results in 100 patients. Ann Thorac Surg 1991; 52:38-45.
8. Morse RP, Rockenmacher S, Pyeritz RE, Sanders SP, Bieber FR, Lin A, MacLeod

P, Hall B, Graham J M Jr. Diagnosis and management of infantile Marfan syndrome. Pediatrics 1990; 86:888–895.

9. Shores J, Berger KR, Murphy EA, Pyeritz RE. Progression of aortic dilatation and the benefit of long-term β-adrenergic blockade in the Marfan syndrome. N Engl J Med 1984; 330:1335–1341.

10. Beighton P, De Paepe A, Danks D, Finidori G, Gedde-Dahl T, Goodman R, Hall J, Hollister DW, Horton W, McKusick VA, Opitz, JM, Pope JM, Pyeritz RE, Rimoin DL, Sillence D, Spranger JW, Thompson E, Tsipouras P, Viljoen D, Winship I, Young I. International nosology of heritable disorders of connective tissue, Berlin 1986. Am J Med Genet 1988; 29:581–594.

11. Abraham PA, Perejda AJ, Carnes WH, Uitto J. Marfan syndrome. Demonstration of abnormal elastin in aorta. J Clin Invest 1982; 70:1245–1252.

12. Nakashima Y. Reduced activity of serum β-glucuronidase in Marfan syndrome. Angiology 1986; 37:576–580.

13. Pulkkinen L, Kainulainen K, Krusius T, Mäkinen P, Schollin J, Gustavsson KH, Peltonen L. Deficient expression of the gene coding for decorin in a lethal form of Marfan syndrome. J Biol Chem 1990; 265:17780–17785.

14. Boucek RJ, Noble NL, Gunja-Smith Z, Butler WT. The Marfan syndrome: a deficiency in chemically stable collagen cross-links. N Engl J Med 1981; 305:988–991.

15. Byers PH, Siegel RC, Peterson KE, Rowe DW, Holbrook KA, Smith LT, Chang Y, Fu JCC. Marfan syndrome: an abnormal a2 chain in type I collagen. Proc Natl Acad Sci USA 1981; 78:7745–7749.

16. Phillips CL, Shrago-Howe AW, Pinnell SR, Wenstrup RJ. A substitution at a non-glycine position in the triple-helical domain of pro-a2(I) collagen chains present in an individual with a variant of the Marfan syndrome. J Clin Invest 1990 86:1723–1727.

17. Appel A, Horwitz AL, Dorfman A. Cell-free synthesis of hyaluronic acid in Marfan syndrome. J Biol Chem 1979; 254:12199–12203.

18. Lamberg SI, Dorfman A. Synthesis and degredation of hyaluronic acid in the cultured fibroblasts of Marfan's disease. J Clin Invest 1973; 52:2428–2433.

19. Blanton SH, Sarfarazi M, Eiberg H, de Groote J, Farndon PA, Kilpatrick MW, Child AH, Pope FM, Peltonen L, Francomano CA, Boileau C, Keston M, Tsipouras P. An exclusion map of Marfan syndrome. J Med Genet 1990; 27:73–77.

20. Kainulainen K, Pulkkinen L, Savolainen A, Kaitila I, Peltonen L. Location of chromosome 15 of the gene defect causing Marfan syndrome. N Engl J Med 1990; 323:935–939.

21. Dietz HC, Pyeritz RE, Hall BD, Cadle RG, Hamosh A, Schwartz J, Meyers DA, Francomano CA. The Marfan syndrome locus: confirmation of assignment to chromosome 15 and identification of tightly linked markers at 15q15-q21.3. Genomics 1991; 9:355–361.

22. Tsipouras P, Sarfarazi M, Devi A, Weiffenbach B, Boxer M. Marfan syndrome is closely linked to a marker on 15q1.5-q2.1. Proc Natl Acad Sci USA 1991; 88:4486–4488.

23. Kainulainen K, Steinmann B, Collins F, Dietz HC, Francomano CA, Child A, Kilpatrick MW, Brock DJH, Keston M, Pyeritz RE, Peltonen L. Marfan syndrome:

no evidence for heterogeneity in different populations and more precise mapping of the gene. Am J Hum Genet 1991; 49:662–667.

24. Sarfarazi M, Tsipouras P, Del Mastro R, et al. A linkage map of 10 loci flanking the Marfan syndrome locus on 15q: results of an international consortium study. J Med Genet 1992; 29:75–80.

25. Sakai LY, Keene DR, Engvall E. Fibrillin, a new 350-kD glycoprotein, is a component of extracellular microfibrils. J Cell Biol 1986; 103:2499–2509.

26. Cleary EG, Gibson MA. Elastin-associated microfibrils and microfibrillar proteins. Int Rev Connect Tissue Res 1983; 10:97–209.

27. Hollister DW, Godfrey M, Sakai LY, Pyeritz RE. Marfan syndrome: immunohistologic abnormalities of the elastin-associated microfibrillar fiber system. N Engl J Med 1990; 323:152–159.

28. Godfrey M, Menashe V, Weleber RG, Koler RD, Bigley RH, Lovrien E, Zonana J, Hollister DW. Cosegregation of elastin-associated microfibrillar abnormalities with the Marfan phenotype in families. Am J Hum Genet 1990; 46:653–660.

29. McGookey-Milewicz D, Pyeritz RE, Crawford ES, Byers PH. Marfan syndrome: defective synthesis, secretion and extracellular matrix formation of fibrillin by cultured dermal fibroblasts. J Clin Invest 1992; 89:79–86.

30. Lee B, Godfrey M, Vitale E, Hori H, Mattei MG, Sarfarazi M, Tsipouras P, Ramirez F, Hollister DW. Linkage of Marfan syndrome and a phenotypically related disorder to two fibrillin genes. Nature 1991; 352:330–334.

31. Maslen CL, Corson GM, Maddox BK, Glanville RW, Sakai LY. Partial sequence of a candidate gene for the Marfan syndrome. Nature 1991; 352:334–337.

32. Pereira L, D'Alessio M, Ramirez F, Lynch JR, Sykes B, Pangilinan T, Bonadio J. Genomic organization of the sequence coding for fibrillin, the defective gene product in Marfan syndrome. Hum Mol Genet 1993; 2:961–968.

33. Corson GM, Chalberg SC, Dietz HC, Charbonneau NL, Sakai LY. Fibrillin binds calcium and is coded by cDNAs that reveal a multidomain structure and alternatively spliced exons at the 5' end. Genomics 1993; 17:476–484.

34. Sakai LY, Keene DR, Glanville RW, Bachinger HP. Purification and partial characterization of fibrillin, a cysteine-rich structural component of connective tissue microfibrils. J Biol Chem 1991; 266:14763–14770.

35. Dietz HC, Cutting GR, Pyeritz RE, Maslen CL, Sakai LY, Corson GM, Puffenberger EG, Hamosh A, Nanthakumar EJ, Curristin SM, Stetten G, Meyers DA, Francomano CA. Marfan syndrome caused by a recurrent de novo missense mutation in the fibrillin gene. Nature 1991; 352:337–339.

36. Magenis RE, Maslen CL, Smith L, Allen L, Sakai LY. Localization of the fibrillin gene to chromosome 15, band 15q21.1. Genomics 1991; 11:346–351.

37. Dietz HC, Pyeritz RE, Puffenberger EG, Kendzior RJ, Corson GM, Maslen CJ, Sakai LY, Francomano CA, Cutting GR. Marfan phenotype variability in a family segregating a missense mutation in the EGF-like motif of the fibrillin gene. J Clin Invest 1992; 89:1674–1680.

38. Cooke RM, Wilkinson AJ, Baron M, Pastore A, Tappin MJ, Campbell ID, Gregory H, Sheard B. The solution structure of human epidermal growth factor. Nature 1987; 327:339–341.

39. Pereira L, Levran O, Ramirez F, Lynch JR, Sykes B, Pyeritz RE, Dietz HC. Diagnosis

of Marfan syndrome: a molecular approach for stratification of cardiovascular risk within families. N Engl J Med 1994; 331:148–153.

40. Dietz HC, McIntosh I, Sakai LY, Corson GM, Chalberg SC, Pyeritz RE, Francomano CA. Four novel FBN1 mutations: significance for mutant transcript level and EGF-like domain calcium binding in the pathogenesis of Marfan syndrome. Genomics 1993; 17:468–475.

41. Kainulainen K, Sakai LY, Child A, Pope MF, Puhakka L, Ryhanen L, Palotie A, Kaitila I, Peltonen L. Two unique mutations in Marfan syndrome resulting in truncated polypeptide chains of fibrillin. Proc Natl Acad Sci USA 1992; 88:5917–5921.

42. Dietz HC, Saraiva JM, Pyeritz RE, Cutting GR, Francomano CA. Clustering of fibrillin (FBN1) missense mutations in Marfan syndrome patients at cysteine residues in EGF-like domains. Hum Mutation 1992; 1:366–374.

43. Dietz HC, Valle D, Francomano CA, Kendzior FJ, Pieritz RE, Cutting GR. The skipping of constitutive exons in vivo induced by nonsense mutations. Science 1993; 254:680–683.

44. Hewett DR, Lynch JR, Smith R, Sykes B. Fibrillin mutation in the Marfan syndrome may disrupt calcium binding of the epidermal growth factor module. Hum Mol Genet 1993; 2:475–477.

45. Godfrey M, Vandemark N, Wang M, Velinov M, Wargowski D, Tsipouras P, Haw J, Becker J, Robertson W, Droste S, Rao VH. Prenatal diagnosis and a donor splice mutation in fibrillin in a family with Marfan syndrome. Am J Hum Genet 1993; 53:472–480.

46. Tynan K, Comeau K, Pearson M, Wilgenbus P, Levitt D, Gasner C, Berg MA, Miller DC, Francke U. Mutation screening of complete fibrillin-1 coding sequence: report of five new mutations, including two in 8-cysteine domains. Hum Mol Genet 1993; 2:1813–1821.

47. Kainulainen K, Karttunen L, Puhakka L, Sakai LY, Peltonen L. Mutations in the fibrillin gene responsible for dominant ectopia lentis and neonatal Marfan syndrome. Nature Genet 1994; 6:64–69.

48. Handford PA, Baron M, Mayhew M, Willis A, Beesley T, Brownlee GG, Campbell ID. The first EGF-like domain from human factor IX contains a high affinity calcium binding site. EMBO J 1990; 9:475–480.

49. Handford PA, Mayhew M, Baron M, Winship PR, Campbell ID, Brownlee GG. Key residues involved in calcium-binding motifs in EGF-like domains. Nature 1991; 351:164–167.

50. Mayhew M, Handford PA, Baron M, Tse AGD, Campbell ID, Brownlee GG. Ligand requirements for calcium binding to EGF-like domains. Prot Engineering 1992; 5:489–494.

51. Aoyama T, Tynan K, Dietz, HC, Francke U, Furthmayer H. Missense mutations impair intracellular processing of fibrillin and microfibril assembly in Marfan syndrome. Hum Mol Genet 1993; 2:2135–2140.

52. Urlaub G, Mitchell PJ, Ciudad CJ, Chasin LA. Nonsense mutations in the dihydrofolate reductase gene affect RNA processing, Mol Cell Biol 1989; 9:2868–2880.

53. Cheng J, Fogel-Petrovic M, Maquat LE. Translation to near the distal end of the penultimate exon is required for normal levels of spliced triosephosphate isomerase mRNA. Mol Cell Biol 1990; 10:5215–5225.

54. Herskowitz I. Functional inactivation of genes by dominant negative mutations. Nature 1987; 329:219-222.
55. Aoyama T, Francke U, Dietz HC, Furthmayer H. Quantitative differences in biosynthesis and extracellular deposition of fibrillin in cultured fibroblasts distinguish five groups of Marfan syndrome patients and suggest distinct pathogenetic mechanisms. J Clin Invest 1994; 94:130-137.
56. Eldadah ZA, Brenn T, Furthmayr H, Dietz HC. Expression of a mutant human fibrillin allele upon a normal human or murine genetic background recapitulates a Marfan cellular phenotype. J Clin Invest 1995; 95:874-880.
57. Tsipouras P, Del Mastro R, Sarfarazi M, Lee B, Vitale E, Child A, Godfrey M, Devereux R, Hewett D, Steinmann B, Viljoen D, Sykes BC, Kilkpatrick M, Ramirez F. Linkage of Marfan syndrome, dominant ectopia lentis, and congenital contractural arachnodactyl to the fibrillin genes on chromosomes 15 and 5. N Engl J Med 1992; 326:905-909.

9

Genetic Risk Factors for Myocardial Infarction

François Cambien
INSERM
Paris, France

Florent Soubrier
INSERM U36
Collège de France
Paris, France

Despite the remarkable therapeutic progress made during the last 20 years, coronary heart disease (CHD) remains one of the major causes of morbidity and mortality in Europe and North America. Myocardial infarction (MI) constitutes a paradigm of multifactorial disease with complex physiopathology. Its frequency increases with age, varies considerably among populations, and is influenced by a large number of environmental and hereditary factors. The contribution of genetic factors to CHD is important, but appears to be stronger on premature than on late-onset forms of the disease. However, the Swedish Twin Registry study has shown that the impact of these factors remains fairly important in patients aged 56 to 75 (1), especially in terms of population-attributable risk since most MIs occur in that age range. Conversely, the contribution of genetic factors to MI occurring after 75 is negligible. The evolution of atherosclerosis, which appears to be a necessary condition for the development of CHD, depends largely on the effect of environmental factors—in particular, diet with a high content of saturated fats and cholesterol—and of genetic factors affecting lipid metabolism. The development of atherosclerosis occurs over several decades, but rupture of the atheroma plaque may considerably accelerate its evolution and eventually lead to acute occlusion of a coronary artery and MI or sudden death. Thrombosis and vasoconstriction contribute synergistically to the acute complications of atheroma (2). Interestingly, these processes, as the response of the vessel wall to injury, which appears to be a major determinant of atheroma (3), are beneficial under normal circumstances; indeed, in the presence of a lesion, they normally both contribute to the prompt termination of the hemorrhage and to the reparation of the injured area.

GENETIC POLYMORPHISMS OF CANDIDATE GENES AND CHD

A candidate gene may be involved in the variable susceptibility to CHD if it exists in functionally different forms. Such polymorphism may be associated with metabolic variability and have detrimental consequences. In contrast to the rare highly deleterious mutations causing monogenic diseases, the known variants predisposing to CHD do not confer a strong increase of risk, but they are quite common and thus carry sizable atrributable risk for the disease in the population. Another consequence of their high frequency is that simultaneous occurrence of several predisposing alleles in an individual is not rare and may lead to high relative risk. Environmental factors are frequently needed for the differential expression of these polymorphic genes, because they affect metabolic pathways that also depend on the availability of substrates provided by the environment. In studying the etiology of CHD, one needs to place strong emphasis on gene–environment interaction. This is particularly logical if the common alleles predisposing to the disease are expressed only in particular environmental contexts.

The patterns of gene–environment interaction may be quite complex. For example, there are gene forms that predispose to atherosclerosis in some populations in which people eat a diet rich in saturated fats and cholesterol. Conversely, other genes predispose to CHD in the presence of atheroma even if they are not involved at all in atheroma formation; this may be the case for gene coding for coagulation factors or proteins involved in cardiovascular repair. Identification of relevant gene–environment interactions is a key to understanding the etiology of CHD.

METHODOLOGICAL ISSUES

Association studies currently constitute the major approach to identifying gene variants predisposing to CHD. Most are directed toward polymorphisms of genes coding for factors involved in the different processes contributing to the development of the disease (candidate genes) in patients with CHD and appropriate controls. This particular type of study raises specific methodological problems. Four are particularly important: 1) the representativeness of the samples of patients and controls, 2) the power of the study, 3) the risk of spurious findings and the need for internal consistency of the results, and 4) the potential bias introduced by selection by death.

1. Ideally, patients and controls should be randomly sampled from the same population during the same period of time. This kind of sampling is possible when CHD registries are available, as for the ECTIM study (see appendix). However, frequently this cannot be realized, and great caution should be exercised in the selection of patients and controls to ensure meaningful comparisions of gene frequencies.

2. The sample size needed for a study must also be considered carefully. In most of the association studies published to date, the samples were too small. This may be due to an inadequate estimate of the minimum relevant relative risk that one wishes to detect. For a common disease such as CHD, and since genetic variants predisposing to this disease are also frequent, a relative risk of 2 or even less may be quite important in terms of attributable risk. Fairly large studies may be needed to identify such relative risks. Formal calculations can easily be done; however, it may be useful to provide general rules defining three levels of studies having increasing power and general nature: 1) to evaluate genetic factors independently, at least 200 patients and 200 controls are needed; 2) to investigate the internal consistency of the results (see below) and study particular subgroups defined a priori, at least 600 patients and 600 controls are needed; and 3) to evaluate a multigenic risk (with possible interaction between factors), at least 2000 patients and 2000 controls are needed. Obviously, situations exist in which much smaller samples may be required. For example, the association between the Apo E e4 allele and Alzheimer's disease has been identified in a large number of small studies because the relative risk is very high. However, the sample sizes given above correspond to the general situation encountered in CHD research. The present trend is to design studies of the third level, to distribute genetic material to a large number of laboratories and to centralize the results to evaluate multigenic risk.

3. Association studies should, by design, allow the possibility of testing the internal consistency of their results: 1) independent, contrasted populations may be included in the same study to verify that the results hold in general; 2) a family history of CHD should be more frequent in carriers than in noncarriers of the detrimental alleles; and 3) when intermediate phenotypes are available (for example,

plasma ACE level when studying the ACE gene), consistent association should be observed between genotype, intermediate phenotype, and disease.

4. Since for CHD the case-fatality rate is high (>30% within the first 3
    months), the frequency of a genetic risk factor in patients surviving
    the disease may be strongly biased downward, if this risk factor
    affects survival. This may be the case for the ACE I/D polymorphism,
    as we will see below. Prospective studies would appear to be more
    appropriate than retrospective case-control studies in such cases;
    however, very large cohorts would have to be followed for a long
    period of time to get a sufficient number of incident cases. It is from
    the conjunction of results collected in different types of studies with
    different endpoints, including surrogate measures of the disease and
    death, that the consistency and strength of a genetic risk factor will
    be established.

Until recently, there were technical limitations to the study of candidate
gene polymorphisms in relation to common diseases, because large numbers
of genes and variants had to be analyzed in large samples of individuals. These
technical constraints are less limiting today, and the possibility of identifying
genes contributing to CHD risk appears to be more dependent on the
availability of appropriate DNA banks from well-characterized patients or
families connected to detailed epidemiological information and intermediate
phenotype exploration. Such a bank has been constituted as part of the ECTIM
study, a large case-control study comparing male patients with MI and randomly selected controls based on different populations with highly contrasting risks of CHD in Europe (see appendix). The results reported in this chapter
are based largely on data collected in that study, and the focus is on the
relationships between MI and genetic polymorphisms of ApoB, ApoE, apo(a),
and ACE.

GENETIC POLYMORPHISMS OF THE ApoB, ApoE, apo(a), AND ACE GENES AND CHD

Apolipoprotein B100

Plasma ApoB level is, to a great extent, genetically determined (4–6), and when
elevated constitutes a major risk factor for the development of atherosclerosis
and CHD (7–9). In the European Atherosclerosis Research Study (EARS), a
large collaborative study with young adults from different populations in
Europe, mean plasma ApoB levels were consistently higher in the offspring
of patients with a premature history of CHD than in controls (Figure 1),

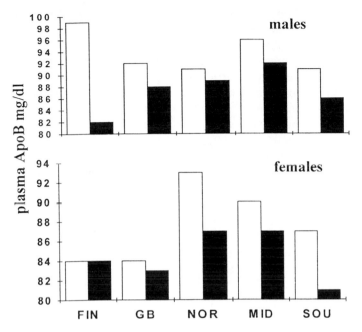

Figure 1 Plasma ApoB levels in offspring of patients with myocardial infarction and controls in the EARS study. FIN: Finland; GB: Great Britain; NOR: northern Europe; MID: central Europe; SOU: southern Europe. White bars: cases; black bars: controls. Case-control difference: $p < 0.01$ in females; $p < 0.001$ in males.

suggesting that plasma ApoB could be an important heritable risk factor for CHD (10).

ApoB is a protein composed of 4536 amino acids coded by a gene on chromosome 2, which comprises 29 exons and covers 43 kilobases. This gene codes for two ApoB forms: one synthesized in the liver, ApoB100, and the other synthesized by the intestine, ApoB48. This latter form is produced posttranscriptionally by an enzymatically mediated editing process of the mRNA, changing a glutamine codon at position 2153 into a stop codon (11). ApoB48 includes the sequence coded by the 5′ part of the ApoB gene, which allows lipoprotein assembly and secretion but lacks the LDL receptor binding domain present in ApoB100. ApoB48 is rapidly cleared from the circulation as a consequence of the rapid catabolism of chylomicrons. In plasma collected in the fasting state, ApoB100 is present in great excess relative to ApoB48. Despite the potential importance of postprandial hypertriglyceridemia as a risk factor for CHD, the possible influence of genetic variability of ApoB48 on triglyceride-rich lipoprotein of intestinal origin has not yet been well investigated.

ApoB100-containing lipoproteins are assembled in hepatocytes, and most ApoB100 is secreted from the liver on triglyceride-rich VLDL particles that also contain ApoE and ApoCs. It appears that the 5' part of the ApoB gene codes for sequences containing structural information necessary for the assembly and secretion of lipoprotein particles. The 3' part, on the other hand, codes for domains involved in lipid and receptor binding. The results of short-term in vitro studies have shown that the secretion of ApoB-containing lipoproteins appears to be modulated co- or posttranslationally. The process of lipoprotien assembly may then be an important step that could possibly be affected by variation in the 5' part of the gene. Importantly, other factors are also needed for the secretion of lipoproteins containing ApoB. Among them, the microsomal triglyceride transfer protein (MTP) catalyses the transport of triglyceride, cholesteryl ester, and phospholipid between phospholipid surfaces. Mutations of MTP may cause abetalipoproteinemia (12). Theoretically, genetic variations in this and similar proteins could affect lipoprotein secretion from hepatic cells. ApoB100 is the only apolipoprotein present in LDL, and it has been hypothesized that functional variability of the ApoB gene—in particular, in regions coding for the LDL receptor binding domain—could interfere with the catabolism of atherogenic lipoproteins.

"Protective" Variants

Several sequence variations of the mature ApoB protein have been described; those characterizing the Ag system are almost neutral, with the exception of the Ag(x/y) polymorphism (13). The frequency of the Ag(x) allele is highly variable among populations; it is frequent in Asians (70–90%) and rare in Europeans (20%). Complete correspondence between the Ag(x/y) protein polymorphism and two polymorphisms changing codons 2712 and 4311 of the ApoB gene has been established in large samples from different populations (14). This observation implies a low frequency of recombination between haplotypes. The particular haplotype characterized by presence of a leucine at position 2712 of the protein and of a serine at position 4311 is associated with lower levels of ApoB-containing lipoproteins in plasma and with a possible protection against CHD (15). Some rare variants of the ApoB gene generate premature stop codons and truncated forms of ApoB; the resulting rare monogenic disorder, hypobetalipoproteinemia (16), is characterized by very low levels of circulating ApoB. Thus, rare mutations with strong effect and frequent polymorphisms with smaller effects, associated with reduced levels of circulating ApoB and LDL and apparently with protection against CHD, are found on the ApoB gene. Low levels of plasma ApoB associated with polymorphisms of the ApoB gene could be the consequence of a less efficient assembly process of VLDL in hepatocytes or of more efficient catabolism.

"Deleterious" Variants

Variants of the ApoB gene associated with increased levels of ApoB and increased risk of CHD have also been identified. The substitution of an arginine to a glutamine at the level of codon 3500, within the region coding for the LDL receptor binding domain of ApoB, causes familial defective ApoB100 (17,18). The frequency of this monogenic trait has been estimated at approximately 1/500 individuals in several studies, but, interestingly, the ApoB3500 mutation has not been found in some populations. Haplotype analysis has shown that this mutation is of unique origin and is generally found in individuals of German descent, suggesting a founder effect. This situation must be distinguised from that observed in the presence of the LDL cholesterol defect leading to familial hypercholesterolemia (19). More than 200 LDL receptor mutations have been described, and even if founder effects have been observed for familial hypercholesterolemia in some populations (French Canadians, Afrikaaners, Finnish, Lebanese), this disease is clearly caused by a large number of independent mutations of the LDL receptor gene. Depending on the type and site of the mutation in the LDL receptor gene, the phenotypic expression may vary. However, in general, familial hypercholesterolemia due to LDL receptor defects is a more severe disease than familial defective ApoB100. In middle-aged individuals of western European origin, the risk of CHD attributable to the ApoB3500 and LDL receptor defects may be estimated at less than 1% and 2%, respectively. This indicates that the contribution of rare highly deleterious mutations to CHD is low and that important genetic contributions to the risk of CHD, if they exist, must be the consequence of frequent variations.

A large number of common polymorphisms of the ApoB gene have been described, and their association with lipid parameters and CHD complications has been studied in several clinical studies. Unfortunately, most of these studies included a small number of patients, and many inconsistent associations were published. Nevertheless, a weak but consistent association has been found between the silent XbaI polymorphism affecting codon 2488 of the ApoB gene and the plasma levels of cholesterol and triglycerides. This polymorphism is probably a marker for one or several functional variants within or near the ApoB gene; however, these functional variants have not been identified so far.

Probably more interesting is the polymorphism characterized by a variable number of tandem repeats located about 300 bp downstream of the ApoB gene (ApoB 3'/VNTR). This highly informative polymorphism is characterized by the presence of at least 15 different alleles composed of 30 to 52 AT-rich repeat units of 30 base pairs. These repeat units are not all identical, and it has been shown that some alleles of identical size differ in their structure and could then be subdivided (20). This structural heterogeneity, however, is apparently not present for the most common alleles, VNTR34, VNTR36, and VNTR48. These alleles are in strong linkage disequilibrium with diallelic polymorphisms of the ApoB gene; for example, ApoB4311-Ser is almost always on VNTR34 and ApoB3500-Gln on VNTR48, and the presence of the ApoB/Xbal cutting site is associated with VNTR36. In the ECTIM study, allele VNTR48 was strongly associated with the risk of MI, with an odds ratio (relative risk) for MI of 1.52 in heterozygotes and homozygotes (VNTR48+) in comparison to VNTR48- individuals ($p < 0.005$). (21). This association was even more striking in overweight individuals in whom the odds ratio for MI was 2.2 ($p < 0.0001$); furthermore, a decreasing trend of the odds ratios, from north to south, suggested that this association might explain part of the higher frequency of MI in Belfast than in Toulouse (Figure 2). Obesity is characterized by an increased secretion of VLDL by hepatocytes, apparently due to the presence of a relative hyperinsulinemia. This effect is not likely to be the consequence of an overexpression of the ApoB gene but possibly results from a more efficient assembly of lipoprotein particles. The VNTR48 allele could, then, be a marker for one or several functional variants affecting the assembly process of lipoproteins in the hepatocyte or their clearance from the circulation by the LDL receptor. These mechanisms could be triggered by an increased input of VLDL. Whether the functional variant(s) preferentially affects the protein assembly of lipoproteins or their catabolism has different implications; in the first instance overefficient alleles might be involved, whereas in the second instance the detrimental allele would be deficient. Identification of the functional variants involved would then be an important step toward understanding how genetic variants of the ApoB gene affect lipoprotein metabolism.

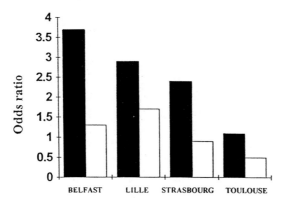

Figure 2 Odds ratios for myocardial infarction associated with the VNTR48 genotypes in the ECTIM study, according to the median of body mass index. Black bars: subjects with a BMI $\geq$ 26 kg/m^2; white bars: subjects with a BMI < 26 kg/m^2.

Interpopulation Differences

Apart from rare forms that probably originated recently, the same alleles of the ApoB gene are found in different populations. However, the frequency of these alleles differs greatly according to population; for example, both alleles of the XbaI polymorphism are found with approximately the same frequency in Caucasians, whereas the XbaI cutting site is very rare in Asians (22). Similarly, long alleles of the VNTR are much less frequent in Asians than in Caucasians, whereas the reverse is true for VNTR34. These differences in haplotype distribution could account, in interaction with environmental factors, for part of the lower mean levels of atherogenic lipoproteins in Asian than in Caucasian populations (22). Detailed studies of haplotype frequencies of the ApoB gene using the Ag system have shown that the differences of haplotype frequencies between human populations are related to the geographical distance between these populations and could be explained by the African origin of humans and their subsequent migration to the Middle East, Europe, Asia, and Australia (23). Although this may be the consequence of random drift, the posible role of natural selection cannot be excluded. It is indeed possible that specific ApoB alleles with different effects on the metabolism of plasma lipids have been preferentially selected in different populations as a consequence of long exposure to highly contrasting diets.

Apolipoprotein E (ApoE)

ApoE is a structural component of chylomicrons, VLDL, and high-density lipoproteins (HDL). Its high affinity for the LDL and remnant receptors and its function in the conversion of VLDL into LDL explain its important role in the metabolism of triglyceride-rich lipoproteins (24).

Two frequent polymorphisms of ApoE exist, one in position 112 of the protein (Cys112 → Arg), the second in position 158 (Arg158 → Cys). These polymorphisms generate three alleles: e2, e3 (the most frequent), and e4. The polymorphism of ApoE has been known for a long time (25), and its clinical and epidemiological impact has been studied in detail. Several methods have been used to identify this polymorphism, first at the level of the protein and, more recently, at the level of the ApoE gene. The frequency of ApoE alleles strongly differs according to population (26) (Figure 3). In particular, e4 is frequent in black populations in Africa and the United States (around 30%), whereas its frequency is low in Asian populations (around 10%). Furthermore, there is a strong decreasing gradient of frequency of e4 from the north (20–25% in Finland) to the south (10–15%) in Europe.

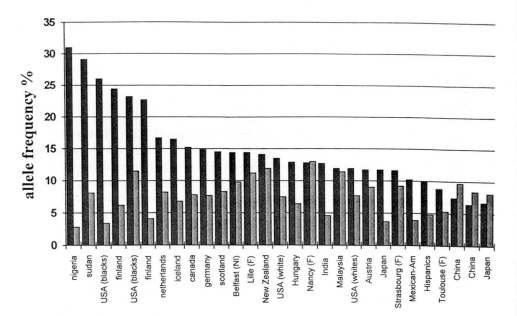

Figure 3 Distribution of ApoE e2 and e4 alleles in different populations. Black bars: ApoE e4 allele; gray bars: ApoE e2 allele.

The ApoE Polymorphism Affects the Metabolism of Plasma Lipoproteins and the Plasma Level of Atherogenic Lipoproteins

The ApoE e2 allele has a much lower affinity than e3 and e4 for the ApoB/E receptors (27); conversely, although the affinity of e3 and e4 for this receptor is similar, they are preferentially associated with HDL and VLDL, respectively (28). This strongly influences the metabolism of dietary fats, which are cleared more rapidly in e4+ than in e2+ individuals, e33 being intermediate (29). In addition, the absorption of dietary cholesterol is more efficient in e4+ than in e33 than in e2+ individuals (30). The higher hepatic cholesterol pool in e4+ individuals would, then, lead to a reduced availability of ApoB/E receptors through a feedback mechanism and to an increased circulating level of LDL cholesterol, and the reverse would be observed in e2+ individuals (31). The delipidation of VLDL induced by lipoprotein lipase (LPL) is also influenced by the ApoE polymorphism; in particular, the delayed formation of LDL from IDL appears to contribute to the lower plasma LDL concentrations associated with e2 (32). The e3 allele is often said to be the normal allele; in fact, this term is inappropriate—the three alleles favor slightly different metabolic pathways and none of these pathways can be considered a priori to be more favorable than another. Even if Apo e4 is associated with increased LDL cholesterol levels in all populations, it is probably the nutritional and genetic contexts that determine whether this allele will predispose to atheroma and its complications. It has also been proposed that ApoE plays an important role in local redistribution of lipids and might be involved in repair mechanisms in different tissues (24). During the process of atherogenesis, and after rupture of the atheroma plaque, local production of ApoE by macrophages could play a role in the redistribution of cholesterol necessary for vascular wall repair. We may speculate that the ApoE polymorphism could affect this process.

The ApoE Polymorphism, LDL Cholesterol, and Vascular Diseases

The results of a large number of studies in different populations with different mean levels of total cholesterol have been very consistent in demonstrating that e4+ individuals have higher mean levels of total or LDL cholesterol than e2+ individuals, e33 subjects being intermediate. These differences were anticipated from the metabolic effects of the polymorphism and, since LDL cholesterol is a risk factor for CHD, it was logical to speculate that consistent associations with the risk of CHD could be observed. Several studies have tested this possibility and the results are not all concordant; in general, however, increased frequency of e4 and decreased frequency of e2 have been found in patients with coronary artery disease or MI in comparison to controls. The PDAY study has provided very interesting information about a large group of young males who died accidentally (33). In this study, the surface extension of atherosclerotic lesions in the thoracic and abdominal aorta was less important in e23 than in e33 and e34 individuals, and there was a tendency for larger lesions in e34 than in e33 individuals. The association was similar in black and white individuals and appeared to be largely independent of serum cholesterol levels. This suggests that the different effects of the three alleles on atherosclerosis extension may be independent of circulating lipid levels; the local process mentioned above could thus be of major importance. In the ECTIM study (34), lower circulating LDL cholesterol and ApoB levels were observed in the presence of e2 than in the presence of e4, and e33 individuals had intermediate levels. The protective effect of e2 and the detrimental effect of e4 on MI risk were also clearly observed. Taking e33 as a reference, the presence of allele e2 was associated with a relative risk of 0.73 ($p < 0.05$) and the presence of allele e4 was associated with a relative risk of 1.33 ($p < 0.02$). Furthermore, the allelic effect appeared to be multiplicative. Figure 4 shows the mean levels of ApoB and the odds ratios for MI according to ApoE genotypes in the ECTIM study. From the frequency of the different alleles and the odds ratios for MI, it was possible to estimate that at least 8% of MI cases in the ECTIM study were attributable to the ApoE polymorphism.

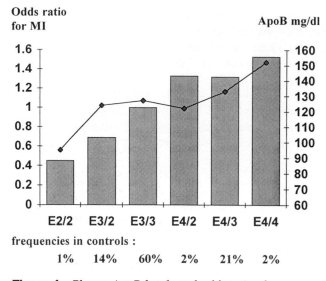

Figure 4   Plasma ApoB levels and odds ratios for myocardial infarction associated with the different ApoE genotypes in the ECTIM study. Bars are odds ratios; points on the curve are mean ApoB levels. Odds ratios and ApoB levels are adjusted for age BMI and population.

An important advance in recent years was the discovery of a strong association between the ApoE polymorphism and Alzheimer's disease (AD) (35). Indeed, several studies have shown that ApoE e4 is at least four times more frequent in Alzheimer's patients than in controls, the attributable risk being greater than 50%. ApoE accumulates in the senile amyloid plaques and in neurofibrillary tangles, two neuropathological charateristics of AD. The β-amyloid peptide is a major component of the senile plaque and ApoE e4 forms preferential associations over e3 and e2 with this peptide. It appears then, that ApoE e4 contributes to two common diseases, AD and CHD, probably through completely different mechanisms. This illustrates the strong pleiotropic effect of the ApoE polymorphism.

Apolipoprotein(a) (Apo(a))

Lp(a) is a macromolecule present in human plasma, constituted of LDL and a large glycoprotein, apo(a). Plasma levels of Lp(a) vary considerably between individuals, but are very stable within individuals and are not easily affected by environmental factors (including drugs). These features suggest that plasma Lp(a) levels may be strongly genetically determined. This hypothesis has been confirmed in several studies; in particular, a study of siblings has shown that more than 90% of the interindividual variability of plasma Lp(a) could be explained by genetic variation at the apo(a) gene locus (36). The apo(a) cDNA sequenced by McLean et al. in 1987 (37) contained a large number of exact or nearly exact repeats of 342 bases, and the deduced protein sequence was highly homologous with that of plasminogen. Plasminogen is the precursor of the proteolytic enzyme plasmin, which is involved in fibrinolysis. Plasminogen contains five sequences of 80–114 amino acids called kringles, followed by a serine protease domain. The kringles are involved in the binding of plasminogen to fibrin, and plasminogen has to be cleaved by tPA at a single arginine residue to form active plasmin. The fourth kringle domain of plasminogen most closely resembles the repeated domain of apo(a). Nucleotide identity between the plasminogen and apo(a) genes ranges from 87% in the 3' untranslated region to 100% at the 5' end, homology for the kringle 4-like repeats being 75–85%. As a consequence of a sequence change affecting the site of activation of plasmin, Lp(a) has no proteolytic activity. However, the possibility remains that it could interfere in vivo with fibrinolysis by acting as a competitor of plasminogen, and it has been proposed that Lp(a) might constitute a link between the lipoprotein and coagulation systems.

The number of kringle 4-like elements in apo(a) is genetically determined and varies from 15 to 40. A strong inverse relationship exists between the number of kringle 4 and the circulating level of Lp(a). However, the repeat polymorphism explains only about 50% of the genetic variability of Lp(a) attributable to the apo(a) locus (38,39).

Lp(a) and Coronary Heart Disease

The mean plasma level of Lp(a) is higher in patients with CHD. Strong evidence exists that the association is causal; however, some important results do not support this causal link. In particular, two prospective studies have failed to demonstrate an association between plasma levels of Lp(a) and coronary heart disease (40,41). Moreover, in the EARS study, mean plasma levels of Lp(a) were similar in offspring of male patients with premature MI and controls (42). Conversely, a prospective study has shown that serum level of apo(a) was predictive of CHD mortality in men (43), and another prospective study has demonstrated an association between levels of Lp(a) and the incidence of CHD in hypercholesterolemic men (44). The discrepancies between these studies could be in part the consequence of differences in the methods used to measure Lp(a) and in duration and storage temperature of the plasma samples before the assay. With a few exceptions, retrospective studies have consistently found higher levels of plasma Lp(a) in CHD patients than in controls. In the ECTIM study, Lp(a) levels higher than 0.3 g/L were associated with a relative risk of MI of 2.5 and with an attributable risk of 16% (9). Furthermore, a positive association exists between plasma levels of Lp(a) and coronary atherosclerosis evaluated by coronary angiography (45) and medial-intimal thickness of the carotid artery evaluated by Doppler technique (46). The causal relationship between high levels of Lp(a) and CHD has been reinforced by the results of studies showing a higher frequency of the short alleles of apo(a) in CHD patients than in controls (47). However, this association was not found in the EARS study (42).

Since more than 90% of the interindividual variability of plasma Lp(a) level is genetically determined, measurement of plasma Lp(a) should provide integrated genetic information, cumulating the effect of the kringle 4-repeat polymorphism and of other, still unknown, polymorphisms of the apo(a) gene. However, it is unclear whether this strong genetic effect also exists in the presence of extensive atherosclerotic lesions or after an MI. It is indeed possible that plasma Lp(a) level increases in particular pathophysiological circumstances; as a consequence, the large differences in Lp(a) levels between patients and controls in retrospective studies could reflect secondary effects and not etiological effects. Prospective studies may also suffer from this drawback; since atherosclerosis in the coronary arteries may simultaneously affect the risk of MI and modify intermediate phenotypes, secondary associations may be expected between these phenotypes and MI. It must also be stressed that even if functional variations in the apo(a) gene are causally linked to the risk of CHD through their effect on atherosclerosis or thrombosis, the length polymorphism may not be the primary factor involved. The mean plasma levels of lp(a) in populations of African origin living in North America are considerably higher than in Caucasians; however, the risk of CHD in these populations is not proportionately greater.

It has also been shown that Lp(a) levels were relatively low in Asians and that the impact of the length polymorphism of apo(a) on Lp(a) level was much weaker in African blacks than in Asians and Caucasians (48). Alleles of apo(a) with the same number of kringle 4 repeats could have occurred at different times during evolution; as a consequence, they may charcterize different alleles possibly having different functional properties (49). There may be functional variations in the regulatory regions of the gene or within its coding sequence that are preferentially linked with alleles of particular sizes. This could explain the interpopulational differences mentioned above and the dissociation between apo(a) size and plasma Lp(a) levels observed within populations.

The Role of Lp(a)

Among the many unresolved questions about Lp(a), the most puzzling concerns its physiological role. The apo(a) and plasminogen genes are located in close proximity on chromosome 6. Studies of sequence homology between the two genes have established that they probably both derive from an ancestral gene that duplicated some 40 million years ago. A more recent conversion between both genes has also been postulated since their 5′ nontranslated region is much more homologous than the 3′ region (37). However, it is also possible that strong selective constraints have resulted in the high conservation of the 5′ regions of both genes. Although the function of Lp(a) is unknown, it is extremely unlikely that the apo(a) gene is the vestige of a useless duplication maintained by random drift. When comparing the apo(a) and plasminogen gene sequences, the remarkable homology of the coding sequences as compared to noncoding ones is striking. This strongly argues against a lack of function of the protein. The relatively recent occurrence (during early primate evolution) of the ancestral gene duplication leading to apo(a) and plasminogen has been challenged by the discovery of apo(a)-like proteins and Lp(a) in hedgehogs (50). This apparently paradoxical observation has not yet been satisfactorily explained.

It is most likely that apo(a) had or is still having an important function. It is possible that the deleterious effects on CHD associated with some forms of apo(a) and with high Lp(a) levels are the consequence of a particular function or of hyperactivity of this particle. As already mentioned, unlike the other proteases to which it is related, Lp(a) has no proteolytic activity. However, in vitro studies have shown that Lp(a) interferes with fibrinolysis (51) and could theoretically induce a prothrombotic state in vivo. Since several clinical studies have failed to demonstrate an association between increased Lp(a) levels and hypofibrinolysis, an in vivo activity and its possible pathophysiological consequences remain to be demonstrated. Proliferation and migration of smooth muscle cells (SMCs) constitute important elements in vascular repair processes and they are also involved in the development of atherosclerosis. High levels of Lp(a) could stimulate the proliferation of vascular SMCs by inhibiting the conversion of latent TGF-β into active TGF-β by plasmin (52). Finally, since Lp(a) can bind to fibrinogen and fibrin, it could constitute a source of cholesterol to injured or wounding sites. This putative function could be highly beneficial in certain situations of low cholesterol intake or synthesis and would furthermore justify the linkage between the apo(a) and LDL molecules (53).

Angiotensin-I-Converting Enzyme (ACE)

In the plasma and on the surface of endothelial cells, ACE converts the inactive decapeptide angiotensin I (AI) into the highly vasoactive and aldosterone-stimulating octapeptide angiotensin II (AII). AII is a powerful coronary vasoconstrictor (54,55) that may modulate the growth of vascular SMCs and induce myointimal hyperplasia after endothelial injury (56). Bradykinin (BK), the other best-known substrate of ACE, is a potent vasodilator and inhibitor of SMC proliferation that by interacting with the BK B2 receptor, induces the release of endothelial factors, including nitric oxide and prostacyclin (57,58). BK is inactivated by ACE; this could account for some of the effects associated with high ACE expression and for part of the beneficial effect of ACE inhibitors (59,60).

The Genetics of Plasma ACE in Humans

Repeated measurement of plasma ACE in adult humans has shown that the level of the enzyme is very stable within individuals but differs strongly from individual to individual and is independent of many environmental, metabolic, and hormonal factors (61). ACE levels in middle-aged men and women are very similar, whereas an important temporary increase is observed during adolescence (62,63).

A low intraindividual, but high interindividual, variability and a lack of association with environmental factors suggested that plasma ACE level is genetically determined. To test this hypothesis, plasma ACE level was measured in a sample of nuclear families composed of healthy individuals (62). Significant correlations were observed between biological relatives but not between spouses, and a major gene transmission was much better supported than a polygenic or a nongenetic transmission. The frequencies of alleles S/s associated with high/low plasma ACE levels were 0.24/0.76. The effect of S was codominant and stronger in offspring than in parents. According to the segregation analysis, the postulated polymorphism S/s accounted for 29% and 75% of the variance of plasma ACE in parents and offspring, respectively. These results strongly suggested that plasma ACE level was determined largely by the effect of a single major gene.

The ACE Gene Is Polymorphic

After the ACE cDNA was cloned, it was possible to elucidate the structure of the gene and to investigate its polymorphism (64) and a frequent insertion/deletion polymorphism (ACE I/D), characterized by the presence/absence of a 287 bp fragment in the sixteenth intron of the gene, was identified by hybridization with a cDNA probe.

The relationship between the ACE I/D polymorphism and plasma ACE level was studied in a sample of healthy adults (65). The concentration of ACE in plasma, measured by RIA, was strongly associated with the genetic polymorphism. In II homozygotes, ID heterozygotes, and DD homozygotes, the mean levels of ACE were 299, 393, and 494 μg/L respectively ($p < 0.001$). The effect of the gene was strictly codominant and accounted for a large part of the interindividual variability of plasma ACE. ACE activity has also been investigated in human circulating mononuclear cells (66). The highest enzyme level was found in T lymphocytes in which ACE activity was approximately 30 times that found in monocytes. As for plasma ACE, the cellular level was very stable within individuals but highly variable between individuals. The correlation between the plasma and T-lymphocyte levels of ACE was 0.42 (p < 0.01), and the mean ACE level in T lymphocytes was greatly increased in the presence of the D allele (p < 0.001). Thus, plasma and cellular ACE levels appear to be largely determined by a polymorphism that probably affects the ACE gene. However, the results of a family study have demonstrated that the ACE I/D polymorphism is only a genetic marker strongly associated with an unknown functional variant (ACE S/s) located within or near the ACE gene (63). In this study, the S/s functional polymorphism and the I/D marker explained 44% and 28%, respectively, of the interindividual variance of plasma ACE levels.

The ACE I/D Polymorphism Is Not Related to Blood Pressure in Humans

As a consequence of the role of ACE in the renin-angiotensin system, it was logical to hypothesize that the ACE gene is a candidate for human hypertension. However, most association studies have found no association between the ACE I/D polymorphism and blood-pressure level or hypertension. A study in pairs of hypertensive siblings using a highly informative marker located close to the ACE gene has also failed to demonstrate a linkage between the ACE locus and hypertension (67). In another study, the distribution of the ACE I/D genotypes did not significantly differ in groups of offspring contrasted for blood pressure and parental history of high blood pressure (68). A lack of association between the polymorphism and blood pressure has also been reported in a Dutch study (69). In a comparison of hypertensives and normotensives of Japanese origin, the distribution of ACE I/D genotypes was also similar in both groups (70). Interestingly, in this study, the frequency of allele D (0.40 in controls) was lower than in Caucasian populations, in which its frequency is generally found to be between 0.53 and 0.57. We have observed in a sample of 150 healthy individuals from northern China that the D allele is effectively less frequent (0.37) in northern Asians than in Caucasians (unpublished result).

Until recently, the only significant association of the ACE I/D polymorphism with hypertension had been found by Zee et al. (71) in a study involving 80 Caucasian hypertensives and 93 normotensives in Australia. In this study, the frequency of the insertion allele (I) was 0.56 in hypertensives, whereas it was 0.41 in normotensives (p < 0.01). In the analysis of the 188 control subjects aged 60 or older included in the ECTIM study, a highly significant association between diastolic blood pressure and the ACE I/D polymorphism was found, higher levels being observed in the presence of the I allele. This association was homogeneous across the four populations studied. After adjusting for population, age, and body mass index, the mean levels of diastolic blood pressure were 77.8 (SEM 1.9), 84.8 (1.4), and 86.3 (2.2) mm Hg in DD, ID, and II, individuals respectively (test for trend: $p < 0.003$). In the younger group, no association was found. This observation, which is in agreement with the results of Zee et al. (71), may appear paradoxical in view of the positive association between the D allele and plasma and cellular ACE. However, given the association between the ACE D allele and CHD risk described below, it is possible that the association between blood pressure and the I allele may result from selection by death of ACE DD hypertensive subjects.

The ACE I/D Polymorphism and MI in the ECTIM Study

In the whole population, the odds ratio (relative risk) for MI was 1.57 for DD vs. II and 1.26 for ID vs. II (test for trend: $p < 0.003$). This association was not significantly heterogeneous across populations, and the frequency of the two alleles was similar in the different control groups (72).

In a low-risk group defined by the absence of hyperlipidemia and obesity (Figure 5), the association between the ACE I/D polymorphism and MI was highly significant and homogeneous across the populations (test for trend: $p < 0.005$). However, the increased risk of MI was present only in DD individuals; the overall odds ratio comparing DD and ID+II individuals was 2.7 ($p < 0.0005$). Conversely, in the high-risk group, the ACE I/D polymorphism was less strongly related to MI ($p < 0.05$).

In the ECTIM study, the parental history of MI was carefully recorded. Figure 6 shows the frequency of parental history of fatal MI in Belfast and France, according to the ACE I/D polymorphism. As expected, parental history of fatal MI was much more frequent in Belfast than in France (about four times). In both countries the frequency of the D allele was higher in those

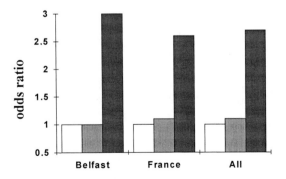

Figure 5   Odds ratios for myocardial infarction associated with the ACE DD genotype in the ECTIM study—low-risk individuals. White bars: II genotype; gray bars: ID genotype; black bars: DD genotype. Adjusted odds ratio: DD/II 2.7; ID/II 1.1. Test for comparison DD/II+ID adjusted for population: $p < 0.0005$.

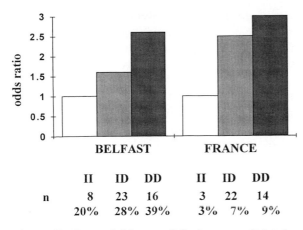

	II	ID	DD	II	ID	DD
n	8	23	16	3	22	14
	20%	28%	39%	3%	7%	9%

Figure 6 Parental history of fatal myocardial infarction in relation to the ACE polymorphism in the control samples of the ECTIM study. Adjusted odds ratio: DD/II 2.7 (1.2-6.4), $p < 0.02$; ID/II 1.9 (0.9-4.1), $p = 0.1$.

having a parental history (73). Taking the II genotype as reference, very similar odds ratios for parental fatal MI associated with the DD and ID genotypes were observed in Belfast and France; the odds ratio adjusted for population being 2.7 ($p < 0.02$) for DD vs. II and 1.9 ($p = 0.1$) for ID vs. II.

Plasma ACE activity was measured in the ECTIM study from frozen plasma samples. A significant interaction between plasma ACE and age on the risk of MI was demonstrated by the strong association between plasma ACE level and MI in patients aged less than 55 (the median age in the study) but not in the older patients. Furthermore, plasma ACE was elevated in the younger cases within each ACE I/D genotype, indicating that this association was independent of the polymorphism (74). In this analysis it was possible to infer the frequency and effect of the ACE S allele in cases and controls by a commingling analysis of the ACE distribution conditioned on the ACE I/D polymorphism. ACE S was more frequent in cases before than after age 55: 45% and 32%, respectively ($p < 0.01$). This might reflect a high mortality in MI patients carrying the D allele.

A Possible Association Between the ACE I/D Polymorphism and Death from MI

A weaker association between the ACE I/D polymorphism and MI in high-risk subjects than in low-risk ones could be the consequence of selection due to a high mortality rate in high-risk DD or ID patients suffering a coronary event. In the ECTIM study, patients were included 3-9 months after the acute event. Several sources of evidence suggest that approximately 30% of the patients who develop an acute coronary event die before arriving at the hospital, and that during the first 3 months after hospitalization about 8% of the patients die from a complication of their MI. Thus, in the populations covered by the ECTIM study, 30–40% of the patients with acute coronary events died before being eligible for the study. Such a selection could strongly affect an association between a risk factor for death and MI, and could lead to an underestimation of the effect of the risk factor. To test this hypothesis, the ACE ID polymorphism was studied from stored autopsy material obtained from patients who had recently died of definite or probable MI in Belfast, and the distribution of ACE I/D genotypes in these individuals was compared with that observed in the random population sample recruited in Belfast for the ECTIM study. Relative to the ACE II genotype, the ACE ID and DD genotypes were associated with odds ratios for MI death of 1.8 and 2.2, respectively (test for trend: $p < 0.01$) (75). Selection by death could also explain the results of a recently published Norwegian study in which the frequency of the ACE D allele was lower in patients recruited several years after an MI compared to controls, whereas it was considerably higher in patients and controls with a parental history of early MI (76,77).

Mechanisms Proposed for the Possible Effect of the ACE Gene Polymorphism on MI

Since allele ACE S is associated with high plasma and cellular levels of ACE, it can be proposed that a chronic increased expression of ACE, by modulating the level of AII and BK at specific sites, affects cardiovascular homeostasis and is responsible for the increased risk of MI. ACE probably acts locally in the coronary arteries and/or in the heart. The paracrine and autocrine effects of vascular or cardiac renin-angiotensin (78) and kallikrein-kinin (79,80) systems could explain the local action of ACE and of ACE inhibitors.

Given the important roles of AII and BK as modulators of cellular growth and of vasomotricity, their respective deleterious and beneficial implications at different stages of the atherosclerotic process and during the acute events leading to MI or sudden death can be postulated. Recent angiographic studies have shown that the increased risk of MI in the presence of ACE D is not the consequence of more severe atherosclerosis (81), but, in the presence of atheroma, ACE D could predispose to plaque rupture and to vasoconstriction of the coronary arteries. It has been shown that the intimal-medial thickness of the carotid artery is greater when plasma ACE levels are high (82). This observation could be important in view of the strong association found in an autopsy study between the intimal thickening of coronary arteries in infants and a family history of CHD (83). High ACE expression in the heart in the presence of ACE D might also increase the local generation of AII and predispose to left ventricular hypertrophy (84), a strong risk factor for CHD

(85). Hence, if ventricular hypertrophy was the main consequence of ACE overexpression, the ACE polymorphism could be a risk factor for MI and sudden cardiac death independent of atherosclerosis development. In this context, it is worth noting that patients with ischemic or idiopathic dilated myocardiopathy are more frequently ACE DD homozygotes than controls (86), and that patients with hypertrophic myocardiopathy have an increased frequency of the ACE D allele, particularly if there is a family history of sudden cardiac death (87).

In patients with coronary artery disease, ACE inhibition attenuates sympathetic coronary vasoconstriction. This may be due to the removal of the facilitating influence of AII on sympathetic modulation of coronary vasomotor tone (88). Furthermore, ACE inhibition has a potent effect to reduce experimental infarct size, probably by increasing collateral flow (89). This may be clinically relevant because viability of the myocardium in patients with MI appears to depend on the presence of collateral blood flow within the infarct bed (90). AII could also interfere with thrombogenesis and contribute to the development of a prothrombic state, since infusion of AII in humans results in a rapid and dose-dependent increase in circulating PAI-1 (91).

ACE D thus appears to influence a large number of phenotypes that may contribute to an increased risk of MI; however, the consequences of a high ACE expression are not necessarily detrimental. Actually, in some circumstances, such as inflammation, cellular growth, wound healing, or angiogenesis, an overefficient ACE gene might even confer benefit. This possibility has been considerably strengthened by the recent observation of an increased frequency of the ACE D allele in French centenarians (92).

SOME SPECULATIONS CONCERNING ALLELES OF CANDIDATE GENES PREDISPOSING TO CHD

Monogenic disorders are generally caused by defective genes, and less frequently by overeffective genes. When considering the different examples given above, it does not appear that alleles predisposing to CHD are functionally deficient. On the contrary, they appear to be overefficient in many respects. It may not be fortuitous that apoE, Lp(a), and ACE may contibute in different ways (local redistribution of cholesterol, cellular growth, vasoconstriction, clot formation, etc.) to vascular repair. We may speculate that the efficient hemostatic and vascular repair mechanisms brought about by evolution to counteract the consequences of injury could become deleterious when they are put into action in the lumen of a coronary artery. Overefficient forms of genes involved in these processes could then contribute to the highly variable individual predisposition to CHD. Some 30 years ago, the thrifty genotype hypothesis was introduced to explain the exploding frequency of diabetes mellitus and obesity observed in some populations undergoing a rapid increase in the availability of food. According to this hypothesis, specific overefficient forms of genes favoring intake, utilization, or storage of nutriments may have been selected in some populations in the past, in situations of shortage or irregular availability of food (93). Analogous hypotheses could also account for the important interpopulational or interindividual variability of lipid metabolism, blood pressure, or vascular repair mechanisms, since the quantity and quality of nutrient fats, the availability of sodium, and physically aggressive environments may have exerted strong selective pressure.

Whether some of the alleles predisposing to CHD have been positively selected by evolution remains speculation, and will be hard to demonstrate. To give support to this hypothesis, it would be of great interest to identify phenotypes that could be favorably affected by these alleles in particular environmental circumstances or at particular stages of development. Aggregation of several variants with similar effects on a single haplotype would also argue against random drift.

MULTIGENIC RISK

Alleles of the ApoB, ApoE, apo(a), and ACE genes contribute significantly to CHD, but several other known or unknown genes are also involved. Table 1 summarizes the available results of the ECTIM study. All the genes listed were selected for study because they were strong candidates, either from a priori physiopathological considerations or because other studies had suggested that they might be involved in CHD. As a consequence, they are not representative of all putative candidates; they are a biased sample that favors the finding of strong effects and minimizes the finding of small or negligible effects. The overall results are compatible with the presence of a large number of functional polymorphisms with negligible effects and a more limited number of functional polymorphisms with significant effects. These latter polymorphisms are those in which we are interested, and an obvious question is what proportion of them do we ignore?

There are a number of cloned candidate genes that have not yet been appropriately explored. This may be less true for genes involved in plasma lipid metabolism than for genes contributing to intracellular lipid metabolism, thrombosis, and vascular homeostasis or repair mechanisms. A large number of genes contributing to CHD may be unknown. In fact, although we know less than 10% of the human genes, the percentage of known genes relevant to CHD may be larger and may differ according to the particular system considered. Circulating proteins are more easily identified than tissular proteins, and proteins present in large amounts are more easily identified than

Table 1 Effects of Several Candidate Polymorphisms on Intermediate Phenotypes and MI risk: Summary of Results of the ECTIM Study

Gene or gene cluster	No. of polymorphisms studied	Effect on intermediate phenotype	Effect on MI risk
ApoAI-CIII-AIV	8	−	−
ApoAII	2	+	−
ApoB	>15[a]	++	++
ApoE	2	++	++
LPL	4	++	+
CETP	1	+++	++
HMG CoA reductase	1[b]	+	−
ACE	1	+++	++
Angiotensinogen	2	++	−
Angiotensin II receptor	2	−	+
Fibrinogen	6	+++	+
Factor VII	2[b]	++	−
PAI-1	2[b]	+	+

[a]Including one minisatellite.
[b]Including one microsatellite.

potent effectors present in very low amounts. Furthermore, the proteins already characterized may be strongly biased toward being more relevant to the pathology than unidentified ones. This is because quantitative or qualitative variation is needed for a particular gene or protein to be involved in the variable predisposition to a disease, and naturally variable components contributing to a pathological trait are searched for in priority and may be more readily identified than nonvariable components.

One factor of progress may be the identification of new relevant domains or the change in focus for atherosclerosis research. For example, endothelial dysfunction and vascular and cardiac hypertrophy have recently been recognized as probable strong contributors to CHD, independent of classic risk factors. This has opened a new area of research and focused attention on new categories of candidate genes for CHD. Thus, even if there are obviously unknown genes contributing to CHD, we may already have access to a sufficient number of genes to be able to characterize a significant part of the genetic component of the disease.

The relevant set of variants for a candidate gene may also be rather limited since these variants are likely to be frequent polymorphisms and not recent rare mutants as for the LDL receptor. It is conceivable that a few tenths of polymorphisms (part of them being still unknown) could account for most of the genetic determination of CHD. If this is true, multigenic risk assessment should be feasible in the future, when most of the relevant forms of genes contributing to CHD have been identified.

CONCLUSION

We are just beginning to uncover the genetic determinants of CHD. It appears that the genotype-phenotype associations are complex as a consequence of pleiotropy, variation with age, selection, and interaction with other factors. Nevertheless, identification of the genes and variants involved in the chronic and acute processes of CHD should considerably improve our understanding of the etiology and mechanisms of this disease. Simultaneous analysis of several alleles predisposing to CHD, moreover, should provide the means by which to identify high-risk individuals and to characterize groups for whom specific therapeutic interventions would be most effective. ACE inhibitors, for example, are powerful cardiovascular drugs; it would be useful to know whether the ACE I/D polymorphism modulates their effects. Adaptation of the therapeutic to the genetic and environmental determinants of the disease may be an important medical challenge for the future. However, it will be important not to interfere with the beneficial effects that overefficient alleles may confer in some situations or at particular ages.

APPENDIX: THE ECTIM STUDY

The ECTIM study is a case-control study specifically designed to identify genetic variants associated with MI. Four populations covered by CHD registries [WHO/MONICA (94)] were targets for study: three populations in France in the regions of Lille, Strasbourg, and Toulouse and one population in Northern Ireland, in the region of Belfast. Of the industrialized countries, France has one of the lowest frequencies of MI; among middle-aged men, this disease is 3 to 4 times more frequent in Northern Ireland than in France. In the ECTIM study, male patients with definite MI aged 35–64 were recruited 3 to 9 months after the acute episode and during a period of $2^{1}/_{2}$ years starting at the end of 1988. Random samples of controls were recruited during the same period and in the same age range as the cases, using electoral rolls in France and the lists of general practioners held by the Central Services Agency in Belfast. The participants were of Caucasian origin; parents of cases and controls had to be born in the same regions; and their four grandparents had to be born in Europe.

REFERENCES

1. Marenberg ME, Risch N, Berkman LF, Floderus B, De Faire U. N Engl J Med 1994; 330:1041–1046.
2. Fuster V, Badimon L, Badimon JJ, Chesebro JH. The pathogenesis of coronary artery disease and the acute coronary syndromes (second of two parts). N Engl J Med 1992; 326:310–318.
3. Ross R. The pathogenesis of atherosclerosis: a perspective for the 1990s. Nature 1993; 362:801–809.
4. Cambien F, Warnet JM, Jacqueson A, Ducimetière P, Richard JL, Claude JR. Relation of parental history of early myocardial infarction to the level of apolipoprotein B in men. Circulation 1987; 76:266–271.
5. Hasstedt SJ, Wu L, Williams RR. Major locus inheritance of apolipoprotein B. in Utah pedigrees. Genet Epidemiol 1987; 4:67–76.
6. Tiret L, Steinmetz J, Herbeth B, Visvikis S, Rakotovao R, Ducimetière P, Cambien F. Familial resemblance of plasma apolipoprotein B: The Nancy Study. Genet Epidemiol 1990; 7:187–197.
7. Sniderman A, Shapiro S, Marpole D, Malcolm I, Skinner B, Kwiterovich PO Jr. Association of coronary atherosclerosis with hyperapobetalipoproteinemia (increased protein but normal cholesterol levels in human low density lipoproteins). Proc Natl Acad Sci USA 1980; 77:604–608.
8. Leitersdorf E, Gottehrer N, Fainaru M, Friedlander Y, Friedman G, Tzivoni D, Stein Y. Analysis of risk factors in 532 survivors of first myocardial infarction hospitalized in Jerusalem. Atherosclerosis 1986; 59:75–93.
9. Parra HJ. Arveiler D, Evans AE, Cambou JP, Amouyel A, Bingham A, McMaster D, Schaffer P, Douste-Blazy P, Luc G, Richard JL, Ducimetière P, Fruchart JC, Cambien F. A case control study of lipoprotein particles in two populations at contrasting risk for coronary heart disease: The ECTIM study. Artérioscler Thromb 1992; 12:701–707.
10. Rosseneu M, Fruchart JC, Bard JM, Nicaud V, Vinaimont M, Cambien F, De Backer G. Plasma apolipoprotein concentrations in young adults with a parental history of premature coronary heart disease and in control subjects—The EARS Study, Circulation 1994; 89:1967–1973.
11. Young SG. Recent progress in understanding Apolipoprotein B. Circulation 1989; 80:219–233.
12. Wettereau JR, Aggerbeck LP, Bouma ME, Eisenberg C, Munck A. Hermier M, Schmitz J, Gay G, Rader DJ, Gregg RE. Absence of microsomal triglyceride transfer protein in individuals with abetalipoproteinemia. Science 1992; 258:999–1001.
13. Berg K, Hannes C, Dahlen G, Frick H, Krishan I. Genetic variation in serum low density lipoproteins and lipid levels in man. Proc Nat Acad Sci USA 1976; 73:937–940.
14. Dunning AM, Renges HH, Xu CF, Peacock R, Brasseur R, Laxer G, Tikkanen MJ, Bütler R, Saha N, Hamsten A, Rosseneu M, Talmud P, Humphries SE. Two amino acid substitutions in apolipoprotein B are in complete allelic association with the antigen group (x/y) polymorphism. Am J Hum Genet 1991; 50:208–221.
15. Moreel JFR, Roizes G, Evans AE, Arveiler D, Cambou JP, Souriau C, Parra HJ, Desmarais E, Fruchart JC, Ducimetière P, Cambien F. The polymorphism Apo

B/4311 in patients with myocardial infraction and controls: the ECTIM study. Hum Genet 1992; 89:169–175.

16. Young SG, Pullinger CR, Zysow BR, Hofmann-Radvani H, Linton MF, Farese RV, Terdiman JF, Synder SM, Grundy SM, Vega GL, Malloy MJ, Kane JP. Four new mutations in the apolipoprotein B gene causing hypobetalipoproteinemia, including 2 different frameshift mutations that yield truncated apolipoprotein B proteins of identical length. J Lipid Res 1993; 34:501–507.

17. Soria LF, Ludwig EH, Clarke HRG, Vega GL, Grundy SM, McCarthy BJ. Association between a specific apolipoprotein B mutation and familial defective ApoB100. Proc Natl Acad Sci USA 1989; 86:587–591.

18. Myant NB. Familial defective apolipoprotein B100. a review, including some somparisons with familial hypercholesterolaemia. Atherosclerosis 1993; 104:1–18.

19. Hobbs, HH, Russel DW, Brown MS, Goldstein JL. The LDL receptor locus in familial hypercholesterolemia: mutational analysis of a membrane protein. Annu Rev Genet 1990; 24:133–170.

20. Desmarais E, Vigneron S, Buresi C, Cambien F, Cambou JP, Roizes G. Variant mapping of the Apo(B) AT rich minisatellite. Dependence on nucleotide sequence of the copy number variations. Instability of non-canonical alleles. Nucleic Acids Res 1993; 21:2179–2184.

21. Cambien F, Evans AE, Cambou JP, Arveiler D, Luc G, Moreel JFR, Jouin E, Desmarais E, Bard JM, Poirier O, Fruchart JC, Tiret L. Body mass is associated with myocardial infraction in individuals carrying the 48 repeats allele of the ApoB 3′ hypervariable region. Am J Hum Genet 1992; 51:A 146.

22. Evans AE, Zhang W, Moreel JFR, Bard JM, Ricard S, Poirier O, Tiret L, Fruchard JC, Cambien F. Polymorphisms of the apolipoprotein B and E genes and their relationship to plasma lipid variables in healthy chinese men. Hum Genet 1993; 92:191–197.

23. Rapacz J, Chen L, Butler-Brunner E, Wu MJ, Hasler-Rapacz JO, Butler R, Schumaker VN. Identification of the ancestral haplotype for apolipoprotein B suggests an African origin of Homo sapiens sapiens and traces their subsequent migration to Europe and the Pacific. Proc Natl Acad Sci USA 1991; 88:1403–1406.

24. Mahley RW. Apolipoprotein E: cholesterol transport protein with expanding role in cell biology. Science 1988; 240:622–630.

25. Uterman G, Pruin N, Steinmetz A. Polymorphism of apolipoprotein E. III. Effect of a single polymorphic gene locus on plasma lipid levels in man. Clin Genet 1979; 15:63–72.

26. Gerdes LU, Klausen IC, Sihm I, Faegerman O. Apolipoprotein E polymorphism in a Danish population compared to findings in 45 other study populations around the world. Genet Epidemiol 1992; 155–167.

27. Weisgraber KH, Innerarity TL, Rall SC, Mahley RW. Abnormal receptor-binding activity of the human E apolipoprotein due to the cystein-arginin interchange at a single site. J Biol Chem 1982; 257:2518–2521.

28. Weisgraber KH. Apolipoprotein E distribution among human plasma lipoproteins; role of cysteine-arginine interchange at residue 112. J Lipid Res 1990; 31:1503–1511.

29. Weintraub MS, Eisenberg S, Breslow JL. Dietary fat clearance in normal subjects is regulated by genetic variation in apolipoprotein E. J Clin Invest 1987; 80:1571–1577.

30. Kesäniemi YA, Ehnholm C, Miettinen TA. Intestinal cholesterol absorption efficiency in man is related to apolipoprotein E phenotype. J Clin Invest 1987; 80:578–581.

31. Davignon J, Gregg RE, Sing CF. ApoE polymorphism and atherosclerosis. Arteriosclerosis 1988; 8:1–21.

32. Demant T, Bedford D, Packard CJ, Shepherd J. Influence of apolipoprotein E polymorphism on apolipoprotein B100 metabolism in normolipidemic subjects. J Clin Invest 1991; 88:1490–1501.

33. Hixson JE and the PDAY Research Group. Apolipoprotein E polymorphisms affect atherosclerosis in young males. Arterioscler Thromb 1991; 11:1237–1244.

34. Luc G, Bard JM, Arveiler D, Evans A, Cambou JP, Bingham A, Amouyel P, Schaffer P, Ruidavets JB, Cambien F, Fruchart JC, Ducimetière P. The impact of apolipoprotein E polymorphism on lipoproteins and risk of myocardial infarction. The ECTIM study. In press.

35. Corder EH, Saunders AM, Strittmatter WJ, Schmechel DE, Gaskell PC, Small GW, Roses AD, Haines JL, Pericak-Vance MA. Gene dose of apolipoprotein E type 4 allele and the risk of Alzheimer's disease in late onset families. Science 1993; 261:921–923.

36. Boerwinkle E, Leffert CC, Lin J, Lackner C, Chiesa G, Hobbs H. Apolipoprotein(a) gene accounts for greater than 90% of the variation in plasma Lp(a) concentration. J Clin Invest 1992; 90:52–60.

37. McLean JW, Tomlinson JE, Kuang WJ, Eaton DL, Chen EY, Fless GM, Scanu AM, Lawn RM. cDNA sequence of human apolipoprotein(a) is homologous to plasminogen. Nature 1987; 300:132–137.

38. Cohen JC, Chiesa G, Hobbs HH. Sequence polymorphisms in the Apolipoprotein(a) gene. J Clin Invest 1993; 91:1630–1636.

39. Kraft HG, Köchl S, Menzel HJ, Sandholzer C, Uterman G. The apolipoprotein(a) gene: a transcribed hypervariable locus controlling plasma lipoprotein(a) concentration. Hum Genet 1992; 90:220–230.

40. Jauhiainen M, Koskinen P, Ehnholm C, et al. Lipoprotein(a) and coronary heart disease risk: a nested case-control study of the Helsinki heart study participants. Atherosclerosis 1991; 89:59–67.

41. Ridker PM, Hennekens CH, Stampfer MJ. A prospective study of lipoprotein(a) and the risk of myocardial infarction. JAMA 1993; 270:2195–2199.

42. Klausen IC, Beisigel U. Menzel HJ, Rosseneu M, Ehnholm C, Tiret L. Apo(a) phenotypes and Lp(a) concentrations in offspring of men under 55 years with myocardial infarction. The EARS study. In press.

43. Wald NJ, Law M, Watt HC, Wu T, Bailey A, Johnson M, Craig WY, Ledue TB, Haddow JE. Apolipoproteins and ischemic heart disease: implications for screening. Lancet 1994; 343:75–79.

44. Schaefer EJ, Lamon-Fava S, Jenner JL, McNamara JR, Ordovas JM, Davis E, Abolafia JM, Lippel K, Levy RI. Lipoprotein(a) levels and risk of coronary heart disease in men. JAMA 1994; 271:999–1003.

45. Dahlen GH, Guyton JR, Attar M, Farmer JA, Kautz JA, Gotto AM Jr. Association of levels of lipoprotein(a), plasma lipids, and other lipoproteins with coronary artery disease documented by angiography. Circulation 1986; 74:758–765.

46. Schreiner PJ, Morrisett, Sharrett R, Patsch W, Tyroler HA, Wu K, Heiss G. Lipoprotein(a) as a risk factor for preclinical atherosclerosis. Arterioscler Thromb 1993; 13:826–833.

47. Sandholzer C, Saha N, Kark JD, Rees A, Jaross W, Dieplinger H, Hoppichler F, Boerwinkle E, Utermann G. Apo(a) isoforms predict risk for coronary heart disease: A study in six populations. Arterioscler Thromb 1992; 12:1214–1226.

48. Sandholzer C, Hallman DM, Saha N, Sigurdsson G, Lackner C, Csaszar A, Boerwinkle E, Utermann G. Effects of the apolipoprotein(a) size polymorphism on the lipoprotein(a) concentration in 7 ethnic groups. Hum Genet 1991; 86:607–614.

49. Perombelon YFN, Soutar AK, Knight BL. Variation in lipoprotein(a) concentration associated with different apolipoprotein(a) alleles. J Clin Invest 1994; 93:1481–1492.

50. Laplaud PM, Beaubatie L, Rall SJ Jr, Luc G, Saboureau M. J Lipid Res 1988; 29:1157–1170.

51. Edelberg JM, Pizzo SV. Lipoprotein(a): link between impaired fibrinolysis and atherosclerosis. Fibrinolysis 1991; 5:135–143.

52. Grainger DJ, Kirschenlohr HL, Metcalfe JC, Weissberg PL, Wade DP, Lawn RM. Proliferation of human smooth muscle cells promoted by lipoprotein(a). Science 1993; 260:1655–1658.

53. Brown MS, Goldstein JL. Teaching old dogma new tricks. Nature 1987; 300:113–114.

54. Whelan RF, Scroop GC, Walsh JA. Cardiovascular action of angiotensin in man. Am Heart J 1969; 77:546–565.

55. Magrini F, Reggiani P, Roberts N, Meazza R, Ciulla M, Zanchetti A. Effects of angiotensin and angiotensin blockade on coronary circulation and coronary reserve. Am J Med 1988; 84(suppl 3A):55–60.

56. Powell JS, Clozel JP, Müller RKM, Kuhn H, Hefti F, Hosang M, Baumgartner HR. Inhibitors of angiotensin-converting enzyme prevent myointimal proliferation after vascular injury. Science 1989; 245:186–188.

57. Pelc LR, Gross GJ, Warltier DC. Mechanism of coronary vasodilatation produced by bradykinin. Circulation 1991; 83:2048–2056.

58. Farhy RD, Ho KL, Carretero OA, Scicli AG. Kinins mediate the antiproliferative effect of ramipril in rat carotid artery. Biochem Biophys Res Comm 1992; 182:283–288.

59. Linz W, Schölkens BA. Role of bradykinin in the cardiac effects of angiotensin-converting enzyme inhibitors. J Cardiovasc Pharmacol 1992; 20(suppl 9):S83–S90.

60. Wiemer G, Schölkens BA, Becker RHA, Busse R. Ramiprilat enhances endothelial autocoid foramtion by inhibiting breakdown of endothelium-derived bradykinin. Hypertension 1991; 18:558–563.

61. Alhenc-Gelas F, Richard J, Courbon D, Warnet JM, Corvol P. Distribution of plasma angiotensin I-converting enzyme levels in healthy men: Relationship to environmental and hormonal parameters. J Lab Clin Med 1991; 117:33–39.

62. Cambien F, Alhenc-Gelas F, Herbeth B, Andre JL, Rakotovao R, Gonzales MF, Allegrini J, Bloch C. Familial resemblance of plasma angiotensin-converting enzyme level: the Nancy study. Am J Hum Genet 1988; 43:774-780.

63. Tiret L, Rigat B, Visvikis S, Breda C, Corvol P, Cambien F, Soubrier F. Evidence, from combined segregation and linkage analysis, that a variant of the angiotensin I-converting enzyme (ACE) gene controls plasma ACE. Am J Hum Genet 1992; 51:197-205.

64. Soubrier F, Alhenc-Gelas F, Hubert C, Allegrini J, John M, Tregcar G, Corvol P. Two putative active centers in human angiotensin I-converting enzyme revealed by molecular cloning. Proc Natl Acad Sci USA 1988; 85:9386-9390.

65. Rigat B, Hubert C, Alhenc-Gelas F, Cambien F, Corvol P, Soubrier F. An insertion-deletion polymorphism in the angiotensin I-converting enzyme gene accounting for half the variance of serum enzyme levels. J Clin Invest 1990; 86:1343-1346.

66. Costerousse O, Allegrini J, Lopez M, Alhenc-Gelas F. Angiotensin I-converting enzyme in human peripheral mononuclear cells: main expression in T lymphocytes under the influence of a genetic polymorphism. Biochem J 1993; 290:33-40.

67. Jeunemaitre X, Lifton RP, Hunt SC, Williams RR, Lalouel JM. Absence of linkage between the angiotensin converting enzyme locus and human essential hypertension. Nature Genet 1992; 1:72-75.

68. Harrap SB, Davidson HR, Connor JM, Soubrier F, Corvol P, Fraser R, Foy CJW, Watt GCM. The angiotensin I converting enzyme gene and predisposition to high blood pressure. Hypertension 1993; 21:455-460.

69. Schmidt S, va Hooft IMS, Grobbee DE, Ganten D, Ritz E. Polymorphism of the angiotensin I converting enzyme gene is apparently not related to high blood pressure: Dutch hypertension and offspring study. J Hypertension 1993; 11:345-348.

70. Higashimori K, Zhao Y, Higaki J, Kamitani A, Katsuya T, Nakura J, Miki T, Mikami H, Ogihara T. Association analysis of a polymorphism of the angiotensin converting enzyme gene with essential hypertension in the Japanese population. Biochem Biophys Res Commun 1993; 191:399-404.

71. Zee RYL, Lou YK, Griffiths LR, Morris BJ. Association of a polymorphism of the angiotensin I-converting enzyme gene with essential hypertension. Biochem Biophys Res Commun 1992; 184:9-15.

72. Cambien F, Poirier O, Lecerf L, Evans A, Cambou JP, Arveiler D, Luc G, Bard JM, Bara L, Ricard S, Tiret L, Amouyel P, Alhenc-Gelas F, Soubrier F. Deletion polymorphism at the angiotensin-converting enzyme gene is a potent risk factor for myocardial infarction. Nature 1992; 359:641-644.

73. Tiret L, Kee F, Poirier O, Nicaud V, Lecerf L, Evans AE, Cambou JP, Arveiler D, Luc G, Amouyel P, Cambien F. Deletion polymorphism in the angiotensin converting enzyme gene is associated with a parental history of myocardial infarction. Lancet 1993; 341:991-992.

74. Cambien F, Costerousse O, Tiret L, Poirier O, Lecerf L, Gonzeles MF, Evans A, Arveiler D, Cambou JP, Luc G, Rakotovao R, Ducimetière P, Soubrier F, Alhenc-Gelas F. Plasma level and gene polymorphism of angiotensin-converting enzyme in relation to myocardial infarction. Circulation 1994; 90:669-676.

75. Evans AE, Poirier O, Kee F, Lecerf L, McCrum E, Crane J, O'Rourke DF, Cambien F. Distribution of the angiotensin converting enzyme gene polymorphism in subjects who die of coronary heart disease. Q J Med 1994; 87:211-214.

76. Bohn M, Berge KE, Bakken A, Erikssen J, Berg K. Insertion/selection (I/D) polymorphism at the locus for angiotensin I-converting enzyme and myocardial infarction. Clin Genet 1993; 44:292-297.

77. Bohn M, Berge KE, Bakken A, Erikssen J, Berg K. Insertion/selection (I/D) polymorphism at the locus for angiotensin I-converting enzyme and parental history of myocardial infarction. Clin Genet 1993; 44:298-301.

78. Paul M, Wagner J, Dzau VJ. Gene expression in the renin-angiotensin system in human tissues. J Clin Invest 1993; 91:2058-2064.

79. Saed GM, Carretero OA, MacDonald RJ, Scicli G. Kallikrein messenger RNA in rat arteries and veins. Circ Res 1990; 67:510-516.

80. Scicli AG, Farhy R, Scicli G. Nolly H. In: Bönner G, Schölkens BA, Scicli AG, eds. The Role of Bradykinin in the Cardiovascular Action of Ramipril. Frankfurt: Hoechst Aktiengesellschaft 1992:17-28.

81. Ludwig EH, Corneli PS, Anderson JL, Marshall HW, Lalouel JM, Ward RH. Angiotensin converting enzyme gene polymorphism is associated with myocardial infarction but not with development of coronary stenosis. Circulation 1995; 91:2120-2124.

82. Bonithon-Kopp C, Ducimetière P, Touboul PJ, Fève JM, Billaud E, Courbon D, Héraud V. Circulation 1994; 89:952-954.

83. Kapprio J, Norio R, Pesonen E, Sarna S. Intimal thickening of the coronary arteries in infants in relation to family history of coronary artery disease. Circulation 1993; 87:1960-1968.

84. Schunkert H, Hense HW, Holmer SR, Stender M, Perz S, Keil U, Lorell BH, Riegger AJ. Association between a deletion polymorphism of the angiotensin converting enzyme gene and left ventricular hypertrophy. N Engl J Med 1994; 330:1634-1638.

85. Kannel WB. Left ventricular hypertrophy as a risk factor: the Framingham experience. J Hypertension 1991; 9(suppl 2):S2-S9.

86. Raynolds MV, Bristow MR, Bush EW, Abraham WT, Lowes BD, Zisman LS, Taft CS, Perryman B. Angiotensin-converting enzyme DD genotype in patients with ischaemic or idiopathic cardimyopathy. Lancet 1993; 342:1073-1075.

87. Marian AJ, Yu QT, Workman R, Greve G, Roberts R. Angiotensin-converting enzyme polymorphism in hypertrophic cardiomyopathy and sudden cardiac death. Lancet 1993; 342:1085-1086.

88. Perondi R, Saino A, Tio RA, Pomidossi G, Gregorini L, Alessio P, Morganti A, Zanchetti A, Mancia G. ACE inhibition attenuates sympathetic coronary vasoconstriction in patients with coronary heart disease. Circulation 1992; 85:2004-2013.

89. Ertl G, Kloner RA, Wayne Alexander R, Braunwald E. Limitation of experimental infarct size by an angiotensin-converting enzyme inhibitor. Circulation 1982; 65:40-48.

90. Sabia PJ, Powers ER, Ragosta M, Sarembock IJ, Burwell LR, Kaul S. An association between collateral blood flow and myocardial viability in patients with recent myocardial infaction. N Engl J Med 1992; 26:1825-1831.

91. Ridker PM, Gaboury CL, Conlin PR, Seely EW, Williams GH, Vaughan DE. Stimulation of plasminogen activator inhibitor in vivo by infusion of angiotensin II. Circulation 1993; 87:1969–1973.
92. Schächter F, Faure-Delanef L, Guénot F, Rouger H, Froguel P, Lesueur-Ginot L, Cohen D. Genetic associations with human longevity at the ApoE and ACE loci. Nature Genet 1994; 6:29–32.
93. Neel JV. Diabetes mellitus: a thrifty genotype rendered detrimental by progress. Am J Hum Genet 1962; 14:353–362.
94. The WHO MONICA Project. A worldwide monitoring system for cardiovascular diseases. World Health Statist Annu 1989:27–149.

10

Using Animal Models to Dissect the Genetics of Complex Traits

John H. Krege and Oliver Smithies
University of North Carolina at Chapel Hill
Chapel Hill, North Carolina

Theodore W. Kurtz
University of California, San Francisco
San Francisco, California

INTRODUCTION

Animal models provide excellent systems for studying environmental and genetic factors that cause complex (multifactorial) diseases such as hypertension, atherosclerosis, stroke, and heart failure. This chapter describes general methods used in generating animal models, and how these recent advances in molecular biology and genetics are providing new ways of investigating the molecular mechanisms of cardiovascular diseases.

A number of single-gene disorders in animals have been discovered that closely parallel human genetic disorders. For example, in animals, as in humans, mutations in the LDL receptor cause hypercholesterolemia (1); mutations in the Factor IX gene cause hemophilia (2); and mutations in the vasopressin gene cause diabetes insipidus (3).

Complex traits (such as lipid metabolism and blood pressure) are determined by the interaction of a wide range of environmental and genetic factors that maintain normal homeostasis. Multiple genetic systems are involved, composed of proteins encoded by several or many genes. We are interested in what happens when these genes go awry. Genetic variants (alleles) may influence disease in several ways. For example, a dysfunctional allele may code for a *qualitative* change, resulting in an abnormal protein, or a *quantitative* change, resulting in reduced or increased production of a normal protein. However, dysfunctional alleles do not always produce detectable changes in phenotype, because the overall system may be able to completely compensate for the adverse effect of the dysfunctional allele. When an altered phenotype does occur, it may be the result of a combination of several factors, genetic and environmental, that in total exceed the capacity of the compensatory mechanisms. Animal models of the types described in this chapter provide important systems in which to precisely manipulate and study environmental and genetic factors that control complex traits such as blood pressure. The study of animal models in cardiovascular research will likely lead to new diagnostic procedures, innovative therapies, and better prevention strategies.

The use of animals to study complex traits can be divided into two general schemes: analytical and synthetic. *Genetic analysis* experiments typically seek to identify potentially deleterious alleles in animals having phenotypic differences (such as in blood pressure) that were generated by selective breeding strategies. The investigator compares the genomes of animals with and without a relevant phenotype and identifies candidate alleles by using special breeding strategies. These candidate genes, once identified through analytical approaches or on *a priori* grounds, can then be studied by a *genetic synthesis* approach. In this type of experiment, the investigator introduces genetic modifications into the genomes of whole animals to determine their impact at the molecular, biochemical, cellular, and physiological levels.

One experimental technique that since its inception in 1980 has proved of great value is the generation of transgenic animals that have *exogenous* DNA sequences introduced into their genomes. The study of animals transgenic for chosen regulatory and coding sequences has revealed important information regarding the mechanisms responsible for the timing, tissue specificity, and level of gene expression (see Ref. 4, for example). Directing the expression of transgenes to specific tissues has also allowed investigators to genetically ablate (5) or stimulate (6) target cells. Models of human diseases, such as amyotrophic lateral sclerosis, have been created through the overexpression of a chosen normal or mutant transgene (7). The study of animals harboring random insertional mutations has led other investigators to discover new genes important in biological functions (see Ref. 8, for example).

An alternative approach to synthesizing genetic diseases is provided by the use of gene targeting in embryonic stem cells. Since its inception in 1989, this technology has been used widely to generate mice having specific mutations in chosen *endogenous* genes. The method has had a great impact in numerous fields of investigation, such as immune function, e.g., generation of defined immune deficiencies (9,10), developmental biology, e.g., genetic control of pattern formation and differentiation (11), neurobiology, e.g., investigation of higher-order functions, such as memory (12,13), tumor biology, e.g., tumor suppressor gene models (14), and human genetic disease models, e.g., the apoE-deficient atherosclerotic mouse (15,16). Since, in principle, targeted mutations can be made in any desired gene, the generation of genetically altered mice will likely continue to increase dramatically as new "targets" of unknown function are revealed by the Human Genome Project. As we discuss later in this chapter, genetically altered animals for studying cardiovascular diseases, including hypertension, atherosclerosis, congenital defects, and other diseases of the heart and vasculature, have already been or are in the process of being developed.

GENETIC ANALYSIS: THE IDENTIFICATION OF CANDIDATE GENES

The ability to generate animal models for hypertension, stroke, cardiac hypertrophy, and atherosclerosis by selective inbreeding constitutes direct evidence that genetic factors can be important determinants of these common cardiovascular disorders. For example, selective breeding protocols have been successfully used to derive a number of rodent strains with hypertension (17), including the spontaneously hypertensive rat (SHR)—the most widely studied animal model of essential hypertension. Okamoto and Aoki (18,19) derived the SHR by measuring blood pressure in a large number of non-inbred Wistar rats and then selectively breeding littermates with the highest blood pressures. After four generations of selective inbreeding, most of the progeny exhibited frank hypertension. Because the majority of the hypertensive phenotype was obtained after only a few generations of selective inbreeding, it is likely that just a few genes are largely responsible for the increased blood pressure of the SHR (18). In some other animal models of hypertension, the selection process resulted in more gradual increases in blood pressure. The derivations of both the BP1 strain of genetically hypertensive mice (20) and the New Zealand strain of genetically hypertensive rats (21) required more than 12 generations of selective breeding. In these models, it is likely that many alleles, each having individually modest effects on blood pressure, are contributing to the hypertension.

Several considerations should be mentioned regarding the use of these types of animal models to understand human disease. First, the method of determining the phenotype may critically influence the genetic analysis (22). During the derivation of the SHR, for example, the selection for hypertension involved tail-cuff measurements. Because tail-cuff blood pressures require the stresses of heating and restraint of the animal (23,24), the selection process may have selected animals more susceptible to stress-induced increases in blood-pressure. Accordingly, the identification of genes contributing to hypertension in the SHR may shed light particularly on genetic susceptibility to increased blood pressure related to stress. Second, the outcome of studies of a candidate gene may be strongly influenced by other environmental or genetic factors; some genes, for example, may be important in producing hypertension or atherosclerosis only under conditions of increased dietary salt (25,26) or fat (27). Finally, the specific alleles that produce a phenotype in rodents may not be present in humans. Even so, genes that are found to cause disease in rodents frequently become important candidates in humans.

Comparisons of Disease and Control Strains

One method to analyze the genetics of cardiovascular diseases involves making systematic comparisons between the genomes of disease strains and their corresponding "controls." In studies of this type, it is hoped that identified genetic differences found in the disease strain are pathological. However, such interstrain comparisons present special challenges in attempts to identify relevant genetic variants, because the disease and control strains usually differ with respect to many genes, not just those affecting the phenotype of interest. During the derivation of the disease strain, the process of selection and inbreeding results not only in the fixation of alleles that affect the phenotype of interest, but also in the random fixation of alleles unrelated to disease pathogenesis. Because of this difficulty, another approach is to compare candidate genes in a disease strain with the corresponding genes from many different control strains (28). The presence of a particular genotype in the disease strain, but not in any of the control strains, provides additional evidence for the involvement of that genotype in disease pathogenesis.

By itself, however, the presence of a unique genetic variant in a disease model does not prove that the variant is involved in disease pathogenesis; it may be a neutral difference that was fixed in the disease strain by chance. In the reverse sense, the presence of a given variant or polymorphism in the control strain does not exclude the possible pathogenic involvement of that variant or polymorphism in the disease of interest; the control strain may have differences in other genes that counterbalance the deleterious variant (29).

Use of Segregating Populations to Map Genes

The genotypic and phenotypic evaluation of segregating populations derived from crosses between a disease strain and a control strain is sometimes very useful for mapping (localizing) genes involved in common cardiovascular disorders. This type of study usually involves crossing two strains with differing phenotypes and determining whether the phenotype of interest (e.g., blood pressure, stroke, heart failure) segregates with the inheritance of a particular genetic marker. The two strains are initially bred to each other to produce first-generation hybrid offspring (F1). The F1 animals are then mated to each other to produce F2 offspring, or with one of the parental strains to produce backcross offspring, and the investigator determines whether the phenotype of offspring inheriting a particular allele at a given candidate locus differs from the phenotype of the offspring inheriting an alternative allele at the same locus (30). Alternatively, the investigator can test the segregating animals with a panel of markers distributed throughout the entire genome. Using these methodologies, studies of segregating populations derived from rodent models of hypertension have recently implicated a number of chromosome regions and candidate genes in the pathogenesis of increased blood pressure (31-37).

Additional investigations are needed to establish that genes implicated by linkage studies are indeed pathogenic. Rapp, in his "Paradigm for Identification of Primary Genetic Causes of Hypertension in Rats" (30) described four criteria for demonstrating a role for a gene in influencing complex traits such as blood pressure:

> 1) a difference in a biochemical or physiological trait between two strains must be demonstrated; 2) the trait must be shown to follow Mendelian inheritance; 3) the genes identified in criterion 2 must co-segregate with an increment in blood pressure which is significantly different from zero; 4) there must be some logical biochemical and/or physiological link between the trait and blood pressure.

Thus, the results of linkage studies must be supported by additional physiological or biochemical data that can be related to mechanisms of disease pathogenesis.

Genetic studies in segregating populations are most useful for the initial identification of relatively large chromosome regions (roughly 10–20 million basepairs of DNA) that contain a genetic variant affecting the phenotype of interest. Because genes that are chromosomally located close together (linked) will usually be inherited together, very large numbers of animals must be analyzed to more definitively localize the pathological genetic variant.

Due to the uniqueness of the individual animals generated in an F2 or backcross study, it is often impossible to share segregating populations among different investigators to confirm the results; it is also difficult to extensively characterize a single group of animals with respect to the wide variety of physiological and biochemical pathways of interest. Recombinant inbred (RI) strains provide an alternative to segregating F2 populations for localizing and identifying important genes (38). Such RI strains provide a reproducible system for studying alleles segregating in different genetic backgrounds, and the resulting animals can be repeatedly studied with different techniques and by different investigators. As shown in Figure 1, the investigator starts by

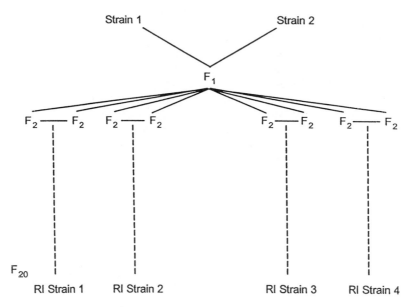

Figure 1 Protocol for derivation of recombinant inbred (RI) strains. Two progenitor strains (strain 1 and strain 2) that differ with respect to blood pressure or some other cardiovascular trait of interest are initially crossed to yield an F1 population. The F1 animals are then intercrossed to yield an F2 population. The F2 animals are paired off at random, and each of these F2 pairs is used to generate a new inbred strain (by repeated brother × sister mating of the offspring for at least 20 generations). This figure depicts the creation of four new RI strains. However, investigators typically create 10–40 new lines when deriving a set of RI strains. The cardiovascular phenotypes of the new RI strains are then correlated with the genotypes that are inherited by these strains.

crossing an inbred disease strain with an inbred control strain to produce F1 offspring, which are then mated to produce F2 offspring. The F2 offspring are paired randomly, and each F2 pair is used to generate a new inbred strain (by repeated brother × sister matings for at least 20 generations). The investigator then determines whether the phenotypes of RI strains inheriting an allele of interest from the disease progenitor strain are different from those RI strains inheriting the allele from the control progenitor strain. Because a great deal of time and effort are required to develop and characterize sets of RI strains, only a few investigators have used this methodology to study the genetics of hypertension. Pravenec et al. (39) have derived a large set of RI strains from the SHR and the normotensive Brown Norway rat. RI strains have been used much more extensively to study the genetics of atherosclerosis in mice (40–42).

Use of Congenic Strains to Map Genes

Once linkage studies have identified a chromosome region containing a genetic variant regulating a cardiovascular phenotype of interest, the next step is to confirm the result and begin to localize the variant to a more restricted chromosome region. This can be accomplished by chromosome "transfer" studies to determine the effects of moving selected chromosome segments from a disease strain into a control strain, or vice versa. By transferring smaller and smaller chromosome segments (as we describe next) between disease and control strains, small chromosome segments containing genes regulating the disease phenotype can be identified.

The transfer of a small chromosomal segment from one (donor) strain into another (recipient) strain results in the creation of a new strain, a congenic strain, that is genetically identical to the recipient strain except for the single chromosome segment coming from the donor. Thus, if the phenotype of the congenic strain differs from that of the recipient progenitor strain, this provides definitive evidence that a locus affecting that phenotype (directly or indirectly) exists within the transferred chromosome segment. Molecular, biochemical, and physiological differences between these congenic strains can then be traced to the differential chromosome segment.

Congenic strains are most commonly derived (see Figure 2) by backcross breeding, i.e., by breeding the "evolving" congenic animals back to the inbred recipient strain. In each backcross cycle, the offspring are genotyped and those carrying the desired (donor) chromosome region are selected for further backcrossing to the recipient strain. After eight to ten backcross generations, the congenic animals that carry the foreign (donor) chromosome segment are genetically very similar (99 to 99.9% identical) to animals of the recipient strain that have not been genetically manipulated (43). Although multiple generations of backcross breeding are required to derive congenic strains, it is possible to accelerate the process by using multiple DNA markers to guide selection of the offspring for use in the next breeding cycle. By genotyping offspring with DNA markers scattered throughout the genome, it is possible to select not only *for* the chromosome segment that one desires to transfer but also *against* other donor strain regions of the genome that one does not intend to transfer. The size of the transferred chromosome segment can be reduced by continued backcross breeding and molecular selection. Congenic strains have been used to study the genetic determinants of atherosclerosis (44,45) and hypertension (39,46–48).

Because the size of chromosome segments transferred by backcross breeding and selection techniques in generating congenic strains is still large at the DNA level (typically 1 or 2 million basepairs), these animals cannot by themselves definitively pinpoint specific mutations contributing to disease

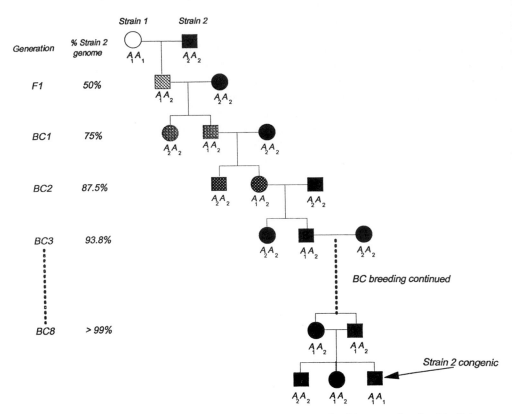

Figure 2   Protocol for derivation of a congenic strain. In this example, the A1 allele from donor strain 1 is transferred onto the genetic background of recipient strain 2. This is accomplished by repeated backcross cycles between offspring that carry the A1 allele and the parental strain 2. After eight backcross (BC) generations, offspring that carry one copy of the A1 allele on the strain 2 background are intercrossed to produce a congenic strain 2 that is homozygous A1A1 (arrow). The intensity of shading reflects the degree of genetic similarity between animals in each backcross generation versus those of the recipient strain 2. For genes that are not linked to the A1 locus, there is greater than 99% probability that the congenic strain derived after eight backcross cycles will be identical to the original parental strain 2. If the new congenic strain differs from parental strain 2 with respect to some cardiovascular phenotype, this provides direct evidence that a gene(s) influencing the phenotype exists in the vicinity of the A1 allele on the transferred chromosome segment.

pathogenesis (unless an unambiguously pathological mutation is discovered). However, the results of physiological and biochemical studies carried out on the congenic strains can implicate candidate genes present within the transferred chromosome segments. Once specific candidate genes have been identified in restricted chromosome regions, other methods can be used to directly test their importance in the genetic control of cardiovascular disease.

GENETIC SYNTHESIS: CHARACTERIZATION OF THE ROLE OF CANDIDATE GENES IN DISEASE

This section describes experiments that provide ways of directly testing the importance of candidate genes in producing a phenotype of interest. Animals can be generated that differ by only a single chosen genetic modification. Any differences in the phenotype of interest in these animals are therefore directly caused by the genetic modification. Additionally, by combining such animals in appropriate ways, the synthesis of complex diseases from their individual components can be attempted.

Creating Synthetic Animal Models

Rats and mice have proved to be of great value in modeling diseases for a variety of reasons. They are relatively inexpensive to breed and house. The physiology, biochemistry, and molecular genetics of humans and rodents are similar (49). Additionally, a great deal of information is available regarding mouse and rat genetics (50–52).

Pronuclear Injection of Single-Cell Embryos

The microinjection of foreign DNA into the pronucleus of a single-cell embryo can result in the random incorporation of *exogenous* sequences into the genome as a "transgene" (Table 1). Alternatively, foreign DNA can be introduced by retroviral infection of embryos (53). Transgenes are typically composed of a desired structural gene along with DNA sequences chosen to control the function of the gene in a useful way. Offspring produced in this manner are screened for successful integration of the transgene into the genome, and for transmission of the transgene to the next generation. This method provides little or no control over the location and the number of copies of transgenes that integrate into the genome. Consequently, it is usually necessary to analyze several independent lines to achieve the desired tissue specificity and expression level. See Mockrin et al. (54), Sigmund (55), and Chapter 11 of this volume for reviews of applications of this technology to the study of hypertension. See Rubin and Smith (56) and Lusis (57) for reviews of its application to the study of atherosclerosis.

Table 1   Two Methods Used in Synthetic Genetic Experiments

Pronuclear injection of single-cell embryos	Gene targeting in embryonic stem cells
Chosen DNA sequences are injected into one-cell embryos and integrate randomly into the genome	Precise chromosomal modifications are made in mouse embryonic stem cells by gene targeting (see Figure 3)
Embryos are implanted into foster mothers for completion of development	Correctly modified ES cells are injected into 3.5-day-old embryos (blastocysts)
Mice (or rats) harboring the transgene (founders) are identified and bred	Blastocysts are implanted into foster mothers for completion of development
Expression of the introduced transgene in tissues is analyzed	Chimeric mice (derived from both the ES cell and recipient blastocyst cells) are bred to generate mice heterozygous and homozygous for the introduced mutations

Gene Targeting in Embryonic Stem Cells

Gene targeting experiments with embryonic stem (ES) cells differ from pronuclear injections because the investigator can make specific precise changes in *endogenous* genes. The desired genetic modifications are first produced in mouse ES cells in culture by means of homologous recombination (58). The correctly modified (gene targeted) cells are then introduced into blastocysts, which are reimplanted into foster mothers (see Table 1). Offspring produced in this fashion are chimeras composed of ES cell–derived and blastocyst-derived cells. Chimeras that transmit the ES cell genome to their offspring can be bred to eventually produce mice homozygous for the mutation introduced into the ES cells in tissue culture. See Maeda (59) and the abovementioned papers (56,57) for reviews of the application of this technology to the study of atherosclerosis.

Gene targeting of this type is currently possible only in the mouse, most often with ES cells from strain 129 (60,61). Since it is desirable, although not always essential, to have a uniform genetic background in which to compare the effects of changing a single gene, the breeding of chimeras and their offspring merits careful attention. Immediate breeding of chimeras with strain 129 mice results in offspring having the mutant gene or its wildtype allele as their sole genetic difference; the animals are otherwise completely strain 129. Unfortunately, strain 129 mice are disease-prone and breed poorly, so it may be difficult to generate the requisite number of mice to conduct complex physiological experiments. One way to circumvent this problem is to cross the ES-derived chimeras to mates from another inbred line, such as C57BL/6. The resulting first-generation (F1) mice are also genetically identical except for their sex chromosomes and the presence or absence of the altered gene, since they have 50% of their chromosomes from strain 129 and 50% from C57BL/6. The F1 animals also show "hybrid vigor" in terms of size, health, and fecundity. The F1 mutant heterozygotes can be interbred to give F2 offspring which will include homozygous mutants, heterozygotes, and wild-type animals. For some purposes, these F2 animals are sufficiently uniform to be used for testing the effects of the mutation. However, they will show systematic heterogeneity in any genes that are linked to the target locus and that differ in strain 129 and strain C57BL/6, and they will show random heterogeneity in genes that are not linked to the target locus (see Ref. 62 for discussion). One way to eliminate the effects of *linked* genes is to breed wild-type F1 mice to generate F2 wild-type mice. The genotype of the wild-type F2 mice at the target locus can be determined. Mice inheriting two strain 129 chromosomal segments containing the wild-type gene that was altered can be used as controls for the homozygous mutant mice. Mice inheriting one strain 129 segment and one strain C57BL/6 segment can be used as controls for the heterozygous F2 mice. One way to eliminate the effects of both linked *and* unlinked genes is to transfer the chromosomal segment containing the strain 129 mutated gene into the C57BL/6 background by backcross breeding, and in parallel to transfer the wild-type 129 gene into the C57BL/6 background. This creates two mutant strains that are effectively C57BL/6, except for having either the targeted or nontargeted strain 129 chromosomal segment.

Modeling Single-Gene Diseases

This section describes and illustrates the use of gene targeting to model human diseases caused by a single defective gene (63).

Cystic fibrosis, the most common single-gene disease in Caucasians, is caused by defects in the cystic fibrosis transmembrane regulator (CFTR) gene (64). These defects produce abnormalities in salt and water exchanges at some epithelial surfaces, resulting in dysfunction of the respiratory, digestive, and reproductive tracts. The mouse CFTR protein is identical to the human protein in more than 78% of its amino acids, but no naturally occurring animal model of cystic fibrosis exists. Several laboratories have therefore produced mouse models of cystic fibrosis by gene targeting (65–69). Two groups disrupted the CFTR gene in very similar ways and demonstrated a phenotype that closely matches some aspects of the human disease (65,67) but that lacks other features. Mice heterozygous for the mutation, like human heterozygotes, show no symptoms, and mating heterozygotes gives homozygotes at the expected frequency. However, about a quarter of the homozygous mice die within 5 days of birth, and most (but not all) of the remainder of the mutant homozygotes die within 30 days of birth because of intestinal blockage (65). The intestinal problems in these mice mimic the meconium ileus that killed about 10% of human homozygous mutant newborns prior to adequate therapy. The intestinal pathology models the inspissation of secretions that is the most fundamental feature of cystic fibrosis, but the mice show no overt lung pathology or evidence of pancreatic dysfunction. Thus, although the cystic fibrosis mouse demonstrates the defective electrolyte transport characteristic of the human disease (70), it does not replicate the human phenotype exactly. Another model (68) has a relatively mild phenotype, most likely because the mutant can still produce a low level of functional protein as a consequence of the specific targeting strategy employed by the investigators (63).

These animal models of cystic fibrosis serve to emphasize several important points regarding gene targeting experiments. First, the production of an animal model of a human genetic disease is possible. Second, the strategy employed to modify a target gene will influence the phenotype obtained. Third, the animal model may have certain, but not all, features of the human disease. However, the differences between the two species are often informative. For example, as indicated above, the cystic fibrosis mouse lacks the pancreatic and pulmonary pathology seen in humans. This absence of pancreatic pathology can be explained by the low level of CFTR gene expression in the normal mouse pancreas (65). But this cannot explain the lack of lung pathology, since the CFTR gene is highly expressed in mouse lung. One explanation for the lack of pathology in the lungs and pancreas of the mouse might be the existence of an alternative plasma membrane chloride conductance that is more strongly expressed in these tissues in the mouse than in humans. In surveying organs that exhibited a range of disease severity, an alternative chloride conductance was indeed found in epithelia of the airways and pancreas from cystic fibrosis mice but not in the small or large intestine (71). If a similar alternative chloride conductance is identified in humans, it may have important clinical implications (71).

Gene targeting can also be used to test special hypotheses regarding the role of individual genes in producing or protecting against disease. For example, Prusiner proposed in 1982 that the causative agent of spongiform encephalopathies, such as scrapie in animals or Creutzfeldt-Jakob disease in humans, was a disease-producing form (a prion) of a normal host protein; prions were postulated to convert the normal protein into the prion form (72). To evaluate the importance of the normal gene in the pathophysiology of scrapie, Bueler and colleagues (73) genetically inactivated the relevant gene (PrP) in mice. They then compared the survival of wild-type mice after inoculation with scrapie with the survival of inoculated mice that lacked the PrP gene. All the wild-type mice died within 6 months, while all mice lacking

PrP remained symptom-free. This experiment elegantly demonstrates the importance of the PrP gene in the etiology of scrapie in mice, and the ability of gene targeting experiments to provide clean answers to complex questions.

Although cardiovascular diseases are usually multifactorial, sometimes a single defective gene can result in an important phenotype, such as congenital heart disease. As part of a systematic study of genes that influence development, Chisaka and Capecchi (11) disrupted the homeobox gene *hox-1.5* (also known as *hoxa-3*). Mice homozygous for this mutation die at or shortly after birth from numerous abnormalities, including defects of the heart and major arteries. They are additionally athymic and aparathyroid, have reduced thyroid and submaxillary tissue, and have throat abnormalities. The constellation of deficiencies is similar to the human congenital disorder DiGeorge's syndrome.

Modeling Quantitative Genetic Disorders

Many human diseases are likely to be caused or influenced by genetic changes that result in under- or overproduction of a qualitatively normal protein product. For example, a genetic deficiency of atrial natriuretic peptide (ANP) may play a role in the etiology of hypertension, as judged by observations that, under conditions of increased dietary sodium intake, plasma ANP levels in some individuals with a hypertensive parent are lower than in individuals with two normotensive parents (74,75). A relative excess of a gene product may also be important in cardiovascular diseases. For example, the human gene encoding angiotensin-converting enzyme (ACE) has two known alleles, one having a small DNA insertion that is absent in the other. While both alleles produce the same protein, the allele lacking the insertion is associated with an increase in serum ACE levels (76) and an increased risk for myocardial infarction (77,78).

To model reduction in the level of a gene product, a gene can be modified or one copy can be functionally eliminated. To functionally eliminate a gene, the gene is usually disrupted by the insertion of a selectable marker gene into the target gene (see Figure 3) in embryonic stem (ES) cells (79), which are then used to generate mice transmitting the disrupted gene to their offspring. Heterozygous mice carrying the disrupted gene (often called a "gene knockout") will typically express a reduced amount of gene product; homozygous animals completely lack the targeted gene product.

Two methods are available to model increases in the level of a gene product. One method involves the random integration of chosen regulatory and coding sequences into the genome to produce greatly increased levels of the encoded protein (described above under "Pronuclear Injection of Single-Cell Embryos"). The second method involves a gene targeting system to increase the number of copies of a chosen gene without changing its regula-

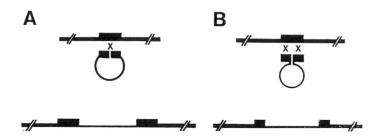

Figure 3 Modifying genes by targeting. (A) Homologous recombination shown with O-type geometry between chromosomal DNA (heavy line at top) containing a target gene (black bar) and a linearized targeting construct (middle) consisting of foreign (plasmid) DNA sequences (light line) plus a length of DNA with the same sequence as the target gene (black bar). O-type recombination *adds* the incoming DNA to the chromosome (bottom), which then contains two copies of the target gene plus the foreign DNA. The target sequence has therefore been duplicated. This strategy can also be used to disrupt a gene by duplicating only *part* of the gene and thereby producing two dysfunctional copies. (B) Recombination with Ω-type geometry between the same chromosomal DNA as in A (top) and a linear DNA targeting construct containing the same sequences as in A but arranged differently (middle). Ω-type recombination *replaces* the target sequences by the incoming DNA. The product of the Ω-type experiment is therefore a chromosome containing the target gene disrupted by the foreign DNA. (From Ref. 86.)

tory or coding sequences (80). A special type of gene targeting (81) is employed to duplicate the entire target gene at its normal chromosomal location along with flanking 5′ and 3′ regulatory sequences. The strategy is the same as the one shown in Figure 3A, except that the two arms of homology need only be long enough to pair with the extreme ends of the target gene. Despite the presence of a gap in the incoming DNA, the procedure results in duplication of the whole of the gene. This gap-repair duplication system has several potential benefits. First, genetic changes of this type produce small, graded increases in the level of the encoded protein that may be similar to those found in human populations (see Ref. 82, for example). Because the duplicated genes are located at the normal chromosomal location, and are controlled by normal endogenous regulatory elements, they are likely to function normally and to be expressed in normal tissues. The duplicated gene does not disrupt other genes. The investigator knows the number of copies of the gene that are produced by the targeting reaction. Detailed knowledge of the regulatory structures of the gene is not necessary, since there is no need to alter them. In short, this technique is likely to provide a useful tool to model physiologically increased levels of gene products. An example of the use of this technique is described below.

Studies of the atrial natriuretic peptide (ANP) gene are presented here as examples of synthetic genetic experiments to study the genetics of cardiovascular diseases. ANP, a peptide produced by the atrium, acts to promote sodium excretion, diuresis, and reduction in arterial blood pressure. Defects in the ANP system have been suggested to be important in the etiology of human essential hypertension. To study the effects of greatly increased levels of ANP on blood pressure, Steinhelper et al. (83) generated mice transgenic for a fusion gene composed of the transthyretin promoter (strongly expressed in the liver) and ANP gene coding sequences. Mice harboring this transgene have circulating ANP levels elevated approximately 10-fold and arterial pressures reduced by approximately 25%. After acute volume expansion, the transgenic mice demonstrate enhanced diuretic and natriuretic responses relative to controls (84). These experiments prove that greatly elevated levels of ANP can produce long-term reduction in blood pressure as well as identifiable differences in kidney function.

To directly test the impact of decreases in the production of ANP, John et al. (85) used gene targeting to generate mice having zero, one, and two copies of the murine *proANP* gene. On an intermediate (2%)-salt diet, the zero-copy mice have blood pressures 8–23 mm Hg higher than two-copy mice, clearly demonstrating that the ANP system has a role in the tonic regulation of blood pressure. On a high (8%)-salt diet, one-copy mice have blood pressures 27 mm Hg higher than controls, indicating that a deficiency in ANP can cause salt-sensitive hypertension.

A Strategy for Studying the Role of Candidate Genes in Complex Diseases

Most common diseases, including hypertension and atherosclerosis, are multifactorial in etiology, with many factors contributing to the disease phenotype. These diseases likely arise from different constellations of genetic and environmental factors, many of which individually may have only a modest effect.

A new strategy for studying the genetics of cardiovascular diseases centers on producing quantitative changes in the level of production of factors that may be important in the disease pathogenesis within the context of the whole animal. A candidate gene, first identified on *a priori* grounds or through genetic analysis of animal models or of human disease populations, is modified in living animals so that its level of production is varied in a graded fashion from absent to increased. The impact of these genetic modifications on the phenotype under investigation (for example, resting blood pressure, blood pressure under high salt, atherosclerotic lesions, etc.) can then be assessed in a dose-response fashion. The magnitude of change in the phenotype with level of expression of the candidate gene will suggest the importance of the gene in producing the phenotype.

To illustrate the use of this strategy, we describe some recent studies of the role of the angiotensinogen gene in blood pressure regulation. Jeune-maitre et al. (82) identified a variant of the angiotensinogen gene that was associated with increased levels of the protein and increased blood pressure in hypertensive patients. To directly determine the impact of quantitative changes in this protein on blood pressure, mice carrying zero, one, two (normal), three, and four copies of the angiotensinogen gene were produced by gene targeting (62,80). As expected, the animals demonstrate graded levels of angiotensinogen production with increasing number of copies of the gene. Defining the angiotensinogen levels of normal two-copy mice as 100%, the respective steady-state plasma angiotensinogen levels of zero-, one-, two-, three-, and four-copy animals are 0%, 35%, 100%, 124%, and 145% (62,80). The blood pressures of these animals show significant and almost linear increases of approximately 8 mm Hg per gene copy number (62). These results establish a direct causal relationship between angiotensinogen genotypes and blood pressures.

The interactions of different genes in producing a phenotype can be studied by combining deliberately engineered variants through breeding. The investigator can then determine whether a chosen combination of genes has significant effects in either amplifying or ameliorating the phenotype. For example, it will be interesting to know whether the combination of a decrease in atrial natriuretic peptide (a blood pressure–decreasing agent) and an increase in angiotensinogen (the precursor of the blood pressure–increasing agent angiotensin II) will amplify the hypertensive phenotype. Animals harboring these relatively subtle quantitative genetic modifications begin to model the sort of functional genetic changes that are likely to occur in humans.

This gene dosage approach appears to be very promising for dissecting the complexities of many multifactorial diseases. The causative element, a genetically changed level of production of a chosen gene product, is never in doubt.

SUMMARY

This chapter describes the use of animals to dissect complex traits. Animals having a disease generated through selective breeding provide a means for identifying some candidate genetic variants that play a role in producing the disease. These genetic variants may be moved into other genetic backgrounds to evaluate whether the disease is transmitted with them, and whether their effects are subsequently modified by additional genetic factors within the new genetic background. Gene targeting experiments allow powerful conclusions of causality, because they allow one to determine the phenotypic effects of precise quantitative and qualitative changes in chosen genes or combinations of genes in animals having a defined genetic background.

New information from animal models such as the ones described in this chapter are likely to have a direct impact on clinical medicine. For example, the demonstration of salt sensitivity in mice deficient in atrial natriuretic peptide suggests that a renewed search for human genetic variants that affect the ANP system may identify hypertensive patients likely to benefit from restriction of dietary salt. Other animal models, such as the apoE-knockout atherosclerotic mice, may be valuable for studying drugs that influence lipid metabolism and atherosclerosis. The systematic study of genes that regulate development, such as *box-1.5*, may shed new light on the genetics, patho-physiology, and diagnosis of congenital heart disease.

The present era of physiological genetics holds great promise for unraveling the etiology of many complex diseases.

ACKNOWLEDGMENTS

This work was supported by grants from the National Institutes of Health to T. W. K. (HL35018) and to O. S. (HL37001, HL49277, GM20069), and to T. W. K. from the American Heart Association National Center, Dallas, Texas; the American Heart Association, California Affiliate, Inc. Burlingame, California; and the Max and Victoria Dreyfus Foundation, New York, New York. John H. Krege is a Howard Hughes Medical Institute Physician Postdoctoral Fellow. The authors thank Eric Bachman, Paul Celmer, Simon John, Jenny Lynch, and Scott Morham for their help.

REFERENCES

1. Yamamoto T, Bishop RW, Brown MS, et al. Deletion in cysteine-rich region of LDL receptor impedes transport to cell surface in WHHL rabbit. Science 1986; 232:1230–1237.
2. Evans JP, Brinkhous JP, Brayer GD. Canine hemophilia B resulting from a point mutation with unusual consequences. Proc Natl Acad Sci 1989; 86:10095–10099.
3. Schmale H, Richter D. Single base deletion in the vasopressin gene is the cause of diabetes insipidus in Brattleboro rats. Nature 1984; 308:705–709.
4. Sigmund CD, Jones CA, Kane CM, Wu C, Lang JA, Gross KW. Regulated tissue- and cell-specific expression of the human renin gene in transgenic mice. Circ Res 1992; 70:1070–1079.
5. Palmiter RD, Behringer RR, Quaife CJ, Maxwell F, Maxwell IH, Brinster RL. Cell lineage ablation in transgenic mice by cell-specific expression of a toxin gene. Cell 1987; 50:435–443.
6. Jacob HJ, Sigmund CD, Shockley TR, Gross KW, Dzau VJ. Renin promoter SV40

T-antigen transgenic mouse: A model of primary renal vascular hyperplasia. Hypertension 1991; 17:1167–1172.

7. Gurney ME, Pu H, Chiu AY, et al. Motor neuron degeneration in mice that express a human Cu, Zn superoxide dismutase mutation. Science 1994; 264:772–775.

8. Vitaterna MH, King DP, Chang AM, et al. Mutagenesis and mapping of a mouse gene, *Clock*, essential for circadian behavior. Science 1994; 264:719–725.

9. Koller BH, Marrack P, Kappler JW, Smithies O. Normal development of mice deficient in beta 2M, MHC class I proteins, and CD8+ T cells. Science 1990; 248:1227–1230.

10. Zijlstra M, Bix M, Simister NE, Loring JM, Raulet DH, Jaenisch R. β2-Microglobulin deficient mice lack CD4-8+ cytolytic T cells. Nature 1990; 344:742–746.

11. Chisaka O, Capecchi MR. Regionally restricted developmental defects resulting from targeted disruption of the mouse homeobox gene *hox-1.5*. Nature 1991; 350:473–479.

12. Huang PL, Dawson TM, Bredt DS, Snyder SH, Fishman MC. Targeted disruption of the neuronal nitric oxide synthase gene. Cell 1993; 75:1273–1286.

13. O'Dell TJ, Huang PL, Dawson TM, et al. Endothelial NOS and the blockade of LTP by NOS inhibitors in mice lacking neuronal NOS. Science 1994; 265:542–546.

14. Donehower LA, Harvey M, Slagle BL, et al. Mice deficient for p53 are developmentally normal but susceptible to spontaneous tumors. Nature 1992; 356:215–221.

15. Zhang SH, Reddick RL, Piedrahita JA, Maeda N. Spontaneous hypercholesterolemia and arterial lesions in mice lacking apolipoprotein E. Science 1992; 258:468–471.

16. Plump AS, Smith JD, Hayek T, et al. Severe hypercholesterolemia and atherosclerosis in apolipoprotein E-deficient mice created by homologous recombination in ES cells. Cell 1992; 71:343–353.

17. Yamori Y. Development of the spontaneously hypertensive rat (SHR) and of various spontaneous rat models, and their implications. In: de Jong W, ed. Handbook of Hypertension. Vol 4. Experimental and Genetic Models of Hypertension. New York: Elsevier, 1984:224–239.

18. Okamoto K, Yamori Y, Ooshima A, et al. Establishment of the inbred strain of the spontaneously hypertensive rat and genetic factors involved in hypertension. In: Okamoto K, ed. Spontaneous Hypertension: Its Pathogenesis and Complications. New York: Springer-Verlag, 1972.

19. Okamoto K, Aoki K. Development of a strain of spontaneously hypertensive rats. Jap Circ J 1963; 27:282.

20. Schlager G. Selection for blood pressure levels in mice. Genetics 1974; 76:537–549.

21. Simpson FO, Phelan EL., Hypertension in the genetically hypertensive strain. In: de Jong W, ed. Handbook of Hypertension. Vol 4. Experimental and Genetic Models of Hypertension. New York: Elsevier, 1984:200–223.

22. Kurtz TW, St Lezin EM. Gene mapping in experimental hypertension. J Am Soc Nephrol 1992; 3:28–34.

23. Bunag RD. Measurement of blood pressure in rats. In: de Jong W, ed. Handbook of Hypertension. Vol. 4. Experimental and Genetic Models of Hypertension. New York: Elsevier, 1984:1–12.

24. Ferrari AU, Daffonchio A, Albergati F, Bertoli P, Mancia G. Intra-arterial pressure alterations during tail-cuff blood pressure measurements in normotensive and hypertensive rats. J Hypertens 1990; 8:909-911.

25. Dahl LK, Love RK. Evidence for relationship between sodium (chloride) intake and human essential hypertension. Arch Intern Med 1954; 94:525.

26. Dahl LK, Heine M, Tassinari L. Role of genetic factors in susceptibility to experimental hypertension due to chronic excess salt ingestion. Nature 1962; 194:480-482.

27. Zhang SH, Reddick R, Burkey B, Maeda N. Diet-induced atherosclerosis in mice heterozygous and homozygous for apolipoprotein E gene disruption. J Clin Inves 1994; 94:937-945.

28. Cicilia GT, Rapp JP, Wang JM, St Lezin E, Ng SC, Kurtz TW. Linkage of 11B-hydroxylase mutations with altered steroid biosynthesis and blood pressure in the Dahl rat. Nat Genet 1993; 3:346-353.

29. Rapp JP, Wang SM, Dene H. Effect of genetic background on cosegregation of renin alleles and blood pressure in Dahl rats. Am J Hypertens 1990; 3:391-396.

30. Rapp JP. A paradigm for identification of primary genetic causes of hypertension in rats. Hypertension 1983; 5(suppl I):I-198-203.

31. Harris EL, Dene H, Rapp JP. SA gene and blood pressure cosegregation using Dahl salt-sensitive rats. Am J Hypertens 1993; 6:330-334.

32. Hilbert P, Lindpaintner K, Beckmann JS, et al. Chromosomal mapping of two genetic loci associated with blood-pressure regulation in hereditary hypertensive rats. Nature 1991; 353:521-529.

33. Jacob HJ, Lindpaintner K, Lincoln SE, et al. Genetic mapping of a gene causing hypertension in the stroke-prone spontaneously hypertensive rat. Cell 1991; 67:213-224.

34. Dubay C, Vincent M, Samani NJ, et al. Genetic determinants of diastolic and pulse pressure map to different loci in Lyon hypertensive rats. Nat Genet 1993; 3:354-357.

35. Rapp JP, Wang SM, Dene H. A genetic polymorphism in the renin gene of Dahl rats cosegregates with blood pressure. Science 1989; 243:542-544.

36. Kurtz TW, Simonet L, Kabra PM, Wolfe S, Chan L, Hjelle BL. Cosegregation of the renin allele of the spontaneously hypertensive rat with an increase in blood pressure. J Clin Invest 1990; 85:1328-1332.

37. Katsuya T, Higaki J, Zhao Y, et al. A neuropeptide Y locus on chromosome 4 cosegregates with blood pressure in the spontaneously hypertensive rat. Biochem Biophys Res Commun 1993; 192:261-267.

38. St Lezin EM, Pravenec M, Kurtz TW. New genetic models for hypertension research. Trends Cardiovasc Med 1993; 3:119-123.

39. Pravenec M, Klir P, Kren V, Zicha J, Kunes J. An analysis of spontaneous hypertension in spontaneously hypertensive rats by means of new recombinant inbred strains. J Hypertens 1989; 7:217-222.

40. Marshall JD, Mu JL, Cheah YC, Nesbitt MN, Frankel WN, Paigen B. The AXB and BXA set of recombinant inbred mouse strains. Mammal Genome 1992; 3:669-680.

41. Paigen B, Ishida BY, Verstuyft J, Winters RB, Albee D. Atherosclerosis suscepti-

bility differences among progenitors of recombinant inbred strains of mice. Arterioclerosis 1990; 10:316-323.

42. Paigen B, Mitchell D, Reue K, Morrow A, Lusis AJ, LeBoeuf RC. Ath-1, a gene determining atherosclerosis susceptibility and high density lipoprotein levels in mice. Proc Natl Acad Sci USA 1987; 84:3763-3767.

43. Festing MFW. Inbred strains in biomedical research. New York: Oxford University Press, 1979.

44. Paigen B, Holmes PA, Morrow A, Mitchell D. Effect of 3-methylcholanthrene on atherosclerosis in two congenic strains of mice with different susceptibilities to methylcholanthrene-induced tumors. Cancer Res 1986; 46:3321-3324.

45. Nishini PM, Naggert JK, Verstuyft J, Paigen B. Atherosclerosis in genetically obese mice: the mutants obese, diabetes, fat, tubby, and lethal yellow. Metab Clin Exper 1994; 43:554-558.

46. St Lezin E, Wong A, Wang JM, et al. Transfer of the Dahl S renin gene into the Dahl R strain does not promote NaCl-sensitivity and may actually cause a decrease in blood pressure [abstr]. Hypertension 1993; 22:421.

47. Ely DL, Turner ME. Hypertension in the spontaneously hypertensive rat is linked to the Y chromosome. Hypertension 1990; 16:277-281.

48. Turner ME, Johnson JL, Ely DL. Separate sex-influenced and genetic components in spontaneously hypertensive rat hypertension. Hypertension 1991; 17:1097-1103.

49. Ganten D. Role of animal models in hypertension research. Hypertension 1987; 9:1-2-4.

50. Searle AG, Peters J, Lyon MF, Evans EP, Edwards JH, Buckle VJ. Chromosome maps of man and mouse. III. Genomics 1987; 1:3-18.

51. Dietrich W, Katz H, Lincoln SE, et al. A genetic map of the mouse suitable for typing intraspecific crosses. Genetics 1992; 131:423-447.

52. Yamada J, Kuramoto T, Serikawa T. A rat genetic linkage map and comparative maps for mouse or human homologous rat genes. Mammal Genome 1994; 5:63-83.

53. Weiher H, Noda T, Gray DA, Sharpe AH, Jaenisch R. Transgenic mouse model of kidney disease: insertional inactivation of ubiquitously expressed gene leads to nephrotic syndrome. Cell 1990; 62:425-434.

54. Mockrin SC, Dzau VJ, Gross KW, Horan MJ. Transgenic animals: New approaches to hypertension research. Hypertension 1991; 17:394-399.

55. Sigmund CD. Major approaches for generating and analyzing transgenic mice: An overview. Hypertension 1993; 22:599-607.

56. Rubin EM, Smith DJ, Atherosclerosis in mice: getting to the heart of a polygenic disorder. Trends Genet 1994; 10:199-203.

57. Lusis AJ. The mouse model for atherosclerosis. Trends Cardiovasc Med 1993; 3:135-143.

58. Koller BH, Smithies O. Altering genes in animals by gene targeting. Annu Rev Immunol 1992; 10:705-730.

59. Maeda N. Gene targeting in mice as a strategy for understanding lipid metabolism and atherogenesis. Curr Opin Lipid 1993; 4:90-94.

60. Evans MJ, Kaufman MH. Establishment of culture of pluripotential cells from mouse embryos. Nature 1981; 292:154-156.

61. Martin GR. Isolation of a pluripotential cell line from early embryos cultured in medium conditioned by teratocarcinoma stem cells. Proc Natl Acad Sci USA 1981; 78:7634-7638.

62. Kim HS, Krege JH, Kluckman KD, et al. Genetic control of blood pressure and the angiotensinogen locus. Proc Natl Acad Sci USA 1995; 92:2735-2739.

63. Smithies O. Animal models of human genetic diseases. Trends Genet 1993; 9:112-116.

64. Worldwide survey of the delta 508 mutation report from the Cystic Fibrosis Genetic Analysis Consortium. Am J Hum Gen 1990; 47:354-359.

65. Snouwaert JN, Brigman KK, Latour AM, et al. An animal model for cystic fibrosis made by gene targeting. Science 1992; 257:1083-1088.

66. Koller BH, Kim HS, Latour AM, et al. Toward an animal model of cystic fibrosis: targeted interruption of exon 10 of the cystic fibrosis transmembrane regulator gene in embryonic stem cells. Proc Natl Acad Sci USA 1991; 88:10730-10734.

67. Colledge WH, Ratcliff R, Foster D, Williamson R, Evans MJ. Cystic fibrosis mouse with intestinal obstruction. Lancet 1992; 340:680.

68. Dorin JR, Dickinson P, Alton EWFW, et al. Cystic fibrosis in the mouse by targeted insertional mutagenesis. Nature 1992; 359:211-215.

69. O'Neal WK, Hasty P, McCray PB, et al. A severe phenotype in mice with a duplication of exon 3 in the cystic fibrosis locus. Hum Mol Genet 1993; 2:1561-1569.

70. Clarke LL, Grubb BR, Gabriel SE, Smithies O, Koller BH, Boucher RC. Defective epithelial chloride transport in a gene-targeted mouse model of cystic fibrosis. Science 1992; 257:1125-1128.

71. Clarke LL, Grubb BR, Yankaskas JR, Cotton CU, McKenzie A, Boucher RC. Relationship of a non-cystic fibrosis transmembrane conductance regulator-mediated chloride conductance to organ-level disease in Cftr(-/-) mice. Proc Natl Acad Sci USA 1994; 91:479-483.

72. Prusiner SB. Novel proteinaceous infectious particles cause scrapie. Science 1982; 216:136-144.

73. Bueler H, Aguzzi A, Sailer A, et al. Mice devoid of PrP are resistant to scrapie. Cell 1993; 73:1339-1347.

74. Ferrari P, Weidmann P, Ferrier C, et al. Dysregulation of atrial natriuretic factor in hypertension-prone man. J Clin Endocrinol Metab 1990; 71:944-951.

75. Weidmann P, Ferrari P, Allemann Y. Developing essential hypertension: a syndrome involving ANF deficience. Can J Physiol Pharmacol 1991; 69:1582-1591.

76. Tiret L, Rigat B, Viskis S, et al. Evidence, from combined segregation and linkage analysis, that a variant of the angiotensin I-converting enzyme (ACE) gene controls plasma ACE levels. Am J Hum Genet 1992, 51:197-205.

77. Cambien F, Poirier O, Lecerf L, et al. Deletion polymorphism in the gene for angiotensin-converting enzyme is a potent risk factor for myocardial infarction. Nature 1992; 359:641-644.

78. Tiret L, Kee F, Poirier O, et al. Deletion polymorphism in angiotensin-converting enzyme gene associated with parental history of myocardial infarction. Lancet 1993; 341:991-992.

79. Thomas KR, Capecchi MR. Site-directed mutagenesis by gene targeting in mouse embryo-derived stem cells. Cell 1987; 51:503-512.

80. Smithies O, Kim HS. Targeted gene duplication and disruption for analyzing quantitative genetic traits in mice. Proc Natl Acad Sci USA 1994; 91:3612–3615.
81. Valancius V, Smithies O. Double-strand gap repair in a mammalian gene targeting reaction. Mol Cell Biol 1991; 11:4389–4397.
82. Jeunemaitre X, Soubrier F, Kotelevtsev YV, et al. Molecular basis of human hypertension: role of angiotensinogen. Cell 1992; 71:169–180.
83. Steinhelper ME, Cochrane KL, Field LJ. Hypotension in transgenic mice expressing atrial natriuretic factor fusion genes [see comments]. Hypertension 1990; 16:301–307.
84. Field LJ, Veress AT, Steinhelper ME, Cochrane K, Sonnenberg H. Kidney function in ANF-transgenic mice: effect of blood volume expansion. Am J Physiol 1991; 260:R1–5.
85. John SWM, Krege JH, Oliver PM, et al. Genetically decreased levels of atrial natriuretic peptide and salt-sensitive hypertension. Science 1995; 267:679–681.
86. Gregg RG, Smithies O. Targeted modification of human chromosomal genes. Cold Spring Harbor Symp Quant Biol 1986; 51(Pt 2):1093–1099.

11

Genetic Dissection of Hypertension in Experimental Animal Models

Howard J. Jacob*
Massachusetts General Hospital
Charleston, Massachusetts

José E. Krieger
University of São Paulo
São Paulo, Brazil

Victor J. Dzau
Stanford University
Stanford, California

Eric S. Lander
Massachusetts Institute of Technology
Cambridge, Massachusetts

Experimental animal models are especially advantageous for genetic studies for several reasons. First, the use of inbred animals eliminates the problem of genetic heterogeneity within a strain and in crosses between strains. Second, large numbers of animals can be produced. Third, experimental animals can be extensively phenotyped. Finally, experimental animals have a shorter lifespan and display the disease phenotype within a few years, unlike patients, who typically show a late age of onset, making collection of patients and their parents harder. The loci identified in the experimental animal models can be used as candidate genes in the genetic dissection of hypertension in man.

Hypertension is a prototypical multifactorial disease and is used in this chapter as an example of how molecular genetics in experimental animal models can be applied to many of the multifactorial disorders outlined in this book. The purpose of this chapter is to: 1) outline the foundation for molecular genetic dissection of complex disease in experimental animal models, 2) outline the successes of molecular genetics in hypertension, and 3) discuss how the results from studies using molecular genetics in experimental animal models are likely to influence our understanding of the etiology of hypertension.

Why Genetic Dissection?

Over the last 30 years, various genetic models of hypertension have been used to study the etiology of hypertension and the regulation of blood pressure.

Current affiliation: Medical College of Wisconsin, Milwaukee, Wisconson.

These models and, in particular, the spontaneously hypertensive rat (SHR) have been extensively studied with physiological, biochemical, and, recently, molecular biological techniques. Since 1966, more than 2000 publications have reported work using the SHR. Most of these involved reports of differences between the SHR and a normotensive control, typically the Wistar-Kyoto (WKY). Since the WKY was derived (seven years later) from the same outbred colony of Wistar rats as the SHR, it was believed that the major differences between the SHR and WKY involve the genes responsible for hypertension. Unfortunately, this assumption is too simplistic. Hypertension did not arise from a spontaneous single- gene mutation in an inbred animal strain; therefore, the differences reported between the two strains could be: 1) the cause of the hypertension, 2) the result of the hypertension, or 3) simply a difference between the strains that has nothing to do with the hypertension. Given that both strains were derived from an outbred colony, the third possibility is the most likely explanation for the majority of the differences between the SHR and WKY. The question is: how are these three possibilities distinguished from one another?

Genetics provides a way to distinguish primary factors causing a complex disease such as hypertension in the SHR. Genes involved in causing the trait will show cosegregation with the trait in pedigrees, whereas genes responsible for secondary responses or unrelated strain differences will not. Still, genetic dissection of hypertension presents a difficult challenge for molecular geneticists, even when using inbred animals for several reasons: 1) hypertension is polygenic (that is, variations in several genes are required to cause expression of the disease), 2) the environment plays a role in expression of the phenotype, and 3) measurements of blood pressure, or the phenotyping, is not standardized and is determined using different methodologies in different laboratories. Molecular genetics brings a promise of identifying the primary genetic components of this complex disease. Once these genes have been identified, it is hoped that etiology of this complex disease will begin to unravel, as the various pieces of the puzzle are put together. For the molecular geneticist, hypertension presents an opportunity to study a prototypical multifactorial trait and to develop techniques to meet the challenges.

GENETIC STUDIES IN EXPERIMENTAL ANIMALS

Why were genetic animal models of hypertension developed? The obvious answer is to study the inheritance of hypertension. In fact, this is not the case, because genetic models were developed before the genetic basis of hypertension was known. Rather, they were developed to be a reliable and reproducible form of hypertension. Prior to the development of the genetically hypertensive (GH) rat in 1958 (1), animal models of hypertension were induced by producing some type of damage to the kidneys via surgical removal of kidneys or renal mass and/or injection of nephrotoxins, ligation of renal arteries, or clipping of renal arteries (2,3). While these experimentally induced models of hypertension provided insight into the regulation of blood pressure, they were not always consistent in producing equivalent levels of hypertension. The GH rat, on the other hand, spontaneously developed hypertension that was relatively uniform in severity. The development of the SHR (4) and the Dahl salt-susceptible strains of rats (5) provided additional reproducible models of hypertension. Interest in genetic models for hypertension research did not develop for nearly 10 years, with the increase in interest roughly parallel to the realization that human hypertension had a polygenetic basis.

Over the last three decades, at least nine different genetically hypertensive rats were developed by selection. Importantly, these nine strains have slightly different etiologies of hypertension, and therefore should enable molecular geneticists to make comparisons to identify the major genes in common. These major genes are more likely to play a role in human hypertension. However, experimental animal models have some clear physiological and pathophysiological differences from humans. Accordingly, it is worth asking how useful these models will be in understanding human disease. The issue of relevance to the human pathobiology is frequently posed in terms of whether there is a 100% correlation between the phenotype of the human and the experimental animal model. It would be extremely unlikely that any of the rat strains developed by simply selecting for animals with the highest blood pressure would be identical to human hypertensives. In fact, it is also unlikely that any particular human pedigree or small set of patients would represent all patients with primary hypertension. In short, there are probably many forms of hypertension, and the purpose of the animal models is to point to mechanisms that potentially can produce hypertension. Human studies can then identify the degree to which one particular mechanism accounts for the overall human incidence. The nine different strains thus provide the unique opportunity to investigate the genetic basis of different "flavors" of hypertension, and this information is likely to help us minimize the effect of the genetic heterogeneity found in human hypertension.

Developing Inbred Strains of Hypertensive Animals

None of the genetic models of hypertension spontaneously arose on an inbred genetic background. They were developed by selecting for high blood pressure in a manner no different from the development of domesticated livestock, animals, and crops. As noted in the previous chapter, the SHR was produced in four generations (4). This might suggest that there is a major gene effect or only a few genes responsible for the hypertension. This would appear to stand in sharp contrast to the 10 generations required to develop the GH (1), which might be expected to harbor many gene effects, each making small contributions to the hypertension. But even these simple conclusions may be unwarranted. Although the question cannot be definitively answered, data from the Kyoto colony suggest that there was a significant incidence of hypertension in this outbred colony even before selection was begun—with 26.4% of the male Wistar rats developing hypertension, and 15.6% of the females (4). So, while the speed of inbreeding is suggestive of only a few genes, many of the "hypertensive" genes may have been present at high frequencies before initiation of selection.

Once the hypertension is "fixed," the next step is to inbreed the strain. By definition, a strain is declared to be inbred after 20 generations of strict brother–sister matings. The standard of 20 generations was chosen to ensure that a strain is homozygous at all loci (genetically defined sites), based on the mendelian assumption that each brother–sister mating "fixes" (makes homozygous) 25% of the genome at each generation. This calculation assumes that each locus is independent and has an equal chance of becoming fixed in any cross; however, loci on the same chromosome are linked and, consequently, fixing at *all* loci will take much longer. The major factor influencing the number of generations is genome length (6). However, 20 generations is sufficient for the vast majority of the genome to become fixed.

QUANTITATIVE GENETICS

Hypertension is the upper end of the blood-pressure distribution curve that is defined by a clinically relevant threshold (7). Consequently, the genetic dissection of hypertension is really the genetic dissection of the genes responsible for "high" blood pressure. The continuous distribution of a quantitative trait is very different from the distribution of a qualitative trait, also called a simple mendelian trait, which follows a simple mode of inheritance (dominant, recessive, or additive). Yet quantitative traits must also follow Mendel's laws of independent assortment and independent segregation. How does a continuous distribution arise from individual genes, each producing a discontinuous variation? There are two explanations: It results from the summation of the effects of multiple genes and from the influence of the environment superimposed on the genetic factors.

Genes Versus Environment

The first step in dissecting quantitative traits is to determine how much of a phenotype is "genetic" and how much is "environmental." For example, the blood-pressure distribution found in an inbred strain of hypertensive rats must be entirely the result of the environment, since the animals are genetically identical. The environmental component consists of variance due to measurement error, uncontrolled factors that influence the level of blood pressure at the time of measurement (e.g., anxiety on the part of the animal), and other potential confounding factors, such as competition for food, establishment of social hierarchies, and perhaps stochastic events during development. Since none of this variation in blood pressure is genetic, it is not possible to study the genes responsible for the hypertension in a single inbred strain.

What if an SHR is mated to a normotensive inbred animal? In this case, the progeny, referred to as the first filial (F1) generation, come from two inbred animals with very different levels in blood pressure and with a very different genetic propensity to hypertension. However, these animals are genetically identical, sharing one chromosome from each parent. Phenotypically, the blood pressure of the F1s should be near the average of the progenitor strains, if the genes are acting additively. The phenotypical variance in the two parental strains and the F1 animals should approximately equal (under our simple assumption) and yield a good estimate of variance caused by the environment (V_E). A second-generation cross is required to carry out linkage analysis. This second cross can be either a backcross (in which F1 animals are mated with one of the parental strains) or an intercross (in which F1 progeny is mated to each other). The progeny from the second generation display greater variability (V_2) as result of genetic segregation and environmental effects. By subtracting V_E (estimated from the F1 or one of the parental strains) from the total variance V_2 found in this progeny, we are left with amount of the variance attributable to genetic factors segregating in the cross (8); see Figure 1.

The ability to distribute the total variance into genetic and environmental components is the major advantage of using inbred animals over outbred animals, or human populations. With the genetic component defined, a search can be conducted to identify the genes producing the effect. The correlation between genotype and the genetic component of the phenotypical variation is the essence of genetic dissection.

Another way to determine the genetic component is to calculate heritability. Heritability (h^2) is a ratio of genetically caused variation to the total variation (6). An h^2 of 1 would indicate that trait is 100% determined by genes. The closer h^2 is to 0, the greater the contribution of the environment to the trait. For genetic models of hypertension, h^2 ranges from 0.30 to 0.92 (see Tables 1 and 2a), demonstrating that a significant proportion of the variance in blood pressure is genetic.

Genetic Background

It is important to realize that genetic studies can distinguish differences only between the two strains studied. Because the selection of strains determines

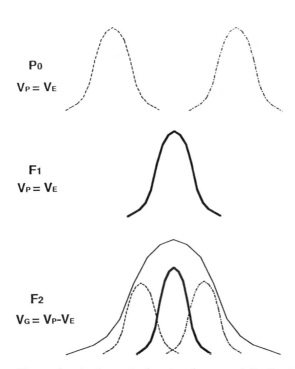

Po

$V_P = V_E$

F1

$V_P = V_E$

F2

$V_G = V_P - V_E$

Figure 1 A schematic showing the normal distribution of a quantitative trait in three generations. The variability in the phenotype (V_P) in the parental generation Po is the result of the environment (V_E). The same holds true for the variability in the phenotype in the F_1 progeny. The variability in the phenotype in the F_2 progeny ($F_1 \times F_1$) is the result of the genetic components (V_G) and V_E; therefore, V_G can be determined by subtracting V_E from V_P.

Table 1   Heritability (h^2) of Rat Genetic Models of Hypertension

Cross	h^2	Ref.
SHR (F14) × Wistar/Imamichi	0.7 (Sex avg. 19 weeks)	9
SHR (F18–F20) × Wistar/Kyoto	0.86 (Sex avg. 20 weeks)	9
SHR (F21) × Wistar/Mishima (F58)	0.92 (Sex avg. 20 weeks)	9
SHR × SHRSP	0.30	10
SS/JR × SR/JR	0.55 (Sex avg.)	11

#: the number of generations of brother–sister mating that precede the animals used in these studies; sex avg.: sex average h^2.
It is important to note that in some cases the differences between sexes were quite large (especially SS/JR × SR/JR). In all cases the cross studied was a backcross.
Source: Modified from Ref. 12.

Table 2a Heritability (h^2) of Blood Pressure at Different Ages

Cross	Week										
	5	7	9	11	13	15	17	19	20	25	30
SHR × WI											
Male	0.51	0.55	0.56	0.60	0.61	0.61	0.63	0.69			
Female	0.43	0.47	0.49	0.55	0.61	0.68	0.73	0.71			
SHR × WKY											
Male						0.88			0.87	0.88	0.82
Female						0.79			0.86	0.87	0.88
SHR × WM											
Male						0.95			0.96	0.95	0.95
Female						0.86			0.88		

WI: Wistar/Imamichi; WKY: Wistar/Kyoto; WM: Wistar/Mishima.
Source: Modified from Ref. 9.

Table 2b Number of Genetic Loci Estimated to Be Responsible for Blood Pressure at Different Ages

Cross	Week										
	5	7	9	11	13	15	17	19	20	25	30
SHR × WI											
Male	0.2	1.6	3.1	4.0	4.2	5.7	6.3	5.4			
Female	0.2	1.2	3.9	3.4	4.8	4.9	4.0	4.0			
SHR × WKY											
Male						2.5			2.6	2.2	2.8
Female						4.1			3.2	4.3	4.0
SHR × WM											
Male						1.6			1.4	2.5	2.4
Female						2.7			3.1		

WI: Wistar/Imamichi; WKY: Wistar/Kyoto; WM: Wistar/Mishima.
Source: Modified from Ref. 9.

the number of genes segregating within the cross, it is meaningless to speak about the "number of genes responsible for hypertension" without specifying the strains studied. The number of genes responsible for a phenotype can be estimated by segregation analysis when the phenotype is qualitative or by a formula developed by Sewall Wright (8) for quantitative traits. For example, Tanase et al. in 1970 (9) used several estimates of heritability to conduct a genetic analysis of hypertension in three crosses (SHR × Wistar/Imamichi, SHR

× WKY, SHR × Wistar/Mishima). The progenies of these crosses were phenotyped over many weeks. Several interesting points arise from this study. Crosses between the SHR and different normotensive strains yield a different estimate of the number of loci, ranging from 1 to 6 (see Table 2b), responsible for the high blood pressure (9). These data demonstrate that the genetic background of the normotensive strain determines how many genes are segregating for high blood pressure in any given cross. Obviously, the cross selected markedly influences the number of loci responsible for high blood pressure, and therefore the difficulty of identifying them using molecular genetics.

Influence of Blood-Pressure Determination

The method and protocol used for estimating blood pressure will affect V_E, and therefore the ability to use genetic dissection. The age at which an animal is phenotyped has an influence on the number of genes contributing to the high blood pressure (Tables 2a and b); this might be referred to as a phenotype window effect. Data presented by Beckett et al. (13) demonstrated that heritability of the genetic components responsible for hypertension in humans similarly changes with age, further supporting the idea of phenotypical windows. A theoretical paper by Nik Schork (personal communication) demonstrated that the window of phenotyping determines which genetic factors can be distinguished, and suggests that longitudinal data may be used to overcome the phenotyping window effect to reveal additional genetic components. However, because of technical limitations in collecting data, longitudinal data have not been used in conjunction with molecular genetics to identify genes responsible for quantitative phenotypes.

The power of quantitative genetics to dissect the genetic factors underlying hypertension is influenced by the accuracy of the measurement of blood pressure. Increased accuracy in the measurement of blood pressure reduces the environmental variance, thereby increasing the contribution of the genetic factors and resulting in a reduction in the number of animals needed for study. Unfortunately, collection of longitudinal data may necessitate the use of the tail-cuff (sphygomanometer) procedure for measuring blood pressure. This method introduces the most variability, owing to the need for restraint and heating of the animal (to maintain vasodilatation of the tail artery). Such measurements may pertain to a very different form of hypertension, compared to measuring blood pressure using other methodologies.

For example, blood pressure of the SHR is lower in animals when measured by radiotelemetry, as compared to animals whose blood pressure is measured using a catheter (the catheter exits the animal at the nape of the neck) or by tail-cuff. These data suggest that animals with telemetry units are less stressed, and therefore have a lower blood pressure. The impact of telemetry on genetic dissection is not known, since there has not been a report of a cross in which blood pressure was measured using this methodology. However, the selection scheme used to develop all the inbred strains of hypertensive rats may have resulted in the selection for a "stress" form of hypertension. In this case, reducing the stress may hamper the genetic dissection of hypertension (discussed in detail later).

Influence of Sex

Most experiments have used a single-cross design. Typically, the male hypertensive animal is crossed with a female normotensive animal. With this simple cross design, one cannot determine whether a phenotypical difference between male and female animals is the result of a sex-linked (gene carried on the sex chromosomes X or Y) gene or genes or of a sex-specific effect (the result of hormonal differences or other factors that distinguish the two sexes). The possibilities can be studied by using a reciprocal-cross design, in which the male hypertensive animal is crossed with the normotensive female *and* the male normotensive crossed with the female hypertensive. A comparison is then made between the male and female F1 progeny of the two crosses. If males and females have different blood pressures but each sex shows the same level of blood pressure independent of the direction of the cross (i.e., whether the father or mother is hypertensive), then the difference is sex-specific. However, if animals of a given sex from the reciprocal crosses are different from each other, then a sex-linked gene, mitochondrial effect, or imprinting effect may be involved.

The sex of the animal also appears to influence the number of genes responsible for the increase in blood pressure, as well as the overall genetic contribution (Tables 2a and b). Unfortunately, very little has been reported about these gender differences; however, investigators may need to adjust the number of animals by gender in each cross to locate the "hypertensive" genes.

The Y-chromosome of the SHR (14) and the X-chromosome of the WKY (15) are reported to carry a "hypertensive" gene. Turner and Ely have extended the Y-chromosome observation by generating a set of congenic animals (see Chapter 10 for details) in which a Y-chromosome from the WKY was placed on an SHR background (SHR-Y^{WKY}) and a Y-chromosome from the SHR was placed on a WKY background (WKY-Y^{SHR}). The blood pressures of these congenic rats were measured and compared to the parental strains. The SHR-Y^{WKY} has a lower pressure than the SHR, and the WKY-Y^{SHR} has a higher blood pressure than the WKY (16). These data would seem to clearly demonstrate the presence of a sex-linked gene on the Y-chromosome. However, another study (17) did not replicate the Y-chromosome effect in the SHR. How can we account for this discrepancy? First, the length of the Y-chromosome varies in different inbred strains, suggesting the existence of different "flavors" of the Y-chromosomes (18); however, the SHR colony was derived from a single male (4). This leaves two possibilities: a mutation arose on the Y-chromosome of one of the SHR substrains or there has been genetic contamination of the SHR, resulting in the introduction of several different Y-chromosomes. The latter case seems most likely, given the overall difficulty of maintaining a colony and the known problems of genetic contamination of the SHR (19–23).

Summary

The design of a genetic study and the ability to use molecular genetic techniques are markedly influenced by: the strains selected, the cross structure, the number of progeny, and the quality of phenotypes measured. For example, the majority of the crosses studied to date are between a hypertensive rat strain and its putative "control" normotensive strain, such as SHR × WKY, MHS × MNS, or SS/JR × SR/JR. This experimental design is somewhat simplistic (assuming the normotensive controls differed only at the hypertensive loci) and fails to exploit the full power of molecular genetic techniques.

TOTAL GENOMIC SEARCH

In 1980, Botstein and colleagues (24) set forth a major concept in mammalian genetics: Differences in DNA sequence can be used as genetic markers. This idea revolutionized molecular genetics by removing the most difficult aspect—finding useful genetic markers. The first DNA-based markers were called restriction fragment length polymorphisms (RFLPs), assayed by whether a restriction enzyme could cut DNA at a specific site. These were very useful but rather cumbersome assays. In 1989, Weber and May (25) proposed the use of dinucleotide repeats, such as $(CA)_n$, as genetic markers. More generally, such simple sequence repeats (SSRs) show a high degree of variation in the length of the repeat. Such polymorphisms are termed simple sequence length polymorphisms (SSLPs) or, for historical reasons, microsatellites. This new class of genetic markers, consisting of mononucleotide, dinucleotide, trinucleotide, and tetranucleotide repeats, are: 1) abundant, with more than 50,000 in a mammalian genome (26), 2) relatively uniformly distributed throughout the genome, 3) easily generated, and 4) assayed by PCR in a manner that can be easily automated.

Complete and dense genetic linkage maps exist for both the mouse and human genomes. For the rat there is now a completed, albeit sparse, genetic map (27). A complete map is one in which *all* genetic markers and their associated linkage groups are assigned to chromosomes.

Lander and Botstein (28) described how, with a complete genetic linkage map, the genome could be scanned simultaneously to look for genetic markers that cosegregate with any phenotype that has a genetic basis. This provides an approach that is more systematic than simply examining candidate genes. In this way, a genome scan would point out which regions contain genes responsible for a specific qualitative or quantitative trait. Genetic dissection has great potential to reveal novel genes responsible for hypertension or any other complex disease phenotype.

The first test of this new strategy came in what might seem to be an unlikely model system for hypertension: the tomato. Paterson et al. (29) reported the first genome scan for quantitative traits—in this case, fruit weight and concentration of soluble solids in the tomato—that could be dissected into discrete genetic loci or quantitative trait loci (QTLs). While this study clearly demonstrated the power of a total genomic search, it was still several years before a similar study could be conducted in a mammalian model, because complete genetic maps were not available.

First Use of a Genomic Scan for Hypertension

The first use of a genomic scan to look for genes responsible for hypertension was reported in 1991 (15,30). A total genome search could not be used previously because there were still very few SSLPs or any other genetic markers available for the rat. The first test case for hypertension involved an intercross between the stroke-prone SHR (SHRSP) and the normotensive control strain WKY, consisting of 115 F2 rats that were arranged and extensively phenotyped by Klaus Lindpaintner, Detlev Ganten, and colleagues (31). We and a group led by Mark Lathrop reported strong genetic evidence that a gene named $Bp1$ that has a major effect on blood pressure maps to rat chromosome 10 with a logarithm of the odds ratio (LOD) score of 5.10 (indicating that the data was $10^{5.10}$-fold more likely to have arisen from linkage of a quantitative trait locus than from random chance). Interestingly, the region showing the effect included the rat gene encoding angiotensin-converting enzyme (ACE) (55,56). The $Bp1$ locus appeared to control more than 20% of the genetic variance for systolic and diastolic blood pressure in sodium-loaded animals, with SHRSP alleles apparently acting dominantly to increase blood pressure (30). We also found significant but weaker evidence for a locus $Bp2$ on chromosome 18 (30), and Hilbert et al. (15) found significant linkage with a marker on the X-chromosome. Interestingly, the hypertensive allele on the X-chromosome came from a WKY parent, demonstrating that even normotensive animals can carry genes that contribute to high blood pressure.

Several studies have corroborated the initial reports. Nara et al. (32) reported that the ACE locus cosegregated blood pressure after a sodium load

for several months in a SHRSP × WKY. Deng and Rapp (33) also reported that
the ACE locus cosegregates with blood pressure after a sodium load in an F2
cross from an SS/JR × MNS. Does this mean that *Bp1* is ACE? The short answer
is no. Inheritance studies alone do not have the resolution to prove that *Bp1*
is ACE, as opposed to any other nearby gene. Work conducted by our group
(34 and unpublished) found only modest differences in DNA sequence and
expression of the ACE gene between the hypertensive and normotensive rat.
The knockout studies reported in Chapter 10 give further support for a role
for ACE in the mouse; however, the role of ACE in the mouse may be different
from that in the rat or human. Despite these encouraging results, the ACE gene
may still be nothing more than a linked genetic marker for the true causative
gene. After these original observations, more than five other candidate genes
were localized within the region containing *Bp1* (see Figure 2). Determining

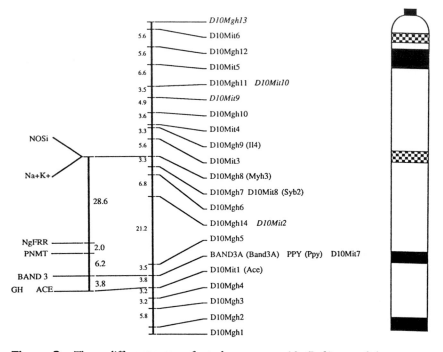

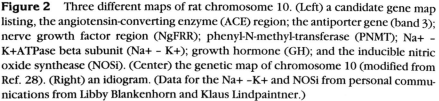

Figure 2   Three different maps of rat chromosome 10. (Left) a candidate gene map
listing, the angiotensin-converting enzyme (ACE) region; the antiporter gene (band 3);
nerve growth factor region (NgFRR); phenyl-N-methyl-transferase (PNMT); Na+ -
K+ATPase beta subunit (Na+ – K+); growth hormone (GH); and the inducible nitric
oxide synthease (NOSi). (Center) the genetic map of chromosome 10 (modified from
Ref. 28). (Right) an idiogram. (Data for the Na+ –K+ and NOSi from personal commu-
nications from Libby Blankenhorn and Klaus Lindpaintner.)

which of these candidate genes, if any, is the causal gene *Bp1* will require additional studies. Definitive proof will require some type of gene transfer or gene targeting approach discussed in depth in Chapter 10.

Results from Partial Genomic Searches

As more genetic markers become available and as investigators continue to develop candidate genes into genetic markers, the rat becomes a better genetic model system. Several groups have used our genetic markers, those developed by Levan, Szpirer and colleagues (35–37), Remmers and colleagues (38,39), and Serikawa and colleagues (40,41) to initiate genomic scans of crosses between hypertensive rats and normotensive rats. These studies have not been complete genome scans and have focused, in large part, on one QTL at a time.

The results of the studies mentioned above are summarized in Table 3. Several points emerge. First, "hypertension-causing" genes are located on a number of rat chromosomes. Second, very few studies have identified QTLs associated with baseline blood pressure, despite there being more than 2 standard deviations between the parental strains. Third, particular QTLs segregate with blood pressure only in particular crosses.

The identification of a large number of QTLs is not really surprising. The blood-pressure regulatory pathway is very complex and has a large number of potential sites that could result in high blood pressure if they are disrupted. Furthermore, the fact that different QTLs are seen in different crosses is not unexpected, since it must be remembered that genetic dissection can only determine genes that are different between the two strains. In addition to the strains used, several other factors are likely to influence the identification of QTLs, including: 1) the methods used for estimating blood pressure (discussed in more detail below), 2) the age of the animal, and 3) how the blood pressure was analyzed.

Why have so few QTLs been identified for baseline blood pressures? There are several potential explanations, all of which have anecdotal evidence. At baseline, the animals are in homeostasis at the genetically predetermined blood pressure. Since blood pressure must be regulated to a fine degree, many compensatory mechanisms are necessarily in place. There may be a large number of genes involved in maintaining the homeostasis, each making only minor contributions to the blood pressure. If so, then a very large number of animals will need to be studied before linkage can be detected. It may be necessary to stress the cardiovascular system in order to see an effect that is predominantly due to a few genes, which could explain why most of QTLs identified are for blood pressure after a salt load of some type. Alternatively, our ability to measure blood pressure has become too "good," in that the level

Table 3   Summary of "Hypertensive" Loci Indentified by Genetic Mapping

Cross[a]	Hypertensive loci[b]	Phenotype[c]	Ref.
SHRSP × WKY	Chr. 10 (ACE), Chr. 18 (?)	SBP after salt	30
	Chr. 10 (ACE), × Chr. (?)	SBP after salt	15
	Chr. 1 (S_A)	SBP after salt	42
SHRSP × WKY	Chr. 10 (ACE)	SBP after salt	32
SHR × WKY	Chr. 1 (S_A)	SBP, DBP	43
SHR × WKY	Chr. 2 (?), Chr. 4 (SPR)	BP	44
SHR × WKY	Chr. 4 (NPY)	SBP, DBP, MAP	45
SHR × WKY	Y-Chr.(?)	SBP	14
SHR × WKY	Chr. 13 (renin)	SBP, DBP	46
SHR × DRY	Chr. 3 (NGF)	MAP	47
RI (SHR × BN)[d]	Chr. 20 (HSP70)	SBP	48
RI (SHR × BN)[d]	Chr. 12 (HSP27)	SBP	e
RI (SHR × BN)[d]	Chr. 13 (renin)	SBP	49
RI (SHR × BN)[d]	Chr. 1 (kallikrien)	SBP	50
SS/JR × SR/JR	Chr. 13 (renin)	SBP after salt	51
SS/JR × MNS	Chr. 2 (GCA), Chr. 10 (ACE)	SBP after salt	33
SS/JR × WKY	Chr. 2 (GCA)	SBP after salt	33
SS/JR × Lew	Chr. 5 (ET2), Chr. 17 (?)	SBP after salt	52
SS/JR × Lew	Chr. 1 (S_A)	SBP after salt	53
LH × LL	Chr. 2 (?)	PP	54
	Chr. 13 (renin)	DBP	
GH × BN	Chr. 2 (?), Chr. 10 (ACE)	SBP	55

[a]Strains of rats—BN: Brown Norway; DRY: Donryu; GH: genetically hypertensive; LEW: Lewis rat; LH: Lyon hypertensive; LL: Lyon low-blood-pressure; MNS: Milan normotensive strain; RI: recombinant inbred strains; SHR: spontaneously hypertensive; SHRSP: stroke-prone spontaneously hypertensive; SR/JR: salt-resistant/John Rapp; SS/JR: salt-susceptible/John Rapp; WKY: Wistar Kyoto.

[b]Candidate genes—ACE: angiotensin-converting enzyme region; ET2, endothelin-2 region; GCA: guanylyl cyclase A/atrial natriuretic peptide receptor region; HSP: heat shock protein region; NGF: nerve growth factor region; NPY: neuropeptide Y region; S_A: subtractive clone A; SPR: substance P receptor region; ?: no known candidate gene.

[c]Phenotypes—DBP: diastolic blood pressure; MAP: mean arterial pressure; PP: pulsatile pressure; SBP: systolic blood pressure.

[d]This methodology, although not as powerful as a cross study, does offer some insight. A problem with this approach is that the same phenotyping information is "scanned" repeatedly, increasing the chance of false linkage, and the segregation of genes is fixed in only 28 strains.

[e]Pavel Hamet, personal communication, 1994.

of anxiety is reduced in the animals. Consider that every hypertensive strain studied was originally selected for using blood pressure measured by tail-cuff, which can be considered a stress model. Our inability to locate the genes responsible for baseline blood pressure may be the result of the animals not being stressed.

In sum, the initial studies utilizing a genetic dissection approach have located several regions in the rat genome that play a role in high blood pressure. However, several points need to be emphasized: 1) the initial studies have not utilized a complete genetic linkage map, 2) many of the crosses studied are quite small (<200 animals), 3) analytical tools have yet to take full account of polygenic interactions—most of the studies to date have simply added the amount of variation accounted for by each locus, and 4) often investigators have not acknowledged that a candidate gene within a QTL may not be the causal gene but rather simply a nearby gene.

Partial Results from a Total Genomic Scan

Using our recently constructed complete genetic linkage map of the laboratory rat (27), we were able to undertake a complete total genome scan. We generated 222 F2 progeny derived from a SHR/NIH × Brown Norway (BN/JK) grandparental cross. The BN strain was selected because it is normotensive and when crossed to the SHR showed the highest rate of polymorphism. More than 300 SSLPs were genotyped on each F2 animal and each animal had 14 different direct phenotypical measurements (see Table 4). Genetic data were analyzed using MAPMAKER/QTL (interval analysis), analysis of variance, and

Table 4 Physiological Measurements in SHR × BNH: Phenotypes Analyzed

Mean arterial pressure (30-minute average)
Systolic blood pressure (30-minute average)
Diastolic blood pressure (30-minute average)
Pulse pressure (30-minute average)
Heart rate (30-minute average)
Baseline MAP before drug administration (30-second average)
MAP after captopril (10 mg/kg)
MAP after captopril and hexamethonium (30 mg/kg)
MAP (30-minute average) after 14 days of salt (1.0% NaCl in water)
SBP (30-minute average) after 14 days of salt (1.0% NaCl in water)
DBP (30-minute average) after 14 days of salt (1.0% NaCl in water)
Pulsatile pressure (30-minute average) after 14 days of salt
HR (30-minute average) after 14 days of salt
Weight (g) before drug administration

a stepwise linear regression. The analyses explicitly take into account polygenic interactions.

    We identified five QTLs linked to SBP after salt loading. The chromo-
somes of interest are chromosomes 2 (with two QTLs), 4, 8, and 16. The most
striking results include: 1) the region containing the ACE locus is not linked
to any of the phenotypes in this particular cross, 2) the region containing the
GCA locus on chromosome 2 shows significant linkage to salt-induced increase in SBP (LOD = 4.5), as previously reported (33), 3) the five loci account
for nearly 40% of the total variance in SBP after a salt load, and 4) the five loci
act in an essentially additive fashion (Figure 3), and with a possible plateau
effect.

    Complete analysis of all 14 phenotypes is not yet finished. However, we
did find at least one QTL for each of these traits. These data illustrate that a
total genomic search has a tremendous amount of power to reveal the genetic
basis of many complex phenotypes.

Problems Identified with Genomic Searches

Although the total genomic search strategy reveals regions of the rat genome
involved in hypertension, there are some problems. The most important is
the inability to replicate studies in some cases. Kreutz et al. (56) report that
the $Bp1$ region did not cosegregate with blood pressure in a second F2 cross
derived from the same SHRSP and WKY colonies in Heidelberg. However, it
appears that the Heidelberg colony had two substrains of WKY, and each may

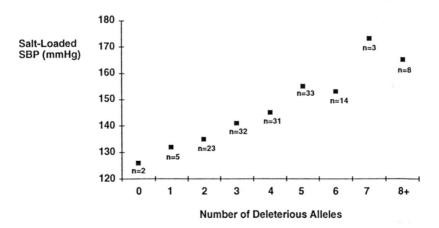

Figure 3 Effect of deleterious alleles. Number of genotypes producing increases in
salt-loaded SBP (NaSBP) possessed by F_2 rat progeny plotted against mean NaSBP value
based on the five loci. n represents the number of rats for each category of putative
NaSBP-increasing alleles.

carry different alleles at *Bp1*. Furthermore, Nara et al. (32) found linkage to the *Bp1* but inferred a recessive mode of inheritance (rather than the previously reported dominant mode) for systolic blood pressure after a salt load in their study. This apparent discrepant result may be accounted for by the different lengths of time the animals were on a salt load.

These two examples point out two major stumbling blocks. First, rat strains with the same name are not necessarily identical. In case of the WKY, the strain was distributed at the 17th generation. The literature has blamed discrepant results on this early distribution, but there are also other likely causes for the existence of different substrains—most importantly, local contamination of the SHR and WKY colonies. Second, the phenotyping protocol and the method used for estimating blood pressure may make cross comparisons difficult. Investigators have used a wide variety of intermediate methodologies between tail-cuff (the most stressful) and direct blood-pressure measurement using radiotelemetry (the least stressful). Furthermore, some investigators have used anesthetized animals and some have used conscious, freely moving animals. Investigators also measure the blood pressure at different ages and, in most cases, for a very short period of time, generally between 30 minutes and 1 hour. Obviously, the methodology affects both the environmental variation and the physiological trait probed (and thus the genetic determinants involved).

These caveats demonstrate an important consideration: Comparison of genetic data from different laboratories with different phenotyping protocols and different ages of phenotyping may produce discrepant results.

UTILITY OF DATA FROM A TOTAL GENOMIC SEARCH

The preceding section illustrates the power of a total genomic search to identify genetic loci responsible for hypertension. In this section, we outline how to use the data generated from the total genomic search.

Candidate Genes Versus Total Genomic Search

Genetic dissection follows two principal approaches. The first is the candidate gene approach, in which a gene known to play a role in the maintenance of blood pressure, such as renin, is investigated as a potential cause of hypertension. As described earlier, the second and newer approach is a total genomic scan. These two approaches should not be considered mutually exclusive; both use cosegregation analysis to determine if a particular genotype of a gene or a genotype of genetic marker is linked with a particular phenotype.

The advantage of the candidate approach is that it focuses on genes demonstrated to play a role in blood-pressure regulation. However, positive cosegregation of a candidate gene with a phenotype is not sufficient evidence to prove the gene is the cause of the phenotype—no matter how much we know about the gene. The causal gene may be an as-yet-unknown gene that resides near the candidate gene.

The total genomic scan indicates only which region of the genome is linked to the hypertension and is independent of any bias due to prior prejudice about which genes are "important." Once linkage is found to a region, the next step is to determine if any candidate genes reside within this genetic locus. Any candidate gene within this region serves as a starting point for future studies, such as those outlined in Chapter 10. However, these studies will need more information. For example, Figure 2 shows that five potential candidate genes for hypertension are all located within a 2 LOD unit confidence interval (15,30) that contains the *Bp1* gene. Which of these candidates is correct, and how do you prove it? To illustrate some of the important issues, let us take a look at the ACE and phenylethanoloamine-N-methyl transferase (PNMT) genes.

ACE Gene

To address the possibility that ACE is *Bp1*, we must first demonstrate either a functional difference in the gene product or a regulatory difference that affects gene expression between the WKY and SHRSP. We have constructed cDNA libraries from the lung of the WKY and the SHRSP, and cloned and characterized the cDNAs for the ACE of both strains. Both WKY and SHRSP ACE cDNAs encode a single polypeptide of 1313 amino acid residues with an estimated molecular weight of 150.9 kDa. Rat ACE cDNA is 80–90% homologous in the nucleotide structure to the mouse, human, and bovine enzyme. The deduced amino acid sequence reveals a high degree of internal homology between two large domains with highly hydrophobic regions and two conserved active sites. Five nucleotide differences (A-620, C-871, C-2514, T-2649, and G-2940 in WKY to G-620, T-871, T-2514, C-2649, and A-2940 in SHRSP) were identified when the nucleotide sequences of the WKY and the SHRSP ACE were compared. One of those differences resulted in an amino acid substitution (Lys-207 in WKY to Arg-207 in SHRSP). Conservative amino acid substitutions, such as that observed in the comparison of the WKY and the SHRSP ACE, might not be expected to have dramatic functional consequences. Consistent with this, the specific activity of the ACE protein was 0.26 units/mg for the WKY versus 0.27 units/mg for the SHRSP/Hei. The K_m values were 3.03 mM (WKY) and 3.17 mM (SHRSP). The V_{max} values were 0.34 mmol/min/mg (WKY) versus 0.34 mmol/min/mg (SHRSP). These data suggest that there are no functionally significant differences between the primary structures of WKY and SHRSP ACE.

To determine if there are differences in the regulatory regions, we cloned and characterized the promoter regions of the ACE gene of both strains. We sequenced a 1.3 kbp fragment from the 5'-flanking region of the WKY and the SHRSP ACE genes. Four nucleotide differences were observed between the promoter region of the two strains, with three of these grouped within 14 bases. In the SHRSP promoter region, a TG deletion at position −270 alters a consensus sequence for the H1 transcription factor and is 17 nucleotides downstream from an AP2 site (Figure 4). These data provide evidence for mutations in the promoter region of SHRSP ACE genes. Studies on whether these differences can alter ACE gene expression have not been completed.

PNMT Gene

We determined the chromosomal assignment and chromosomal location of the rat PNMT gene to chromosome 10, within the confidence interval of BP1. To examine whether there are any differences in the PNMT gene between WKY and SHRSP that can result in an abnormality of the primary structure and/or the aberrancy in the transcriptional regulation of the gene, we set out to clone the PNMT genes of both strains. Entire exon regions, 1077 bp 5'-flanking region, and 256 bp 3'-end region of both PNMT genes were sequenced. Intron–exon boundaries were also sequenced. Intron sizes were determined by restriction mapping and/or sequencing. There were no nucleotide differences in the regions sequenced between WKY and SHRSP. These data suggest that the PNMT gene itself is an unlikely candidate gene contributing to the hypertension in the SHRSP; however, a regulatory region further upstream is still a possibility.

Unfortunately, neither of these two studies shed much light on which gene is *Bp1*. However, knowing in advance that this region contains a "hypertensive" locus at least provided additional rationale for initiating these time-consuming studies.

Identification of the Causal Gene Corresponding to a QTL

Cloning mammalian genes based solely on position remains a daunting task even for genes controlling simple mendelian traits. Fewer than 75 genes have

```
WKY    GGCGGCCAGAGGGCACGGTTGGGCCGCAGCACTGTGTTTGCAGCCCGC
SHRSP  CGCGGCCAGAGGGCACGGTTGGGCCGCAGCAC--TGTTTGCAGCCGGC
         AP2                              H1-Box
```

Figure 4 Nucleotide differences in promoter region of ACE gene between WKY and SHRSP. 5'-flanking regions of the rat ACE genes (from positions −311 to −263 in WKY) are aligned. Putative consensus sequences (AP2 and H1-Box) are boxed. (−) indicates a deletion. Bold character indicates different nucleotides between the WKY and the SHRSP.

been cloned in this manner, and none of them is for quantitative traits. Nonetheless, genomic mapping tools are improving and the task is becoming more tractable. In anticipation of continued improvement, we briefly outline an approach to cloning a QTL based on position:

1. Confirm the linkage by constructing a second cross using the original parental strains and phenotyping protocol.
2. Once linkage is confirmed, genetically transfer a small region containing the QTL allele from the hypertensive strain into the normotensive strain and vice versa by successive backcrosses to create congenic strains, which differ only in the region of one QTL (cf. Ref. 57).
3. Using congenic strains, map the QTL more closely by arranging additional crosses and looking for recombinant animals (i.e., animals that have inherited different parental alleles on either side of the region). A recombinant animal allows the geneticist to reduce the size of the region containing the causal gene.
4. When the region has been adequately narrowed, construct a physical map (clone the entire region, which is usually between 1 million and 2 million base pairs in length), and identify genes in the region. The mutations are likely to be subtle rather than gross gene disruption, making the mutations difficult to distinguish from sequence variation.
5. To prove that a candidate gene corresponds to a QTL, use homologous recombination to construct gene knockouts in strains carrying a dominant allele (which can then be tested for loss of dominance at the QTL) or to substitute the allele in strain A into strain B and demonstrate that the QTL effect has been transferred as well. To prove a candidate gene with a recessive mode of inheritance, a transgenic approach could be used to determine if adding the "disease-free" allele reduces blood pressure. Transgenic "rescue" experiments may require the use of congenic animals carrying both recessive alleles, rather than the hypertensive rat model, because the complex polygenic environment may prevent the expression of an altered phenotype.

Starting with a plausible candidate gene may accelerate the process, but does not alter the fundamental logic. Taking ACE as an example, it will be important to look in the ACE gene and protein for structural and functional differences between SHRSP and WKY, but this alone cannot provide rigorous genetic proof that ACE is the QTL in question. Rather, it will still be important to introgress the SHRSP allele of ACE onto a WKY genetic background and to demonstrate by homologous gene replacement that an isogenic SHRSP strain with a gene knockout at the ACE locus lacks the dominant effect of the SHRSP *Bp1* allele or that an isogenic WKY strain carrying the SHRSP allele at ACE shows the dominant effect of the SHRSP *Bp1* allele. The knock-out of the ACE gene in mouse demonstrates the power of approach for hypertension research (Chapter 10).

Clearly, a major problem with step 5 is the fact that, at present, homologous recombination and gene knockout studies are limited to mice. It will be important to extend this technology to the rat. Iannaccone et al. (58) recently reported the production of a chimeric rat using a rat embryonic stem (ES) cell. Unfortunately, these chimeric rats failed to produce germ cells containing the ES cell genome. However, these experiments illustrate that ES cell technology may soon be available for the rat.

Homology Mapping

As outlined above, genetic mapping techniques have begun to identify regions in the rat genome that contain genes that contribute to complex disease. Unfortunately, the molecular genetic tools required to identify the gene and then prove that the gene is causal have not been developed in the rat (59). However, in many cases it will be possible to begin the hunt for the gene in another mammalian species using a technique referred to as conserved synteny (gene order) mapping—or, more appropriately, homology mapping—which takes advantage of the order of homologous genes tending to be conserved in different mammalian species.

For example, we are interested in identifying a gene (*Iddm1*) responsible for lymphopenia and insulin-dependent diabetes in the BB rat. We mapped the gene to within 0.7 cM (~1 million base pairs) of neuropeptide Y (Npy) (60). Little was known about this region in the rat, and the tools, namely large insert libraries, required for positional cloning of an unknown gene were not available for the rat. Therefore, we turned to the mouse, for which large insert libraries were available and a large number of genes and genetic markers had been genetically mapped.

The first step was to determine if the region surrounding Npy in the mouse was likely to carry the Iddm1 gene. We used Npy as an anchor point for looking for which genes reside near Npy in the mouse and human. We designed primers from the mouse or human gene sequences near Npy that would work in the rat and then mapped these genes in the rat. There is exceptional conservation of order for homologous genes between rat and mouse at the *Iddm1* locus (Figure 5). Therefore, we could use the mouse genetic and physical mapping reagents to begin the positional cloning.

A similar strategy can be used to determine if a homologous region (to a "hypertensive" region in the rat) in the human contributes to hypertension.

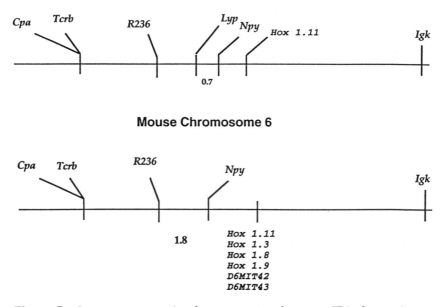

Figure 5 Synteny conservation between rat and mouse. This figure shows the synteny conservation between this region of rat chromosome 4 and mouse chromosome 6. D6MIT#s are mouse SSLPs, and R236 is a rat SSLP. Gene names are as follows: *Cpa*, carboxypeptidase; *Tcrb*, T-cell receptor beta; *Hox* 1.2–1.9, housekeeping genes; *Npy*, neuropeptide Y; *Lyp*, lymphopenia/*Iddm 1*; *Igk*, immunoglobulin kappa.

Once the region of homology has been established between rat and human, existing human genetic markers can be used to test if this region cosegregates with hypertension. This approach allows candidate regions from the rat to be tested in humans long before the gene has been identified.

Summary

Over a decade ago, John Rapp (12) set four criteria to be fulfilled before accepting that a genetic locus controls blood pressure: 1) a biochemical and/or physiological difference between two strains must be established, 2) the locus must segregate in a mendelian fashion, 3) the loci identified must account for a significant portion of the blood pressure, and 4) there must be some logical link between the locus and its control of blood pressure. This last criterion, since we can identify hypotheses to meet it, has led to the continued focus on candidate genes. These criteria need to be further refined to reflect the difference between linkage to region and the actual identification of the causal gene. For linkage to a region, only criteria 1, 2, and 3 need to be fulfilled. However, all four criteria need to be fulfilled before a gene can be determined to be causal. It is crucial to recognize that identification of the causal gene remains a major challenge even after the genetic dissection, since this approach cannot determine biological function.

Molecular genetics today offers several advantages for current research and several promises for tomorrow's understanding of hypertension and other complex diseases. First, genetic analysis in experimental animal models continues to define and improve the approaches used to study complex disease by providing investigators with locations known to harbor a disease-causing gene. Second, once the molecular genetic infrastructure (genetic maps and physical maps) is in place and the genetic loci identified, there will be a need to study the physiology. These physiological studies will come in many forms—most of which will focus on an animal with a single disease gene or with the "protective" gene knocked out by homologous recombination. Third, genetic maps and physical mapping tools are already enabling investigators to initiate these studies.

CLINICAL IMPLICATIONS

Genetic analysis of experimental animal models provides: 1) understanding of the multifactorial basis of hypertension and 2) genetic pointers that highlight regions to be studied in humans long before the genes are cloned, thereby revealing clues about the genetic basis of hypertension in people.

The detailed physiology known about the animal models combined with this new information about the genetic underpinnings is likely to provide a more complete understanding of the primary causes of hypertension. Once primary causes have been determined, the potential for clinical diagnosis, defining the "flavor" of hypertension, and the potential for new and novel therapies become a possibility. Certainly any hope of prevention requires an understanding of the etiology of hypertension. The molecular genetic approaches outlined in this chapter represent one powerful approach that builds on our experimental and clinical understanding of hypertension to obtain a better understanding of its etiology.

REFERENCES

1. Smirk FH, Hall WH. Inherited hypertension in rats. Nature 1958; 182:727–728.
2. Barger AC. The Goldblatt Memorial Lecture. Part I. Experimental renovascular hypertension. Hypertension 1979, 1:447–455.
3. Goldblatt H, Lynch J, Hanzal RF, Summerville WW. Studies on experimental

hypertension: production of persistent elevation of systolic blood pressure by means of renal ischemia. J Exp Med 1934; 59:347–379.

4. Okamoto K, Aoki K. Development of a strain of spontaneously hypertensive rats. Jpn Circ J 1963; 27:282–293.

5. Dahl LK, Heine M, Tassinari L. Role of genetic factors in susceptibility to experimental hypertension due to chronic excess salt ingestion. Nature 1962; 194:480–482.

6. Falconer DS. Introduction to Quantitative Genetics. Essex, England: Longman Scientific & Technical, 1989.

7. Group HDaF-UPC. The hypertension detection and follow-up program: A progress report. Circ Res 1977; 40:I-106–I-109.

8. Wright S. The genetics of quantitative variability. In: The Genetics of Human Populations: A Treatise in Four Volumes. Chicago: University of Chicago Press, 1968:373–420.

9. Tanase H, Suzuki Y, Ooshima A, Yamori Y, Okamoto K. Genetic analysis of blood pressure in spontaneously hypertensive rats. Jpn Circ J 1970; 34:1197–1212.

10. Yamori Y, Ikeda K, Ooshima A, Fukase M. Inheritance of hypertension in stroke-prone spontaneously hypertensive rats. In: Prophylactic Approach to Hypertension Diseases. Yamori Y, Lovenberg W, Freis ED, eds. New York: Raven, 1979:121–125.

11. Knudsen KD, et al. Effects of chronic salt ingestion: inheritance of hypertension in the rat. J Exp Med 1970; 132:976–1000.

12. Rapp JP. Genetics of experimental and human hypertension. In: Hypertension. 2nd ed. Genest J, Kuchel O, Hamet P, Cantin M, eds. New York: McGraw-Hill, 1983:534–555.

13. Beckett LA, Rosner B, Roche AF, Guo S. Serial changes in blood pressure from adolescence into adulthood. Am J Epidemiol 1992; 135:1166–1177.

14. Ely DL, Turner ME. Hypertension in the spontaneously hypertensive rats is linked to the Y-chromosome. Hypertension 1990; 16:282–289.

15. Hilbert P, et al. Chromosomal mapping of two genetic loci associated with blood pressure regulations in hereditary hypertensive rats. Nature 1991; 353:521.

16. Ely DL, Daneshvar H, Turner ME, Johnson ML, Salisbury RL. The hypertensive Y chromosome elevates blood pressure in F_{11} normotensive rats. Hypertension 1993; 21:1071–1075.

17. Vincent M, Kaiser MA, Orea V, Lodwick D, Samani NJ. Hypertension in the spontaneously hypertensive rat and the sex chromosomes. Hypertension 1994; 23:161–166.

18. Levan G, Fredga K. Cytogenetic markers. In: Genetic Monitoring of Inbred Stains of Rats. Hedrich HJ, ed. Stuttgart: Fischer Verlag, 1990:42–58.

19. Kurtz TW, Morris RC. Biological variability in Wistar-Kyoto: implications for research with the spontaneously hypertensive rat. Hypertension 1987; 10:127–131.

20. Kurtz TW, Morris RC. Biological variability in the Wistar- Kyoto and spontaneously hypertensive rats. Hypertension 1988; 11:106.

21. Kurtz TW, Montano M, Chan L, Kabra P. Molecular evidence of genetic heterogeneity in Wistar-Kyoto rats: Implications for research with the spontaneously hypertensive rat. Hypertension 1989; 13:188–192.

22. Samani NJ, Swales JD, Jefferys A, et al. DNA fingerprinting of spontaneously hypertensive and Wistar-Kyoto rats: Implications for hypertension research. J Hypertension 1989; 7:809–816.

23. St Lezin EM, et al. Genetic contamination of Dahl SS/Jr rats: Impact on studies of salt-sensitive hypertension. Hypertension 1994; 23:786–790.

24. Botstein D, White RL, Skolnick M, Davis RW. Construction of a genetic linkage map in man using RFLP's. Am J Hum Genet 1980; 32:314.

25. Weber JL, May PE. Abundant class of human DNA polymorphisms which can be typed using PCR. Am J Hum Genet 1989; 44:388.

26. Weber JL. Informativeness of human (dC-dA)n* (dG-dT)n polymorphisms. Genomics 1990; 7:524–530.

27. Jacob HJ, et al. Genetic linkage map of the laboratory rat, *Rattus norvegicus.* Nature Genet 1995; 9:63–69.

28. Lander ES, Botstein D. Mapping complex genetic traits in humans: new methods using a complete RFLP linkage map. Cold Spring Harb Symp Quant Biol 1986; 1:49–62.

29. Paterson AH, et al. Resolution of quantitative traits into Mendelian factors by using a complete linkage map of restriction fragment length polymorphisms. Nature 1988; 335:721–726.

30. Jacob HJ, et al. Genetic mapping of a gene causing hypertension in the stroke-prone spontaneously hypertensive rat. Cell 1991; 67:213–224.

31. Lindpaintner K, Takahashi S, Ganten D. Structural alterations of the renin gene in the stroke-prone spontaneously hypertensive rat: examination of genotype-phenotype correlations. J Hypertension 1990; 8:763–773.

32. Nara Y, et al. Blood pressure cosegregates with a microsatellite of angiotensin I converting enzyme (ACE) in F2 generation from a cross between original normotensive Wistar-Kyoto rat (WKY) and stroke-prone spontaneously hypertensive rat (SHRSP). Biochem Biophys Res Commun 1991; 181:941–946.

33. Deng Y, Rapp JP. Cosegregation of blood pressure with angiotensin converting enzyme and atrial natriutetic peptide receptor genes using Dahl salt-sensitive rats. Nature Genet 1992; 1:267–272.

34. Krieger JE, et al. Evidence for mutation in the promoter region of the ACE gene of SHR-SP vs. WKY (abstr). Hypertension 1992; 20:412.

35. Lenfant C, Roccella EJ. National High Blood Pressure Education Program (editorial). J Am Optom Assoc 1986; 57:347–348.

36. Levan G, et al. The gene map of the norway rat (*Rattus norvegicus*) and comparative mapping with mouse and man. Genomics 1991; 10:699–718.

37. Levan G, Klinga-Levan K, Szpirer C, Szpirer J. Gene map of the rate (*Rattus norvegicus*). In: O'Brien SJ, ed. Locus Maps of Complex Genomes. 6th ed. Cold Spring Harbor, NY: Cold Spring Harbor Press, 1992.

38. Goldmuntz E, et al. Genetic map of 12 polymorphic loci on rat chromosome 1. Genomics 1993; 16:761–764.

39. Zha H, et al. Linkage map of 10 polymorphic markers on rat chromosome 2. Cytogenet Cell Genet 1994; 63:117–123.

40. Serikawa T, et al. Rat gene mapping using PCR-analyzed microsatellites. Genetics 1992; 131:701–721.

41. Yamada J, Kuramoto T, Serikawa T. A rat genetic linkage map and comparative maps for mouse or human homologous rat genes. Mammal Genome 1994; 5:63-83.

42. Lindpaintner K, et al. Molecular genetics of the S_A-gene: cosegregation with hypertension and mapping to rat chromosome 1. J Hypertension 1993; 11:19-23.

43. Samani N, et al. A gene differentially expressed in the kidney of the spontaneously hypertensive rat cosegregates with increased blood pressure. J Clin Invest 1993; 92:1099-2005.

44. Nakajima S, Rioseco N, Ma L, Printz M. Candidate gene loci for hypertension in LaJolla colony SHR(lj) and WKY(lj) on chromosomes 2 and 4 (abstr). J Hypertension 1994; 12:S66.

45. Katsya T, et al. A neuropeptide Y locus on chromosome 4 cosegregates with blood pressure in the spontaneously hypertensive rat. Biochem Biophys Res Comm 1993; 192:261-267.

46. Sun L, McArdle S, Chun M, Wolff DW, Pettinger WA. Cosegregration of the renin gene with an increase in mean arterial blood pressure in the F2 rats of SHR-WKY cross. Clin Exp Hypertension 1993; 15:797-805.

47. Kapuscinski M, Charchar F, Mitchell G, Harrap S. The nerve growth factor gene and blood pressure in the spontaneously hypertensive rat (abstr). J Hypertension 1994; 12:S191.

48. Hamet P, et al. Restriction fragment length polymorphism of hsp70 gene, localized in the RT1 complex, is associated with hypertension in spontaneously hypertensive rats. Hypertension 1992; 19:611-614.

49. Pravenec M, et al. The rat renin gene: assignment to chromosome 13 and linkage to the regulation of blood pressure. Genomics 1991; 9:466-472.

50. Pravenec M, et al. Cosegregation of blood pressure with a kallikrein gene family polymorphism. Hypertension 1991; 17:242-246.

51. Rapp JP, Wang SM, Dene H. A genetic polymorphism in the renin gene of Dahl rats cosegregates with blood pressure. Science 1989; 243:542-544.

52. Deng AY, Dene H, Pravenec M, Rapp JP. Genetic mapping of two new blood pressure quantitative trait loci in the rat by genotyping endothelin system genes. J Clin Invest 1994; 93:2701-2709.

53. Harris EL, Dene H, Rapp JP. SA gene and blood pressure cosegregation using Dahl salt-sensitive rats. Am J Hypertension 1993; 6:330-334.

54. Dubay C, et al. Genetic determinants of diastolic and pulse pressure map to different loci in Lyon hypertensive rats. Nature Genet 1993; 3:354-357.

55. Harris E, Phelan E, Grigor M. The GCA allele of the New Zealand genetically hypertensive (GH) rat cosegregates with blood pressure in male but not female rats (abstr). J Hypertension 1994; 12:S66.

56. Kreutz R, Hubner N, Jacob H, Ganten D, Lindpaintner K. Evidence that the hypertension-associated locus of rat chromosome 10, BP-SP1, is not identical with the angiotensin-converting enzyme gene locus (abstr). Hypertension 1994; 24: 373.

57. Paterson AH, DeVerna JW, Lanini B, Tanksley SD. Fine mapping of quantitative trait loci using selected overlapping recombinant chromosomes, in an interspecies cross of tomato. Genetics 1990; 124:735-742.

58. Iannaccone PM, Taborn GU, Garton RL, Caplice MD, Brenin D. Pluripotent embryonic stem cells from the rat are capable of producing chimeras. Dev Biol 1994; 163:288-292.
59. Frankel WN. Of rats, mice and men? Nature Genet 1995; 9:3-4.
60. Jacob HJ, et al. Genetic dissection of autoimmune type I diabetes in the BB rat. Nature Genet 1992; 2:56-60.

12

Hypertension Research with Genetically Altered Animals

Linda J. Mullins and John J. Mullins
Centre for Genome Research
University of Edinburgh
Edinburgh, Scotland

INTRODUCTION

The generation of genetically modified animals by transgenic technology is an increasingly important tool in the field of hypertension. Transgenesis affords the opportunity to analyze gene function and regulation in vivo, alter the level of expression of specific genes, ablate specific cell types, generate tissue-specific tumors from which differentiated cell lines can be derived, and generate and refine disease models.

In this chapter, we first outline the methodology of transgenesis, by both microinjection and embryonic-stem-cell technology, pointing out the general considerations for experimental design. Second, we highlight the applications of transgenesis to hypertension research, reviewing the most important transgenic lines that have been generated to date. Finally, we consider the future possibilities for hypertension research.

METHODOLOGY OF TRANSGENESIS

Microinjection

Generation of Transgenic Animals

Transgenesis, in the simplest terms, may be defined as the introduction of exogenous DNA into the genome, such that it is stably maintained in a heritable manner. The technique by which transgenic animals are produced is schematically outlined in Figure 1. The preliminary stage involves the isolation of fertilized eggs for microinjection. In mice, this is achieved by the superovulation of young virgin females (approximately 4–5 weeks of age),

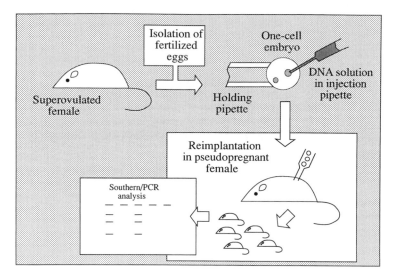

Figure 1   The generation of transgenics by microinjection.

which are injected with a source of follicle-stimulating hormone (pregnant mare's serum gonadotropin) 60 hours prior to mating. In the case of rats, a purified preparation of follicle-stimulating hormone, administered by osmotic minipump, is required to induce superovulation of healthy oocytes. Human chorionic gonadotropin is administered, intraperitoneally, 12 hours prior to pairing with proven stud males. The following day, fertilized eggs are recovered from donor females. Between 20 and 30 mouse zygotes, and 50 to 100 rat zygotes, can be isolated per female, assuming superovulation has been successful, and depending on the strain used.

The single-cell embryos are immobilized on a holding pipette, and DNA is injected into one of the pronuclei, typically the male pronucleus. Embryos that have been successfully injected may be incubated overnight, developing to the two-cell stage. Whether at the one-cell or the two-cell stage, embryos are reimplanted into the oviduct of anesthetized pseudopregnant females (experienced mothers that have been mated the previous night with vasectomized or genetically infertile males). The females are allowed to recover from the anesthetic, and the pregnancy is continued to term. For a more detailed description, see Hogan et al. (1).

Testing the Progeny

At weaning, progeny are tested for incorporation of the transgene. A number of techniques are available, including polymerase chain reaction (PCR) (2) or Southern blot hybridization analysis (3) of genomic DNA isolated from tail biopsies or PCR analysis of whole blood (4). The intensity of the signal, on Southern blot analysis, gives an indication of the number of transgene copies that have been integrated. This can range from a fraction of a copy per cell (suggesting that the transgene is not present in all cells—see below) to tens of copies per cell.

Mosaicism and its Consequences

If the transgenic DNA failed to integrate into the genome prior to the first cell division, it may not be present in all cells of the developing embryo. This is indicated when, upon analysis, the copy number is found to be less than one per cell, and results in a mosaic transgenic animal. If the transgene is not integrated in cells that subsequently contribute to the germline, then the transgene will, by definition, not be inherited by subsequent offspring. If, however, a proportion of the germline cells carry the transgene, then it will be passed on to some of the F1 generation through the germline, at a frequency that reflects the percentage of mosaicism in the germ cells. Careful characterization of these progeny will allow the transgenic line to be rescued.

Multiple Insertion Sites and Their Segregation

The possibility always exists that there may be more than one transgene insertion site in a founder (G_0) animal. This can be discerned from the complexity of the Southern blot analysis, since each insertion site will yield two unique restriction fragments flanking the transgene array (Figure 2). Unless the insertion sites are closely linked, they will act as independent genetic loci, and will segregate in Mendelian fashion. Thus, careful analysis of the progeny should allow each individual insertion event to be derived as a unique transgenic line.

Concatomer Arrays and Insertion Site Effects

The transgene integrates into the genome in an entirely random fashion, often as a head–tail concatomer. Expression of the transgene is not necessarily copy-number-related, however, and may be very low despite the presence of a high copy number. This is due to the fact that expression can be affected by sequences flanking the site of integration, for example, if it integrates into an area of the genome that is actively repressed/silenced. Alternatively, expression of the transgene may be so high that it causes unexpected phenotypic alterations to the organism. Finally, the transgene may integrate within an endogenous gene, altering or insertionally inactivating that gene function and presenting as a totally unexpected phenotype (5–7).

ES Cells

An alternative route to transgenesis involves the introduction of foreign DNA into embryonic stem (ES) cells. These cells are derived from the inner cell

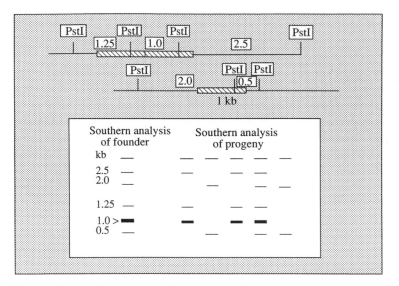

Figure 2 Resolution of multiple unlinked transgene insertion sites by Southern hybridization analysis. Where present, internal transgene-specific bands (>) will reflect transgene copy number.

mass of the developing blastocyst, and are passaged on feeder layers, or in the presence of differentiation-inhibiting activity (DIA) (8), to maintain their undifferentiated state. At present, ES-cell technology is limited to the mouse, most notably cells derived from strain 129 Ola, and much work is being done to isolate ES cells from other mouse strains and also other species. Obviously the rat would be the species of choice for hypertension and cardiovascular research, and a recent report of rat chimeras generated from embryo-derived cells gives promise for the development of germline-competent ES cells from this species (9). [For a more detailed description of ES cell technology see Hooper (10)].

Homologous Recombination

Transgenes introduced by microinjection (or retroviral infection) integrate in a random fashion, such that the endogenous copy of the gene remains intact and structurally unaltered. However, its expression may be affected by or mask that of the transgene, such that phenotypic analysis of the transgenic line may be complicated. On the other hand, with the recent advances in gene targeting it is now possible to select ES cells in which endogenous genes have been specifically modified by homologous recombination (11). Following selection, targeted clones are rapidly identified by Southern or PCR analysis (Figure 3).

General Design of Vectors

Although gene targeting strategies are becoming increasingly elegant, the design of vectors has been somewhat empirical. As our understanding of the factors underlying the frequency of gene targeting improves, the approaches used are becoming more standard. Since homologous recombination in mammalian cells is relatively rare compared to random integration, sophisticated selection protocols are required to enrich for cells in which the desired event has occurred. One of the key factors in efficient homologous recombination is the use of isogenic DNA (i.e., isolated from the same strain as that from which ES cells were derived) to prepare the targeting vector. It appears that heterology in nonisogenic DNA sequences significantly reduces recombination efficiency (12,13). Progress is now being made toward defining the extent of homology required between targeting vector and target locus for high-fidelity recombination to occur (13,14).

Gene Knockout

Two basic types of vector are used for homologous recombination: insertion and replacement vectors (Figure 4) (15). Using insertion constructs, the normal structure of the targeted gene is interrupted by the insertion of the

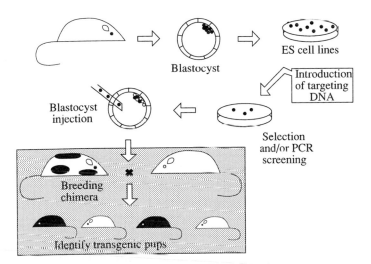

Figure 3   The generation of transgenics by homologous recombination in ES cells.

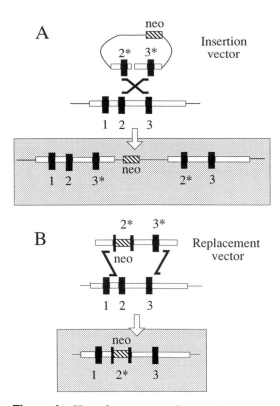

Figure 4 Homologous recombination by (A) insertion vector and (B) replacement vector. (Open boxes = homologous DNA; hatched box = the neomycin resistance gene sequences.)

targeting vector. Using replacement vectors, a double-crossover event across two regions of homology flanking nonhomologous sequences (such as a positive selectable marker) leads to the replacement of endogenous sequences by the sequences present in the construct.

Since integration can occur both specifically and at random, a number of strategies have been devised for both positive selection of homologous recombination events and negative selection against random integration (Figure 5). Positive selection is achieved by interrupting homologous sequences in the targeting vector with a selectable marker, such as the bacterial neomycin-resistance gene neo^r (using G418 selection). Any additional vector sequences outside the region of homology will be lost. By placing the herpes simplex virus-1 thymidine kinase (HSV-1-tk) gene in the nonhomologous region of the targeting vector, any cells retaining this gene through random

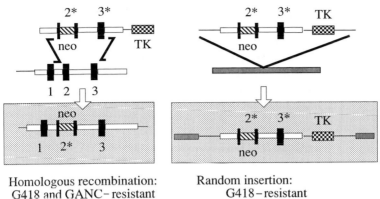

Homologous recombination: Random insertion:
G418 and GANC–resistant   G418–resistant
                                                GANC–sensitive

Figure 5 The positive–negative selection strategy, selecting for homologous recombinants and against random integration events. (Open boxes = homologous DNA; hatched box = the neomycin resistance gene; checked box = thymidine kinase gene; gray box = nonhomologous DNA.)

integration events will be killed in the presence of appropriate synthetic nucleosides such as gancyclovir (GANC). This approach was used to inactivate the *Wnt*-1 proto-oncogene (16,17) and the retinoblastoma gene (18–20). Toxigenic genes can equally be used as a negative selection (21).

Homologous integration can be indirectly selected if expression of the selectable marker is dependent on correct gene targeting. Here, the marker, lacking its translational start, is fused, in frame, to coding sequences of the targeted gene. It is essential that the target gene be active, or inducible in the ES cell, for this approach to work. The strategy was used successfully to sequentially inactivate both alleles of the *pim*-1 proto-oncogene (22). A more flexible and efficient strategy involves the incorporation of an internal ribosome-entry site (IRES) upstream of the selectable marker *Bgeo* (a lacZ-neo^r fusion gene) (23). This makes production of the selectable marker and/or reporter protein independent of the reading frame generated by the fusion, but it still requires detectable expression of the target gene in ES cells.

In/Out Strategies

All the above techniques will generate null mutations through insertional inactivation of the target gene. By using the so-called "hit-and-run" or double-replacement strategies, more subtle gene alterations can be introduced into the target gene. The hit-and-run procedure introduces a site-specific mutation into a nonselectable gene by a two step recombination event (24). In the first step, the vector, containing the desired mutation within sequences homologous to the target gene together with selectable markers for monitoring the integration and reversion events, integrates into the target gene. The resultant duplication can then be resolved to yield clones carrying the desired mutation. The double-replacement strategy requires two targeting constructs. The first introduces a functional *hprt* minigene into the target gene (25), while the second removes the *hprt* gene and replaces it with a subtly altered target gene. Selection both for and against the incorporation of *hprt* can be achieved by growth in HAT medium or in the presence of 6-thioguanine, respectively.

Germline Transmission/Homozygosity

Foreign DNA can be introduced into the ES cells by a number of means: electroporation, transfection, or microinjection. The genetically altered cells are selected appropriately, and are then reintroduced into a host blastocyst, and can contribute to all tissues of the developing embryo following reimplantation into a pseudopregnant female. The pups resulting from such manipulation will, by definition, be chimeras, since cells harboring the transgene constitute only a proportion of the inner cell mass of the blastocyst. However, provided that transgene-containing cells have contributed to the germline and their karyotype is normal, then a suitable breeding strategy will allow the establishment of a transgenic line and, if viable, the generation of homozygotes (26). In some cases, generation of chimeras using homozygous ES cells lacking gene functions essential to early development may be informative when breeding to homozygosity fails because of early lethality. Homozygosity of the ES cells can be achieved either by two rounds of homologous recombination, using a different selection marker for each round (22), or by selecting heterozygous cells at higher concentrations of the drug to which the cells are resistant (15,27).

An example of germline transmission not being reported was the targeted integration of the Ren-1D locus, using a Ren-1D targeting vector (28). Interestingly, although the ES cells contained both Ren-1 and Ren-2 genes, only Ren-1 was targeted, again confirming the importance of homology between targeting construct and target gene. Since renin expression has not been confirmed in ES cells, the selectable marker was placed under the control of the phosphoglycerate kinase (PGK) promoter.

General Considerations

Choice of Species

The mouse has traditionally been the species of choice for transgenic research. A transgenic animal program requires significant animal-breeding facilities, to ensure the regular production of large numbers of single-cell embryos for microinjection as well as of pseudopregnant females to receive the injected embryos, and to maintain breeding colonies for the various transgenic lines generated from each microinjection series. For some physiological studies, however, such as cardiovascular research, neurobiology, and pharmacology, rat transgenics may be preferable, because of size constraints with the mouse (29) or the historical use of this species within a particular discipline. Other features to be taken into consideration are listed in Table 1.

If larger animals such as pigs are to be used, it is practical to carry out initial transgenic studies with a given construct in rodents prior to costly trials in large domestic animals. It must be noted, however, that the response of one species to a gene construct may vary from that of another (29,30).

Breeding Considerations

It is worth considering the respective pros and cons of using inbred versus outbred strains as a source of embryos. Outbred or F1 crossbred animals are generally superior to inbred strains, yielding higher numbers of eggs on superovulation, having larger litters, and being generally better mothers. However, the complex genetic background of the resultant transgenic line may lead to complications in subsequent analysis. This has been amply demonstrated with the TGR(mRen)27 rats that carry the mouse renin transgene on a Sprague Dawley (Hannover) background. When the transgene was bred onto a Sprague Dawley (Edinburgh) background, the penetrance of malignant hypertension increased significantly. In contrast, crossing such animals with inbred Lewis rats suppressed the development of this exaggerated phenotype (31).

The introduction of Transgenes onto an inbred background, on the other hand, provides for control animals that are essentially congenic with the transgenic line. Factors such as characteristics of endogenous gene versus the transgene may play a part in the decision to use inbred strains rather than F1 hybrids.

Table 1   Mouse Versus the Rat as a Tool for Transgenesis

	Mouse	Rat
Gestation time	19−21 days	20−23 days
Litter size	5−12	6−15
Onset of fertility	6−8 weeks	9−14 weeks
Superovulation	20−30 eggs	50−100 eggs
Husbandry costs	Low	High
Genetic characterization	High	Low
Cryopreservation	Yes	Yes
Sample size (e.g., blood)	Low	High

Size of Construct

Interestingly, homologous recombination was recently employed in murine zygotes, to reconstruct a large functional gene from microinjected DNA fragments. Pieper et al. (32) injected three overlapping genomic DNA fragments, which together constituted the human serum albumin (hSA) gene (a 33 kb segment of DNA). A significant proportion of resulting transgenic mice contained the correctly reconstituted hSA gene. The limits of this coinjection procedure have yet to be determined. However, the method could be superseded by more recent experiments in which P1 clones (33) and yeast artificial chromosomes (YACs) (34) have been microinjected.

Experimental Design

Gene Expression

Increasing the expression of a gene may be informative in evaluating the role that the gene product plays in normal development. This can be achieved by placing the gene under the control of a strong, heterologous promoter (35), by linking the transgene to a housekeeping gene promoter (such as that of the PGK gene) rendering its expression constitutive, or by using an inducible promoter [such as that of metallothionine (36)].

To identify and define control elements in and around a gene that affect its tissue-specific and developmental pattern of expression, one could design a series of constructs with nested deletions around the promoter region, the 3′ end of the gene, and, if necessary, within introns. By analyzing expression patterns of the transgenes in the resultant transgenic lines, one could build up a map of the sequences that are absolutely required for correct gene expression. In practice, such onerous endeavors are rarely undertaken.

If expression of the transgene is likely to be masked or affected by expression of its endogenous counterpart, then it is possible to link the promoter sequences to a reporter gene, such as the SV40 T antigen (37), CAT (38), or β-galactosidase (lacZ) (39). One can then ascertain whether the promoter directs expression to the appropriate cell types. In the former example, the resultant tumors arising in the target cells have proved to be invaluable in generating new cell lines that retain highly differentiated phenotypes (40,41). In the latter case, X-gal staining of the embryos/tissue slices readily identifies any promoter-directed sites of β- galactosidase expression.

Reduction of Gene Function

Abolishing or reducing the function of a gene can be equally informative with respect to the role that the gene plays in vivo. There are a number of strategies by which this can be achieved. By introducing a gene encoding antisense RNA under the promoter sequences of its endogenous counterpart, its expression is limited to those sites where it can reduce the amount of targeted gene product. The mechanism by which antisense inhibition occurs remains obscure, and its general efficacy has yet to be proven. There are, however, a growing number of examples in the literature of its application (42–44).

The second strategy involves the use of a ribozyme: an RNA molecule with enzymic activity that is capable of cleaving specific target RNA molecules. If such a sequence is placed within an appropriate antisense sequence, one would predict a much more efficient inhibition of targeted gene expression. Ribozyme-mediated destruction of RNA has been demonstrated in tissue culture (45–48) and, more recently, in transgenics, in which ribozyme-mediated reduction of β-microglobulin mRNA levels has been demonstrated (49).

Gene knockout experiments are an extreme case in which, in homozygosity, gene expression can be abolished as described above. If progeny are viable, such animals can be invaluable in understanding the developmental and physiological role of specific genes and in revealing functional redundancy, where complete knockout of one gene product appears to have no phenotypic effect. Often an underlying compensatory increase in a closely related gene product can be demonstrated (50).

Cell Ablation

To answer questions about the lineage, fate, or function of a cell, it can be informative to observe the effects of removing that cell type in the developing embryo or later in the adult (51). Although this might be difficult or impossible to achieve by physical means, it can be done by introducing genes encoding cytotoxins, such as the catalytic subunits of diphtheria toxin (DT-A) (52,53) or ricin (RT-A) (54), under appropriate cell-specific promotors. The selective ablation of cell types by toxins is potentially very powerful, since the toxic gene products may act at very low concentrations. However, it is important that the toxin be confined to the cells in which it is expressed, or damage to neighboring cells may occur. Obviously, some promotor-toxigene constructs are likely to be lethal to the developing embryo if the ablated cells are essential for viability.

To give the researcher more control over the degree of cell ablation or timing of the event during development, a number of strategies have been devised. The first is the use of an attenuated DT-A gene (55). An alternative strategy makes ablation dependent on the administration of drugs. This has been achieved by introducing the HSV-1-*tk* transgene under the control of appropriate promoter elements. Cells expressing the gene are rendered susceptible to drugs such as gancyclovir (56). The power of this strategy is that the timing and degree of cell ablation are controlled by the investigator. Evidence suggests that both actively dividing and nondividing cells may be susceptible to drug-induced ablation, indicating a wide application for this strategy. With these more sophisticated methods of regulation, it should be possible to use promoters that are active in cells essential to development.

Locus Control Regions

The locus control region, a regulatory element, plays an important role in controlling the expression of certain genes. Such regions have been characterized for the human globin gene locus (57), the chicken β-globin gene (58), the chicken lysozyme gene (59), and, more recently, the red-green visual pigment genes (39). Importantly, these elements appear to confer position-independent, tissue-specific expression on the transgene. Identification of transcription factor binding sites and regulatory elements within these locus control elements may give the researcher a greater degree of control over construct design-limiting expression of the transgene to specific cells, yet ensuring good expression levels.

APPLICATIONS OF TRANSGENESIS TO HYPERTENSION RESEARCH

Study of Genes Known to Be Involved in Blood-Pressure Homeostasis

Essential hypertension is a quantitative trait, resulting from the combined phenotypes of a number of genetic loci interacting with environmental influences. This occurs on a background of complicated and sophisticated regulatory and feedback circuits. One might imagine that alteration of any gene function known to be involved in blood-pressure control might lead to a perturbation in blood pressure. The list of potential candidates is almost endless, including members of the renin angiotensin system (RAS), the kallikrein-kinin system, aldosterone, endothelin, atrial natriuretic factor (ANF), nitric oxide synthase, vasopressin, catecholamines, the sodium hydrogen antiporter, sodium potassium ATPase, and others. The contribution of specific genes and allelic variants can be investigated through genetic manipulation, in the whole animal, by both traditional genetics and transgenesis.

Identification of New Candidates Through Gene Mapping

Since this topic is covered in detail elsewhere in the book, we shall allude only briefly to the identification of new candidate genes linked to hypertension through genetic analyses of hypertension in rat strains. The ability to follow, through the use of a panel of polymorphic markers, the inheritance pattern of the entire genome in all progeny from an experimental cross allows the analysis of quantitative trait loci (60) to localize genes that contribute significantly to the phenotypic variation. Candidate loci, implicated in this way, include those for angiotensinogen (61) and angiotensin converting enzyme (ACE) (62,63).

However, demonstration of potential linkage in one species does not necessarily signal its contribution to pathogenesis in another, since no linkage was demonstrated for ACE (64) in human populations (tested by sib pair analysis) or for angiotensinogen in the SHRSP$_{HD}$ rat (65), and such linkage studies can depend on the strain used (66). This may be because the causative loci are in fact unidentified genes lying close to the putative candidate genes, or because the animal models reflect subclasses of human hypertensives.

A SURVEY OF TRANSGENIC EXPERIMENTS TO DATE

Renin

Microinjection of a 24 kb fragment spanning the mouse Ren-2 gene (including 5 kb of upstream and 9 kb of downstream sequences) resulted in the generation of a number of transgenic mice, each of which exhibited transgene expression in the correct spectrum of tissues (67). Quantitatively, the levels of expression deviated considerably from normal, although no adverse effect on blood pressure was observed. (A closely related fragment used by Tronik et al. (68) gave near wild-type expression levels in the kidney and submandibular gland of transgenic mice.)

When the 24 kb transgene was introduced into rats (29), it was found to cause fulminant hypertension, with adult transgenic animals exhibiting blood pressures typically 80 to 100 mm Hg above those observed in non-transgenic littermates. The hypertension was characterized by low kidney and plasma renin levels and high levels of circulating prorenin, the latter being transgene-derived and produced by the adrenal gland (69). Further transgenic models are currently being investigated to determine the exact mechanism by which the Ren-2 transgene exerts its phenotypic effect. The hypertension may be due, in part, to enhanced kinetics of the reaction between mouse renin and rat angiotensinogen (70).

Genomic clones encoding the human renin gene have been successfully microinjected, yielding both transgenic mice (71) and rats (72). Again, expression was seen in the correct spectrum of tissues, but blood-pressure levels were unaltered as compared to nontransgenic littermates, despite high levels of secretion of the transgene products into the circulation, due to the species specificity of the renin angiotensinogen reaction.

Transgenic mice containing the mouse renin promoter fused to SV40 T antigen were used to define the promoter sequences in the mouse Ren-2 renin gene capable of directing cell specificity (37). Expression of an oncoprotein in a cell predisposes it to immortalization and/or transformation. The transgenic mice therefore developed tumors in tissues that would normally express renin.

From one such kidney tumor, a renin-expressing cell line exhibiting many characteristics of juxtaglomerular cells was isolated (37).

Angiotensinogen

A "high AII" hypertensive phenotype was generated in three lines of transgenic mice overexpressing the rat angiotensinogen gene (73). When angiotensinogen was expressed under the metallothionein promoter, no increase in blood pressure was seen, although plasma angiotensinogen levels were lower in this transgenic line (74).

Expression of the human angiotensinogen gene (under its native regulatory sequences) was found predominantly in the liver, although with high levels in the kidney (75). Blood-pressure levels were not elevated despite high levels of circulating transgene product.

By using targeted gene disruption, and a special form of gap-repair gene targeting, that tandemly duplicated the whole of the gene, together with 5′ and 3′ flanking sequences, animals were generated with one, two, or three copies of the angiotensinogen gene (76). The animals were otherwise genetically identical. Progressively higher levels of gene product were detected in the animals with increasing numbers of gene copies. This model system addresses the hypothesis that genetically determined elevation of angiotensinogen might predispose to hypertension. This may be a generally applicable strategy for exploring the effects of gene dosage on a complex quantitative phenotype. However, it is possible that phenotypic effects may be subtle, if the gene product is normally in excess, or because compensatory changes occur in some related system.

Through the crossbreeding of individual lines of transgenic mice carrying either the human renin or human angiotensinogen genes under the control of their native promoters, a double transgenic line exhibiting elevated blood pressure was produced (77). This mouse line represents a model for high human renin hypertension and will facilitate pharmacological studies on the human RAS system.

Angiotensin Converting Enzyme

Angiotensin converting enzyme exists in two different isoforms: the somatic isozyme and the germinal-specific form. The latter is generated via a germinal-specific promoter, located within intron 12 of the gene encoding the former. Transgenic studies (78) have localized the germinal-specific promoter to a 91 bp portion of the immediate upstream sequence. This encompasses a highly conserved cyclic AMP-responsive element, which is thought to confer germinal-specific expression through interaction with a testis-specific transactivating factor.

Atrial Natriuretic Factor (ANF)

Transgenic mice, carrying the mouse transthyretin promoter upstream of the mouse ANF structural gene, chronically expressed high levels of ANF in the liver (35,79). Phenotypically, they were found to exhibit long-term hypotension, with blood pressures typically 20 to 30 mm Hg below those of nontransgenic siblings. The expected natriuretic and diuretic effects of ANF were not observed, suggesting that the animals are able to compensate for the renal effects of hormone overexpression.

Transgenic mice containing the ANF promoter fused to SV40 T Ag developed unilateral right atrial tumors composed of differentiating cardiomyocytes. The atrial tumors could be propagated as transplantable tumor lineages in syngeneic mice, and cardiomyocyte lines were established from this source. The cells were highly differentiated and expressed numerous cardiac-specific proteins. They also displayed spontaneous electrical and contractile activities (80).

Endothelin

The endothelin-1 (ET-1) gene was disrupted by homologous recombination to generate mice deficient in ET-1 (81). Homozygotes died of respiratory failure at birth, with craniofacial abnormalities, while heterozygotes (which produced lower levels of ET-1 than wild-type mice) developed elevated blood pressure. This suggests that ET-1 plays a physiological role in cardiovascular homeostasis. The hypertensive effect is surprisingly at odds with previous data suggesting that a sustained increase in plasma ET-1 levels is positively correlated with blood pressure (82,83). The blood-pressure elevation seen in ET-1 heterozygotes may be caused by changes in central cardiorespiratory control, but further investigation is required before the underlying mechanism is determined.

Kallikrein

Transgenic mice carrying the human tissue kallikrein gene under the control of the mouse metallothionein promoter were found to have significantly lowered blood pressure compared with control mice (84). The effect could be reversed by aprotinin administration. Metal ion induction of transgene expression failed to lower the blood pressure further, suggesting that above a critical concentration of tissue kallikrein, compensatory mechanisms came into play.

Sodium-Protein Antiporter

Overexpression of the sodium-proton antiporter, under the human elongation factor 1α (EF1α) promotor, impaired urinary excretion of Na^+ and Cl^-. Elevation of blood pressure was observed following excessive salt intake (85). This is the first evidence that enhancement of Na^+/H^+ exchange could be a cause of essential hypertension.

Vasopressin

When an 8.2 kb genomic fragment spanning the rat vasopressin gene was introduced into the mouse germline (86), the transgene was expressed in a tissue-specific manner, and the animals demonstrated appropriate osmotic regulation of the transgene-derived mRNA. Water metabolism was normal, even in animals homozygous for the transgene.

IMMORTALIZATION OF CELL LINES

A concern in experiments involving the tissue-specific expression of onco-proteins such as SV40 Tag is that the transformation event leading to immortalization may alter normal cellular physiology. It is very important that any cell line emulate the in vivo counterpart. To this end, conditionally immortalizing genes, such as those encoding temperature-sensitive mutants of SV40 TAg, have been used to limit the effects of the transformation process. The T-antigen mutant ts A58 (tsA58) is capable of immortalizing cells at the permissive temperature of 33°C in the presence of γ-interferon when expression is driven by the H-2K promoter (87). Such proliferating cell lines are capable of differentiation after inactivation of the immortalizing gene at the nonpermissive temperature (39°C) (Figure 6).

Cell lines derived from transgenic animals, which expressed TAg at high levels, escaped conditionality. However, one transgenic line proved to be viable, exhibiting lower TAg expression. This "immortomouse" proved to be a multipotential source of conditionally immortal cell lines. Immortalized myoblasts cultured from the leg muscle (88) exhibited typical myogenic morphology in the permissive conditions, but readily formed myotubes when grown at high density in nonpermissive conditions. On reinjection into leg muscle, the cells were able to form new muscle.

Conditionally immortalized astrocytes have been isolated from the brain (89) and epithelial cells from the colon and small intestine (90). It would appear that the greatest challenge is to identify appropriate growth conditions

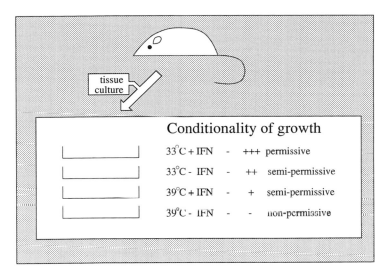

Figure 6 Summary of the conditionality of the immortalizing gene, tsA58, under various growth conditions.

for the particular cell type of interest. The promise of this approach for cardiovascular research is obvious, permitting genetic alterations or reporter constructs to be introduced into immortalized cells and the cells subsequently reintroduced into animal hosts.

CONCLUSION

Transgenic animal models have proved to be an important tool in cardiovascular and hypertension research. Given the range of potential applications for transgenic technology, it is clear that very precise questions about the effect of altering gene expression on metabolic or physiological pathways can now be addressed in vivo.

One can investigate the developmental and tissue-specific expression of a gene through the appropriate use of reporter constructs. One can question the importance of a candidate gene in predisposition to hypertension, by increasing or decreasing its expression in the presence or absence of its endogenous counterpart.

Some caution must be exercised when crossing species barriers—one must be aware of potential interactions between the foreign gene and transcription factors that may cause inappropriate tissue expression. Likewise, the transgene product may or may not interact with components of the endogenous blood-pressure control system, due to species specificity. This can be used to advantage, as in the case of human renin and angiotensinogen transgenic mouse lines, which facilitate pharmacological investigation of the human RAS in vivo. Another point to bear in mind is that candidate genes linked to hypertension in the rat or mouse model may not always reflect the situation in essential hypertension in humans.

Despite these cautionary notes, transgenic experiments are expanding our understanding of gene regulation in the cardiovascular system, and highlight important questions about aspects of blood pressure homeostasis, requiring the interaction between molecular biology, pharmacology, physiology, and neurobiology.

FUTURE PROSPECTS

Transgenesis is having, and will continue to have, far-reaching effects on the fields of animal-model production, gene-therapy strategies, new therapeutic drug treatments, and the commercial production of biologically important molecules. Advances in transgenic technology will see the tissue-directed site-independent integration of transgenes, whose expression is under the direct control of the researcher. Progress in antisense and ribozyme technology may allow precise control of endogenous gene expression.

With the ability to subtly alter, duplicate, or disrupt genes at will, ES cell technology presents a unique opportunity to control the level of expression of endogenous genes. Although this technology is limited to the mouse at present, it is hoped that in time it will be extended to the rat, and other important species. The potential use of traditional hypertensive strains for transgenic experimentation should not be overlooked. One might envision that, for example, the SHR rat could be studied directly through suitable transgenic manipulation, and the prospect of SHR ES cells would extend such analyses.

With further developments in gene mapping and QTL analysis, one can foresee the identification of a whole range of factors that predispose to hypertension. The introduction of multiple, subtle changes to several factors in a given transgenic line may lead to the generation of a range of animal models mimicking the spectrum of essential hypertension subclasses seen in humans, and open the possibility of improved treatment and prevention.

REFERENCES

1. Hogan B, Costantini F, Lacy E. Manipulating the Mouse Embryo—A Laboratory Manual. Cold Spring Harbor, NY: Cold Spring Harbor Laboratories Press, 1986.
2. Mullis KB, Faloona F. Specific synthesis of DNA in vitro via a polymerase-catalysed

chain reaction. In: Wu R, ed. Methods in Enzymology. Vol 155. New York: Academic Press, 1987:335–350.

3. Southern EM. Detection of specific sequences among DNA fragments separated by gel electrophoresis. J Mol Biol 1975; 98:503–517.

4. Ivinson AJ, Taylor GR. PCR in genetic diagnosis. In: McPherson MJ, Quirke P, Taylor GR, eds. PCR—A Practical Approach. New York: Oxford Press, 1991:15–27.

5. Woychik RP, Stewart TA, Davis LG, D'Eustachio P, Leder P. An inherited limb deformity created by insertional mutagenesis in a transgenic mouse. Nature 1985; 318:36–40.

6. Ratty AK, Fitzgerald LW, Titeler M, Glick SD, Mullins JJ, Gross KW. Circling behaviour exhibited by a transgenic insertional mutant. Mol Brain Res 1990; 8:355–358.

7. Karls U, Muller U, Gilbert DJ, Copeland NG, Jenkins NA, Harbers K. Structure, expression, and chromosomal location of the gene for the β subunit of brain-specific Ca^{2+}/calmodulin-dependent protein kinase II identified by transgene integration in the embryonic lethal mouse mutant. Mol Cell Biol 1992; 12:3644–3652.

8. Smith AG, Nichols J, Robertson M, Rathjen PD. Differentiation inhibiting activity (DIA/LIF) and mouse development. Dev Biol 1992; 151:339–351.

9. Iannaccone PM, Taborn GU, Garton RL, Caplice MD, Brenin DR. Pluripotent embryonic stem cells from the rat are capable of producing chimeras. Dev Biol 1994; 163:288–292.

10. Hooper ML. Embryonal stem cells: Introducing planned changes into the animal germline. In: Evans HJ, ed. Modern Genetics. Vol 1. Switzerland: Harwood Academic Press, 1992.

11. Evans MJ. Potential for genetic manipulation of mammals. Mol Biol Med 1989; 6:557–565.

12. Deng C, Capecchi MR. Reexamination of gene targeting frequency as a function of the extent of homology between the targeting vector and the target locus. Mol Cell Biol 1992; 12:3365–3371.

13. te Riele H, Maandag ER, Berns A. Highly efficient gene targeting in embryonic stem cells through homologous recombination with isogenic DNA constructs. Proc Natl Acad Sci USA 1992; 89:5128–5132.

14. Thomas KR, Deng C, Capecchi MR. High-fidelity gene targeting in embryonic stem cells by using sequence replacement vectors. Mol Cell Biol 1992; 12:2919–2923.

15. Mortensen RM. Double knock-outs: production of mutant cell lines in cardiovascular research. Hypertension 1993; 22:646–651.

16. Thomas KR, Capecchi MR. Targeted disruption of the murine int-1 protooncogene resulting in severe abnormalities in midbrain and cerebellar development. Nature 1990; 346:847–850.

17. McMahon AP, Bradley A. The Wnt-1 (int-1) proto-oncogene is required for development of a large region of the mouse brain. Cell 1990; 62:1073–1085.

18. Lee EY-HP, Chang C-Y, Hu N, Wang Y-CJ, Lai C-C, Herrup K, Lee W-H, Bradley A. Mice deficient for Rb are nonviable and show defects in neurogenesis and hematopoiesis. Nature 1992; 359:288–294.

19. Jacks T, Fazeli A, Schmitt EM, Bronson RT, Goodell MA, Weinberg RA. Effects of an *Rb* mutation in the mouse. Nature 1992; 359:295-300.

20. Clarke AR, Maandag ER, van Roon M, van der Lugt NMT, van der Valk M, Hooper ML, Berns A, te Riele H. Requirement for a functional *Rb-1* gene in murine development. Nature 1992; 359:328-330.

21. Yagi T, Ikawa Y, Yoshida K, Shigetani Y, Takeda N, Mabuchi I, Yamamoto T, Aizawa S. Homologous recombination of c-fyn locus of mouse embryonic stem cells with use of diphtheria toxin A-fragment gene in negative selection. Proc Natl Acad Sci USA 1990; 87:9918-9922.

22. Riele H, Maandag ER, Clarke A, Hooper M, Berns A. Consecutive inactivation of both alleles of the pim-1 proto-oncogene by homologous recombination in embryonic stem cells. Nature 1990; 348:649-651.

23. Mountford P, Zevnik B, Duwel A, Nichols J, Li M, Dani C, Robertson M, Chambers I, Smith A. Dicistronic targeting constructs: reporters and modifiers of mammalian gene expression. Proc Natl Acad Sci USA 1994; 91:4303-4307.

24. Hasty P, Ramirez-Solis R, Krumlauf R, Bradley A. Introduction of a subtle mutation into the *Hox-2.6* locus in embryonic stem cells. Nature 1991; 350:243-246.

25. Ratcliff R, Evans MJ, Doran J, Wainwright BJ, Williamson R, Colledge WH. Disruption of the cystic fibrosis transmembrane conductance regulator gene in embryonic stem cells by gene targeting. Transgen Res 1992; 1:177-181.

26. Nichols J, Evans EP, Smith AG. Establishment of germ-line-competent embryonic stem (ES) cells using differentiation inhibiting activity. Development 1990; 110:1341-1348.

27. Mortensen RM, Conner DA, Chao S, Geisterfer LA, Seidman JG. Production of homozygous mutant ES cells with a single targeting construct. Mol Cell Biol 1992; 12:2391-2395.

28. Miller CC, McPheat JC, Potts W. Targeted integration of the *Ren-1D* locus in mouse embryonic stem cells. Proc Natl Acad Sci USA 1992; 89:5020-5024.

29. Mullins JJ, Peters J, Ganten D. Fulminant hypertension in transgenic rats harbouring the mouse Ren-2 gene. Nature 1990; 344:541-544.

30. Hammer RE, Maika SD, Richardson JA, Tank J-P, Taurog JD. Spontaneous inflammatory disease in transgenic rats expressing HLA-B27 and human β_2m: an animal model of HLA-B27-associated human disorders. Cell 1990; 63:1099-1112.

31. Whitworth CE, Fleming S, Cumming AD, Morton JJ, Burns NJT, Williams BC, Mullins JJ. Spontaneous development of malignant phase hypertension in transgenic Ren-2 rats. Kidney Int 1994; 46:1528-1532.

32. Pieper FR, de Wit ICM, Pronk ACJ, Kooiman PM, Strijker R, Krimpenfort PJA, Nuyens JH, de Boer HA. Efficient generation of functional transgenes by homologous recombination in murine zygotes. Nucl Acids Res 1992; 20:1259-1264.

33. Linton MF, Farese RV Jr, Chiesa G, Grass DS, Chin P, Hammer RE, Hobbs HH, Young SG. Transgenic mice expressing high plasma concentrations of human apolipoprotein B100 and lipoprotein (a). J Clin Invest 1993; 92:3029-3037.

34. Schedl A, Beermann F, Thies E, Montoliu L, Kelsey G, Schutz G. Transgenic mice generated by pronuclear injection of a yeast artificial chromosome. Nucl Acid Res 1992; 12:3073-3077.

35. Steinhelper ME, Cochrane KL, Field LJ. Hypotension in transgenic mice express-
     ing atrial natriuretic factor fusion genes. Hypertension 1990; 16:301–307.
36. Behringer RR, Cate RL, Froelick GJ, Palmiter RD, Brinster RL. Abnormal sexual
     development in transgenic mice chronically expressing mullerian inhibiting
     substance. Nature 1990; 345:167–170.
37. Sigmund CD, Okuyama K, Ingelfinger J, Jones CA, Mullins JJ, Kane-Haas C, Kim
     U, Wu C, Kenny L, Rustum Y, Dzau VJ, Gross KW. Isolation and characterization
     of renin-expressing cell lines from transgenic mice containing a renin promoter
     viral oncogene fusion construct. J Biol Chem 1990; 265:19916–19922.
38. Lui ML, Olson AL, Moye-Rowley WS, Buse JB, Bell GI, Pessin JE. Expression and
     regulation of the GLUT4/muscle-fat facilitative glucose transporter gene in trans-
     genic mice. J Biol Chem 1992; 267:11673–11676.
39. Wang Y, Macke JP, Merbs SL, Zack DJ, Klaunberg B, Bennett J, Gearhart J, Nathans
     J. A locus control region adjacent to the human red and green visual pigment
     genes. Neuron 1992; 9:429–440.
40. Windle JJ, Weiner RI, Mellon PL. Cell lines of the pituitary gonadotrope lineage
     derived by targeted oncogenesis in transgenic mice. Mol Endocrinol 1990;
     4:597–603.
41. Mellon PL, Windle JJ, Goldsmith PC, Padula CA, Roberts JL, Weiner RI. Im-
     mortalization of hypothalamic GnRH neurons by genetically targeted tumorigen-
     esis. Neuron 1990; 5:1–10.
42. Munir MI, Rossiter BJF, Caskey CT. Antisense RNA production in transgenic mice.
     Somat Cell Mol Genet 1990; 16:383–394.
43. Pepin M-C, Pothier F, Barden N. Impaired type II glucocorticoid-receptor func-
     tion in mice bearing antisense RNA transgene. Nature 1992; 355:725–728.
44. Simons M, Edelman ER, DeKeyser J-L, Langer R, Rosenberg RD. Antisense c-myb
     oligonucleotides inhibit arterial smooth muscle cell accumulation in vivo. Nature
     1992; 359:67–70.
45. Cotten M, Birnstiel ML. Ribozyme-mediated destruction of RNA *in vivo*. EMBO
     J 1989; 8:3861–3866.
46. Cameron FH, Jennings PA. Specific gene suppression by engineered ribozymes
     in monkey cells. Proc Natl Acad Sci USA 1989; 86:9139–9143.
47. Sarver N, Cantin EM, Chang PS, Zaia JA, Ladne PA, Stephens DA, Rossi JJ.
     Ribozymes as potential anti-HIV-1 therapeutic agents. Science 1990; 247:1222–
     1225.
48. Dropulic B, Lin NH, Martin MA, Jeang K-T. Functional characterization of a U5
     ribozyme: intracellular suppression of human immunodeficiency virus type 1
     expression. J Virol 1992; 66:1432–1441.
49. Larsson S, Hotchkiss G, Andang M, Nyholm T, Inzunza J, Jansson I, Ahrlund-Rich-
     ter L. Reduced β2- microglobulin mRNA levels in transgenic mice expressing a
     designed hammerhead ribozyme. Nucl Acid Res 1994; 22:2242–2248.
50. Rudnicki MA, Braun T, Hinuma S, Jaenisch R. Inactivation of Myo-D in mice leads
     to up-regulation of the myogenic HLH gene Myf-5 and results in apparently
     normal muscle development. Cell 1992; 71:383–390.
51. O'Kane CJ, Moffat KG. Selective cell ablation and genetic surgery. Curr Op Genet
     Devel 1992; 2:602–607.

52. Palmiter RD, Behringer RR, Quaife CJ, Maxwell F, Maxwell IH, Brinster RL. Cell lineage ablation in transgenic mice by cell-specific expression of a toxin gene. Cell 1987; 50:435-443.

53. Breitman ML, Clapoff S, Rossant J, Tsui L-C, Glode LM, Maxwell IH, Bernstein A. Genetic ablation: Targeted expression of a toxin gene causes micropthalmia in transgenic mice. Science 1987; 238:1563-1565.

54. Landel CP, Zhao J, Bok D, Evans GA. Lens-specific expression of recombinant ricin induces developmental defects in the eyes of transgenic mice. Genes Dev 1988; 2:1168-1178.

55. Breitman ML, Rombola H, Maxwell IH, Klintworth GK, Bernstein A. Genetic ablation in transgenic mice with an attentuated diphtheria toxin A gene. Mol Cell Biol 1990; 10:474-479.

56. Borrelli E, Heyman RA, Arias C, Sawchenko PE, Evans RM. Transgenic mice with inducible dwarfism. Nature 1989; 339:538-541.

57. Collis P, Antoniou M, Grosveld F. Definition of the minimal requirements within the human beta-globin gene and the dominant control region. EMBO J 1990; 9:233-240.

58. Reitman M, Lee E, Westphal H, Felsenfeld G. Site-independent expression of the chicken beta^A-globin gene in transgenic mice. Nature 1990; 348:749-752.

59. Bonifer C, Vidal M, Grosveld F, Sippel AE. Tissue-specific and position-independent expression of the complete gene domain for chicken lysozyme in transgenic mice. EMBO J 1990; 9:2843-2848.

60. Lander ES, Botstein D. Mapping Mendelian factors underlying quantitative traits using RFLP linkage maps. Genetics 1989; 121:185-199.

61. Jeunemaitre X, Soubrier F, Kotelevtsev Y, Lifton RP, Williams CS, Charru A, Hunt SC, Hopkins PN, Williams RR, Lalouel J-M, Corvol P. Molecular basis of human hypertension: role of angiotensinogen. Cell 1992; 71:169-180.

62. Hilbert P, Lindpaintner K, Beckmann JS, Serikawa T, Soubrier F, Dubay C, Cartwright P, de Gouyon B, Julier C, Takahashi S, Vincent M, Ganten D, Georges M, Lathrop GM. Chromosomal mapping of two genetic loci associated with blood-pressure regulation in hereditary hypertensive rats. Nature 1991; 353:521-529.

63. Jacob HJ, Lindpaintner K, Lincoln SE, Kusumi K, Bunker RK, Mao Y-P, Ganten D, Dzau VJ, Lander ES. Genetic mapping of a gene causing hypertension in the stroke-prone spontaneously hypertensive rat. Cell 1991; 67:213-224.

64. Jeunemaitre X, Lifton RP, Hunt SC, Williams RR, Lalouel J-M. Absence of linkage between the angiotensin converting enzyme locus and human essential hypertension. Nature Genet 1992; 1:72-75.

65. Hubner N, Kreutz R, Takahashi S, Ganten D, Lindpaintner K. Unlike human hypertension, blood pressure in a hereditary hypertensive rat strain shows no linkage to the angiotensinogen locus. Hypertension 1994; 23:797-801.

66. Rapp JP, Dene H, Deng AY. Seven renin alleles in rat and their effects on blood pressure. J Hypertens 1994; 12:349-355.

67. Mullins JJ, Sigmund CD, Kane-Haas C, Wu C, Pacholec F, Zeng Q, Gross KW. Studies on the regulation of renin genes using transgenic mice. Clin Exp Hyper Theory Prac 1988; A10:1157-1167.

68. Tronik D, Dreyfus M, Babinet C, Rougeon F. Regulated expression of the Ren-2 gene in transgenic mice derived from parental strains carrying only the Ren-1 gene. EMBO J 1987; 6:983–987.

69. Bachmann S, Peters J, Engler E, Ganten D, Mullins J. Transgenic rats carrying the mouse renin gene—morphological characterization of a low-renin hypertension model. Kidney Int 1992; 41:24–36.

70. Tokita Y, Franco-Saenz R, Reimann EM, Mulrow PJ. The hypertension in the transgenic rat TGR(mRen-2)27 may be due to enhanced kinetics of the reaction between mouse renin and rat angiotensinogen. Hypertension 1994; 23:422–427.

71. Takaori K, Kim S, Fukamizu A, Sagari M, Hsoi M, Katsuki M, Murakami K, Yamamoto K. Biochemical characteristics of human renin expressed in transgenic mice. Clin Sci 1993; 84:21–29.

72. Ganten D, Wagner J, Zeh K, Bader M, Michel JB, Paul M, Zimmermann F, Ruf P, Hilgenfeldt U, Ganten U, Kaling M, Bachmann S, Fukamizu A, Mullins JJ, Murakami K. Species specificity of renin kinetics in transgenic rats harboring the human renin and angiotensinogen genes. Proc Natl Acad Sci USA 1992; 89:7806–7810.

73. Kimura S, Mullins JJ, Bunneman B, Metzger R, Hilgenfeldt U, Zimmerman F, Jacob H, Froce X, Ganten D, Kaling M. High blood pressure in transgenic mice carrying rat angiotensinogen gene. EMBO J 1992; 11:821–827.

74. Ohkubo H, Kawakami H, Kakehi Y, Takumi T, Arai H, Yokota Y, Iwai M, Tanabe Y, Masu M, Hata J, Iwao H, Okamoto H, Yookoyama M, Nomura T, Katsuki M, Nakanishi S. Generation of transgenic mice with elevated blood pressure by introduction of the rat renin and angiotensinogen genes. Proc Natl Acad Sci USA 1990; 87:5153–5157.

75. Takahashi S, Fukamizu A, Hasegawa T, Yokoyama M, Nomura T, Katsuki M, Murakami K. Expression of the human angiotensinogen gene in transgenic mice and transfected cells. Biochem Biophys Res Comm 1991; 180:1103–1109.

76. Smithies O, Kim H-S. Targeted gene duplication and disruption for analyzing quantitative genetic traits in mice. Proc Natl Acad Sci USA 1994; 91:3612–3615.

77. Fukamizu A, Sugimura K, Takimoto E, Sugiyama F, Seo M-S, Takahashi S, Hatae T, Kajiwara N, Yagami K-I, Murakami K. Chimeric renin-angiotensinogen system demonstrates sustained increase in blood pressure of transgenic mice carrying both human renin and human angiotensinogen genes. J Biol Chem 1993; 268: 11617–11621.

78. Howard T, Balogh R, Overbeek P, Bernstein KE. Sperm-specific expression of angiotensin-converting enzyme (ACE) is mediated by a 91-base-pair promoter containing a CRE-like element. Mol Cell Biol 1993; 13:18–27.

79. Koh GY, Klug MG, Field LJ. Atrial natriuretic factor and transgenic mice. Hypertension 1993; 22:634–639.

80. Steinhelper ME, Lanson NA Jr, Dresdner KP, Delcarpio JB, Wit AL, Claycomb WC, Field LJ. Proliferation in vivo and in culture of differentiated adult atrial cardiomyocytes from transgenic mice. Am J Physiol 1990; 259:H1826–H1834.

81. Kurihara Y, Kurihara H, Suzuki H, Kodama T, Maemura K, Nagai R, Oda H, Kuwaki T, Cao W-H, Kamada N, Jishage K, Ouchi Y, Azuma S, Toyoda Y, Ishikawa T, Kumada M, Yazaki Y. Elevated blood pressure and craniofacial abnormalities in mice deficient in endothelin-1. Nature 1994; 368:703–710.

82. Mortensen LH, Pawloski CM, Kanagy NL, Fink GD. Chronic hypertension produced by infusion of endothelin in rats. Hypertension 1990; 15:729–733.

83. Yokokawa K, Tahara H, Kohno M, Murakawa K, Yasunari K, Nakagawa K, Hamada T, Otani S, Yanagisawa M, Takeda T. Endothelin-secreting tumor. J Cardiovasc Pharmacol 1991; 17:S398–S401.

84. Wang J, Xiong W, Yang Z, Davis T, Dewey MJ, Chao J, Chao L. Human tissue kallikrein induces hypotension in transgenic mice. Hypertension 1994; 23:236–243.

85. Kuro-o M, Hanaoka K, Noguchi T, Hiroi Y, Takewaki S, Hayasaka M, Aikawa M, Miyagishi A, Yazaki Y, Nabeshima Y, Nagai R. Overexpression of sodium-proton antiporter causes salt-sensitive hypertension in transgenic mice. J Hypertens 1994; 12:S63.

86. Grant FD, Reventos J, Kawabata S, Miller M, Gordon JW, Majzoub JA. Transgenic mouse models of vaspressin expression. Hypertension 1993; 22:640–645.

87. Jat PS, Noble MD, Ataliotis P, Tanaka Y, Yannoutsos N, Larsen L, Kioussis D. Direct derivation of conditionally immortal cell lines from an H-$2K^b$-tsA58 transgenic mouse. Proc Natl Acad Sci USA 1991; 88:5096–5100.

88. Noble M, Groves AK, Ataliotis P, Morgan J, Peckham M, Partridge T, Jat PS. Biological and molecular approaches to the generation of conditionally immortal neural cells. Neuroprotocols 1993; 3:189–199.

89. Groves AK, Entwistle A, Jat PS, Noble M. The characterisation of astrocyte cell lines that display properties of glial scar tissue. Dev Biol 1993; 159:87–104.

90. Whitehead RH, VanEeden PE, Noble MD, Ataliotis P, Jat PS. Establishment of conditionally immortalized epithelial cell lines from both colon and small intestine of adult H-$2K^b$-tsA58 transgenic mice. Proc Natl Acad Sci USA 1993; 90:587–591.

13

Genetically Altered Mouse Models of Lipoprotein Disorders and Atherosclerosis

Jan L. Breslow, Andrew Plump, and Marilyn Dammerman
The Rockefeller University
New York, New York

INTRODUCTION

In humans, lipoprotein disorders are strongly associated with coronary heart disease susceptibility. Several types of abnormal lipoprotein patterns are commonly observed in heart attack victims, including elevated LDL cholesterol; reduced HDL cholesterol, usually with increased triglycerides; elevated chylomicron remnant and IDL cholesterol levels; and elevated levels of Lp(a) particles. These patterns are caused by environmental and genetic factors that alter the synthesis, processing, or catabolism of lipoprotein particles. Over the last decade, genes have been isolated that code for proteins that directly interact with plasma lipids. There are approximately 17 such lipoprotein transport proteins, including apolipoproteins that coat lipoprotein particles, lipoprotein-processing proteins, and lipoprotein receptors (Table 1). The genes coding for these proteins have all turned out to be single-copy in the human genome. They have been sequenced, mapped, and used as candidate genes to identify mutations underlying lipoprotein phenotypes associated with coronary heart disease susceptibility (1). The lipoprotein transport genes have also been used to make transgenic and knockout animals, principally mice, which have provided new insights into how these genes are expressed and the functions they serve in an intact organism. Either singly or combined through crossbreeding, these genetically altered mice are now being used to make animal models of human lipoprotein disorders associated with coronary heart disease susceptibility (Table 2) (2–4). In at least one instance, a mouse has been produced that develops diffuse fibroproliferative atherosclerotic lesions very much like those seen in humans.

Table 1   Lipoprotein Transport Proteins

Apolipoproteins	Apo A-I, Apo A-II, Apo A-IV
	Apo B
	Apo C-I, Apo C-II, Apo C-III
	Apo E
	Apo(a)
Processing proteins	Lipoprotein lipase (LPL)
	Lecithin:cholesterol acyltransferase (LCAT)
	Cholesterol ester transfer protein (CETP)
Receptors	LDL receptor
	Chylomicron remnant receptor
	Scavenger receptor

INCREASED LDL CHOLESTEROL LEVELS

Increased levels of LDL cholesterol are a significant risk factor for coronary heart disease in humans (5). There is a 2 to 3% change in the risk of heart disease for each 1 mg/dl change in LDL cholesterol levels. LDL particles are a constituent of the endogenous (nondietary) fat transport pathway and are

Table 2   Mouse Models of Lipoprotein Disorders Associated with Coronary Artery Disease

Lipoprotein pattern	Mouse	Ref.
↑ LDL cholesterol	LDL receptor deficient	12
	Apo B transgenic	16–18
↓ HDL cholesterol and ↑ VLDL triglyceride	Apo A-I deficient	36–38
	Apo CIII transgenic	43,44
	Apo CI transgenic	45
	Apo CII transgenic	46
	CETP transgenic	50,53,54,78
	Apo A-I, CETP transgenic	51
	Apo A-I, Apo CIII, CETP transgenic	52
↑ IDL and chylomicron remnant cholesterol	Apo E deficient	67–70,80,81
	Apo $E3_{Leiden}$ transgenic	65
	Apo $E4_{Arg142-Cys}$ transgenic	66
↑ Apolipoprotein(a)	Apo(a) transgenic	79
	Apo(a), Apo B transgenic	17,18

formed via the action of lipases on precursor particles. Dietary carbohydrate or fat reaching the liver that is not required for energy or synthetic purposes is converted into triglycerides, packaged with apolipoproteins, and secreted as VLDL particles. LPL, present on the capillary endothelium mainly in adipose tissue and skeletal muscle, hydrolyzes VLDL core triglycerides, using surface apo CII as a cofactor. This results in conversion of VLDL to IDL particles. The fatty acids thus liberated are re-esterified to form triglycerides in adipose tissue or are oxidized to generate energy in muscle. IDL is cleared from plasma by the LDL receptor, which binds apo E on the IDL surface. IDL particles that escape clearance via this route are subject to further triglyceride hydrolysis by hepatic lipase to form cholesterol ester-enriched LDL particles. The LDL surface contains a single molecule of apo B, which is recognized by LDL receptors. Approximately 70% of LDL is cleared by the LDL receptor, which is expressed primarily in the liver.

The mouse is a poor model for elevated LDL cholesterol levels. On a chow diet, mouse LDL cholesterol levels are approximately 10 mg/dl; this can only be doubled by feeding a Western-type diet. This contrasts with average human LDL cholesterol levels of 140 mg/dl, with the 95th percentile of the human LDL cholesterol distribution at 200 mg/dl. Several inherited disorders in humans are recognized to cause high levels of LDL cholesterol, and these have suggested genetic manipulations in the mouse that are being used to produce animals with elevated LDL cholesterol levels.

The first of these conditions is familial hypercholesterolemia (FH) (6,7). This is an autosomal dominant disorder due to a defective LDL receptor gene on human chromosome 19. Reduction in functional cell surface LDL receptor molecules impairs LDL and IDL uptake, resulting in decreased LDL catabolism and enhanced conversion of IDL to LDL. FH heterozygotes have LDL cholesterol levels approximately double those in unaffected family members. They suffer from premature coronary heart disease, with 25% of the males dying of myocardial infarction by the age of 50, compared to less than 5% in the general population. Approximately 75% of heterozygous familial hypercholesterolemics also develop tendon xanthomas. FH homozygotes have sixfold elevations in LDL cholesterol levels, with total cholesterol levels in the 600 to 1000 mg/dl range. Coronary heart disease is often apparent before age 10, and most untreated homozygotes suffer fatal myocardial infarction before age 20. In addition to tendon xanthomas, homozygotes develop planar cutaneous xanthomas. Clearly genetic modifications changing the number of functional LDL receptors should be a powerful way to alter mouse LDL cholesterol levels.

LDL receptor transgenic and knockout mice have been made. LDL receptor transgenic mice were made with a human cDNA construction driven by the mouse metallothionein-I promoter (8). After heavy metal induction, transgenic mice cleared injected radiolabeled LDL eight to ten times faster than control mice, and the plasma concentrations of LDL receptor ligands, apo B and apo E, declined by more than 90%. The increase in LDL clearance was due primarily to increased liver removal of LDL, mainly by parenchymal cells. These LDL receptor transgenic mice were also used to suggest that Lp(a) particles are cleared from plasma by LDL receptors (9). This issue remains controversial because other lines of evidence suggest that LDL receptors do not play a physiological role in Lp(a) clearance.

LDL receptor transgenic mice were also made with a human minigene driven by the transferrin promoter, resulting in animals with chronically elevated LDL receptor levels (10). When these transgenic mice were challenged with a diet of 1.25% cholesterol, 15% fat, and 0.5% sodium cholate, there was no increase in IDL or LDL levels, and only slightly increased VLDL levels, whereas the levels of all three of these lipoprotein fractions were significantly increased in control mice. Thus, it appears that unregulated expression of LDL receptors can affect diet response. This suggests one possible mechanism for the variation in dietary response to fats and cholesterol observed in humans.

These LDL receptor transgenic mice have also been used to study receptor sorting to the surfaces of epithelial cells (11). In the transgenic mice, LDL receptors appropriately localize to the basolateral surface of hepatocytes and intestinal epithelial cells, and to the apical surface of renal tubular epithelial cells. A signal present in the coding sequence of the transgene apparently interacts in a tissue-specific manner to control cell membrane sorting.

LDL receptor knockout mice have also been created (12). On a chow diet these mice had a twofold increase in total cholesterol, due mainly to an eightfold increase in IDL and LDL cholesterol levels. Metabolic studies showed reduced clearance of apo B–containing lipoproteins. When fed a high-fat diet with 0.2% cholesterol and 10% coconut oil, the LDL receptor-deficient mice had a further threefold increase in the IDL plus LDL cholesterol levels, whereas this diet did not increase the levels of these lipoproteins in control mice. These studies indicate that decreased expression of LDL receptors causes increased diet response. The LDL receptor knockout mice were also used as a model for gene replacement therapy. Intravenous administration of an adenovirus vector containing the human LDL receptor cDNA driven by the cytomegalovirus promoter caused high levels of hepatic LDL receptor expression, increased clearance of apo B–containing lipoproteins, and a return to normal of the IDL plus LDL lipoprotein fractions. A full report on atherosclerosis susceptibility in the LDL receptor-deficient mice has not yet appeared, but one would expect them to develop lesions—if not on chow, then on a high-cholesterol and/or high-fat diet. The LDL receptor-deficient mouse is a good model for the human disorder FH. For some types of studies, particularly those involving genetics or requiring a large number of animals, the LDL receptor-deficient mouse is preferable to the previously described animal models of FH: the WHHL rabbit and the LDL receptor-deficient rhesus monkey.

There are two other inherited conditions in humans characterized by increased LDL cholesterol levels that suggest an alternative approach to making a mouse with high levels of LDL cholesterol. The first of these is familial defective apo B, which is due to a missense mutation in the apo B gene on human chromosome 2 (13,14). The resulting amino acid substitution, which replaces an arginine at amino acid residue 3500 of the 4536 amino acid long apo B polypeptide, disrupts apo B binding to the LDL receptor and impairs LDL uptake. Heterozygosity for this disorder increases LDL cholesterol levels by at least 50% relative to unaffected family members. In general, familial defective apo B may be milder than familial hypercholesterolemia, but patients do sometimes present with premature coronary heart disease and tendon xanthomas. The second condition is familial combined hyperlipidemia, a complex disorder of unknown etiology, which is the most common genetic hypercholesterolemia and a frequent cause of premature coronary heart disease (5,6,15). This disorder is characterized by elevations of LDL cholesterol and VLDL triglyceride within the same family. Affected subjects may have one or both abnormalities, and the lipid profile may vary over time. Xanthomas are uncommon. Metabolic studies have shown overproduction of VLDL and apo B. This leads to elevated VLDL and hypertriglyceridemia in some family members, while in others with more efficient lipolysis the consequence is elevated LDL. These two disorders suggest that production of a receptor binding defective form of apo B or overproduction of apo B in mouse liver would elevate mouse LDL cholesterol levels. Because human apo B is recognized poorly by mouse LDL receptors, both strategies can be realized by creating transgenic mice that express human apo B in the liver.

In humans, full-length apo B, B100, is produced in the liver, whereas truncated apo B, B48, is produced in the intestine by an mRNA editing mechanism. In rodent liver, apo B mRNA editing also occurs and mouse liver produces both B48 and B100. Transgenic mice expressing a human apo B minigene driven by the mouse transthyretin promoter have been made (16). These mice have very low levels of human apo B in plasma (<1% of normal), but the human apo B mRNA transcripts produced in the transgenic mouse liver undergo editing at an efficiency comparable to endogenous mouse apo B mRNA. This indicates that the apo B mRNA editing process can occur across species. Recently, human apo B transgenic mice have been made by microinjecting a large genomic piece of DNA derived from a phagemid vector containing the entire 43 kb gene and 15 to 20 kb of 5' and 3' flanking sequence (17,18). Transgenic lines were obtained with varying amounts of human apo B in plasma up to physiological levels without diminution of mouse apo B.

Human apo B mRNA was found mainly in the liver and mRNA editing occurred. Transgenic lines with high levels of human apo B had a high ratio of B100/B48 in plasma, whereas the opposite was true for lines with low levels. High-expression human apo B transgenic mice had increased total cholesterol due to an increase in LDL cholesterol. The LDL was relatively enriched in triglycerides compared to human LDL, perhaps due to the lack of CETP in mouse plasma. A major difference between mouse and human lipoprotein patterns is the low LDL cholesterol level in the mouse. The human apo B transgenic mice are more similar to humans in this respect, although the absolute levels of LDL cholesterol are still low compared to those in hypercholesterolemic humans.

Gene targeting has been used to alter the endogenous mouse apo B gene (19). An apo B gene that contained a premature stop codon was created in mouse embryonic stem cells used to create altered mice. Homozygosity for the mutant allele reduced VLDL cholesterol, IDL cholesterol, and LDL cholesterol. These mice are a model for a human disorder called hypobetalipoproteinemia, caused by apo B gene mutations that result in the production of abnormally truncated apo B. However, these mice also display findings not associated with human hypobetalipoproteinemia, such as low HDL cholesterol and central nervous system abnormalities (exencephalus and hydrocephalus). It is not clear whether these findings are related to the truncated apo B; they may represent other mutations occurring in the embryonic stem cell line that gave rise to these mice.

REDUCED HDL CHOLESTEROL AND ELEVATED TRIGLYCERIDES

Reduced HDL cholesterol is the most common lipoprotein abnormality associated with coronary heart disease (20). Each mg/dl decrease in HDL cholesterol is associated with a 4% increase in risk. Low HDL cholesterol is often found along with other lipoprotein abnormalities, including high levels of triglycerides in VLDL, increased levels of IDL, and dense LDL. There are two major HDL particle-size classes in plasma, HDL3 and the larger HDL2 (21). Nascent HDL particles produced by the liver and small intestine consist primarily of complexes of phospholipid and apo A-I, and remodeling of these particles occurs as they circulate in plasma. HDL particles attract excess free cholesterol from extrahepatic tissues and from other types of lipoprotein particles. The cholesterol is esterified by the enzyme lecithin:cholesterol acyltransferase, using apo A-I as a cofactor, and the resulting cholesterol ester enters the HDL core, enlarging the particle. HDL particles may also become smaller as a result of the action of cholesterol ester transfer protein (CETP), which exchanges the cholesterol ester in HDL for triglycerides in VLDL and

IDL. Hepatic lipase can then hydrolyze HDL triglycerides, reducing HDL size. Excess cholesterol in peripheral tissues can thus be transferred from HDL to other lipoprotein particles, which are cleared from plasma by hepatic receptors. This process, termed reverse cholesterol transport, may account in part for the protective effect of HDL on atherosclerosis susceptibility.

HDL cholesterol levels are strongly correlated with the levels of its major apolipoprotein, apo A-I (21). Transgenic animals have been made with the human apo A-I gene under the control of its natural flanking sequences (22–25). The apo A-I gene is in a cluster of apolipoprotein genes on human chromosome 11q23 consisting, in order, of the apo A-I, apo CIII, and apo A-IV genes (Figure 1). The apo CIII gene is transcribed in the opposite orientation from the other two genes. The apo A-I gene is expressed primarily in liver and intestine and human apo A-I transgenes extending from as little as 256bp 5' to 80bp 3' of the gene achieved high-level liver expression, whereas this construction and others extending from 5kb 5' to 4kb 3' of the gene failed to give intestinal expression. Subsequent experiments revealed a region approximately 6kb 3' to the apo A-I gene between –0.2kb and –1.4kb 5' of the apo CIII gene required for apo A-I intestinal expression. Whether expressed from the liver alone or liver plus intestine, in several transgenic mouse lines and a line of transgenic rats, apo A-I overexpression was found to selectively increase HDL cholesterol levels. Human apo A-I expression in the mouse also resulted in decreased levels of mouse apo A-I, with some transgenic lines having 80–90% of plasma apo A-I of the human variety (24,26). Coincident with the expression of human apo A-I in the mouse, changes were also noted in the physical properties of HDL. Mouse HDL normally consists of a single major size distribution of particles of approximately 10 nm in diameter. In human apo A-I transgenic mice, there are two major size distributions of

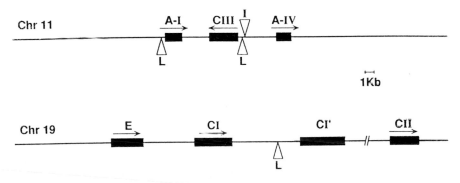

Figure 1   Tissue-specific enhancer elements in the apo A-I, C III, and A-IV and the apo E, C-I, C-I', and C-II gene clusters. L = liver; I = intestine. (From Ref. 4.)

particles, with diameters of approximately 10.3 and 8.8 nm. This corresponds to the two major size distributions of HDL particles in human plasma, HDL_{2b} and HDL_{3a}. The transgenic mouse studies show that the structure of apo A-I is an important determinant of HDL particle size distribution, and for the first time suggest an explanation for HDL subspeciation in humans.

Human apo A-I transgenic mice have served as a model system to examine the mechanisms whereby diet and drugs alter HDL cholesterol and apo A-I levels. A high-fat Western-type diet, which increases HDL cholesterol and apo A-I levels in humans, was mimicked in transgenic mice (27). In these animals the main metabolic effect was to increase HDL cholesterol ester and apo A-I transport rates, without an increase in apo A-I mRNA levels. In contrast, probucol decreases HDL cholesterol and apo A-I levels, and this could also be reproduced in transgenic mice (28). In probucol-treated mice the HDL cholesterol ester fractional catabolic rate increased but the apo A-I transport rate decreased. The latter was not accompanied by a change in apo A-I mRNA levels. Thus, over the wide range of apo A-I levels observed in the high-fat feeding and probucol experiments, there were no changes observed in the apo A-I fractional catabolic rates or in the apo A-I mRNA levels. Two interesting implications can be drawn from these studies. The first is that over the physiological range of apo A-I levels there is no saturable apo A-I or HDL receptor. The second is that previously unrecognized, potent, post-mRNA levels of regulation exist that regulate apo A-I production in relevant clinical situations.

Human apo A-I transgenic mice have been used to test possible non-lipid-transport functions of HDL. *Trypanosoma brucei* fatally infect livestock, but humans are resistant, apparently because human HDL lyses the organism. Human apo A-I is fully trypanolytic, whereas cattle and sheep apo A-I are not. Plasma from human apo A-I transgenic mice was found to be less trypanolytic than human plasma in vitro, and the transgenic mice were fully susceptible to infection with this organism (29). The difference between the lytic activity of human apo A-I in human and mouse plasma was shown to be due to inhibition by mouse HDL apolipoproteins. Thus, it appears that mouse apo A-I can antagonize the trypanolytic effect of human apo A-I. In another study, human apo A-I transgenic mice were used to confirm an in vitro observation suggesting that HDL might be a natural component of plasma that neutralizes endotoxin (30). In this study, after injection of lipopolysaccharide, transgenic mice with a twofold increase in HDL cholesterol levels had better survival than control mice. This was accompanied by an increased transfer of lipopolysaccharide from the peritoneum to plasma and a marked reduction in endotoxin-induced increase in plasma TNF-α levels. Thus, HDL may indeed provide natural protection against endotoxin, and protection may vary with plasma HDL concentration.

While it is clear that low HDL cholesterol levels are an important risk factor for coronary heart disease in the general population, there is little understanding of mechanism. Three theories have been proposed: low HDL reflects decreased reverse cholesterol transport from peripheral tissues—including the artery wall—to the liver, where it can be excreted; low HDL results in decreased protection of the blood vessel wall; and low HDL is merely reflective of elevated levels of atherogenic lipoproteins such as VLDL, IDL, and dense LDL. Further studies of the physiological consequences of low HDL would be aided by the creation of an animal model. One method for accomplishing this is suggested by the existence of a small number of patients with mutations of the apo A-I gene that preclude synthesis of the protein (31–35). In the homozygous state these patients are characterized by very low to undetectable HDL cholesterol levels, and most of them develop planar xanthomas and coronary heart disease between the ages of 25 and 50. To mimic these patients and cause a low-HDL state, apo A-I gene knockout mice were created by gene targeting in embryonic stem cells (36,37). Apo A-I deficient mice maintained on a chow diet had a 75% reduction in total and HDL cholesterol levels. HDL cholesterol ester flux was diminished sevenfold, but tissue levels of free and esterified cholesterol were normal, except in the adrenal, which was deficient in cholesterol ester. Hepatic levels of LDL receptor and HMG CoA reductase mRNA were normal, whereas cholesterol 7-α hydroxylase mRNA was decreased by over 50%. In spite of the marked effects on HDL cholesterol levels and HDL cholesterol ester flux, it remains to be determined whether apo A-I deficiency causes a decrease in reverse cholesterol transport of peripheral tissue cholesterol to the liver for excretion. Histological examination of blood vessels from the apo A-I deficient mice on a chow diet and after 20 weeks of a diet with 1.25% cholesterol, 15% fat, and 0.5% sodium cholate failed to show atherosclerosis (37,38). This suggests that apo A-I deficiency and low HDL cholesterol levels themselves are not sufficient to cause atherosclerosis in mice, which are normally atherosclerosis-resistant. It will be necessary to breed the trait of apo A-I deficiency onto mice susceptible to atherosclerosis (e.g., the apo E knockout mice discussed below) to determine whether apo A-I can act as a modifier of atherosclerosis.

The second most abundant HDL protein is apo A-II, and transgenic mice expressing human apo A-II have been produced (39). Unlike the human apo A-I transgenics, these animals did not have elevated HDL cholesterol levels, nor did they have diminished levels of mouse apo A-I or apo A-II. The human apo A-II transgenic mice had normal levels of VLDL, IDL, and LDL cholesterol and total triglycerides. The lack of effect of excess apo A-II production on HDL cholesterol levels is compatible with clinical studies that fail to show a correlation between plasma apo A-II and HDL cholesterol levels, and the relatively normal HDL cholesterol levels reported in an apo A-II deficient patient (40). The human apo A-II transgenics did show an alteration of HDL particle size, with the appearance of a population of particles 8.0 nm in diameter along with the normal-sized mouse HDL particles. The protein composition of the smaller HDL consisted almost entirely of human apo A-II, whereas the larger particles consisted of mouse apo A-I and human apo A-II. Thus, human apo A-II appears to affect the quality of HDL particles. Recently, a transgenic mouse line expressing mouse apo A-II was derived with a two- to threefold elevation of apo A-II levels (41). In contrast to human apo A-II transgenic mice, these animals had a twofold increase in HDL cholesterol levels and a two- to threefold increase in non-HDL cholesterol and triglyceride levels. Mouse apo A-II transgenics also had larger HDL particles. The marked differences in the lipoprotein profiles between mouse and human apo A-II transgenics may be due to the different physical properties of the two proteins. For example, in plasma mouse apo A-II is monomeric whereas human apo A-II is homodimeric. In addition, the two proteins are only about 60% identical in their amino acid sequences. As with apo A-I, this is another example of important species differences in in vivo apolipoprotein behavior, and provides a unique perspective on apolipoprotein function.

As previously noted, low HDL cholesterol levels are often found together with high triglycerides. The principal triglyceride-rich lipoproteins are VLDL and chylomicrons. The initial step in the metabolism of these particles is triglyceride hydrolysis, which is carried out by LPL, a molecule that resides on endothelial surfaces, mainly of muscle and adipose tissue. LPL requires apo CII as a cofactor. Elevated triglycerides are common in the population, with 10% of middle-aged men having triglycerides over 250 mg/dl, yet the known genetic causes, homozygosity for mutations in LPL or its cofactor apo CII, are quite rare, each with a frequency of less than one in a million.

Transgenic mice have been made that overexpress LPL in a tissue-non-specific manner utilizing a human LPL cDNA driven by the chicken beta actin promoter (42). These mice expressed high levels of LPL mRNA in heart, skeletal muscle, and adipose tissue, but also in the lungs, brain, spleen, and aorta, with low levels in other tissues. Transgenic mice had fivefold higher LPL activity in adipose tissue and 1.7-fold higher activity in postheparin plasma than controls. The overexpression of LPL activity resulted in a 75% reduction of triglyceride levels, and a 1.4-fold increase in HDL_2 cholesterol. Metabolic studies indicated rapid clearance of VLDL and dietary fat, as measured by a vitamin A–fat-tolerance test. In the transgenic mice no increase was observed in VLDL after sucrose feeding, and the development of hypercholesterolemia was suppressed after high-cholesterol feeding. The LPL transgenic mice confirm that LPL has a strong influence on triglyceride levels in vivo.

The mouse normally has low levels of triglycerides compared with humans, and overexpression of LPL can diminish these levels even further.

Transgenic techniques have also been used to make models of hypertriglyceridemia. In the course of making transgenic mice to study the *cis* acting regions responsible for the tissue-specific expression of the apo A-I gene, a DNA construction was used that contained the apo A-I gene plus the neighboring apo CIII gene, which codes for a protein in VLDL and HDL (3). These mice were found to have massive hypertriglyceridemia, whereas mice made with the apo A-I gene alone had normal triglyceride levels. Subsequently, several transgenic mouse lines were made with only the apo CIII gene, and triglyceride levels were proportional to apo CIII gene expression as measured by human apo CIII plasma concentrations (43). In one transgenic line there was a single copy of the transgene and only 30–40% extra apo C-III in plasma, yet these mice had more than twice normal triglyceride levels. The human apo CIII transgenic mice were the first animal model of primary hypertriglyceridemia.

The mechanism of the hypertriglyceridemia has been studied in the human apo CIII transgenic mice (44). These animals accumulate VLDL that is slightly larger than normal. The VLDL composition is appropriately triglyceride-rich, but there is altered apolipoprotein content with increased apo CIII and diminished apo E. The transgenic mice also have higher plasma-free fatty acid levels. Metabolic studies indicate that the primary abnormality is a decreased VLDL fractional catabolic rate, with a small increase in the production rate of VLDL triglycerides but not apo B. In vitro, the transgenic VLDL showed decreased LDL receptor-mediated uptake by tissue culture cells but normal lipolysis by purified lipoprotein lipase. Thus, the hypertriglyceridemia appears to be due to a prolonged VLDL residence time but not with the accumulation of remnant particles. This implies decreased in vivo lipolysis and tissue uptake, presumably secondary to altered surface apolipoprotein composition.

There are two other C apolipoproteins in VLDL and HDL: apo CI and apo CII. Transgenic mouse models of hypertriglyceridemia have been made with both. The human apo CI transgenics (45) were mildly hypertriglyceridemic, and, surprisingly, human apo CII transgenics (46) were found to be as hypertriglyceridemic as the human apo CIII transgenic mice, even though apo CII is an activator of LPL. As with the human apo CIII transgenics, the human apo CII transgenic mice accumulated VLDL of almost normal size but with an increased ratio of apo C to apo E. These mice also had delayed VLDL clearance without increased production. The human apo CII transgenic mouse VLDL had markedly decreased binding to heparin-sepharose, suggesting that apo CII-rich, apo E-poor VLDL may be less accessible to cell-surface lipases or receptors within glycosaminoglycan matrices. The human apo CII transgenic mice are a model of primary hypertriglyceridemia and suggest a more complex role for apo CII in the metabolism of triglycerides than previously thought. Apo CII deficiency is known to cause hypertriglyceride-mia. It now appears that overproduction of apo CII might cause hypertri-glyceridemia as well.

Hypertriglyceridemia is common in humans, but the known genetic abnormalities are quite rare. These transgenic mouse experiments prove that apo CIII and apo CII overexpression can cause hypertriglyceridemia and suggest that apo CIII and/or apo CII gene expression may regulate triglyceride levels in humans. Evidence for involvement of the apo CIII gene in human hypertriglyceridemia has come from association studies. These have repeat-edly shown that apo CIII alleles with an SstI cutting site in the 3′ untranslated region are more common in affected Caucasian hypertriglyceridemics than in controls. More recently, five new sites of genetic variation have been identi-fied in the apo CIII gene promoter, and haplotyping utilizing these and the SstI site has revealed three classes of apo CIII alleles: susceptible, neutral, and protective with regard to hypertriglyceridemia (47). Further efforts are under-way to identify the causative mutations and prove that apo CIII expression influences human triglyceride levels. In addition, a polymorphism of the apo CII gene has recently been associated with hypertriglyceridemia (48). If this line of research proves successful, the use of transgenic animals to provide clues to the genes underlying complex human traits may become an important paradigm in human genetics.

In the transgenic mouse models thus far described, HDL cholesterol and triglyceride levels have been altered relatively independently of each other. This is because the mouse lacks CETP activity in plasma. In humans, it is thought that hypertriglyceridemia actually causes low HDL cholesterol levels through the CETP-mediated exchange of HDL cholesterol esters for triglycer-ides, with subsequent HDL triglyceride hydrolysis by hepatic lipase. CETP deficiency has been described in humans, causing elevated HDL cholesterol levels and reduced levels of non-HDL cholesterol (49). CETP mutations occur frequently in Japanese and in this population are a common cause of high HDL cholesterol levels. Transgenic techniques have been used to induce CETP activity in mouse plasma. A human CETP minigene driven by the mouse metallothionein-I promoter was used to make a transgenic line with human-like levels of activity in plasma, which could be doubled by feeding zinc (50). After zinc treatment, compared to control mice, these mice had lower levels of HDL cholesterol and apo A-I—35% and 24%, respectively—and smaller HDL particles (mean diameter of 10 nm as compared to 9.7 nm). These effects of CETP were less than expected based on studies comparing normal and CETP-deficient humans, suggesting functional differences between mouse and human HDL. In subsequent studies, CETP was found to be more potent in mice expressing the human apo A-I transgene (51). In these experiments human CETP transgenic mice were crossed with human apo A-I transgenic mice. After zinc induction, compared to the human apo A-I transgenic mice, the doubly transgenic mice had a more pronounced reduction in HDL cholesterol and apo A-I levels, 66% and 42%, respectively, with even smaller HDL particles (mean diameter of 10.4, 8.8, and 7.4 nm as compared to 9.7, 8.5, and 7.3 nm). In the doubly transgenic mice it was also found that 100% of the CETP was HDL-associated versus 22% in the singly transgenic animals. Thus, CETP overexpression can reduce HDL cholesterol levels and particle size, an effect that is more dramatic on a human apo A-I background. This implies a specific interaction of human CETP with human apo A-I or human apo A-I-containing particles.

The CETP-mediated exchange of triglycerides for HDL cholesterol ester is driven by the level of VLDL, which is quite low in the mouse. Therefore, through crossbreeding, the effect of CETP was studied in hypertriglyceridemic human apo CIII transgenic mice coexpressing human apo A-I (52). In these mice, human CETP gene expression reduced HDL cholesterol and apo A-I to very low levels, with a dramatic reduction in HDL particle size. This mimics the high triglyceride–low HDL cholesterol phenotype in humans, which is the most common lipoprotein disorder associated with susceptibility to coronary heart disease. The human apo A-I, apo CIII, CETP transgenic mice are the first animal model of this disorder. These animals provide insights into which genes may cause this abnormal phenotype in humans and present opportunities to study the mechanisms of the relationship between this lipoprotein abnormality and atherosclerosis susceptibility.

A cynomologus monkey CETP cDNA driven by the mouse metallothionein-I promoter was used to make mice with very high levels of CETP (53). The monkey CETP transgenic mice showed a strong inverse correlation of CETP activity with HDL cholesterol and apo A-I levels as well as HDL size, and a positive correlation of CETP activity with apo B levels and the size of apo B–containing lipoproteins. The monkey CETP mice were also more diet-responsive than control animals. Recently, human CETP transgenic mice were also found to show a gene-dosage-dependent effect on apo B levels (54). The experiments with human and monkey CETP transgenic mice confirm the proposed role of CETP in lipoprotein metabolism deduced from other systems and will make it possible to test the effect of CETP on cholesterol homeostasis, particularly reverse cholesterol transport, and atherosclerosis.

ELEVATED CHYLOMICRON REMNANTS AND IDL CHOLESTEROL

Chylomicron remnants and IDL are atherogenic cholesterol ester–rich particles that are not normally present in large amounts in fasting plasma or commonly measured in coronary heart disease risk factor assessment. Chylomicron remnants and IDL are normally cleared from plasma by hepatic chylomicron remnant and LDL receptors, which recognize apo E on the surface of these particles (55). There are three common apo E alleles: E3 (Caucasian frequency 77%), E4 (15%), and E2 (8%). These specify six common apo E phenotypes: E3/3 (frequency 59%), E4/4 (2%), E2/2 (1%), E4/3 (23%), E3/2 (12%), and E4/2 (2%). Mature apo E is 299 amino acids long; E4 differs from E3 by a Cys112Arg substitution and E2 differs from E3 by an Arg158Cys substitution. Apo E2 is defective in receptor binding, whereas apo E3 and E4 bind receptors normally. Individuals with type III hyperlipoproteinemia, who have increased plasma levels of chylomicron remnants and IDL particles due to impaired catabolism, generally have the E2/2 phenotype. These patients are susceptible to premature coronary heart disease, strokes, and peripheral vascular disease. Type III hyperlipoproteinemia can also be caused by heterozygosity for other rare mutations of apo E. In addition to contributing to type III hyperlipoproteinemia, the apo E phenotype can affect LDL cholesterol levels. Individuals with the E4/3 genotype have mean LDL cholesterol levels 5 to 10 mg/dl higher than subjects with the E3/3 genotype, while individuals with E3/2 have LDL cholesterol levels 10 to 20 mg/dl lower than E3/3 subjects (56). In an autopsy study of more than 500 young male trauma victims, E3/2 was associated with reduced atherosclerosis relative to E3/3 (57). Recently, studies of the elderly have revealed a decreased frequency of the E4 allele, suggesting that this locus influences longevity (58). In a surprising finding, the E4 allele has been found to be a risk factor for familial late onset and sporadic Alzheimer's disease (59,60). Clearly, mice genetically altered in their apo E gene complement should provide interesting models for human disease.

Human apo E transgenic mice have been made utilizing gene constructions containing natural flanking sequences (61,62). The apo E gene is in a cluster with two other apolipoprotein genes on human chromosome 19q13 (Figure 1). The cluster consists of, in order, the apo E, apo CI, and apo CII genes. The genes are transcribed in the same orientation, and there is an apo CI pseudogene between the apo CI and apo CII genes. The endogenous apo E gene is expressed primarily in the liver, with expression—but at lower levels—in most body tissues. Human apo E transgenes extending from 5kb 5' to 2kb 3' of the gene gave low-level liver expression, but high-level kidney expression. Studies have localized a region 11kb 3' to the apo E gene, between the apo CI gene and the apo CI pseudogene, that is required for high-level liver expression. This region also suppresses kidney expression, and may control the liver expression of the other two apolipoprotein genes in this cluster, apo CI and apo CII. The initial human apo E transgenic mice were made with constructions that lacked the liver control element; as a result, the animals produced had relatively low levels of apo E expression with no significant effect of transgene expression on lipoprotein levels. Recently, transgenic mice have been made with the rat apo E gene driven by the metallothionein promoter (63,64). After zinc induction, these animals had a fourfold increase in apo E levels accompanied by a significant decrease in VLDL and LDL cholesterol levels. Metabolic studies indicated a several-fold increase in the clearance rate of radiolabeled VLDL and LDL. In addition, these animals were resistant to diet-induced hypercholesterolemia. These studies indicate that apo E overexpression lowers fasting levels of atherogenic lipoproteins and decreases diet response.

Transgenic mice have also been made with mutant forms of apo E that can cause dominantly inherited type III hyperlipoproteinemia, including $E3_{Leiden}$ (tandem duplication of apo E amino acids 120 to 126) (65) and $E4_{Arg142Cys}$ (66). In each case the phenotype was similar, with increased levels of cholesterol and triglycerides in the VLDL and IDL lipoprotein fractions. The $E3_{Leiden}$ mice were shown to be extremely responsive to dietary cholesterol. The $E3_{Leiden}$ and the $E4_{Arg142Cys}$ mice appear to be reasonable phenocopies of human type III hyperlipoproteinemia and will be useful models to study the genetic and environmental factors that influence the expression of this disease.

Apo E knockout mice with a true null mutation have also been created (67-69). Homozygous deficient animals are viable and fertile. On a chow diet, which is very low in cholesterol (0.01%) and low in fat (4.5%), they have cholesterol levels of 400 to 500 mg/dl. Most of this is in the VLDL plus IDL lipoprotein fractions. When the homozygous apo E knockout mice are fed a Western-type diet, which has moderate amounts of cholesterol (0.15%) and fat (20%), they respond with cholesterol levels of approximately 1800 mg/dl, also mostly in the VLDL plus IDL lipoprotein fractions (69). On both diets triglyceride levels are minimally elevated. The lipoprotein particles that accumulate in the apo E deficient mice are similar in size to normal VLDL, but are cholesterol ester–enriched, similar to β- VLDL (70). Metabolic studies indicate a severe defect in lipoprotein clearance from plasma, as expected from the known function of apo E as a ligand for lipoprotein receptors. The β-VLDL in the apo E deficient mice are probably remnants of intestinally derived lipoproteins. Heterozygous apo E knockout mice have diminished plasma apo E levels, normal fasting lipoprotein levels, and slightly delayed postprandial lipoprotein clearance. Thus, half-normal apo E expression in the mouse is nearly sufficient for normal lipoprotein metabolism. As discussed below, the accumulation of atherogenic β-VLDL in the apo E knockout mice is sufficient to produce human-like atherosclerotic lesions.

ELEVATED LIPOPROTEIN(a)

In case-control studies, elevated levels of Lp(a) have been found to be an independent risk factor for coronary heart disease. Lp(a) consists of a large glycoprotein, apo(a), disulfide bonded to the apo B moiety of LDL. Apo(a) resembles plasminogen, containing domains of plasminogen-like kringle 4, in multiple copies, and of plasminogen-like kringle 5 and protease, in single copies. The protease domain of apo(a) is unable to degrade fibrin. Apo(a) alleles specify proteins that differ in size due to variation in the number of kringle 4–like domains. In humans levels of Lp(a) vary greatly from less than 0.1 to greater than 200 mg/dl, and are almost entirely genetically determined by the apo(a) gene locus on human chromosome 6q25. The larger apo(a) forms are associated with lower plasma Lp(a) levels and vice versa. Perhaps due to its resemblance to LDL, Lp(a) is a tightly bound constituent of the atherosclerotic plaque. In addition, through its plasminogen-like properties, Lp(a) may also participate in thrombogenic processes. In vitro, it has been shown that Lp(a) can compete with plasminogen for binding to endothelial surfaces and in this manner interfere with the assembly of the fibrinolytic system on vascular surfaces. The physiological function of Lp(a) is unknown.

Lp(a) has a limited phylogenetic distribution and is found only in humans, Old World primates, and hedgehogs. Mice do not express apo(a), and transgenic animals have been made with a human apo(a) gene construction consisting of a cDNA containing 17 kringle 4s, kringle 5, and protease coding regions driven by the transferrin promoter (71). Expression was achieved in all tissues analyzed, while the gene is normally expressed only in liver. Mean plasma levels equivalent to 9 mg/dl of Lp(a) were achieved, but, in contrast to humans, apo(a) was found in the lipoprotein-free fraction. Infusion of human LDL into these transgenic mice resulted in binding of apo(a) to these lipoproteins, indicating the formation of Lp(a). In addition, when apo(a) transgenic mice were crossbred to human apo B transgenic mice, Lp(a) was also formed (72,73). These experiments suggest that human apo(a) can bind to human apo B but not mouse apo B, perhaps due to the lack of conservation of a crucial cysteine in the mouse protein. These animals and others with higher levels of Lp(a) will be useful in defining the metabolism and function of this protein and how it participates in the atherosclerotic process, and in designing effective pharmacological means for lowering Lp(a).

MOUSE MODELS OF ATHEROSCLEROSIS

The mouse is the best mammalian system for the study of genetic contributions to disease. This is because of the easy breeding, short generation time, and availability of inbred strains, many of which have interesting heritable phenotypes. Unfortunately, the mouse is highly resistant to atherosclerosis, which makes it difficult to use to identify the genes controlling this complex disease. In an attempt to overcome this problem and produce atherosclerotic lesions, mice have been fed an unphysiological diet consisting of 1.25%

cholesterol, 15% fat, and 0.5% cholic acid. This diet—which contains 10–20 times the amount of cholesterol of a human diet and an unnatural constituent, cholic acid—is toxic when fed for a long time to mice. However, it does produce a cholesterol level of 200 to 300 mg/dl, with the increase in cholesterol in the non-HDL lipoprotein fractions. This is in contrast to a chow diet, which produces cholesterol levels of 60 to 80 mg/dl, mostly in HDL. When certain strains of mice are fed the high-cholesterol diet for 4–5 months, they develop foam cell lesions at the base of the aorta in the region of the aortic valves, whereas others do not. In this model, crosses between resistant and susceptible mouse strains have been used to identify atherosclerosis-susceptibility loci (74).

Transgenic techniques have now been used to further exploit this model (Table 3). In these studies, lipoprotein transport genes have been introduced into one of the susceptible strains, C57BL/6, to evaluate their effect on diet-induced atherosclerosis. In one study (75), human apo A-I gene expression reduced aortic sinus foam cell lesion area. These results suggest that apo A-I expression, with its attendant increase in HDL-C levels, can protect against atherosclerosis. In a second study, human apo A-I and human apo A-II transgenic mice were crossed, and there was less protection when both genes were expressed than when the human apo A-I gene was expressed alone (76). Since human apo A-II gene expression does not increase HDL-C levels but rather increases the apo A-I plus A-II and decreases the apo A-I-only HDL particles, this experiment suggests that not all HDL particles are equally antiatherogenic. In other studies, the overexpression of mouse apo A-II (77) or monkey CETP (78) in this model increased atherosclerosis susceptibility. Finally, in a variation on this type of study, the apo(a) transgene was introduced into outbred mice that do not develop atherosclerosis (79). The apo(a) transgenic mice developed aortic sinus lesions. As previously noted, Lp(a) does not form in these mice. This implies that apo(a) is itself atherogenic.

The homozygous apo E knockout mice have provided a new model of atherosclerosis (68,69). Although these animals are outbred, representing a mixture of C57BL/6 and 129 strain genetic backgrounds, they develop widespread fibroproliferative atherosclerotic lesions when fed a chow diet (80,81). This is in contrast to the model cited above in which an extremely high-cholesterol, cholic-acid diet is required to produce foam cells localized to the base of the aorta (74). In the apo E deficient mice, lesions occur at the base of the aorta, proximal coronaries, and at all the major branch points of vessels coming off the aorta, including the carotid, intercostal, mesenteric, renal, and femoral arteries (Figure 2C). Histologically the lesions progress from subintimal foam cells at 5 to 10 weeks of age to fibrous plaques with smooth muscle cell caps and necrotic cores at 10 to 20 weeks of age (Figure 2A and B). Some lesions show fibrous plaques flanked by foam cells at the shoulder areas. Other

Table 3 Mouse Models of Atherosclerosis and the Effect of Lipoprotein-Modifying Genes on Lesion Formation

Mouse Model[a]	Transgene	Non-HDL cholesterol[b]	HDL cholesterol[b]	Atherosclerosis
C57BL/6 diet-induced	None	↑↑	↓	Aortic root foam cells
	Apo A-I	↑↑	↑	↓↓↓[c]
	Apo A-II (human)	↑↑	↓	—[c]
	Apo A-I, Apo A-II	↑↑	↑↓	↓[c]
	CETP	↑↑	↑↓	↑[c]
Hybrid diet-induced	None	↑↑	↓	No lesions
	Apo(a)	↑↑	↓	Aortic root foam cells
Apo E deficient	None	↑↑↑↑	↓	Diffuse fibroproliferative
	Apo A-I	↑↑	↑	↓↓↓[d]

[a]C57BL/6 are inbred mice fed an atherogenic high-cholesterol, high-fat, cholic-acid-containing diet; the hybrid genetic mice are mixtures of C57BL/6 and a resistant strain also fed the atherogenic diet; apo E deficient mice were fed a low-fat, chow diet.
[b]As compared to chow-fed mice.
[c]As compared to C57BL/6 diet-induced mice.
[d]As compared to apo E deficient mice.

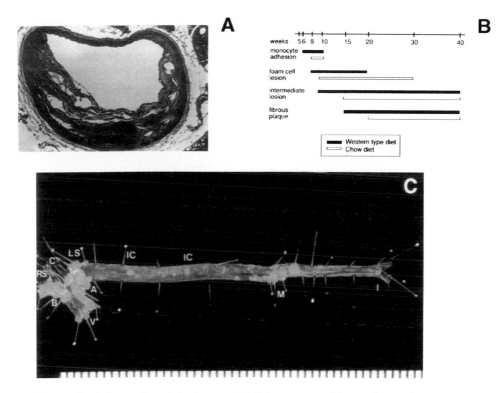

Figure 2   Atherosclerosis in the apo E deficient mouse. (A) An advanced atherosclerotic fibrous plaque from the carotid artery of an apo E deficient mouse. (B) The progression of lesion formation in chow- and Western-type-diet-fed apo E deficient mice. (C) An aorta from an apo E deficient mouse stained for lipid with Oil Red O. (A from Ref. 83, B and C from Ref. 80.)

lesions have medial necrosis with occasional aneurysm formation. Lesion formation can be accelerated and lesion size increased by feeding a Western-type diet containing 0.15% cholesterol and 20% fat (Figure 2B). In addition, lesions can be inhibited by the overexpression of human apo A-I by a mechanism due primarily to an increase in HDL cholesterol levels (82). The foam cell lesions are enriched in oxidized epitopes of lipoproteins, and the plasma of these mice contains very high levels of autoantibodies to oxidized lipoproteins (83). Thus, the single genetic lesion causing apo E absence and severe hypercholesterolemia is sufficient to convert the mouse from a species that is highly resistant to one that is highly susceptible to atherosclerosis. These animals should be of great assistance in studies of diets, genes, and drugs that influence atherosclerosis and greatly accelerate research in this field.

REFERENCES

1. Breslow JL. Lipoprotein transport gene abnormalities underlying coronary heart disease susceptibility. Annu Rev Med 1991; 42:357-371.
2. Breslow JL. Transgenic mouse models of lipoprotein metabolism and atherosclerosis. Proc Natl Acad Sci USA 1993; 90:8314-8318.
3. Breslow JL. Insights into lipoprotein metabolism from studies in transgenic mice. Annu Rev Physiol 1994; 56:797-810.
4. Breslow JL. Lipoprotein metabolism and atherosclerosis susceptibility in transgenic mice. Curr Opin Lipidol 1994; 5:175-184.
5. Bierman EL. Atherosclerosis and other forms of arteriosclerosis. In: Wilson JD, Braunwald E, Isselbacher KJ, et al., eds. Harrison's Principles of Internal Medicine. 12th ed. New York: McGraw-Hill, 1991:992-1001.
6. Brown MS, Goldstein JL. The hyperlipoproteinemias and other disorders of lipid metabolism. In: Wilson JD, Braunwald E, Isselbacher KJ, et al., eds. Harrison's Principles of Internal Medicine. 12th ed. New York: McGraw-Hill, 1991:1814-1825.
7. Goldstein JL, Brown MS. Familial hypercholesterolemia. In: Scrivner CR, Beaudet AL, Sly WS, Valle D, eds. The Metabolic Basis of Inherited Disease. 6th ed. Vol I. New York: McGraw-Hill, 1989:1215-1250.
8. Hofmann SL, Russell DW, Brown MS, Goldstein JL, Hammer RE. Overexpression of low density lipoprotein (LDL) receptor eliminates LDL from plasma in transgenic mice. Science 1988; 239:1277-1281.
9. Hofmann SL, Eaton DL, Brown MS, McConathy WJ, Goldstein JL, Hammer RE. Over-expression of human low density lipoprotein receptors leads to accelerated catabolism of Lp(a) lipoprotein in transgenic mice. J Clin Invest 1990; 85:1542-1547.
10. Yokode M, Hammer RE, Ishibashi S, Brown MS, Goldstein JL. Diet-induced hypercholesterolemia in mice: Prevention by overexpression of LDL receptors. Science 1990; 250:1273-1275.
11. Pathak RK, Yokode M, Hammer RE, Hofmann SL, Brown MS, Goldstein JL, Anderson RGW. Tissue-specific sorting of the human LDL recepor in polarized epithelia of transgenic mice. J Cell Biol 1990; 111:347-359.
12. Ishibashi S, Brown MS, Goldstein JL, Gerard RD, Hammer RE, Herz J. Hypercholesterolemia in low density lipoprotein receptor knockout mice and its reversal by adenovirus-mediated gene delivery. J Clin Invest 1993; 92:883-893.
13. Breslow JL. Lipoprotein transport gene abnormalities underlying coronary heart disease susceptibility. Annu Rev Med 1991; 42:357-371.
14. Zannis VI, Kardassis D, Zanni EE. Genetic mutations affecting human lipoproteins, their receptors, and their enzymes. In: Harris H, Hirschhorn K, eds. Advances in Human Genetics. Vol 21. New York: Plenum Press, 1993:145-319.
15. Kane JP, Havel RJ. Disorders of the biogenesis and secretion of lipoproteins containing the B apolipoproteins. In: Scrivner CR, Beaudet AL, Sly WS, Valle D, eds. The Metabolic Basis of Inherited Disease. 6th ed. Vol I. New York: McGraw-Hill, 1989:1139-1164.
16. Chiesa G, Johnson DF, Yao Z, Innerarity TL, Mahley RW, Young SG, Hammer RH, Hobbs HH. Expression of human apolipoprotein B100 in transgenic mice. J Biol Chem 1993; 268:23747-23750.

17. Linton MF, Farese RV Jr, Chiesa G, Grass DS, Chin P, Hammer RE, Hobbs HH, Young SG. Transgenic mice expressing high plasma concentrations of human apolipoprotein B100 and lipoprotein(a). J Clin Invest 1993; 92:3029-3037.

18. Callow MJ, Stoltzfus LJ, Lawn RM, Rubin EM. Expression of human apolipoprotein B and assembly of lipoprotein(a) in transgenic mice. Proc Natl Acad Sci USA 1994; 91:2130-2134.

19. Homanics GE, Smith TJ, Zhang SH, Lee D, Young SG, Maeda N. Targeted modification of the apolipoprotein B gene results in hypobetalipoproteinemia and developmental abnormalities in mice. Proc Natl Acad Sci USA 1993; 90:2389-2393.

20. Genest JJ Jr, Martin-Munley SS, McNamara JR, et al. Familial lipoprotein disorders in patients with premature coronary artery disease. Circulation 1992; 85:2025-2033.

21. Eisenberg, S. High density lipoprotein metabolism. J Lipid Res 1984; 25:1017-1058.

22. Walsh A, Ito Y, Breslow JL. High levels of human apolipoprotein A-I in transgenic mice result in increased plasma levels of small high density lipoprotein (HDL) particles comparable to human HDL3. J Biol Chem 1989; 264:6488-6494.

23. Walsh A, Azrolan N, Wang K, Marcigliano A, O'Connell A, Breslow JL. Intestinal expression of the human apo A-I gene in transgenic mice is controlled by a DNA region 3′ to the gene in the promoter of the adjacent convergently transcribed apo C-III gene. J Lipid Res 1993; 34:617-623.

24. Rubin EM, Ishida BY, Clift SM, Krauss RM. Expression of human apolipoprotein A-I in transgenic mice results in reduced plasma levels of murine apolipoprotein A-I and the appearance of two new high density lipoprotein size subclasses. Proc Natl Acad Sci USA 1991; 88:434-438.

25. Swanson ME, Hughes TE, St. Denny I, France DS, Paterniti JR Jr, Tapparelli C, Gfeller P, Burki K. High level expression of human apolipoprotein A-I in transgenic rats raises total serum high density lipoprotein cholesterol and lowers rat apolipoprotein A-I. Transgenic Res 1992; 1:142-147.

26. Chajek-Shaul T, Hayek T, Walsh A, Breslow JL. Expression of the human apolipoprotein A-I gene in transgenic mice alters high density lipoprotein (HDL) particle size distribution and diminishes selective uptake of HDL cholesteryl esters. Proc Natl Acad Sci USA 1991; 88:6731-6735.

27. Hayek T, Ito Y, Azrolan N, Verdery RB, Aalto-Setälä K, Walsh A, Breslow JL. Dietary fat increases high density lipoprotein (HDL) levels both by increasing the transport rates and decreasing the fractional catabolic rates of HDL cholesterol ester and apolipoprotein (apo) A-I. J Clin Invest 1993; 91:1665-1671.

28. Hayek T, Chajek-Shaul T, Walsh A, Azrolan N, Breslow JL. Probucol decreases apolipoprotein A-I transport rate in control and human apolipoprotein A-I transgenic mice. Arterioscler Thromb 1991; 11:1295-1302.

29. Owen JS, Gillett MPT, Hughes TE. Transgenic mice expressing human apolipoprotein A-I have sera with modest trypanolytic activity in vitro but remain susceptible to infection by Trypanosoma brucei. J Lipid Res 1992; 33:1639-1646.

30. Levine DM, Parker TS, Donnelly TM, Walsh A, Rubin AL. In vivo protection against endotoxin by plasma high density lipoprotein. Proc Natl Acad Sci USA 1993; 90:12040-12044.

31. Breslow JL. Familial disorders of high density lipoprotein metabolism. In: Scrivner CR, Beaudet AL, Sly WS, Valle D, eds. The Metabolic Basis of Inherited Disease. 6th ed. Vol I. New York: McGraw-Hill, 1989:1251-1266.

32. Schaefer EJ, Heaton WH, Wetzel MG, Brewer HB Jr. Plasma apolipoprotein A-I absence associated with a marked reduction of high density lipoproteins and premature coronary artery disease. Arteriosclerosis 1982; 2:16-26.

33. Hiasa Y, Maeda T, Mori H. Deficiency of apolipoproteins A-I and C-III and severe coronary heart disease. Clin Cardiol 1986; 9:349-352.

34. Matsunga T, Hiasa Y, Yanagi Y, Maeda T, Hattori N, Yamakawa K, Yamanouchi Y, Tanaka I, Obara T, Hamaguchi H. Apolipoprotein A-I deficiency due to a codon 84 nonsense mutation of the apolipoprotein A-I gene. Proc Natl Acad Sci USA 1991; 88:2793-2797.

35. Lackner KJ, Dieplinger H, Nowicka G, Schmitz G. High density lipoprotein deficiency with xanthomas. J Clin Invest 1993; 92:2262-2273.

36. Williamson R, Lee D, Hagaman J, Maeda N. Marked reduction of high density lipoprotein cholesterol in mice genetically modified to lack apolipoprotein A-I. Proc Natl Acad Sci USA 1992; 89:7134-7138.

37. Plump AS, Hayek T, Walsh A, Breslow JL. Diminished HDL cholesterol ester flux in apo A-I-deficient mice. Circulation 1993; 88:I-422.

38. Li H, Reddick RL, Maeda N. Lack of apo A-I is not associated with increased susceptibility to atherosclerosis in mice. Arterioscler Thromb 1993; 13:1814-1821.

39. Schultz JR, Gong EL, McCall MR, Nichols AV, Clift SM, Rubin EM. Expression of human apolipoprotein A-II and its effect on high density lipoproteins in transgenic mice. J Biol Chem 1993; 267:21630-21636.

40. Deeb SS, Takata K, Peng RL, Kajiyama G, Albers JJ. A splice-junction mutation responsible for familial apolipoprotein A-II deficiency. Am J Hum Genet 1990; 46:822-827.

41. Hedrick CC, Castellani LW, Warden CH, Puppione DL, Lusis AJ. Influence of mouse apolipoprotein A-II on plasma lipoproteins in transgenic mice. J Biol Chem 1993; 268:20676-20682.

42. Shimada M, Shimano H, Gotoda T, Yamamoto K, Kawamura M, Inaba T, Yazaki Y, Yamada N. Overexpression of human lipoprotein lipase in transgenic mice. J Biol Chem 1993; 268:17924-17929.

43. Ito Y, Azrolan N, O'Connell A, Walsh A, Breslow JL. Hypertriglyceridemia as a result of human apolipoprotein CIII gene expression in transgenic mice. Science 1990; 249:790-793.

44. Aalto-Setälä K, Fisher EA, Chen X, Chajek-Shaul T, Hayek T, Zechner R, Walsh A, Ramakrishnan R, Ginsberg HN, Breslow JL. Mechanism of hypertriglyceridemia in human apo CIII transgenic mice: Diminished VLDL fractional catabolic rate associated with increased apo CIII and reduced apo E on the particles. J Clin Invest 1992; 90:1889-1900.

45. Simonet WS, Bucay N, Pitas RE, Lauer SJ, Taylor JM. Multiple tissue-specific

elements control the apolipoprotein E/C-I gene locus in transgenic mice. J Biol Chem 1991; 265:8651–8654.

46. Shachter NS, Hayek T, Leff T, Smith JD, Rosenberg DW, Walsh A, Ramakrishnan R, Ginsberg HN, Breslow JL. Overexpression of apolipoprotein CII causes hypertriglyceridemia in transgenic mice. J Clin Invest 1994; 93:1683–1690.

47. Dammerman M, Sandkuijl LA, Halaas J, Chung W, Breslow JL. An apo CIII haplotype protective against hypertriglyceridemia is specified by novel promoter polymorphisms and a known 3′ untranslated region polymorphism. Proc Natl Acad Sci USA 1993; 92:2262–2273.

48. Hegele RA, Connelly PW, Maguire GF, et al. An apolipoprotein CII mutation, CII$_{Lys19-Thr}$, identified in patients with hyperlipidemia. Dis Markers 1991; 9:73–80.

49. Tall AR. Plasma cholesteryl ester transfer protein. J Lipid Res 1993; 34:1255–1274.

50. Agellon LB, Walsh A, Hayek T, Moulin P, Jiang XC, Shelanski SA, Breslow JL, Tall AR. Reduced high density lipoprotein cholesterol in human cholesteryl ester transfer protein transgenic mice. J Biol Chem 1991; 266:10796–10801.

51. Hayek T, Chajek-Shaul T, Walsh A, Agellon LB, Moulin P, Tall AR, Breslow JL. An interaction between the human cholesteryl ester transfer protein (CETP) and apolipoprotein A-I genes in transgenic mice results in a profound CETP-mediated depression of high density lipoprotein cholesterol levels. J Clin Invest 1992; 90:505–510.

52. Hayek T, Azrolan N, Verdery RB, Walsh A, Chajek-Shaul T, Agellon LB, Tall AR, Breslow JL. Hypertriglyceridemia and cholesteryl ester transfer protein interact to dramatically alter high density lipoprotein levels, particle sizes, and metabolism. J Clin Invest 1993; 92:1143–1152.

53. Marotti KR, Castle CK, Murray RW, Rehberg EF, Polites HG, Melchior GW. The role of cholesteryl ester transfer protein in primate apolipoprotein A-I metabolism: Insights from studies with transgenic mice. Arterioscler Thromb 1992; 12:736–744.

54. Jiang X-C, Masucci-Magoulas L, Ma J, Lin M, Walsh A, Breslow JL, Tall A. Down regulation of LDL receptor mRNA in CETP transgenic mice: mechanism to explain accumulation of lipoprotein B particle. Circulation 1993; 88:I-421.

55. Mahley R. Apolipoprotein E: Cholesterol transport protein with an expanding role in cell biology. Science 1988; 240:622–630.

56. Davignon J, Gregg R, Sing C. Apolipoprotein E polymorphism and atherosclerosis. Arteriosclerosis 1988; 8:1–21.

57. Hixson JE and the PDAY Research Group. Apolipoprotein E polymorphisms affect atherosclerosis in young males. Arterioscler Thromb 1991; 11:1237–1244.

58. Kervinen K, Savolainen MJ, Salokannel J, Hynninen A, Heikkinen J, Ehnholm C, Koistinen MJ, Kesaniemi YA. Apolipoprotein E and B polymorphisms—longevity factors assessed in nonagenarians. Atherosclerosis 1994; 105:89–95.

59. Strittmatter WJ, Saunders AM, Schmechel D, Pericak-Vance M, Enqhild J, Salvesen GS, Roses AD. Apolipoprotein E: high-avidity binding to beta-amyloid and increased frequency of type 4 allele in late-onset familial Alzheimer disease. Proc Natl Acad Sci USA 1993; 90:1977–1981.

60. Corder EH, Saunders AM, Strittmatter WJ, Schmechel DE, Gaskell PC, Small GW, Roses AD, Haines JL, Pericak-Vance MA. Gene dose of apolipoprotein E type 4 allele and the risk of Alzheimer's disease in late onset families. Science 1993; 261:921-923.

61. Simonet WS, Bucay N, Lauer SJ, Wirak DO, Stevens ME, Weisgraber KH, Pitas RE, Taylor JM. In the absence of a downstream element, the apolipoprotein E gene is expressed at high levels in kidneys of transgenic mice. J Biol Chem 1990; 265:10809-10812.

62. Smith JD, Plump AS, Hayek T, Walsh A, Breslow JL. Accumulation of human apolipoprotein E in the plasma of transgenic mice. J Biol Chem 1990; 265:14709-14712.

63. Shimano H, Yamada N, Katsuki M, Shimada M, Gotoda T, Harada K, Murase T, Fukazawa C, Takaku F, Yazaki Y. Overexpression of apolipoprotein E in transgenic mice: Marked reduction in plasma lipoproteins except high density lipoprotein and resistance against diet-induced hypercholesterolemia. Proc Natl Acad Sci USA 1992; 89:1750-1754.

64. Shimano H, Yamada N, Katsuki M, Yamamoto K, Gotoda T, Harada K, Shimada M, Yazaki Y. Plasma lipoprotein metabolism in transgenic mice overexpressing apolipoprotein E. J Clin Invest 1992; 90:2084-2091.

65. van der Maagdenberg AMJ, Hofker MH, Krimpenfort PJA, De Bruijn I, van Vlijmen B, van der Boom H, Havekes LM, Frants RR. Transgenic mice carrying the apolipoprotein E3-Leiden gene exhibit hypercholesterolemia. J Biol Chem 1993; 268:10540-10545.

66. Fazio S, Lee Y, Sheng Z, Rall SC Jr. Type III hyperlipoproteinemic phenotype in transgenic mice expressing dysfunctional apolipoprotein E. J Clin Invest 1993; 92:1497-1503.

67. Piedrahita JA, Zhang SH, Hagaman JR, Oliver PM, Maeda N. Generation of mice carrying a mutant apolipoprotein E gene inactivated by gene targeting in embryonic stem cells. Proc Natl Acad Sci USA 1992; 89:4471-4475.

68. Zhang SH, Reddick RL, Piedrahita JA, Maeda N. Spontaneous hypercholesterolemia and arterial lesions in mice lacking apolipoprotein E. Science 1992; 258:468-471.

69. Plump AS, Smith JD, Hayek T, Aalto-Setälä K, Walsh A, Verstuyft JG, Rubin EM, Breslow JL. Severe hypercholesterolemia and atherosclerosis in apolipoprotein E-deficient mice created by homologous recombination in ES cells. Cell 1992; 71:343-353.

70. Plump AS, Forte TM, Eisenberg S, Breslow JL. Atherogenic β-VLDL in the apo E-deficient mouse: Composition, origin, and fate. Circulation 1993; 88:I-2.

71. Chiesa G, Hobbs HH, Koschinsky ML, Lawn RM, Maika SD, Hammer RE. Reconstitution of lipoprotein(a) by infusion of human low density lipoprotein into transgenic mice expressing human apolipoprotein(a). J Biol Chem 1992; 267:24369-24374.

72. Linton MF, Farese RV Jr, Chiesa G, Grass DS, Chin P, Hammer RE, Hobbs HH, Young SG. Transgenic mice expressing high plasma concentrations of human apolipoprotein B100 and lipoprotein(a). J Clin Invest 1993; 92:3029-3037.

73. Callow MJ, Stolzfus LF, Lawn RM, Rubin EM. Expression of human apolipoprotein

B and assembly of lipoprotein(a) in transgenic mice. Proc Natl Acad Sci USA 1994; 91:2130-2134.

74. Paigen B, Mitchell D, Reue K, Morrow A, Lusis A, LeBoeuf RC. Ath-1, a gene determining atherosclerosis susceptibility and high density lipoprotein levels in mice. Proc Natl Acad Sci USA 1987; 84:3763-3767.

75. Rubin EM, Krauss RM, Spangler EA, Vestuyft JG, Clift SM. Inhibition of early atherogenesis in transgenic mice by human apolipoprotein AI. Nature 1991; 353:265-267.

76. Schultz JR, Verstuyft JG, Gong EL, Nichols AV, Rubin EM. Protein composition determines the anti-atherogenic properties of HDL in transgenic mice. Nature 1993; 365:762-764.

77. Warden CH, Hedrick CC, Qiao J-H, Castellani LW, Lusis AJ. Atherosclerosis in transgenic mice overexpressing apolipoprotein A-II. Science 1993; 261:469-472.

78. Marotti KR, Castle CK, Boyle TP, Lin AH, Murray RW, Melchior GW. Severe atherosclerosis in transgenic mice expressing simian cholesteryl ester transfer protein. Nature 1993; 364:73-74.

79. Lawn RM, Wade DP, Hammer RE, Chiesa G, Verstuyft JG, Rubin EM. Atherogenesis in transgenic mice expressing human apolipoprotein(a). Nature 1992; 360: 670-671.

80. Nakashima Y, Plump AS, Raines EW, Breslow JL, Ross R. Apo E-deficient mice develop lesions of all phases of atherosclerosis throughout the arterial tree. Arterioscler Thromb 1994; 14:133-140.

81. Reddick RL, Zhang SH, Maeda N. Atherosclerosis in mice lacking ApoE. Arterioscler Thromb 1994; 14:141-147.

82. Plump AS, Scott CJ, Breslow JL. Human apolipoprotein A-I gene expression increases high density lipoproteins and suppresses atherosclerosis in the apolipoprotein E-deficient mouse. Proc Natl Acad Sci USA 1994; 91:9607-9611.

83. Palinski W, Ord VA, Plump AS, Breslow JL, Steinberg D, Witzum JL. Apolipoprotein E-deficient mice are a model of lipoprotein oxidation in atherogenesis: Demonstration of oxidation-specific epitopes in lesions and high titers of autoantibodies to malondialdehyde-lysine in serum. Arterioscler Thromb 1994; 14:606-616.

14

Homologous Recombination and Growth Factors

Marcia M. Shull, Ronald J. Diebold, Michael Eis, Greg Boivin, Ingrid L. Grupp, and Thomas Doetschman
University of Cincinnati College of Medicine
Cincinnati, Ohio

INTRODUCTION

Growth factors, polypeptide signaling molecules that play a prominent role in intercellular communication, are critical regulators of cell growth, proliferation, differentiation, and survival (reviewed in Ref. 1). Growth factors include both soluble and membrane-anchored polypeptides acting in an autocrine or paracrine manner (1,2). Cytokines that specifically regulate hematopoiesis and immune cell function represent a special functional class of growth factors that will not be covered in this chapter. Growth factors exert their effects by binding to specific cell-surface receptors, triggering a cascading network of events that transduce the extracellular signal to the nucleus, resulting in altered gene expression and the characteristic growth factor response (reviewed in Refs. 3–6). Activation or modulation of the activity of various intracellular signal-transducing molecules, often by phosphorylation/dephosphorylation, is a central theme in the mechanism by which growth factors effect a response (reviewed in Refs. 3 and 6).

Growth Factors and Growth Factor Receptor Families

Growth factors and their receptors can be organized into families based on structural homologies (reviewed in Refs. 2 and 6). Growth factor receptor families (and their ligands) that may be important in cardiovascular physiology and cell biology include (2,6):

> The platelet-derived growth factor (PDGF) family—PDGF, macrophage colony-stimulating factor-1, vascular endothelial growth factor, and mast cell growth factor/steel factor

The epidermal growth factor (EGF) receptor family—EGF, transforming growth factor-α (TGF-α), amphiregulin, heparin binding EGF-like factor, and schwannoma-derived growth factor

The insulin receptor family—insulin, insulin-like growth factor-I, and insulin-like growth factor-II

The transforming growth factor-β (TGF-β) family—TGF-βs, bone morphogenetic proteins, inhibin, activin, and Müllerian inhibiting substance

The fibroblast growth factor (FGF) family—FGF-1/acidic FGF, FGF-2/basic FGF, FGF-3/int-2, FGF-4/hst/K-FGF, FGF-5, FGF-6, and keratinocyte growth factor

The neurotrophin receptor family—nerve growth factor, brain-derived neurotrophic factor, neurotrophin-3, and neurotrophin-4)

In addition to these classic growth factors, the vasoactive peptides—angiotensin II and endothelin—appear to be locally produced in the heart and vascular system and function as cardiovascular growth factors (7–9).

Growth Factors in Cardiovascular Biology

A wealth of studies of growth factor expression and function suggest that many of the factors listed above may play important roles in cardiac and vascular system morphogenesis; in cardiovascular disease processes, either as components mediating the pathology of the disease or as agents ameliorating the disease process; and in the characteristic and often deleterious responses to therapeutic intervention in cardiovascular disorders.

Growth Factors in Cardiac and Vascular System Morphogenesis

Growth factors may play a major role in ontogeny of the cardiovascular system, as evidenced by the temporal and spatial patterns of expression of growth factors and their receptors during embryogenesis and during early postnatal development. The biological activities of growth factors in in vitro models of cardiac and vascular morphogenesis also support the concept that these factors function in development of the cardiovascular system. In particular, the TGF-βs, FGFs, insulin-like growth factors, and neurotrophins may participate in cardiovascular morphogenesis.

The TGF-βs are multifunctional growth factors that influence proliferation and differentiation of numerous cell types and regulate production and degradation of extracellular matrix (reviewed in Ref. 10). Expression of TGF-β isoforms has been extensively studied during embryogenesis, and, in particular, during cardiogenesis (11–17). In the 7.0–7.5-day mouse embryo, TGF-β1 and -β2 mRNAs are expressed in the cardiogenic plate region, the location of presumptive myocardial cells (11,16). By 8.0 days postcoitus (p.c.), two distinct cell layers, the endocardium and myocardium, separated by an acellular matrix, the cardiac jelly, are present in the primitive heart tube (see Ref. 16). At this time, TGF-β1 mRNA is expressed in the endocardial, and TGF-β2 in the myocardial, cells of the primitive heart tube (11,16). At approximately 8.5 days p.c., pronounced cardiac jelly persists only in the atrioventricular junction and outflow tract (see Ref. 11). At about 9 days p.c., mesenchymal cells derived from the endocardium begin invading the underlying cardiac jelly in the atrioventricular canal and proximal outflow tract region, forming the cardiac cushion tissue (see Refs. 11,13). This tissue will contribute to formation of the valves and membranous septa (see Refs. 11,13,16). During this period, TGF-β1 mRNA expression in the heart is limited to the endothelium of the ventricular trabeculae and to endothelial cells overlying the cardiac jelly and cushion of tissue of the outflow tract and atrioventricular junction (11,13,16). During this same period, TGF-β2 mRNA is localized and up-regulated in the myocardium under the cardiac jelly and cushion tissue of the atrioventricular junction and outflow tract (11,13).

At approximately 10.5 days p.c., the atrioventricular cushions begin to fuse, contributing to atrial and ventricular septation, and at this time TGF-β1 and TGF-β2 mRNAs are limited to this region: TGF-β1 is localized to the endothelium and TGF–β2 to the myocardium (11,13,16). By 11.5 days p.c., TGF-β2 in the AV and outflow tract is restricted to a narrow band of cells adjacent to the cushion tissue, perhaps contributing to the formation of the atrioventricular conduction system (11). High levels of TGF-β1 mRNA persist in the endothelia of heart valves up to 1 week postpartum (16). TGF-β3 expression in the embryonic heart is limited, and has been detected only in cells around the outflow tract at 8.5–9.0 days p.c., in the AV cushions at 11.5 days, and in mesenchyme of the valves at 14.5–16.5 days (11,13,19).

Immunohistochemical studies have confirmed that TGF-β proteins are expressed in the developing heart (11,12,14–18). TGF-β1 immunoreactivity is first detected in the mouse heart at 8.5 days p.c. (12). Intense immunostaining for TGF-β1 protein is observed in the mouse heart (intracellularly in the myocardium and extracellularly in the endocardium and cardiac jelly) at all stages examined, from 9.0 to 10.5 days gestation, with the intensity of staining intensifying as heart development progresses (12). TGF-β1 immunoreactivity persists in the adult heart in the valve leaflets and the base of the cusps (16). TGF-β2 immunostaining is initially detected in the heart at 8.25 days p.c. (11,12). TGF-β2 immunoreactivity, which is localized to the myocardium, increases from day 8.5 to 12.5 and persists in the adult (11). TGF-β2 immunostaining is also detected in endothelial cells of the dorsal aortae at 9.5 days p.c. (12). Immunoreactivity for TGF-β3 has been detected in developing mouse heart, in particular, in the ventricular myocardium at 9.0 days p.c., in the pericardium at 10.0 days, and in the ventricles and atria at 12.5 days (12,18). All three TGF-β isoform mRNAs and proteins are expressed in blood vessels during embryonic development (13,16,18–20). Based on these localization studies, it has been suggested that the TGF-βs may play major roles in cardiac morphogenetic processes, particularly in inductive interactions involved in cardiac cushion formation and in subsequent septation and valve formation (11–14,16).

The TGF-βs may also modulate postnatal cardiac growth and development (21). Neonatal growth of the rodent heart consists of three phases: enlargement due to cardiomyocyte hyperplasia in the fetal period and first few days of the neonatal period; a transition from hyperplastic to hypertrophic growth, with karyokinesis without cytokinesis, at approximately 6 to 14 days postpartum; and an increase in heart mass resulting from hypertrophic growth occurring at about 14 to 21 days postpartum (22). TGF-β1 mRNA and immunoreactive protein are expressed in the neonatal and adult myocardium, with transcript levels becoming near maximal at about 7 to 14 days postpartum, the period of transition from hyperplastic to hypertrophic growth (21,23). TGF-β1 transcripts continue to be expressed into adulthood (21,23). Thus, it has been proposed that TGF-β1 may participate in regulation of neonatal myocyte maturation, including termination of myocyte proliferation and maintenance of the nonproliferative state of the mature cardiomyocyte (21).

In vitro studies substantiate a role of TGF-βs in cardiogenesis and vasculogenesis (reviewed in Refs. 24 and 25). TGF-β1 modulates the expression of cardiac-specific genes in cultured neonatal cardiomyocytes (reviewed in Ref. 24). In primary cultures of fetal and neonatal cardiomyocytes, TGF-β1 inhibits stimulation of DNA synthesis in response to IGF-I and IGF-II (21). TGF-β1 enhances cardiac differentiation of precardiac mesodermal explants from axolotl (26). TGF-β1, -β2, and -β3 maintain the beating rate of neonatal cardiac myocytes in serum-free medium and prevent inhibition of beating of these cells induced by IL-1β (27). Anti-TGF-β antibodies and antisense TGF-β3 oligodeoxynucleotides inhibit the epithelial–mesenchymal transformation of cells in the atrioventricular canal that give rise to the valves and membranous septa (28,29). TGF-β1 inhibits endothelial cell proliferation and migration, although it can stimulate endothelial formation of branching tubular structures in vitro and stimulates angiogenesis in vivo (reviewed in Ref. 25). Thus, both localization studies and studies of biological activities implicate the TGF-βs in development of the cardiovascular system.

Fibroblast growth factors induce proliferation and differentiation of cells of mesodermal and neuroectodermal origin (referenced in Ref. 30). FGF-1 and FGF-2 mRNAs and proteins are expressed in developing and adult vertebrate heart (31–39). FGF-1 and FGF-2 proteins are expressed in a superimposable pattern in embryonic (11–20 days gestation) and postnatal (1–35 days) rat heart at all stages examined (31–33). FGF-1 and FGF-2 immunoreactive proteins are present in cardiac myocytes from the earliest stages (day 11 p.c.) through the postnatal period (31,32). FGF-1 and FGF-2 are observed in mesenchymal cells of the cardiac cushions from the time when the cushion tissue is first recognizable and in endothelium and connective tissue of the valves from the time of morphogenesis throughout the postnatal period (31,32). Capillary endothelium exhibits immunostaining throughout development and in the postnatal period (31,32,34). The medial layer of the aorta and large arteries develops during embryonic life in the rat; immunostaining of the smooth-muscle cells in these vessels is intense during early development, but declines at later stages, although it persists throughout the postnatal period examined (up to 5 weeks) (31,34). In contrast, the media of the intramural coronary arteries develops during the neonatal period in the rat, and in these vessels immunostaining of the smooth-muscle cells is intense immediately after birth, then diminishes after the first week of life (31).

Transcripts for FGF-1 and the FGF receptor, the product of the *flg* gene, which are relatively abundant in the fetal rat heart, decrease dramatically during the perinatal period of transition from hyperplastic to hypertrophic growth, but then reappear, primarily in capillary endothelial cells, during the second to third and fifth to seventh weeks after birth (33). This reappearance of Flg and FGF-1 corresponds to the period of ventricular remodeling characterized by extracellular matrix formation and capillary angiogenesis which is necessary for oxygenation of the adult myocardium (33). Based on these studies, it has been suggested that FGF-1 and FGF-2 play an important role in morphogenesis of heart and vessels, in particular, regulation of cardiomyocyte proliferation, formation and maintenance of valvular tissue, differentiation of ventricular capillary endothelium, and development of smooth muscle of large and small arteries (31–34).

Studies of the biological activities of the FGFs also support the concept that these molecules may be involved in cardiovascular system development. FGF-1 and FGF-2 stimulate endothelial cell proliferation and migration (reviewed in Refs. 37, 40, and 41). Both factors stimulate angioblast outgrowth and vascular cord formation in ES cell-derived embryoid bodies (41). FGF-1 stimulates proliferation of cultured neonatal cardiac myocytes (42). In an in vitro cardiogenic system, oligodeoxynucleotides complementary to FGF-2 mRNA inhibit the proliferation and contractility of cardiomyocytes arising from cultured chick embryo precardiac lateral plate mesoderm (30). All of these studies provide data that are consistent with a role of FGF-1 and FGF-2 in cardiac and vascular system morphogenesis.

The insulin-like growth factors (IGFs), polypeptides that are structurally similar to proinsulin, exert insulin-like metabolic effects when present at high concentrations and exhibit effects on cell proliferation and differentiation when present at nanomolar concentrations (reviewed in Ref. 43). Transcripts for both IGF-I and IGF-II are expressed in rodent heart during embryogenesis, although IGF-II is the predominant form expressed during prenatal development (44–49). IGF-I mRNA is expressed in the midgestation rat embryo in the epicardium and in areas of tissue remodeling, specifically mesenchymal or fibroblast cells of the truncus arteriosus, endocardial cushions, and forming valves (44). High levels of IGF-II mRNA are detected in the heart and vasculature during development (44–47). In the early mouse embryo (embryonic day 7.5), IGF-II mRNA is first observed in the anterior proximal mesoderm, the location of presumptive heart cells (45). By embryonic day 8, intense hybridization, as well as immunostaining, is observed in the heart (45). IGF-II mRNA levels are particularly high in the developing vascular system of the central nervous system, including the choroid plexus, organum vasculosum, adventitia of the cerebral arteries, and vessels of the trigeminal ganglia (44,46,47). During the neonatal period in rodents, transcripts for both IGF-I and IGF-II are detected in the ventricle, although levels of IGF-II are higher than those of IGF-I; for both IGF-I and IGF-II, transcript levels are highest during the first few days postpartum, corresponding to the final phase of cardiomyocyte proliferation (50). Both IGF-I and IGF-II mRNAs have been detected in adult rat heart, although IGF-II expression in the adult is primarily confined to the choroid plexus and leptomeninges (see Refs. 45 and 51). Transcripts for the type 1 IGF receptor are of relatively low abundance in the heart during embryonic development (44).

In in situ hybridization analyses of IGF type 2 receptor (cation-independent mannose-6-phosphate receptor) expression during rodent embryogenesis, the strongest hybridization signals were observed in the heart and major vessels (52,53). Transcripts are detected in the embryonic mouse heart at day 9.5, the earliest stage examined, and persist in the heart through embryonic day 17.5 (52). Levels of IGF type 2 receptor protein, determined by immunohistochemistry, become intense in rodent heart and vessels by midgestation (45,53). Expression of mRNAs for the type 1 and type 2 IGF receptors are also detected in the neonatal heart, with the highest levels occurring during the first few postnatal days (48,50,54). Based on these observations, it has been suggested that the IGFs may participate in cardiogenesis and vasculogenesis in the embryo and may influence growth and maturation of the heart during the neonatal period of transition from hyperplastic to hypertrophic growth (44,45,50,54).

Neurotrophins are growth factors produced in target tissues that influence the survival and differentiation of neurons in those tissues (reviewed in Ref. 55). Some members of the neurotrophin family of growth factors and their receptors are expressed in developing and adult heart (56–61). In developing rodent heart, nerve growth factor (NGF) mRNA increases rapidly from embryonic day 17 (the earliest stage examined), reaching maximal levels 14 days after birth, then stabilizing at the relatively high levels observed in the adult (56). NGF protein is detected as early as embryonic day 12 in the mouse (the earliest stage examined) and increases to reach maximal prenatal levels at embryonic day 14 (57). A transient peak of NGF protein occurs between 4 and 14 days after birth, after which NGF levels remain relatively stable up to 1.5 years of age (57). In a study of NGF mRNA expression in adult human tissues, heart exhibited the highest levels among the tissues examined (62). Transcripts for the low-affinity NGF receptor, p75^{LNGFR}, are also abundant in adult heart (62). In the rodent embryonic and early neonatal period, immunoreactivity for p75^{LNGFR} is intense in the adventitia surrounding blood vessels (58). Transcripts for the high-affinity NGF receptor, p140^{trkA}, and the related receptor, p145^{trkB}, are limited primarily to cells of the peripheral and central nervous system, while p145^{trkC} expression has been detected in vessels of the central nervous system (61,63,64). The observed developmental pattern of NGF expression correlates with the time course of initial sympathetic innervation, subsequent sympathetic neuron differentiation, and cessation of sympathetic cell death in the heart (see Ref. 56). Thus, NGF in the heart may influence proper levels of sympathetic innervation and sympathetic differentiation in this tissue during development (56,57).

Growth Factors and Cardiovascular Disease

Expression of growth factors in the adult heart and vascular system and the altered patterns of expression that are frequently observed during pathological processes suggest that these molecules may play a role in normal cardiovascular physiology and in cardiovascular disease. In some cases, perturbed growth factor or growth factor receptor expression may be a factor in the etiology of the disease, while in other situations altered expression may be secondary to the disease process but may exacerbate or mitigate the disease phenotype. Several representative examples of growth factor expression in cardiovascular disease are described in the following paragraphs.

Cardiac hypertrophy occurring in disease situations such as hypertension represents an adaptive response of the heart to increased hemodynamic load and is characterized by an increase in size of the mature cardiomyocyte and, in rodents, by induction of the fetal program of cardiomyocyte contractile protein gene expression (reviewed in Refs. 42 and 65–68). It has been proposed that growth factors may mediate this characteristic response to load (reviewed in Refs. 24 and 65–68). The roles of the TGF-βs and FGFs in this process have recently been reviewed (24,65–67). TGF-β1 and FGF-2 in vitro induce the fetal pattern of gene expression characteristic of hemodynamic overload, and expression of both growth factors is up-regulated during hypertrophy induced in vivo by aortic constriction or as a result of infarction, suggesting that these factors are involved in the hypertrophic response (see Refs. 24, 42, 66, and 68).

Locally produced angiotensin II may play a major inductive role in load-induced hypertrophy. Angiotensin II promotes hypertrophy of cultured neonatal rat cardiac myocytes, an effect mediated by the angiotensin II type 1 receptor; mechanical stretch of cultured cardiac myocytes causes release of angiotensin II, which mediates the stretch-induced hypertrophic response of these cells; and angiotensin converting enzyme inhibitors prevent cardiac hypertrophy in both animal models of ventricular hypertrophy and in patients with myocardial infarction (see Refs. 7 and 8 and references therein). These results suggest that angiotensin II is a critical mediator of the stretch-induced hypertrophic response (7,8).

The elevation of serum IGF-I levels observed in acromegalic patients and the common occurrence of cardiac hypertrophy in these patients has implicated IGF-I in the development of cardiac hypertrophy (referenced in Refs. 69 and 70). In several rodent models of hypertension, IGF-I mRNA and protein levels increased significantly in ventricular muscle undergoing pressure overload hypertrophy (70,71). Furthermore, treatment of cultured neonatal rat cardiomyocytes with IGF-I induces hypertrophy, as evidenced by an increase in cell-surface area and induction of muscle-specific gene expression (69). Based on these observations, it has been proposed that IGF-I may participate in the initiation of ventricular hypertrophy in response to pressure overload (69–71).

Growth factor expression during cardiac ischemia and infarction has been extensively investigated (reviewed in Refs. 24 and 65–67). TGF-β1, FGF-2, and IGF-I are elevated in ischemic or infarcted heart, especially at the border of the infarcted area (24,66,72–77). TGF-β1 appears to prevent cardiac damage resulting from reperfusion of ischemic myocardium, possibly by inhibiting tumor necrosis factor release (78,79). Collateral vessel growth after coronary artery occlusion may ameliorate damage due to ischemia (see Ref. 75). Both FGF-2 and vascular endothelial growth factor increase collateral artery growth in experimentally induced ischemia or infarction in dogs, suggesting that these growth factors may be useful therapeutic agents in treating human ischemia (75,80).

Atherosclerosis is a major contributing factor in infarction (see Ref. 81). Injury to the endothelium and abnormal proliferation of intimal smooth-muscle cells are prominent events in the development of atherosclerosis (discussed in Ref. 81). Growth factors (including PDGF, TGF-β1, FGF-2, IGF-I, IGF-II, and EGF) are present in many of the cell types involved in formation of the atherosclerotic plaque, including endothelial cells, smooth-muscle cells, macrophages, and platelets (67,81–84). Furthermore, these factors exhibit biological activities that may be relevant to plaque formation, including stimulation or inhibition of endothelial and smooth-muscle cell proliferation, stimulation of cell migration, modulation of vascular smooth-muscle cell gene expression, regulation of extracellular matrix production, and stimulation or suppression of production of tissue plasminogen activator (reviewed in Refs. 67,81,82,84). Perturbations in the balance between stimulatory and inhibitory actions of growth factors may be involved in the development of atherosclerotic lesions (81,83,84).

Vascular complications including atherosclerosis and proliferative retinopathy are common in diabetes mellitus (see Ref. 85). Since FGF-2 stimulates endothelial proliferation and angiogenesis, the expression of this growth factor in diabetes is of interest. FGF-2 mRNA levels are elevated in several tissues of rats with streptozocin-induced diabetes relative to the levels observed in control animals (85), and FGF-2 protein is increased in the vitreous of diabetic patients with proliferative retinopathy, particularly those with active neovascularization (40). These observations are consistent with an involvement of FGF-2 in the development of vascular complications occurring in diabetes (40,85).

Nerve growth factor may play a role in the pathogenesis of hypertension (86–89 and references therein). Enhanced sympathetic innervation and elevated sympathetic activity of resistance arteries is observed in the spontaneously hypertensive rat as compared to its normotensive control (reviewed in Refs. 86–89). Since sympathetic activity exerts a trophic effect on the vasculature, it has been proposed that the peripheral sympathetic nervous system plays a primary role in the observed vessel wall thickening and development of hypertension in these animals (86,89). Density of sympathetic innervation correlates with the production of NGF in the target organ (see Ref. 86). NGF content is elevated in the mesenteric vasculature of the spontaneously hypertensive rat compared to the normotensive control (86–88). Sympathectomy of neonatal SHR rats using anti-NGF antibodies and guanethidine prevents smooth-muscle hyperplasia in the reactive muscular arteries and completely prevents the development of hypertension in these animals (89). From these observations, it has been suggested that elevated NGF may be responsible for the heightened sympathetic innervation of the vasculature of the SHR animal and may directly contribute to the development of hypertension in this animal model (86–89).

Growth factor involvement in more obscure cardiovascular disorders has also been suggested. For example, carcinoid heart disease, characterized by subendocardial fibrotic lesions which are usually located in the atrial wall and valves, is a serious complication of carcinoid tumor, a neurendocrine tumor of the gastrointestinal tract (90). Enhanced expression of TGF-β1 and TGF-β3 proteins in fibroblasts of the carcinoid plaques relative to fibroblasts in normal subendocardium suggests that the TGF-bs may be involved in extracellular matrix protein production and deposition in the carcinoid lesions (90).

Growth Factors in Response to Therapeutic Intervention

Therapeutic intervention, including both surgical procedures and drug therapy, has dramatically increased the survival of patients suffering from various cardiovascular diseases. Altered growth factor expression as a result of these interventions may play a role in the subsequent success or failure of these therapies.

Growth factors may be involved in both the survival and rejection of cardiac allografts (91,92). TGF-β administration tends to prolong the survival of rodent heart allografts (91,93). On the other hand, a major complication of cardiac transplantation is cardiac allograft vasculopathy characterized by smooth-muscle proliferation and occlusion of the donor coronary arteries (see Ref. 92), and evidence suggests that growth factors, such as TGF-β, TGF-α, FGF-2, or PDGF, produced by the donor endothelium in response to allogeneic lymphocytes may induce smooth-muscle proliferation, contributing to the coronary artery occlusion (92).

Artery restenosis, characterized by proliferation of vascular smooth-muscle cells and synthesis of extracellular matrix resulting in neointimal thickening, is a common complication following balloon angioplasty (see Ref. 94). Given the stimulatory effects of growth factors on smooth-muscle cell mitogenesis and extracellular matrix production, the role of these factors in repair of arterial injury and restenosis is of interest. In a rat model of balloon catheter–induced arterial injury, expression of TGF-β1 mRNA and protein is induced in the neointima of the injured artery (94,95). The temporal pattern of TGF-β1 mRNA expression correlates with the initiation and maintenance of smooth-muscle cell proliferation and the stimulation of extracellular matrix synthesis (94). Furthermore, administration of TGF-β1 to rats following arterial injury elicits an increase in smooth-muscle cell proliferation in the neointima (94). Elevated TGF-β1 mRNA and protein levels have also been detected in restenotic lesions from patients treated by balloon angioplasty (96). Expression of IGF-I mRNA and immunoreactivity is induced in the vessel wall following catheter-induced injury (97,98), and PDGF is produced by platelets and intimal smooth-muscle cells following arterial injury (reviewed in Refs. 97 and 98). Taken together, these observations suggest that growth factors, including TGF-β1, IGF-I, and PDGF, may participate in the initiation and perpetuation of arterial repair and restenosis subsequent to balloon angioplasty.

The wealth of information concerning patterns of growth factor and growth factor receptor expression in the cardiovascular system during embryogenesis and during normal and pathological conditions in the adult, as well as the numerous studies of the biological activities of growth factors in vitro and in vivo, suggest that these factors play a cardinal role in the development and functioning of the cardiovascular system. In the following sections, we discuss the targeted disruption of growth factors, growth factor receptors, and growth factor signaling molecules as they relate to the cardiovascular system, with particular emphasis on TGF-βI.

TARGETED DISRUPTION OF GROWTH FACTORS AND GROWTH FACTOR RECEPTORS

To date, several reports of animals exhibiting a null mutation in genes encoding growth factors or their receptors have been published. These include TGF-β1, α-inhibin, TGF-α, int-2, IGF-I, IGF-II, the IGF type 1 receptor, NGF, brain-derived neurotrophic factor (BDNF), p75^{LNGFR}, trkB, Wnt-1, and ciliary neurotrophic factor (CNTF) (99–116). Based on localization of expression (or lack thereof) and biological activity, several of these growth factors would not be expected to show a cardiovascular phenotype. Among these are the int-1 and int-2 gene products, α-inhibin, and CNTF; not unexpectedly, no cardiovascular-related phenotypes are reported in animals lacking these factors (101,104,114–116). Other growth factors, such as the neurotrophins, although not generally considered cardiovascular growth factors, may exhibit effects on the cardiovascular system by affecting innervation of heart and vasculature or influencing cardiovascular function as a result of modulation of sympathetic or parasympathetic activity. Growth factors or growth factor receptors that are expressed in cardiovascular tissues or exhibit biological activity relevant to cardiovascular function that have been targeted include TGF-b1, TGF- a, IGF-I, IGF-II, IGF type 1 receptor, NGF, BDNF, the low-affinity receptor for NGF, and TrkB (99,100,102,103,105–111, 113). Phenotypes of animals lacking these growth factors or receptors are described below.

Targeted Disruption of Transforming Growth Factor- β1

The TGF-βs are multifunctional growth factors exhibiting pleiotropic effects on growth and differentiation of numerous cell types. Studies of expression in the developing heart and biological activities in in vitro systems of cardiogenesis and in studies of cardiovascular system function suggest that these factors are cardinal regulators of cardiovascular system morphogenesis and function. To more fully understand the role of TGF-β1 in the context of the whole animal, the corresponding gene has been ablated in murine embryonic stem cells via homologous recombination and animals homozygous for the null allele have been generated (99,100). Surprisingly, at birth TGF-β1-deficient neonates are indistinguishable from their littermates. However, the actual number of live-born homozygous mutants is approximately one-half the expected number, assuming that the TGF-β1 null allele is recessive, as indicated by the otherwise normal phenotype of the heterozygous animals. The occurrence of live-born mutants may suggest that TGF-β1 does not play an essential role in development and calls into question interpretations based on expression analysis. However, it has been cautioned that this conclusion may be premature because this factor may be accessible to the developing embryo or neonate via the placenta or milk (99).

Mice homozygous for the null allele invariably develop a wasting syndrome that culminates in death around the time of weaning, at approximately 3–4 weeks of age (99,100). Histological and immunohistochemical analysis reveals multiorgan inflammation with the heart being a primary target of infiltrating leukocytes, which consist primarily of T cells (Figure 1). Of all mutants examined, 90% exhibited varying degrees of cardiac inflammation, and some of the older mutants (>30 days) exhibited fibrosis. The mutant phenotype is not observed in wild-type or heterozygous littermates. Thus, it

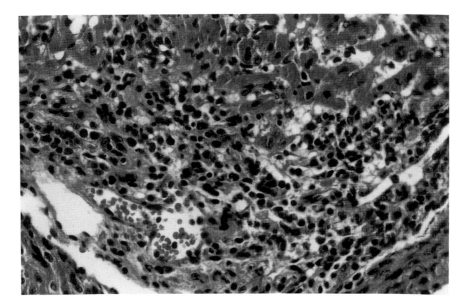

Figure 1 Histological analysis of TGF-β1-deficient heart ventricle. Inflammatory cells consist mostly of lymphocytes (identified immunohistochemically as T cells). A small number of neutrophils and macrophages are also present. Tissues were stained with hematotoxylin and eosin. Magnification 400×.

appears that the cardioprotective role of TGF-β1 is a vital physiological function.

The cellular and molecular mechanisms that give rise to cardiac inflammation at such a young age in the TGF- β1 null animals are not presently clear. The proinflammatory cytokines, tumor necrosis factor-α (TNF-α), interleukin-1β (IL-1β), and interferon-γ, are elevated in some tissues in the mutants (99) and may result in increased expression of major histocompatibility complex (MHC) and cell adhesion molecules (CAMs) in the vasculature (117,118). This in turn would enhance leukocyte activation and extravasation similar to that observed in cardiac transplant rejection and in reperfusion injury following ischemia. MHC II expression is elevated in the cardiac compartment of the mutant animals (Figure 2). However, IFN-γ levels may not be elevated prior to or at the time of increased MHC II expression (119), suggesting that elevated MHC II levels may occur independently from inflammatory cytokines. The cardiac inflammatory phenotype suggests that the TGF-β1-deficient mouse

Figure 2 Immunohistological analysis of TGF-β1 heart showing elevated expression of MHC II antigen (arrowheads). Sections were stained with biotinylated antimurine-IA^k antibodies followed by avidin-conjugated horseradish peroxidase. 3,3'-diaminobenzidine (DAB) was used as substrate. Sections were counterstained with hematotoxylin and eosin. Magnification 200×.

may be an appropriate model for testing therapeutic compounds for prevention of reperfusion injury and graft rejection.

Physiological analyses have been carried out on cardiac tissue derived from the TGF-β1 mutants. Contractility measurements on cardiac muscle strips from wild- type mutant mice reveal a depressed response to β–adrenergic stimulation (Figure 3). It has been shown previously that TNF-α and IL-1 are potent inhibitors of cardiac contractile responsiveness to β-adrenergic stimulation (120). Consequently, the impaired cardiac contractility observed in TGF- β1-deficient animals may be mediated, at least in part, by inflammatory cytokines.

Previous in vitro studies have shown that addition of TGF- β1 to cultured rat neonatal cardiomyocytes induces a switch from adult to fetal isoforms of several cardiac- specific genes, a response characteristic of pressure-overload hypertrophy (reviewed in Ref. 66). Expression of cardiac-specific genes previously shown to be regulated by TGF-β1 in vitro has been examined in the TGF-β1 mutants in an attempt to further define potential cardiac abnor-

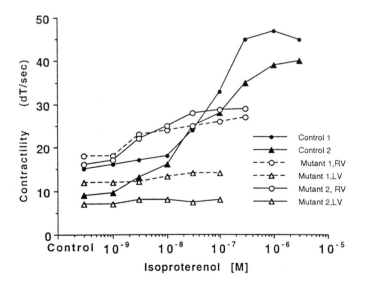

Figure 3   Isolated strips from the right ventricles (RV) or left ventricles (LV) of control or mutant animals were mounted in a bath chamber (70 ml) containing oxygenated balanced salt solution. One end of the strip is fixed to an electrode to stimulate the strip and the other end is connected to a force transducer (164). Increasing concentrations of isoproterenol are added to the bath and contractility is measured as tension an change in tension per second. In normal tissue, increasing amounts of β-adrenergic agonist resulted in increased constractility. Strips isolated from mutant hearts displayed little response to β-adrenergic stimulation.

malities independent of inflammation. RNA dot blot analysis reveals that the normal fetal to adult isoform switch for both cardiac myosin heavy chain (Figure 4) and actin (data not shown) occurs at birth in the TGF-β1-deficient animals. Although these results suggest that TGF-β1 is not essential for regulating the developmental program of these genes, assuming it is not provided transplacentally, they do not rule out the possibility that TGF-β1 may participate in regulating isoform switching.

Targeted Disruption of Transforming Growth Factor-α

Transforming growth factor-α (TGF-α), which is structurally related to EGF and interacts with the EGF receptor, is mitogenic for epithelial and mesenchymal cells (see Refs. 102 and 103). TGF-α is a potent angiogenic factor and is expressed in many solid tumors, suggesting a role in tumor angiogenesis (121). Mice homozygous for disruption of the TGF-α gene are viable and fertile, the primary defect being a waviness of the whiskers and fur and eye abnormalities (102,103). No obvious cardiovascular abnormalities were reported. Since it has been suggested that TGF-α may participate in angiogenesis associated with tumor growth, studies of angiogenic processes in these animals—including extent of neovascularization in experimentally induced tumors in the homozygous mutant animals, formation of tubular networks by cultured endothelial cells isolated from the mutants, and angiogenesis/vasculogenesis in embryoid bodies derived from the targeted ES cells—would be of interest.

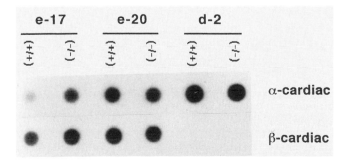

Figure 4 RNA dot blot analysis showing appropriate myosin heavy-chain isoform switch at birth in TGF-β1-null mutants. RNA was isolated from ventricles of two hearts each of mutant (-/-) and control (+/+) animals at the time points indicated: embryonic day 17 (e-17), embryonic day 20 (e-20), and postnatal day 2 (d-2). 10 μg of total RNA was applied to nitrocellulose membrane and probed with 32P labeled probes specific to the α- and β-cardiac myosin heavy-chain genes.

Targeted Disruption of Insulin-Like Growth Factors I and II and the IGF Type 1 Receptor

As indicated in the introduction, both IGF-I and IGF-II are expressed in embryonic and neonatal rodent heart during development, and IGF-II expression is high in the developing vascular system of the central nervous system. Animals homozygous for targeted disruption of IGF-I exhibit growth deficiency and variable survival (105,106). Depending on genetic background, some animals survive to adulthood, while others die shortly after birth (105). Surviving adult mutants exhibit dwarfism and infertility (105,106). Animals homozygous for IGF-II disruption also exhibit a growth deficiency, being viable and fertile proportionate dwarfs (107,108). Animals in which the IGF type 1 receptor gene has been ablated exhibit a more severe growth deficiency and die at birth due to respiratory failure (105). No specific cardiac or vascular system defects were mentioned. Given the expression of IGF-I in cardiac cushion and forming valve tissue during embryogenesis, expression of IGF-II in the developing vascular system of the CNS, and expression of both IGF-I and -II during the neonatal phase of hyperplastic to hypertrophic cardiac growth, a careful examination of the developing cardiovascular system in these animals would be warranted.

Targeted Disruption of Neurotrophins and Their Receptors

Neurotrophins influence neuronal growth, differentiation, and survival (reviewed in Ref. 55). Nerve growth factor, the prototypic neurotrophin, influences the survival of sympathetic, neural crest-derived sensory, and forebrain cholinergic neurons (see Ref. 109). NGF is expressed in embryonic, neonatal, and adult heart (56,57,62). Animals homozygous for a disruption in the NGF gene are born alive but fail to thrive and die within several weeks of life, probably as a result of abnormal feeding behavior secondary to sensory nervous-system defects (109). Histological analysis revealed that homozygous mutant animals exhibit neuronal cell loss in sympathetic and sensory ganglia (109). No other obvious abnormalities in major organs were observed (109). Specific cardiovascular effects, such as altered sympathetic innervation of the heart or perturbed function of the heart or vasculature, were not mentioned.

The low-affinity NGF receptor p75[LNGFR] has been ablated in mice by homologous recombination (111). As mentioned in the introduction, p75[LNGFR] is abundant in adult heart. Animals homozygous for disruption of the gene encoding p75[LNGFR] are viable and fertile but exhibit decreased cutaneous sensory innervation (111). In these animals, sympathetic innervation of the heart was examined and appeared qualitatively normal (112).

Brain-derived neurotrophic factor (BDNF) is a neurotrophin supporting survival of certain neuronal populations in the brain and sensory neurons derived from neural crest (see Ref. 110). Relatively high levels of BDNF mRNA are detected in adult rat heart, although expression of BDNF and its receptor, TrkB, is predominant in the central and peripheral nervous system (60,110). Mice homozygous for *BDNF* gene ablation, although born alive, usually die within 48 hours, although some survive for up to 25 days (110). Histological analysis revealed a profound deficiency in the number of cranial and spinal sensory neurons, including a loss of neurons in the petrosal and nodose ganglia (110). Sensory neurons in the petrosal-nodose ganglia transmit information from the internal organs, including heart and great vessels, to the central nervous system (see Ref. 110). This transmission is important in regulation, via the parasympathetic system, of visceral organ function, including heart rate and blood pressure (see Ref. 110). Thus, the postnatal lethality in the BDNF-deficient animals may be due in part to the loss of neurons in these ganglia (110).

Animals homozygous for disruption of the gene encoding the BDNF receptor, *trk*B, develop to birth but fail to thrive and die during the first postnatal week, apparently as a result of abnormal feeding behavior (113). In these animals, reductions in the number of facial motor neurons and lumbar spinal motor neurons, and reductions in the number of neurons in the trigeminal and dorsal root ganglia are observed; no specific cardiovascular effects were mentioned (113).

TARGETED DISRUPTION OF MOLECULES INVOLVED IN GROWTH FACTOR SIGNAL TRANSDUCTION

Growth factors elicit their characteristic responses by binding to specific cell-surface receptors. In many cases, this binding activates an intrinsic tyrosine or serine-threonine kinase activity of the receptor molecule, triggering a cascading network of events that transduce the signal from the cell surface to the nucleus (3,5,6,122). Proteins involved in the transduction pathway include protein kinases and phosphatases, phospholipases, and phosphatidyl inositol kinases (3,5,123). A major signal-transduction pathway utilized by many different growth factors is a kinase cascade referred to as the Ras pathway, which involves the Ras protein and a number of serine-threonine kinases including Raf, mitogen-activated protein kinases (MAPKs), and ribosomal S6 kinases (4,5,124). Transduction of a signal through this pathway ultimately results in phosphorylation and activation of nuclear transcription factors that regulate gene expression, including Fos, Jun, and Myc (4,124). Some growth factors may also utilize a more direct Ras-independent signaling pathway in which ligand binding triggers direct tyrosine phosphorylation of latent cytoplasmic transcription factors which then translocate to the nucleus and activate transcription (125–128). Proteins involved in cell-cycle control, including cyclins, cyclin-dependent kinases, and the retinoblastoma gene product, are also intracellular targets of growth factor–mediated responses (122,129–132).

Targeted disruption of genes encoding several molecules that may be involved in growth-factor signal transduction has been reported. These include genes encoding several nonreceptor kinases (*src, fyn, lck, csk,* and *abl*), genes encoding transcription factors (c-*myc*, N-*myc, fos,* and *jun*), and the retinoblastoma gene (133–152). For three of these, effects on the cardiovascular system were notable (137,138,141–143,145,153).

Members of the *myc* family of nuclear proto-oncogenes are induced by mitogenic stimuli and encode putative transcription factors (see Ref. 153). While c-*myc* is widely expressed in both the embryo and in the adult, N-*myc* is expressed to a greater extent in the embryo than in the adult and in a more restricted pattern (see Refs. 141, 142, and 145). Studies suggest that c-*myc* and possibly N-*myc* play a role in regulating cardiomyocyte proliferation and differentiation (reviewed in Refs. 65 and 66). Both c-*myc* and N-*myc* are down-regulated during cardiac development, coincident with the transition from hyperplastic to hypertrophic growth (see references in 65 and 66). Furthermore, in transgenic mice engineered to overexpress c-*myc* in the heart throughout development, cardiac enlargement occurs as a result of ventricular myocyte hyperplasia during fetal development (154).

Targeted disruption of c-*myc* and N-*myc* support the hypothesis that these factors are involved in cardiac development. Homozygosity for a null mutation in c-*myc* results in embryonic lethality at 9.5–10.5 days of gestation (141). Homozygous mutant embryos are retarded in development and exhibit pathological abnormalities, including enlarged heart, dilated and fluid-filled pericardium, delayed neural-tube closure, and delayed embryo turning (141). Based on this phenotype, it has been proposed that cells lacking c-*myc* can proliferate and populate the embryo initially, but are unresponsive to the growth factor–induced burst in proliferation necessary for subsequent normal development (141).

Targeted disruption of N-*myc* also results in embryonic lethality, occurring at 10.5–12.5 days of gestation (142,143). Homozygous mutant embryos are reduced in size but appear to be at the same developmental stage as their wild-type littermates (142,143). Histological abnormalities were observed in the heart, central nervous system, genitourinary system, and lung (142,143, 145). The embryonic lethality may be a consequence of abnormal cardiovascular system development (142,143). In one report, homozygous mutant animals were anemic, bled easily, had distended aortas, and exhibited reduced myocardial thickness (142). In another report, cardiac morphogenesis was dramatically and severely affected (143). Cardiac abnormalities observed included absent or underdeveloped septation and valve formation, reduced ventricular trabeculation, and dilatation of the anterior cardinal vein (143). Based on these observations, it has been proposed that the N-*myc* gene product may not be essential for initial organogenesis, but may function in subsequent development by regulating continued proliferation or differentiation (143). With respect to cardiac morphogenesis, the aberrant septation and valve formation in these animals suggest that N-*myc* may be involved in the process by which signals from the myocardium elicit an epithelial-mesenchymal transformation of cells from the endocardium, which is required for development of the valves and membranous septa (143).

Csk (C-terminal Src kinase) is a kinase that has been shown to associate with the PDGF receptor (see Ref. 155). Csk negatively regulates the activity of Src, a protein tyrosine kinase that also associates with and is activated by the PDGF receptor (see Ref. 155). Animals homozygous for a null mutation in the *csk* gene die during embryogenesis between days 9.5 and 10.5 (137,138). Biochemical studies revealed that activity of Src kinases is elevated in the mutants (137,138). Homozygous mutant mice exhibit growth retardation and neural-tube defects (137,138). With respect to the cardiovascular system, two observations are relevant. Although normal beating of the heart was observed in the 9.5-day mutant embryo, circulation in the yolk sac and in the embryo was "essentially absent" (137,138). Blood vessels of the first branchial arch were enlarged (137). The implications of these observations as related to the function, if any, of *csk* or *src* in the cardiovascular system are unknown.

DETECTION OF CARDIAC PHENOTYPES

Complex organisms require an intricate set of regulatory mechanisms that govern developmental and homeostatic processes. Maintenance of homeostasis during periods of injury and stress is essential to survival of the organism. The mild or apparent lack of phenotype of some knockout animals may be attributed to the existence of compensatory mechanisms or redundancies in these developmental and homeostatic processes (however, see Ref. 156 for an alternative interpretation). Furthermore, since laboratory animals are usually maintained in relatively stress-free conditions, some phenotypes may not be readily apparent.

In genetically altered mice, valuable information concerning potential cardiac phenotypes can be derived from descriptive studies in which morphological and molecular events occurring throughout cardiogenesis are investigated. However, if a particular genetic alteration is expected to exhibit a cardiac phenotype, yet none is detected, one must ensure that subtle effects have not been overlooked. While more detailed morphogenetic studies may reveal such effects, functional analyses may be required to better understand them as well as to expose others. Phenotypes detected by functional analyses would probably be relevant to diseases of the adult heart.

To detect subtle defects in cardiac function, experimental approaches are required in which the heart and vasculature are subjected to stresses characteristic of the natural setting. These approaches include both cell/tissue-culture methods and whole-organ studies. Cell or tissues from animals in which a gene of interest has been ablated or altered can be manipulated in ways to elicit predictable responses. In vitro studies using cardiac cells from genetically altered animals will provide numerous opportunities for manipulation at the molecular, biochemical, and physiological levels in order to determine whether the genetic defect alters the normal response. For example, neonatal rat cardiac cells grown on a deformable substrate represent an in vitro model of load-induced cardiac hypertrophy (157,158). Using cardiomyocytes from growth factor–deficient mice in this system, effects of growth-factor absence on parameters such as morphology, rate of beating, protein synthesis, and induction of immediate early and contractile protein genes can be examined. As a specific example, angiotensin II mediates the stretch-induced hypertrophic response of neonatal rat cardiomyocytes in this in vitro model of load-induced hypertrophy; since angiotensin also induces TGF- β1 expression in cultured neonatal rat cardiac cells, it has been proposed that the hypertrophic response may involved production of TGF-β1 (7,8). Using cardiac cells from TGF-β1–deficient animals, this hypothesis can be tested directly.

In vitro studies using tissues explanted from growth factor–deficient animals may also reveal information concerning growth-factor function. For example, in a cardiac explant system, TGF-β has been shown to be essential for the epithelial-mesenchymal cell transformation required for valve and membranous septa formation (28,29). This study can be repeated using explants from TGF-β- null mice to confirm the role of the TGF-βs in this process and to further define which TGF-β isoforms are involved.

Physiological study of the isolated, intact mouse heart, following experimental manipulation of either the whole animal or the isolated heart, has become a valuable technique for investigating the cardiac phenotype of genetically altered animals. Although in the past the mouse has not been the small animal of choice for physiological studies on the heart, the tools supplied by over half a century of mouse genetics and the recent breakthrough in gene-manipulation capabilities in the mouse make it imperative that cardiac physiology also be directed toward the mouse. Whole-heart preparations can be used to assess contractility in normal, hypo-, and hyperdynamic mouse hearts (159), and in mice with genetic variations in muscle-specific genes

(160). The work-performing mouse-heart preparation sets preloads, after-loads, and heart rates at constant levels to provide an objective basis for heart-to-heart comparisons. If analysis is required on smaller hearts, such as those from neonates, measurements can be taken on muscle strips (161). Mechanical parameters can also be investigated on isolated mouse cardiomy-ocytes, thereby eliminating influences of chamber geometry, non-muscle cells, or extracellular matrix (162). Finally, mechanisms underlying pressure-overload hypertrophy can now be examined in the mouse after aortic banding, and changes ranging from gene expression to myocyte mechanics can be characterized in the whole heart or the isolated myocyte (162,163).

SUMMARY

The ability to ablate or alter a specific genetic locus in embryonic stem cells and to generate mice carrying the genetic alteration in the homozygous state provides a powerful tool for studying the role of growth factors, their receptors, and signaling molecules in cardiovascular-system development and function. This is especially so when descriptive studies of the phenotype resulting from the gene modification are combined with molecular, biochem-ical, and physiological studies conducted using cells, tissues, or intact organs from the genetically altered animals. In particular, the ability to experimentally manipulate and conduct physiological studies on the intact mouse heart in vitro (primarily working-heart but also Langendorff preparations) and in vivo (aortic banding) should enhance our understanding of the function of growth factors in cardiac physiology in both normal and disease states.

ACKNOWLEDGMENTS

We thank Drs. J. Robbins and J. Lessard for assistance with myosin and actin dot blots, G. Grupp for the analysis of contractility data, and I. Ormsby for maintenance of the TGF-β1 mouse colony. Supported by grants HL41496, HL46826, HD26471, and AHA (SW9309I).

REFERENCES

1. Cross M, Dexter TM. Growth factors in development, transformation and tumor-igenesis. Cell 1991; 64:271–280.
2. Massague J, Pandiella A. Membrane-anchored growth factors. Annu Rev Biochem 1993; 62:515–541.
3. Ullrich A, Schlessinger J. Signal transduction by receptors with tyrosine kinase activity. Cell 1990; 61:203–212.

4. Chao MV. Growth factor signaling: where is the specificity? Cell 1992; 68:995–997.
5. Roberts TM. A signal chain of events. Nature 1992; 360:534–535.
6. Fantl WJ, Johnson DE, Williams LT. Signalling by receptor tyrosine kinases. Annu Rev Biochem 1993; 62:453–481.
7. Sadoshima J, Xu Y, Slayter HS, Izumo S. Autocrine release of angiotensin II mediates stretch-induced hypertrophy of cardiac myocytes *in vitro*. Cell 1993; 75:977–984.
8. Sadoshima J, Izumo S. Molecular characterization of angiotensin II–induced hypertrophy of cardiac myocytes and hyperplasia of cardiac fibroblasts. Circ Res 1993; 73:413–423.
9. Shubeita HE, McDonough PM, Harris AN, Knowlton KU, Glembotski CC, Brown JH, Chien KR. Endothelin induction of inositol phospholipid hydrolysis, sarcomere assembly, and cardiac gene expression in ventricular myocytes. J Biol Chem 1990; 265:20555–20562.
10. Roberts AB, Sporn MB. The transforming growth factor- βs. In: Sporn MB, Roberts AB, eds. Peptide Growth Factors and Their Receptors. I. Berlin: Springer-Verlag, 1990:419–472.
11. Dickson MC, Slager HG, Duffie E, Mummery CL, Akhurst RJ. RNA and protein localisations of TGFβ2 in the early mouse embryo suggest an involvement in cardiac development. Development 1993; 117:625–639.
12. Mahmood R, Flanders KC, Morris-Kay GM. Interactions between retinoids and TGF-βs in mouse morphogenesis. Development 1992; 115:67–74.
13. Millan FA, Denhez F, Kondaiah P, Akhurst RJ. Embryonic gene expression patterns of TGFβ1, β2 and β3 suggest different developmental functions *in vivo*. Development 1991; 111:131–144.
14. Choy M, Armstrong MT, Armstrong PB. Transforming growth factor-β1 localized within the heart of the chick embryo. Anat Embryol 1991; 183:345–352.
15. Jakowlew SB, Dillard DJ, Winokur TS, Flanders KC, Sporn MB, Roberts AB. Expression of transforming growth factor- βs 1–4 in chicken embryo chondrocytes and myocytes. Dev Biol 1991; 143:135–148.
16. Akhurst RJ, Lehnert SA, Faissner A, Duffie E. TGF beta in murine morphogenetic processes: the early embryo and cardiogenesis. Development 1990; 108:645–656.
17. Heine UI, Munoz EF, Flanders KC, Ellingsworth LR, Lam H-YP, Thompson NL, Roberts AB, Sporn MB. Role of transforming growth factor-β in the development of the mouse embryo. J Cell Biol 1987; 105:2861–2876.
18. Pelton RW, Saxena B, Jones M, Moses HL, Gold LI. Immunohistochemical localization of TGFβ1, TGFβ2, and TGFβ3 in the mouse embryo: expression patterns suggest multiple roles during embryonic development. J Cell Biol 1991; 115:1091–1105.
19. Pelton RW, Dickinson ME, Moses HL, Hogan BLM. In situ hybridization analysis of TGFβ3 RNA expression during mouse development: comparative studies with TGFβ1 and β2. Development 1990; 110:609–620.
20. Gatherer D, ten Dijke P, Baird DT, Akhurst RJ. Expression of TGF-β isoforms during first trimester human embryogenesis. Development 1990; 110:445–460.

21. Engelmann GL, Boehm KD, Birchenall-Roberts MC, Ruscetti FW. Transforming growth factor-beta 1 in heart development. Mech Develop 1992; 38:85–98.
22. Clubb FJ, Bishop SP. Formation of binucleated myocardial cells in the neonatal rat. Lab Invest 1984; 50:571–577.
23. Thompson NL, Flanders KC, Smith JM, Ellingsworth LR, Roberts AB, Sporn MB. Expression of transforming growth factor-β1 in specific cells and tissues of adult and neonatal mice. J Cell Biol 1989; 108:661–669.
24. MacLellan WR, Brand T, Schneider MD. Transforming growth factor-β in cardiac ontogeny and adaptation. Circ Res 1993; 73:783–791.
25. Roberts AB, Sporn MB. Regulation of endothelial cell growth, architecture, and matrix synthesis by TGF-β. Am Rev Respir Dis 1989; 140:1126–1128.
26. Muslin AJ, Williams LT. Well-defined growth factors promote cardiac development in axolotl mesodermal explants. Development 1991; 112:1095–1101.
27. Roberts AB, Roche NS, Winokur TS, Burmester JK, Sporn MB. Role of transforming growth factor-β in maintainance of function of cultured neonatal cardiac myocytes. J Clin Invest 1992; 90:2056–2062.
28. Potts JD, Runyan RB. Epithelial-mesenchymal cell transformation in the embryonic heart can be mediated, in part, by transforming growth factor β. Dev Biol 1989; 134:392–401.
29. Potts JD, Dagle JM, Walder JA, Weeks DL, Runyan RB. Epithelial-mesenchymal transformation of embryonic cardiac endothelial cells is inhibited by a modified antisense oligodeoxynucleotide to transforming growth factor β3. Proc Natl Acad Sci USA 1991; 88:1516–1520.
30. Sugi Y, Sasse J, Lough J. Inhibition of precardiac mesoderm cell proliferation by antisense oligodeoxynucleotide complementary to fibroblast growth factor-2 (FGF-2). Dev Biol 1993; 157:28–37.
31. Spirito P, Fu Y-M, Yu Z-X, Epstein SE, Casscells W. Immunohistochemical localization of basic and acidic fibroblast growth factors in the developing rat heart. Circulation 1991; 84:322–332.
32. Fu Y-M, Spirito P, Yu Z-X, Biro S, Sasse J, Lei J, Ferrans VJ, Epstein SE, Casscells W. Acidic fibroblast growth factor in the developing rat embryo. J Cell Biol 1991; 114:1261–1273.
33. Engelmann GL, Dionne CA, Jaye MC. Acidic fibroblast growth factor and heart development. Circ Res 1993; 72:7–19.
34. Gonzalez AM, Buscaglia M, Ong M, Baird A. Distribution of basic fibroblast growth factor in the 18-day rat fetus: localization in the basement membranes of diverse tissues. J Cell Biol 1990; 110:753–765.
35. Parlow MH, Bolender DL, Kokan-Moore NP, Lough J. Localization of bFGF-like proteins as punctate inclusions in the preseptation myocardium of the chicken embryo. Dev Biol 1991; 146:139–147.
36. Kardami E, Fandrich RR. Basic fibroblast growth factor in atria and ventricles of the vertebrate heart. J Cell Biol 1989; 109:1865–1875.
37. Consigli SA, Joseph-Silverstein J. Immunolocalization of basic fibroblast growth factor during chicken cardiac development. J Cellu Physiol 1991; 146:379–385.
38. Speir E, Tanner V, Gonzalez AM, Farris J, Baird A, Casscells W. Acidic and basic fibroblast growth factors in adult rat heart myocytes. Circ Res 1992; 71:251–259.

39. Casscells W, Speir E, Sasse J, Klagsbrun M, Allen P, Lee M, Calvo B, Chiba M, Haggroth L, Folkman J, Epstein SE. Isolation, characterization, and localization of heparin-binding growth factors in the heart. J Clin Invest 1990; 85:433–441.

40. Sivalingam A, Kenney J, Brown GC, Benson WE, Donoso L. Basic fibroblast growth factor levels in the vitreous of patients with proliferative diabetic retinopathy. Arch Ophthalmol 1990; 108:869–872.

41. Doetschman T, Shull M, Kier A, Coffin JD. Embryonic stem cell model systems for vascular morphogenesis and cardiac disorders. Hypertension 1993; 22:618–629.

42. Parker TG, Packer SE, Schneider MD. Peptide growth factors can provoke "fetal" contractile protein gene expression in rat cardiac myocytes. J Clin Invest 1990; 85:507–514.

43. Humbel R. Insulin-like growth factors I and II. Eur J Biochem 1990; 190:445–462.

44. Bondy CA, Werner H, Roberts CT, LeRoith D. Cellular pattern of insulin-like growth factor-I (IGF-I) and type I IGF receptor gene expression in early organogenesis: comparison with IGF-II gene expression. Molec Endocrinol 1990; 4:1386–1398.

45. Lee JE, Pintar J, Efstratiadis A. Pattern of the insulin-like growth factor II gene expression during early mouse embryogenesis. Development 1990; 110:151–159.

46. Stylianopoulou F, Efstratiadis A, Herbert J, Pintar J. Pattern of insulin-like growth factor II gene expression during rat embryogenesis. Development 1988; 103:497–506.

47. Beck F, Samani NJ, Penschow JD, Thorley B, Tregear GW, Coghlan JP. Histochemical localization of IGF-I and -II mRNA in the developing rat embryo. Development 1987; 101:175–184.

48. Soares MB, Turken A, Ishii D, Mills L, Episkopou V, Cotter S, Zeitlin S, Efstratiadis A. Rat insulin-like growth factor II gene. J Molec Biol 1986; 192:737–752.

49. Wood TL, Streck RD, Pintar JE. Expression of the IGFBP-2 gene in post-implantation rat embryos. Development 1992; 114:59–66.

50. Engelmann GL, Boehm KD, Haskell JF, Khairallah PA, Ilan J. Insulin-like growth factors and neonatal cardiomyocyte development: ventricular gene expression and membrane receptor variations in normotensive and hypertensive rats. Molec Cell Endocrinol 1989; 63:1–14.

51. Murphy LJ, Bell GI, Friesen HG. Tissue distribution of insulin-like growth factor I and II messenger ribonucleic acid in the adult rat. Endocrinol 1987; 120:1279–1282.

52. Matzner U, von Figura K, Pohlmann R. Expression of the two mannose 6-phosphate receptors is spatially and temporally different during mouse embryogenesis. Development 1992; 114:965–972.

53. Senior PV, Byrne S, Brammar WJ, Beck F. Expression of the IGF-II/mannose-6-phosphate receptor mRNA and protein in the developing rat. Development 1990; 109:67–73.

54. Werner H, Woloschak M, Adamo M, Shen-Orr Z, Roberts CT, LeRoith D. Developmental regulation of the rat insulin-like growth factor I receptor gene. Proc Natl Acad Sci USA 1989; 86:7451–7455.

55. Raffioni S, Bradshaw RA. The receptors for nerve growth factor and other neurotrophins. Annu Rev Biochem 1993; 62:823–850.

56. Clegg DO, Large TH, Bodary SC, Reichardt LF. Regulation of nerve growth factor mRNA levels in developing rat heart ventricle is not altered by sympathectomy. Dev Biol 1989; 134:30–37.

57. Korsching S, Thoenen H. Developmental changes of nerve growth factor levels in sympathetic ganglia and their target organs. Dev Biol 1988; 126:40–46.

58. Yan Q, Johnson EM. An immunohistochemical study of the nerve growth factor receptor in developing rats. J Neurosci 1988; 8:3481–3498.

59. Heuer JG, Fatemie-Nainie S, Wheeler EF, Bothwell M. Structure and developmental expression of the chicken NGF receptor. Dev Biol 1990; 137:287–304.

60. Maisonpierre PC, Belluscio L, Squinto S, Ip NY, Furth ME, Lindsay RM, Yancopoulos GD. Neurotrophin-3: a neurotrophic factor related to NGF and BDNF. Science 1990; 247:1446–1451.

61. Merlio JP, Ernfors JP, Jaber M, Persson H. Molecular cloning of rat *trkC* and distribution of cells expressing messenger RNAs for members of the *trk* family in the rat central nervous system. Neuroscience 1992; 51:513–532.

62. MacGrogan D, Saint-Andre J-P, Dicou E. Expression of nerve growth factor and nerve growth factor receptor genes in human tissues and in prostatic adenocarcinoma cell lines. J Neurochem 1992; 59:1381–1391.

63. Martin-Zanca D, Barbacid M, Parada LF. Expression of the *trk* proto-oncogene is restricted to the sensory cranial and spinal ganglia of neural crest origin in mouse development. Genes Develop 1990; 4:683–694.

64. Klein R, Martin-Zanca D, Barbacid M, Parada LF. Expression of the tyrosine kinase receptor gene *trk*B is confined to the murine embryonic and adult nervous system. Development 1990; 109:845–850.

65. Parker TG, Schneider MD. Growth factors, proto-oncogenes, and plasticity of the cardiac phenotype. Annu Rev Physiol 1991; 53:179–200.

66. Parker TG. Molecular biology of cardiac growth and hypertrophy. Herz 1993; 18:245–255.

67. Schneider MD, Parker TG. Cardiac myocytes as targets for the action of peptide growth factors. Circulation 1990; 81:1443–1456.

68. Schneider MD, McLellan WR, Black FM, Parker TG. Growth factors, growth factor response elements, and the cardiac phenotype. Basic Res Cardiol 1992; 87(suppl 2):33–48.

69. Ito H, Hiroe M, Hirata Y, Tsujino M, Adachi S, Shichiri M, Koike A, Nogami A, Marumo F. Insulin-like growth factor-I induces hypertrophy with enhanced expression of muscle specific genes in cultured rat cardiomyocytes. Circulation 1993; 87:1715–1721.

70. Donohue TJ, Dworkin LD, Lango MN, Fliegner K, Lango RP, Bernstein JA, Slater WR, Catanese VM. Induction of myocardial insulin-like growth factor-I gene expression in left ventricular hypertrophy. Circulation 1994; 89:799–809.

71. Wahlander H, Isgaard J, Jennische E, Friberg P. Left ventricular insulin-like growth factor I increases in early renal hypertension. Hypertension 1992; 19:25–32.

72. Wunsch M, Sharma HS, Markert T, Bernotat-Danielowski S, Schott RJ, Kremer P, Bleese N, Schaper W. In situ localization of transforming growth factor $\beta 1$ in

porcine heart: enhanced expression after chronic coronary artery constriction. J Molec Cell Cardiol 1991; 23:1051-1062.

73. Thompson NL, Bazoberry F, Speir EH, Casscells W, Ferrans VJ, Flanders KC, Kondaiah P, Geiser AG, Sporn MB. Transforming growth factor beta-1 in acute myocardial infarction in rats. Growth Factors 1988; 1:91-99.

74. Casscells W, Bazoberry F, Speir E, Thompson N, Flanders K, Kondaiah P, Ferrans VJ, Epstein SE, Sporn M. Transforming growth factor-β1 in normal heart and in myocardial infarction. Ann NY Acad Sci 1990; 593:148-160.

75. Yanagisawa-Miwa A, Uchida Y, Nakamura F, Tomaru T, Kido H, Kamijo T, Sugimoto T, Kaji K, Utsuyama M, Kurashima C, Ito H. Salvage of infarcted myocardium by angiogenic action of basic fibroblast growth factor. Science 1992; 257:1401-1403.

76. Sharma HS, Wunsch M, Brand T, Vedouw PD, Schaper W. Molecular biology of the coronary vascular and myocardial responses to ischemia. J Cardiovasc Pharm 1992; 20(suppl 1):S23-S31.

77. Reiss K, Kajstura J, Capasso JM, Marino TA, Anversa P. Impairment of myocyte contractilily following coronary artery narrowing is associated with activation of the myocyte IGF1 autocrine system, enhanced expression of late growth related genes, DNA synthesis, and myocyte nuclear mitotic division in rats. Exp Cell Res 1993; 207:348-360.

78. Lefer AM, Tsao P, Aoki N, Palladino MA. Mediation of cardioprotection by transforming growth factor-β. Science 1990; 249:61-64.

79. Lefer AM. Mechanisms of the protective effects of transforming growth factor-β in reperfusion injury. Biochem Pharm 1991; 42:1323-1327.

80. Haglund K. Angiogenesis for heart disease? J NIH Res 1994; 6:29-32.

81. Ross R. The pathogenesis of atherosclerosis—an update. N Engl J Med 1986; 314:488-500.

82. Ferns GAA, Motani AS, Anggard EE. The insulin-like growth factors: their putative role in atherogenesis. Artery 1991; 18:197-225.

83. Barrett TB, Benditt EP. Platelet-derived growth factor gene expression in human atherosclerotic plaques and normal artery wall. Proc Natl Acad Sci USA 1988; 85:2810-2814.

84. Libby P, Warner SJC, Salomon RN, Birinyi LK. Production of platelet-derived growth factor-like mitogen by smooth muscle cells from human atheroma. N Engl J Med 1988; 318:1493-1498.

85. Karpen CW, Spanheimer RG, Randolph AL, Lowe WL. Tissue-specific regulation of basic fibroblast growth factor mRNA levels by diabetes. Diabetes 1992; 41:222-226.

86. Zettler C, Rush RA. Elevated concentrations of nerve growth factor in heart and mesenteric arteries of spontaneously hypertensive rats. Brain Res 1993; 614:15-20.

87. Falckh PH, Harkin LA, Head RJ. Nerve growth factor mRNA content parallels altered sympathetic innervation in the spontaneously hypertensive rat. Clin Exp Pharm Physiol 1992; 19:541-545.

88. Ueyama T, Hamada M, Hano T, Nishio I, Masuyama Y, Furukawa S. Increased nerve growth factor levels in spontaneously hypertensive rats. J Hypertens 1992; 10:215-219.

89. Lee RMKW, Triggle CR, Cheung DWT, Coughlin MD. Structural and functional consequence of neonatal sympathectomy on the blood vessels of spontaneously hypertensive rats. Hypertension 1987; 10:328-338.

90. Waltenberger J, Lundin L, Oberg K, Wilander E, Miyazono K, Heldin C-H, Funa K. Involvement of transforming growth factor- β in the formation of fibrotic lesions in carcinoid heart disease. Am J Pathol 1993; 142:71-78.

91. Waltenberger J, Wanders A, Fellstrom B, Miyazono K, Heldin C-H, Funa K. Induction of transforming growth factor-β during cardiac allograft rejection. J Immunol 1993; 151:1147-1157.

92. Wagner CR, Morris TE, Shipley GD, Hosenpud JD. Regulation of human aortic endothelial cell-derived mesenchymal growth factors by allogeneic lymphocytes in vitro. J Clin Invest 1993; 92:1269-1277.

93. Wallick SC, Figari IS, Morris RE, Levinson AD, Palladino MA. Immunoregulatory role of transforming growth factor β (TGF-β) in development of killer cells: comparison of active and latent TGF-β1. J Exp Med 1990; 172:1777-1784.

94. Majesky MW, Lindner V, Twardzik DR, Schwartz SM, Reidy MA. Producton of transforming growth factor β1 during repair of arterial injury. J Clin Invest 1991; 88:904-910.

95. Madri JA, Reidy MA, Kocher O, Bell L. Endothelial cell behavior after denudation injury is modulated by transforming growth factor-β1 and fibronectin. Lab Invest 1989; 60:755-764.

96. Nikol S, Isner JM, Pickering JG, Kearney M, Leclerc G, Weir L. Expression of transforming growth factor-β1 is increased in human vascular restenosis lesions. J Clin Invest 1992; 90:1582-1592.

97. Cerek B, Fishbein MC, Forrester JS, Helfant RH, Fagin JA. Induction of insulin-like growth factor I messenger RNA in rat aorta after balloon denudation. Circ Res 1990; 66:1755-1760.

98. Hansson H-A, Jennische E, Skottner A. Regenerating endothelial cells express insulin-like growth factor-I immunoreactivity after arterial injury. Cell Tiss Res 1987; 250:499-505.

99. Shull MM, Ormsby I, Kier AB, Pawlowski S, Diebold RJ, Yin M, Allen R, Sidman C, Proetzel G, Calvin D, Annunziata N, Doetschman T. Targeted disruption of the mouse transforming growth factor- β1 gene results in multifocal inflammatory disease. Nature 1992; 359:693-699.

100. Kulkarni AB, Huh C-G, Becker D, Geiser A, Lyght M, Flanders KC, Roberts AB, Sporn MB, Warn JM, Karlsson S. Transforming growth factor β1 null mutation in mice causes excessive inflammatory response and early death. Proc Natl Acad Sci USA 1993; 90:770-774.

101. Matzuk MM, Finegold MJ, Su J-GJ, Hsueh AJW, Bradley A. α-Inhibin is a tumor suppressor gene with gonadal specificity in mice. Nature 1992; 360:313-319.

102. Mann GB, Fowler KJ, Gabriel A, Nice EC, Williams RL, Dunn AR. Mice with a null mutation of the TGFα gene have abnormal skin architecture, wavy hair, and curly whiskers and often develop corneal inflammation. Cell 1993; 73:249-261.

103. Luetteke NC, Qiu TH, Peiffer RL, Oliver P, Smithies O, Lee DC. TGFα deficiency results in hair follicle and eye abnormalities in targeted and wave-1 mice. Cell 1993; 73:263-278.

104. Mansour SL, Goddard JM, Capecchi MR. Mice homozygous for a targeted disruption of the proto-oncogene *int-2* have developmental defects in the tail and inner ear. Development 1993; 117:13–28.
105. Liu J-P, Baker J, Perkins AS, Robertson EJ, Efstratiadis A. Mice carrying null mutations of the genes encoding insulin-like growth factor I (*Igf-1*) and type 1 IGF receptor (*Igf1r*). Cell 1993; 75:59–72.
106. Baker J, Liu J-P, Robertson EJ, Efstratiadis A. Role of insulin-like growth factors in embryonic and postnatal growth. Cell 1993; 75:73–82.
107. DeChiara TM, Robertson EJ, Efstratiadis A. Parental imprinting of the mouse insulin-like growth factor II gene. Cell 1991; 64:849–859.
108. DeChiara TM, Efstratiadis A, Robertson EJ. A growth-deficiency phenotype in heterozygous mice carrying an insulin-like growth factor II gene disrupted by targeting. Nature 1990; 345:78–80.
109. Crowley C, Spencer SD, Nishimura MC, Chen KS, Pitts-Meek S, Armanini MP, Ling LH, McMahon SB, Shelton DL, Levinson AD, Phillips HS. Mice lacking nerve growth factor display perinatal loss of sensory and sympathetic neurons yet develop basal forebrain cholinergic neurons. Cell 1994; 76:1001–1011.
110. Jones KR, Farinas I, Backus C, Reichardt LF. Targeted disruption of the BDNF gene perturbs brain and sensory neuron development but not motor neuron development. Cell 1994; 76:989–999.
111. Lee K-F, Li E, Huber J, Landis SC, Sharpe AH, Chao MV, Jaenisch R. Targeted mutation of the gene encoding the low affinity NGF receptor p75 leads to deficits in the peripheral sensory nervous system. Cell 1992; 69:737–749.
112. Lee K-F, Bachman K, Landis S, Jaenisch R. Dependence on p75 for innervation of some sympathetic targets. Science 1994; 263:1447–1449.
113. Klein R, Smeyne RJ, Wurst W, Long LK, Auerbach BA, Joyner AL, Barbacid M. Targeted disruption of the *trk*B neurotrophin receptor gene results in nervous system lesions and neonatal death. Cell 1993; 75:113–122.
114. McMahon AP, Bradley A. The *Wnt-1* (*int-1*) proto- oncogene is required for development of a large region of the mouse brain. Cell 1990; 62:1073–1085.
115. Thomas KR, Capecchi MR. Targeted disruption of the murine *int-1* proto-ncogene resulting in severe abnormalities in midbrain and cerebellar development. Nature 1990; 346:847–850.
116. Masu Y, Wolf E, Holtmann B, Sendtner M, Brem G, Thoenen H. Disruption of the CNTF gene results in motor neuron degeneration. Nature 1993; 365:27–32.
117. Dustin ML, Rothlein R, Bhan AK, Dinarello CA, Springer TA. Induction by IL 1 and interferon-γ: tissue distribution, biochemistry, and function of a natural adherence molecule (ICAM-1). J Immunol 1986; 137:245–254.
118. Munro JM, Pober JS, Cotran RS. Tumor necrosis factor and interferon-γ induce distinct patterns of endothelial activation and associated leukocyte accumulation in skin of Papio anubis. Am J Path 1989; 135:121–133.
119. Geiser AG, Letterio JJ, Kulkarni AB, Karlsson S, Roberts AB, Sporn MB. Transforming growth factor β*1* (TGF-β1) controls expression of major histocompatibility genes in the postnatal mouse: aberrant histocompatibility antigen expression in the pathogenesis of the TGF-β1 null mouse phenotype. Proc Natl Acad Sci USA 1993; 90:9944–9948.

120. Gulick T, Chung MK, Pieper SJ, Lange LG, Schreiner GF. Interleukin 1 and tumor necrosis factor inhibit cardiac myocyte β-adrenergic responsiveness. Proc Natl Acad Sci USA 1989; 86:6753-6757.

121. Schreiber AB, Winkler ME, Derynck R. Transforming growth factor-α: a more potent angiogenic mediator than epidermal growth factor. Science 1986; 232:1250-1253.

122. Lin HY, Lodish HF. Receptors for the TGF-β superfamily: multiple peptides and serine/threonine kinases. Trends Cell Biol 1993; 3:14-19.

123. Pazin MJ, Williams LT. Triggering signaling cascades by receptor tyrosine kinases. Trends Biol Sci 1992; 17:374-378.

124. Egan SE, Weinberg RA. The pathway to signal achievement. Science 1993; 365:781-783.

125. Fu X-Y, Zhang J-J. Transcription factor p91 interacts with epidermal growth factor receptor and mediates activation of the c-fos gene promoter. Cell 1993; 74:1135-1145.

126. Silvennoinen O, Schindler C, Schlessinger J, Levy DE. Ras-independent growth factor signaling by transcription factor tyrosine phosphorylation. Science 1993; 261:1736-1739.

127. Sadowski HB, Shuai K, Darnell JE, Gilman MZ. A common nuclear signal transduction pathway activated by growth factor and cytokine receptors. Science 1993; 261:1739-1744.

128. Ruff-Jamison S, Chen K, Cohen S. Induction by EGF and interferon-γ of tyrosine phosphorylated DNA binding proteins in mouse liver nuclei. Science 1993; 261:1733-1736.

129. Hunter T. Braking the cycle. Cell 1993; 75:839-841.

130. Sherr CJ. Mammalian G_1 cyclins. Cell 1993; 73:1059-1065.

131. Sherr CJ. The ins and outs of RB: coupling gene expression to the cell cycle clock. Trends Cell Biol 1994; 4:15-18.

132. Ewen ME, Sluss HK, Whitehouse LL, Livingston DM. TGFβ inhibition of Cdk4 synthesis is linked to cell cycle arrest. Cell 1993; 74:1009-1020.

133. Soriano P, Montgomery C, Geske R, Bradley A. Targeted disruption of the c-src proto-oncogene leads to osteopetrosis in mice. Cell 1991; 64:693-702.

134. Stein PL, Lee H-M, Rich S, Soriano P. $pp59^{fyn}$ mutant mice display differential signaling in thymocytes and peripheral T cells. Cell 1992; 70:741-750.

135. Appleby MW, Gross JA, Cooke MP, Levin SD, Qian X, Perlmutter RM. Defective T cell receptor signaling in mice lacking the thymic isoform of $p59^{fyn}$. Cell 1992; 70:751-763.

136. Molina TJ, Kishihara K, Siderovski DP, van Ewijk W, Narendran A, Timms E, Wakeham A, Paige CJ, Hartmann K-U, Veillette A, Davidson D, Mak TW. Profound block in thymocyte development in mice lacking $p56^{lck}$. Nature 1992; 357:161-164.

137. Imamoto A, Soriano P. Disruption of the csk gene, encoding a negative regulator of src family tyrosine kinases, leads to neural tube defects and embryonic lethality in mice. Cell 1993; 73:1117-1124.

138. Nada S, Yagi T, Takeda H, Tokunaga T, Nakagawa H, Ikawa Y, Okada M, Aizawa S. Constitutive activation of src family kinases in mouse embryos that lack csk. Cell 1993; 73:1125-1135.

139. Tybulewicz VLJ, Crawford CE, Jackson PK, Bronson RT, Mulligan RC. Neonatal lethality and lymphopenia in mice with a homozygous disruption of the c-abl proto-oncogene. Cell 1991; 65:1153–1163.

140. Schwartzberg PL, Stall AM, Hardin JD, Bowdish KS, Humaran T, Boast S, Harbison ML, Robertson EJ, Goff SP. Mice homozygous for the abl^{m1} mutation show poor viability anf depletion of selected B and T cell populations. Cell 1991; 65:1165–1175.

141. Davis AC, Wims M, Spotts GD, Hann SR, Bradley A. A null c-myc mutation causes lethality before 10.5 days of gestation in homozygotes and reduced fertility in heterozygous female mice. Genes Dev 1993; 7:671–682.

142. Stanton BR, Perkins AS, Tessarollo L, Sassoon DA, Parada LF. Loss of N-myc function results in embryonic lethality and failure of the epithelial component of the embryo to develop. Genes Dev 1992; 6:2235–2247.

143. Charron J, Malynn BA, Fisher P, Stewart V, Jeannotte L, Goff SP, Robertson EJ, Alt FW. Embryonic lethality in mice homozygous for a targeted disruption of the N-myc gene. Genes Dev 1992; 6:2248–2257.

144. Sawai S, Shimono A, Hanaoka K, Kondoh H. Embryonic lethality resulting from disruption of both N-myc alleles in mouse zygotes. New Biologist 1991; 3:861–869.

145. Moens CB, Auerbach AB, Conlon RA, Joyner AL, Roussant J. A targeted mutation reveals a role for N-myc in branching morphogenesis in the embryonic mouse lung. Genes Dev 1992; 6:691–704.

146. Johnson RS, Spiegelman BM, Papaioannou V. Pleiotropic effects of a null mutation in the c-fos proto-oncogene. Cell 1992; 71:577–586.

147. Wang Z-Q, Ovitt C, Grigoriadis AE, Mohle-Steinlein U, Ruther U, Wagner EF. Bone and haematopoietic defects in mice lacking c-fos. Nature 1992; 360:741–745.

148. Johnson RS, van Lingen B, Papaioannou VE, Spiegelman BM. A null mutation at the c-jun locus causes embryonic lethality and retarded cell growth in culture. Genes Dev 1993; 7:1309–1317.

149. Hilberg F, Aguzzi A, Howells N, Wagner EF. c-Jun is essential for normal mouse development and hepatogenesis. Nature 1993; 365:179–181.

150. Lee EY-HP, Chang C-Y, Hu N, Wang Y-CJ, Lai C-C, Herrup K, Lee W-H, Bradley A. Mice deficient for Rb are nonviable and show defects in neurogenesis and haematopoiesis. Nature 1992; 359:288–294.

151. Jacks T, Fazeli A, Schmitt EM, Bronson RT, Goodell MA, Weinberg RA. Effects of an Rb mutation in the mouse. Nature 1992; 359:295–300.

152. Clarke AR, Maandag ER, van Roon M, van der Lugt NMT, van der Valk M, Hooper ML, Berns A, te Riele H. Requirement for a functional Rb-I gene in murine development. Nature 1992; 359:328–330.

153. Davis A, Bradley A. Mutation of N-myc in mice: what does the phenotype tell us? BioEssays 1993; 15:273–275.

154. Jackson T, Allard MF, Sreenan CM, Doss LK, Bishop SP, Swain JL. The c-myc proto-oncogene regulates cardiac development in transgenic mice. Mol Cell Biol 1990; 10:3709–3716.

155. Cooper JA, Howell B. The when and how of Src regulation. Cell 1993; 73:1051–1054.

156. Erickson HP. Gene knockouts of c-*src*, transforming growth factor β1, and tenascin suggest superfluous, nonfunctional expression of proteins. J Cell Biol 1993; 120:1079–1081.

157. Sadoshima JI, Jahn L, Takahashi T, Kulik TJ, Izumo S. Molecular characterization of the stretch-induced adaptation of cultured cardiac cells. J Biol Chem 1992; 267:10551–10560.

158. Komuro I, Katoh Y, Kaida T, Shibazaki Y, Kurabayashi M, Hoh E, Takaku F, Yazaki Y. Mechanical loading stimulates cell hypertrophy and specific gene expression in cultured rat cardiac myocytes. J Biol Chem 1991; 266:1265–1268.

159. Grupp IL, Subramaniam A, Hewett TE, Robbins J, Grupp G. Comparison of normal, hypodynamic, and hyperdynamic mouse hearts using isolated work-performing heart preparations. Am J Physiol 1993; 265:H1401–1410.

160. Hewett TE, Grupp IL, Grupp G, Robbins J. α-Skeletal actin is associated with increased contractility in the mouse heart. Circ Res 1994; 74:740–746.

161. Capasso JM, Robinson TF, Anversa P. Alterations in collagen cross-linking impair myocardial contractility in the mouse heart. Circ Res 1989; 65:1657–1664.

162. Dorn GW, Robbins J, Ball N, Walsh RA. Myosin heavy chain regulation and myocyte contractile depression after LV hypertrophy in aortic banded mice. Am J Physiol 1994; 267:H400–405.

163. Rockman HA, Ross RS, Harris AN, Knowlton KU, Steinhelper ME, Field LJ, Ross J, Chien KR. Segregation of atrial-specific and inducible expression of an atrial natriuretic factor transgene in an *in vivo* murine model of cardiac hypertrophy. Proc Natl Acad Sci USA 1991; 88:8277–8281.

164. Grupp IL, Grupp G. Isolated heart preparations perfused or superfused with balanced salt solutions. In: Schwartz A, ed. Methods in Pharmacology. Vol 5. New York: Plenum, 1984:111–128.

15

Gene Targeting of Adhesion Molecules in the Vasculature

Denisa D. Wagner
Harvard Medical School
Boston, Massachusetts

Richard O. Hynes
Massachusetts Institute of Technology
Cambridge, Massachusetts

INTRODUCTION

Specific adhesion processes direct and modulate all the dynamic interactions among vascular cells. Nowhere else in the adult organism do the cells have the capacity to change their interactions so rapidly and completely as in the blood vessels, and nowhere else are defective interactions so immediately life-threatening. For example, during normal hemostasis, free-flowing platelets have to adhere rapidly at the site of damage and then to each other to form an effective platelet plug; otherwise, blood loss cannot be contained and hemorrhage and death follow. Just as dangerous as lack of platelet adhesion is excessive platelet aggregation. This results in thrombosis, which may lead to stroke or heart attack. Similarly, deficiency in leukocyte adhesion leads to life-threatening infections, while excessive leukocyte adhesion—for example, during reperfusion after a period of ischemia—may cause tissue destruction.

Lymphocytes use similar adhesion processes in their normal circulation between lymphoid organs and peripheral tissues. These various adhesive processes are diagrammed in Figure 1, which also depicts the role of cell migration and adhesion in angiogenesis, the formation of new blood vessels, an important process in wound healing. Finally, it is important to remember that cell adhesion plays major roles in development, including vascular development. Defects in vasculogenesis or heart development are of considerable clinical importance. Understanding of the many adhesive interactions in the development and function of the vascular system, and of the precise roles a particular adhesion molecule may play, will lead to the development of therapies that will allow one to modulate these processes as required by the clinical circumstances.

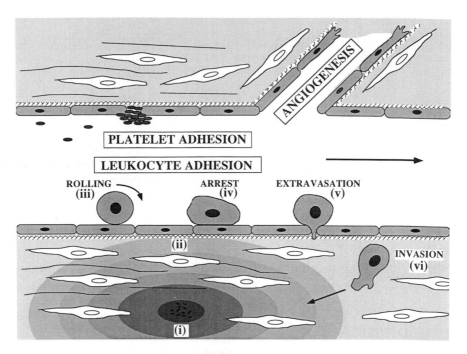

Figure 1   Three major processes in the vasculature involving cell adhesion. *Platelet adhesion* to exposed basement membranes or other extracellular matrices and aggregation of platelets with each other. Both processes involve integrin adhesion receptors as well as other molecules (see Figures 2 and 3). *Leukocyte adhesion* to the vessel wall at sites of infection or inflammation proceeds in several steps. (i) Bacteria at a site of infection (or other stimuli) cause release of chemokines and cytokines, with several effects on surrounding cells. (ii) Cytokines activate the endothelial cells lining adjacent blood vessels, causing them to express several novel cell-surface molecules that promote leukocyte adhesion. (iii) Selectins expressed on the endothelial cells bind carbohydrates on the leukocytes and cause these cells to roll along the wall. This slows the passage of the leukocytes but does not cause arrest. (iv) Cytokines or molecules expressed on endothelial cell surfaces activate the leukocytes. A key consequence of this activation is that leukocyte integrins—present but inactive on resting circulating leukocytes—become active. They then bind to counterreceptors of the immunoglobulin superfamily expressed on the endothelial cells; this interaction causes strong cell adhesion and arrest. (v) The leukocytes then move through the endothelial layer, leading to extravasation. This is thought to involve changes in adhesive interactions of a variety of cell-adhesion receptors, probably including selectins, integrins, cadherins, and Ig superfamily molecules. (vi) The leukocytes then migrate through the interstitial matrix and respond to chemotactic factors released by the bacteria or by cells surrounding the site of infection. The migration involved in this invasion is thought to involve integrin–matrix interactions. *Angiogenesis*, the process of generation of new blood vessels, involves migration of endothelial cells through interstitial matrices. This requires changes in cell–cell adhesion between adjacent endothelial cells, cell–matrix adhesion leading to migration, and, probably, degradation of extracellular matrix at the leading edge of the extending blood vessel.

404

ADHESION MOLECULES IN THE VASCULAR SYSTEM

Important changes in the adhesion of cells to their neighbors, to cells of different types, and to extracellular matrices are central to many physiological and pathological events in the vasculature. These various adhesive processes are mediated by a large number of adhesive molecules (1,2). These include both adhesive proteins of the extracellular matrix and cell-surface adhesion receptors. Fortunately, these molecules fall into a limited number of families. Four major families of transmembrane adhesion receptors are known to mediate adhesion in blood vessels (Figures 2 and 3). Only one of these families, the selectins, is specific to the vasculature.

Selectins

Selectins are unusual in that they mediate binding to a carbohydrate ligand rather than to another protein structure as the other receptors do (3,4). Selectins have a lectin domain located farthest from the cell membrane followed by an epidermal growth factor–like domain, also thought to contribute to the binding specificity. Between these two domains and the membrane are several segments homologous with domains that are also present in proteins binding complement. The functions of these latter domains in the selectins are unknown, but the possibility that they promote complement fixation cannot be excluded. The proteins span the membrane once and contain a short cytoplasmic tail.

The letter prefix of a selectin is determined by the cell type in which the selectin was first identified (L = leukocytes; E = endothelium; P = platelets). L-selectin is known as the homing receptor on lymphocytes for high endothelial venules of peripheral lymph nodes. L-selectin is also found on neutrophils and may be involved in neutrophils' extravasation at sites of inflammation. The ligand for L-selectin on the high endothelium of peripheral lymph nodes is a mucin-like molecule called CD34 (5). Another soluble mucin, glyCAM-1 (6), can modulate lymphocyte homing by binding to L-selectin and thus block the leukocyte adhesion to CD34 (experimental evidence for this is presented below). The ligand for L-selectin on other endothelial cells is not known.

E-selectin is synthesized by endothelial cells after they have been activated with inflammatory cytokines and is not found on resting endothelium. E-selectin is a receptor for neutrophils, monocytes, and a subset of T lymphocytes. The carbohydrate ligand appears to be sialyl-Lewis^x, which is found on both proteins and lipids.

The largest selectin, P-selectin, is found in endothelial cells and in platelets. It is different from the other two selectins in that it is stored in granules. It becomes rapidly expressed on the cell surface only after cellular

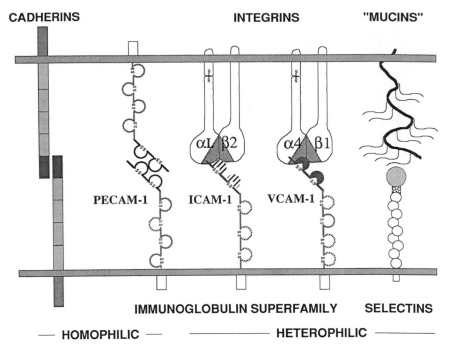

Figure 2 Major classes of cell-surface receptors involved in cell–cell adhesion events in the vasculature. *Integrins* are $\alpha\beta$ heterodimers expressed widely on cells. Each integrin has specificity for a limited number of extracellular matrix proteins (Figure 3) or Ig superfamily receptors. *Cadherins* mediate Ca^{2+}-dependent adhesion between like cells via homophilic (like with like) interactions; e.g., cadherins mediate adhesions between endothelial cells. *Immunoglobulin superfamily* molecules can also mediate homophilic interactions. For example, PECAM-1, expressed on both endothelial cells and on transmigrating leukocytes, may mediate adhesion between them. Ig superfamily molecules can also act as counterreceptors for integrins in heterophilic interactions; e.g., integrins of the β_2 subfamily can bind to ICAM-1 (shown) or other ICAMs and $\alpha 4$ integrins can bind to VCAM-1, one form of which is shown. These integrins are expressed on various leukocytes and the ICAMs and VCAM-1 are up-regulated on activated endothelial cells. Thus, these heterophilic integrin–Ig superfamily interactions are central in leukocyte–endothelium interactions. *Selectins* are another family of adhesion molecules that mediate heterophilic interactions by binding to carbohydrate groups on highly glycosylated proteins sometimes called, collectively, *mucins*. Shown is P-selectin, expressed on activated platelets or endothelial cells, binding to a counterreceptor on a leukocyte. E-selectin on endothelium functions in a similar fashion, whereas L-selectin is expressed on the leukocytes and binds a counterreceptor on the surfaces of endothelial cells.

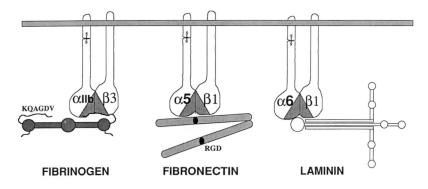

Figure 3 Three representative examples of integrin-mediated cell–matrix adhesion. (Left) $\alpha_{IIb}\beta_3$, expressed solely on platelets, binds preferentially to fibrinogen and recognizes a short peptide sequence, KQAGDV, in the γ chain. With lower affinity it can also bind to RGD sequences elsewhere in the fibrinogen molecule and in other proteins. $\alpha_{IIb}\beta_3$-fibrinogen binding leads to cross-linking of platelets by the bivalent fibrinogen molecule and platelet aggregation. $\alpha_{IIb}\beta_3$ exemplifies an important property shared by all or most integrins. On resting circulating platelets, it is in an inactive state and does not bind soluble fibrinogen. After activation of platelets (e.g., by thrombin or collagen) the $\alpha_{IIb}\beta_3$ integrin switches to an active conformation in which it can bind its ligands. (Center) $\alpha_5\beta_1$ is expressed by many cells, including endothelial cells, platelets, and activated lymphocytes. It binds the RGD sequence in fibronectin and mediates cell–matrix adhesion. (Right) $\alpha_6\beta_1$ shares one subunit with $\alpha_5\beta_1$ but has a different adhesive specificity; it binds to a specific domain of laminin, a major constituent of basement membranes, and does not bind fibronectin. $\alpha_6\beta_1$-laminin interactions are probably important in adhesion of endothelial cells to the basement membrane. $\alpha_6\beta_1$ is also expressed by platelets and activated lymphocytes. In the latter cell type, it can mediate cell–matrix adhesion and can also act as a costimulatory receptor for T-cell activation, as can $\alpha_5\beta_1$.

activation accompanied by degranulation. P-selectin is a receptor for neutrophils, monocytes, and some other subsets of leukocytes. The known ligand for P-selectin is also a sialylated mucin, called PSGL-1 (7). Selectin-mediated adhesion is thought to be weak and transient. This would explain how they are involved in such interactions as leukocyte rolling on the vessel wall. The rolling contact with the endothelium allows the leukocyte to check the vessel wall for signs of inflammation indicated by the presence of soluble activators and the expression of new adhesion receptors. These receptors, members of the immunoglobulin superfamily, mediate stronger adhesion, leading to arrest and extravasation (Figure 1). The counterreceptors on the leukocytes that mediate these higher-affinity interactions are ubiquitous molecules called integrins (1,2,8,9).

Integrins

Integrins are heterodimers of α and β subunits, each of which spans the membrane; both subunits contribute to ligand binding (8). There are many different integrins divided into subfamilies, depending on the kind of β subunit that they contain. Integrins can bind to secreted adhesion molecules such as fibrinogen, von Willebrand factor and fibronectin found in plasma, or those that are deposited in the extracellular matrix (basement membrane—see Figure 3). As mentioned above, integrins also mediate binding to adhesion receptors from the immunoglobulin superfamily, such as ICAM-1, -2, and -3 and VCAM-1. These interactions are very important in the vascular system (see Figures 1 and 2). It is the leukocyte's β_2 integrins that mediate the strong adhesion to endothelium expressing Ig superfamily molecules in many areas of inflammation. Other integrins of the α_4 type can also contribute to the adhesion of lymphocytes, monocytes, and eosinophils by binding to VCAM-1, another immunoglobulin superfamily molecule expressed on activated endothelial cells.

An important feature of the integrins is that their affinity for their ligands can change rapidly and drastically, depending on the state of cellular activation (8). An activating signal reaching the cytoplasmic domains of the integrins somehow propagates across the membrane, causing changes in the conformation of the extracellular domains and leading to a change in ligand affinity. This process is best documented for the platelet integrin GPIIb-IIIa (α_{IIb} β_3) (10,11). In the blood, platelets circulate singly because their integrin receptors are in the resting/low-affinity state. Upon platelet activation, within seconds, the GPIIb-IIIa integrins become competent for binding fibrinogen, which then cross-links the platelets, leading to aggregation (Figures 1 and 3). Similarly, stimulation of leukocytes with, for example, bacterially derived peptides leads to a rapid increase in the affinity of their integrins for the immunoglobulin superfamily of adhesion receptors (1,2,8, 9).

Other members of the immunoglobulin superfamily of adhesion receptors, that are also built from several immunoglobulin-like domains, engage in homotypic interactions (adhesion between identical molecules). One example of these homotypic interactions, important in blood vessels, is that of PECAM-1, which is distributed in contact areas between endothelial cells and likely contributes to the strength of cohesiveness of the endothelial sheet (12). PECAM-1 is also present on leukocytes, and there are several experimental results pointing to the possibility that it may be involved in leukocyte transmigration (13). This may again be accomplished through homotypic interactions between the PECAM-1 on the endothelial cell and that on the transmigrating leukocyte.

Cadherins

The last major family of adhesion receptors are the cadherins (14,15). Their role in cardiovascular pathology has been the least studied. Many cadherins, like selectins, are designated by a letter indicating the tissue in which they were first described. Their interactions appear to be homophilic and between like cells. Each molecule has four homologous calcium-binding domains, a single transmembrane domain, and a conserved cytoplasmic tail that mediates association with specific components of the cellular cytoskeleton. Cadherins probably play a major role in embryonic development because their appearance and disappearance correlate temporally and spatially with major morphogenetic events. In adult tissues, cadherins likely contribute to maintenance of correct intercellular interactions and to fortification of adhesion among cells of a particular cellular layer. Recently, an endothelial-specific cadherin was described that is distributed in areas of contact between the endothelial cells (16). Experimental evidence implicates all four families of adhesion receptors in mediating adhesion of endothelial cells. The mechanics of this interaction must be very sophisticated and highly regulated, since the monolayer has to be strong enough to withstand the high shearing force of streaming blood, and at the same time the junctions must be easily openable, to allow quick passage of leukocytes.

GENE TARGETING TECHNOLOGY

Many cardiovascular and blood diseases are due to the absence of, or a defect in, a single gene. If the gene is known, an animal model of that disease can be obtained by engineering mice defective in that particular gene. Other diseases, such as atherosclerosis or cancer, are much more complex and are the results of several genetic and/or environmental factors. In investigating the role a particular gene product plays in such diseases, the generation of an animal lacking one gene product may prove to be very informative. In acute disease models, the involvement of an adhesion receptor can be investigated through the injection of an inhibitory antibody or a soluble form of the receptor that acts as an inhibitor. This is technically not possible in chronic diseases such as arthritis or chronic inflammation. In these cases, gene ablation may prove to be the most direct way to study the role of an adhesion molecule in such processes.

Inactivating a desired gene without affecting the rest of the animal's genetic background can now be accomplished by gene targeting technology (17,18). This revolutionary new technique is the result of combining major advances in fields as diverse as embryology and molecular biology. The most important breakthrough was the obtaining of embryonic stem (ES) cell lines that could be passaged in culture long enough to allow genetic manipulation, without losing their original totipotency. This means that these cells can be modified in vitro and, after injection into a blastocyst, they contribute to all organs including the germ cells. Thus, the modified gene can be transmitted to future generations.

To accomplish the gene inactivation, first the mouse gene is cloned and the exons encoding important regions of the protein are located. These exon(s) are then best deleted and replaced by a positive selection gene (providing neomycin resistance, for example). This drug-resistance gene, surrounded by 2–4 kb of genomic sequence on each side, forms the basis for the "targeting vector" to be transfected into the ES cells. A deletion that includes the signal sequence is a good strategy, as this will likely prevent the translocation of any residual protein product through the secretory pathway, thus preventing its expression. However, other strategies of deletion or insertion have also worked.

Most investigators include at one or both ends of the targeting vector the negative selection gene, thymidine kinase. This gene is lost during a double crossover homologous recombination event, while it would be retained through a random insertion of the targeting construct, which occurs via the ends of a DNA molecule. A double selection procedure can then be used, which enriches for ES clones that 1) incorporated the transfected DNA and are therefore neomycin-resistant and 2) lost the thymidine kinase gene since they are resistant to a toxic metabolite produced by this enzyme from gancyclovir added to the medium. The selected ES clones are checked for the correct homologous recombination event and injected into recipient blastocysts, which are implanted into a pseudopregnant female. The genetic transmission of the ES cell genotype is monitored by the coat color of the offspring, as the ES cells encode a dominant coat-color gene. The resulting heterozygous mutant animals can then be intercrossed to produce homozygotes. If the gene targeted is necessary for embryonic development, no homozygous progeny result (see later discussion), but, in many cases, homozygous animals are viable.

It is important to check that the resulting homozygous mutant animals are truly deficient in the gene product, i.e., do not contain the targeted protein or an RNA transcript of the correct size. A targeting construct comprising mostly the 3′ end of the gene may result in the production of a truncated protein with its N-terminal domains intact. This would be especially disturbing in the case of a single-chain transmembrane adhesion receptor where the N-terminal domains contain most of the binding activity, and are known to be active in a soluble form, independently of the rest of the molecule. The production of a truncated soluble molecule would therefore complicate the interpretation of the phenotype of such an animal, as the phenotype would be the consequence of both the absence of the normal receptor and the presence of a soluble adhesion inhibitor.

Using these approaches, mutations have now been made in representative members of all the families of adhesion receptors (Figures 2 and 3) and in several extracellular matrix proteins. In the following sections, we review the results obtained and discuss their relevance to vascular biology.

TARGETED MUTATIONS IN GENES FOR ADHESION PROTEINS

Extracellular Matrix Molecules

The first extracellular matrix (ECM) molecule "knocked out" was actually an accident: a retroviral insertion in the gene of $\alpha_1(I)$ collagen inactivated the gene (19). This mutation produced embryonic lethality because of rupture of the vasculature. This result proves the importance of type I collagen for vascular integrity, and results on human mutations in collagen genes conform with this conclusion (20; Chapter 7 in this volume). Embryonic lethality, caused at least in part by vascular defects, also occurs when the gene for fibronectin is knocked out (21). Fibronectin is a widely expressed ECM protein found in, among other places, the basement membranes of blood vessels and in the developing heart (22). When the gene for fibronectin is eliminated, there are major defects in vascular development. The extraembryonic yolk sac vasculature fails to develop properly and, depending on genetic background, the embryonic blood vessels and heart either do not form or are severely deformed. Therefore, the mutations in collagen I and fibronectin clearly demonstrate their roles in vascular development and integrity.

In contrast with results on collagens and fibronectin, knockouts of two other extracellular matrix proteins, tenascin-C and thrombospondin-1 produce viable homozygous mutant animals (23; J. Lawler, H. B. Rayburn, and R. O. Hynes, unpublished). These animals appear largely normal, although ongoing work will likely uncover subtle defects. This is particularly surprising for the thrombospondin-1 knockout since this protein is a major released protein of the platelet α granules and is a prominent constituent of endothelial basement membranes. This is a good example of the sort of unexpected result that can arise because of *redundancy* or *compensation*. In the case of redundancy, two gene products are expressed in a given situation and, on elimination of one of them, the other suffices. Compensation arises when expression of a given gene is eliminated and another gene up-regulates to compensate for the missing function. Presumably one or another of these phenomena accounts for the absence of marked phenotypic defects in the tenascin-C and thrombospondin-1 knockouts. It may be relevant that each of these genes is a member of a multigene family.

Integrins

Genes for several integrin subunits have been altered by gene targeting. Mutation of the gene encoding one subunit of $\alpha_5\beta_1$, a fibronectin receptor integrin (see Figure 3), produces embryonic lethality (24). The phenotype is distinct from, and milder than, that of the fibronectin-null mutation discussed above. This shows that there must be other fibronectin receptors. However, the α_5-null mutation also leads to defects in the developing blood vessels of the yolk sac and the embryo itself. For example, the dorsal aortae are distended and frequently leaky. Consequently, the embryos die midgestation, in part because of these vascular defects although, as for the fibronectin-null mutant, there are other developmental defects as well. Since α_5 integrin is expressed in many endothelial cells, these results are not surprising, but they do confirm the crucial importance of $\alpha_5\beta_1$ integrin–fibronectin interactions for integrity of the lining of blood vessels.

Targeting of a second integrin gene encoding the α_4 subunit produced some surprising results (J. T. Yang, H. B. Rayburn, and R. O. Hynes, in preparation). There are two integrins ($\alpha_4\beta_1$ and $\alpha_4\beta_7$) containing this subunit. Both these integrins function as fibronectin receptors, recognizing a different site from that bound by $\alpha_5\beta_1$, but both can also bind to VCAM-1, an immunoglobulin superfamily member (Figure 2). α_4 Integrin–VCAM-1 interactions are known to be important for adhesion of various white blood cells to inflamed endothelium (Figures 1 and 2 and discussion above). The defects caused by the α_4 knockout revealed other, previously unsuspected roles for these integrins. Two defects were observed. In the first, the allantois failed to fuse with the chorion, leading to failure of placentation. The second defect was in the heart; the endocardium and the forming cardiac blood vessels within it failed to develop properly, leading to death by hemorrhage into the pericardial cavity. Both these defects probably reflect α_4 integrin–VCAM-1 interactions necessary for these morphogenetic processes. While these were completely unexpected, it has now become clear that α_4 integrin and VCAM-1 (as well as fibronectin) are expressed in the involved tissues and, furthermore, that a knockout of the VCAM-1 gene produces similar defects (see "Immunoglobulin Superfamily Adhesion Receptors" below).

Although these embryonic defects provide important information about development, including that of the heart and vasculature, the associated lethality precludes analysis of any consequences for function in the mature animal. There are potential ways around this problem, which we discuss later, under "Future Prospects." At this point we can summarize the results by saying that the targeted mutations confirm the essential nature of integrin–matrix interactions for a variety of developmental processes, including several involving the cardiovascular system.

A final targeted mutation in a different integrin subunit, β_2, brings up some other issues. This mutation generated by an insertional event produces a hypomorphic mutant (25). That is, the normal protein is produced at a greatly reduced level as a consequence of cryptic promoter and splice sites within the inserted prokaryotic sequences. Nonetheless, this is an interesting mutation, which can serve as a model of a human disease. Defects in the human gene for the integrin β_2 subunit lead to the genetic disease leukocyte adhesion deficiency, or LAD I (26,27). As the name suggests, this disease is characterized by defects in adhesion of leukocytes, particularly granulocytes. Patients have elevated granulocyte counts and do not form pus, because their neutrophils cannot adhere to the vessel wall and extravasate at sites of infection (cf. Figure 2). This experiment of nature demonstrates the importance of β_2 integrins for leukocyte adhesion; LAD I patients suffer from recurrent infections and defects in wound healing. The severity varies with the particular mutation—in severe cases, the β_2 subunit is absent; in mild cases it is reduced in level and/or altered (27,28). The β_2 hypomorphic mutation in mice is a model for the milder form of LAD I. The mutant mice are viable but show somewhat elevated circulating granulocytes and lymphocytes, defective recruitment of neutrophils to a peritoneal inflammation, and delayed rejection of cardiac transplants. These results confirm the importance of β_2 integrins in the extravasation of neutrophils and in the rejection of foreign tissues. It will be interesting to see the phenotype of a true β_2-null mutation.

Cadherins

As we mentioned earlier, cadherins are important for adhesions between like cells, including endothelial cells. Two cadherins have been knocked out. N-cadherin is expressed in heart and in endothelial cells, among other places (14,15). An N-cadherin-null mutation is a homozygous embryonic lethal, which shows major defects in cardiogenesis (G. Radice, H. B. Rayburn, M. Takeichi, and R. O. Hynes, unpublished results). Basically, the myocardium falls apart early in heart development because of loss of adhesions between the cells. So, this mutation, like some of the ones discussed above, provides information about the importance of this cadherin for heart development but tells us little about its roles in the mature animal.

The other cadherin mutation is an example of one that produces a *less* severe defect than expected. Although P-cadherin is strongly expressed in the deciduum and placenta, knockout of the gene yields viable, fertile, homozygous P-cadherin-null mice (G. Radice, H. Rayburn, M. Takeichi, and R. O. Hynes, unpublished results). This result suggests redundancy or compensation by another cadherin.

Selectins and Their Ligands

In contrast to the developmental defects produced by many mutations in ECM molecules, integrins, and cadherins, mutations of genes encoding selectins produce viable mice with defined defects.

The selectins represent the best example of the fast progress of discovery in the field of adhesion. It was only in 1989 that these molecules were independently cloned and their relatedness realized (3,4). Now, animals engineered to be deficient in each selectin exist. These mice will be very useful in deciphering in which pathological condition a particular selectin is most involved, leading to therapies specifically targeted to that selectin or combination of selectins. Selectins are likely to contribute to many disease states, where excessive leukocyte adhesion and infiltration are a problem. Some examples of these are: reperfusion injury, arthritis, atherosclerosis, frostbite, transplant rejection, asthma, diabetes, cystic fibrosis, and many other conditions in which chronic inflammation or autoimmunity is involved.

Mice lacking any one selectin have no obvious gross phenotype and are fertile. The severity of their phenotype after challenge varies, indicating that the functions of selectins are not fully overlapping. Results from the P- and L-selectin knockout mice have been published (28,33); those from E-selectin-deficient animals have been reported through personal communication by M. Labow/B. Wolitzky.

P-selectin is a platelet and endothelial protein. Except for lack of binding of activated platelets to neutrophils, a known role of P-selectin (29), no major defects in platelet function have yet been observed in the P-selectin-deficient platelets, and the platelet counts in these animals are normal. As in the β_2-deficient mice (see "Integrins," above) and the ICAM-1-deficient animals (see below, "Immunoglobulin Superfamily Adhesion Receptors"), the numbers of peripheral neutrophils are elevated but those of lymphocytes are not (28). Since there do not appear to be more myelocytic precursors in the bone marrow of these mice, and since injected radiolabeled neutrophils have a longer half-life in the P-selectin-deficient mice (R. Johnson, unpublished observation), it is likely that P-selectin is involved in the normal mechanism of neutrophil clearance. The major difference we observed between wild-type and P-selectin-deficient animals was the complete absence of leukocyte-rolling in mesenteric venules of the P-selectin-deficient animals, monitored in live, anesthetized mice, the intestines of which were exteriorized on the stage of a microscope equipped with a TV camera and videorecorder. The very act of exteriorization causes some mast cell degranulation (30), so the endothelial linings of the mesenteric venules are effectively activated. In wild-type mice, this leads to rolling of leukocytes along the venule wall, which can readily be recorded on videotape. In contrast, in P-selectin-deficient mice there are absolutely no rolling leukocytes, even after deliberately induced degranulation (28). Antibodies to L-selectin have been shown to decrease the numbers of rolling leukocytes significantly (31,32), and decreased rolling is also seen in L-selectin-deficient animals. In these animals, the inhibitory effect of the absence of L-selectin on leukocyte-rolling is not complete but becomes somewhat more pronounced after the mesentery is externalized for longer periods (33). Both selectins therefore appear to contribute to the rolling of leukocytes, with P-selectin being absolutely crucial for baseline rolling, i.e., before the vessel becomes inflamed. Our most recent results indicate that P- selectin is important even in inflamed vessels, where rolling cells are numerous. Although no longer completely absent, the numbers of rolling leukocytes are still drastically reduced in inflamed P-selectin-deficient vessels in comparison with wild-type vessels treated the same way (R. Johnson, unpublished observation).

As for leukocyte extravasation, a 2-hour delay in neutrophil extravasation into the peritoneum is seen in both P- and L-selectin-deficient animals after injection of thioglycollate, again as if both molecules contribute cooperatively (28,33). In contrast, only a small decrease in neutrophil recruitment was seen in the E-selectin-deficient animals (M. Labow and B. Wolitzky, personal communication). In the preliminary studies performed, the E-selectin-deficient animals showed only very mild defects. It is possible that the function of E-selectin is restricted to certain organs or types of vessels or to certain inflammatory states. The E-selectin-deficient animals should help to pinpoint the situations and places in which this selectin comes into play.

P-selectin-deficient animals have been available since 1992, and were therefore already submitted to several experimental protocols. Interestingly, defects were seen both in acute phenomena, such as lack of rapid recruitment of phagocytes to the site of a skin wound, and in chronic phenomena, such as reduction in contact hypersensitivity response (M. Subramaniam, unpublished observation). This may reflect the fact that P-selectin can be rapidly expressed from storage sites in acute situations and that its synthesis is also up-regulated by inflammatory cytokines, which may lead to a direct surface expression in chronic situations (34). In addition, after surface expression, P-selectin returns to storage granules from which it can be re-expressed, possibly in a chronic manner (35).

Recently, the first of the selectin mucin ligands was eliminated by gene targeting: the soluble ligand for L-selectin, glyCAM-1 (6). The lymph nodes in the knockout mice were much larger and hypercellular, indicating that, in the absence of glyCAM-1, lymphocyte-homing is more efficient (L. Lasky and M.

Moore, personal communication). The role of this soluble ligand may therefore be to modulate the function of the homing receptor L-selectin.

Immunoglobulin Superfamily Adhesion Receptors

ICAM-1, composed of five immunoglobulin-like domains, is found both on leukocytes and endothelial cells. It binds to the β_2 integrins LFA-1 ($\alpha_L\beta_2$), expressed on all leukocytes, through its domains 1 and 2, and to Mac-1 ($\alpha_M\beta_2$), expressed on monocytes and granulocytes, through domain 3 (2,9). ICAM-1 on antigen-presenting cells was shown to be important for stimulation of T-cell responses in vitro (36). As discussed above, ICAM-1 on endothelial cells, together with other adhesion receptors of the immunoglobulin family, mediates adhesion to leukocytes (see Figures 1 and 2). After treatment of cultured endothelial cells with inflammatory cytokines, ICAM-1 expression increases up to 40-fold, resulting in much enhanced binding of both lymphocytes and neutrophils to the endothelial cells (2,9). Considering these and other indications of the important role this molecule may play in inflammatory processes and in immune responses, it is not surprising that ICAM-1 was chosen by two laboratories to be the first immunoglobulin adhesion receptor to be disrupted through gene targeting. This was accomplished by inserting a neomycin-resistance cassette in the fourth (37) or fifth exon (38) encoding the third and fourth immunoglobulin domain, respectively. Although these manipulations clearly prevented the formation of the intact transmembrane receptor, whether a soluble truncated form was generated by the mutant animals was not examined.

Animals homozygous for ICAM-1 gene disruption were grossly normal and fertile. Both neutrophil and lymphocyte counts were elevated several-fold in comparison with wild-type animals, indicating that these animals may have a defect in leukocyte extravasation. This assumption was confirmed by the observation of a twofold reduction in neutrophil recruitment in the deficient animals in response to the injection of the inflammatory irritant thioglycollate (38). The fact that neutrophils could still extravasate in large numbers shows that other endothelial molecules can serve as ligands for the β_2 integrins. Both groups have further found that ICAM-1 deficiency played a prominent role in the contact hypersensitivity response and in the ability of spleen cells to stimulate proliferation of ICAM-1 positive allogeneic T cells in the mixed lymphocyte reaction.

In addition, Xu and colleagues (37) have made an observation that is likely to lead to improved treatment of patients with septic shock. They demonstrated that the absence of ICAM-1 provides a significant protection from lethality in mouse models of both gram-negative and gram-positive septic shock. Although both wild-type and ICAM-1-deficient animals responded to a high dose of endotoxin by an initial rapid decline in neutrophil count and infiltration of neutrophils into the liver, by 24 hours the ICAM-1-deficient animals showed significantly reduced infiltration of neutrophils into the liver. Xu et al. proposed that the protective effect in this model is due to decreased neutrophil adhesion to the liver endothelium. This could be tested by translating wild-type bone marrow into ICAM-1-deficient mice. If the interpretation is correct, these mice should have the same level of protection as animals deficient in ICAM-1 in both leukocyte and endothelium. The gram-positive septic shock is mediated by exotoxins, which act as superantigens, thus producing T-cell activation. The protective effect of ICAM-1 deficiency in this model was due to a reduction in T-cell activation. This resulted in decreased production of inflammatory cytokines, which are the primary mediators in both types of shock. From these observations it appears that in the gram-positive shock the leukocyte ICAM-1 is the primary culprit.

As mentioned briefly above, under "Integrins," another Ig superfamily molecule, VCAM-1 has also been knocked out. This has been done by two groups (M. Cybulsky and M. Labow, personal communication). The defects seen in these mutant mice are similar to those discussed above for the α_4 integrin knockout, namely, failure of allantois–chorion fusion during placentation and defects in heart development, including failure of development of the epicardium. These results are consistent with roles for α_4 integrin–VCAM-1 interactions in both processes, an unexpected result.

INTERPRETATIONS AND SIGNIFICANCE

The targeted mutations discussed in the preceding sections represent only the beginnings of the application of this approach to questions concerning adhesion in the vasculature. There are more genes that need to be targeted, the mutations already made need further analysis, and extensions of the current approaches (see "Future Prospects" below) will undoubtedly be applied. Nonetheless, one can already see the value of this approach.

The mutations made and analyzed so far fall into several broad categories. First, there are those that show unexpectedly mild, or virtually no, phenotypes. Examples are the two ECM molecules, tenascin-C and thrombospondin-1, and P-cadherin. As mentioned, the absence of marked defects in these homozygous mutant mice suggests either redundancy or compensation by some other gene product(s). Since all three are members of multigene families, the most obvious suggestion is that homologous genes can overlap in function and obscure the roles of individual genes. Genes are not maintained during evolution unless they confer some selective advantage, but a small increase in fitness (as low as 1%) can provide sufficient selective pressure. The analyses to date would not have revealed such subtle defects, and further analyses of these mutants are in order. Even the absence of an obvious phenotype is informative; it tells one that, despite the fact that a given gene is expressed somewhere (e.g., in megakaryocytes and endothelial cells in the case of thrombospondin-1), this does not mean that it subserves some key function one might want to attribute to it and that one needs to look elsewhere for the molecules that are crucial.

The second broad group of mutants comprises those that cause embryonic lethality. While this precludes analyses of physiological functions in adults (barring further analyses to be discussed under "Future Prospects"), valuable information can be elicited by analyses of the nature of the embryonic defects. As we discussed earlier, several of these mutations have demonstrated roles for ECM proteins (collagen I, fibronectin), integrins (α_4 and α_5), VCAM-1, and N-cadherin in development of the heart and/or vasculature. Some of the defects observed resemble cardiac defects seen in humans; others are more severe, and equivalent defects in humans would cause early spontaneous abortion. These mutations have already provided fundamental insights and surprising new results pertaining to cardiac development. Again, the strong defects produced by null mutations presage more subtle defects likely to results from more subtle mutations (see the following section, "Future Prospects").

The most readily interpretable group of mutants includes the viable mutants with measurable defects. The most progress here has been made in studies of genes of restricted expression. Because of this restriction, mutations in these genes do not cause embryonic lethality; viable mice are born and can be analyzed for defects in their physiology. The three selectin mutants are a good case in point. A great deal remains to be done, but already one can tell that this three-gene family is not fully redundant; mutations in any of the three produce effects on leukocyte behavior, although to date the deficits due to loss of E-selectin have proven to be rather mild as compared with those exhibited by L-selectin mutants and, even more so, by P-selectin mutants. This was unexpected—perhaps because E-selectin was discovered by its function in neutrophil-endothelial cell adhesion and L-selectin was discovered as a lymphocyte homing receptor, while P-selectin was discovered as an activation antigen on platelets, it had been commonly believed that E- and L-selectin were likely to be most important for leukocyte-vessel wall adhesion. In fact, it appears that P-selectin is at least as important or even more important. The phenotype of P-selectin-null mice indicates that P-selectin functions in both acute and chronic inflammation. In contrast, its role on activated platelets remains less clear.

The current status of studies on the three selectin mutants clearly reveals the importance of the in vivo analyses made possible by the mutations. Neither in vitro studies nor antibody-blocking studies in vivo had given an accurate and complete picture of selectin functions. It seems likely that selectins play complementary roles, partially overlapping, partially synergistic, and partially independent. Further studies of the mutant mice challenged in different ways, combined with studies of the detailed patterns of expression of each selectin in individual cases of inflammation and so forth, will be necessary to determine which of them is most important in given situations. Such dissection of their roles will be crucial for planning therapeutic interventions using blocking agents such as carbohydrate mimetics designed to block selectin–ligand interactions.

The knockout of glyCAM-1 is a good example of the value of this approach in choosing targets for therapy. This protein, a ligand for L-selectin that is expressed in high endothelial venules, had been suggested as a molecule targeting lymphocytes to particular lymph nodes (6). However, the fact that the glyCAM-1 knockout mice show *increased* rather than *decreased* numbers of lymphocytes in lymph nodes clearly changes the perspective; glyCAM-1 now appears to be a negative regulator of lymphocyte homing, not a counterreceptor for homing.

Mutations in the receptors involved in integrin–Ig superfamily interactions between leukocytes and endothelial cells are still under development. The published β_2 mutation (25) is a hypomorph; a null mutation will provide additional information. The α_4 and VCAM-1 mutations are embryonic lethals and ways around this difficulty must be developed (see the following section, "Future Prospects"). The ICAM-1 mutation needs to be supplemented by mutations in ICAM-2 and ICAM-3, alternative ligands for β_2 integrins. However, the existing mutations already confirm the importance of β_2 integrins and ICAM-1 for leukocyte function. Of course, LAD I patients also implicate β_2 integrins, but they are rare patients and there are severe limitations on the experiments that can be done. One can anticipate with confidence that a variety of β_2 mutations will be made in mice, reflecting the range of mutations seen in human LAD I patients. Furthermore, mutations in each of the three α subunits (α_L of LFA-1, α_M of Mac-1/CR3, and α_x of p150,95/CR4) will also be made, as will the various ICAM mutants. This, like the situation discussed above for selectins, should allow us to determine exactly which receptor pairs are most important in given pathologies, the extent of overlap in function, and which molecules offer the best targets for therapeutic intervention.

The β_2 integrin/LAD I situation brings up another point. This is a good example of the way in which one can create animal models of a human disease. Another leukocyte adhesion deficiency syndrome, LAD II, is caused by a defect in fucose metabolism that ablates the ligands for some (perhaps all) selectins (39). One should be able to mimic this aspect of the disease by double or triple mutants in the selectins, which will, in any case, be informative. Other human vascular diseases for which mouse models should soon be forthcoming include Glanzmann's thrombasthenia (β_3 and probably also α_{IIb} integrin subunits) and Bernard-Soulier syndrome (glycoprotein Ib of platelets; see Ref. 40) and von Willebrand disease (von Willebrand factor; Ref. 41). Mutations in these genes can be expected to be viable because severe mutations in the human genes allow viability. The mutant mice will then be susceptible to detailed analyses not possible with the rare human patients. The existing mutant mice should produce much additional information as they are tested further for their capacities. Intercrossing to produce mice doubly mutant in different genes should also be very informative.

FUTURE PROSPECTS

In addition to the obvious future extensions mentioned in the preceding section, we should mention some likely technological advances that will render the targeted mutagenesis approach even more powerful. We relate them to specific examples having to do with genes involved in vascular cell adhesion.

As we have mentioned several times, embryonic lethality precludes analyses of the role of the gene in question in later stages, including adulthood. There are several potential ways around this problem. The most straightforward is the use of mosaic or chimeric mice. If the gene product in question is cell-autonomous—for example, a cell-surface receptor—then it is possible with available methods to generate mice that are a mixture of normal and mutant cells. The easiest way to do this is to take the heterozygous mutant ES cells (see "Gene Targeting Technology" above) and convert them to homozygosity, either by targeting the second allele with a targeting construct harboring a different selectable marker or, alternatively, by increasing the selection pressure with a higher level of G418 drug (42). This, with a reasonably high frequency, selects for cells in which the wild-type allele is converted to mutant form including a second copy of the selectable marker. In either case, the homozygous mutant ES cells can then be injected into wild-type blastocysts and the resulting mice are chimeras composed of both mutant and normal cells. This approach has been used to good effect in analyses of the immune system (43), and, in this way, it is possible to generate chimeric mice containing α_4-integrin-null blood cells (J. T. Yang, A. Arroyo, H. Rayburn, and R. O. Hynes, unpublished results). This should allow analyses of the role of α_4 integrins in leukocyte behavior.

Another way around embryonic lethality may be to make more subtle mutations in the gene so that, rather than ablating it completely, the level of expression is reduced or, by mutations in the promoter, eliminated in only certain cells. Subtle mutations in the coding region of genes to produce mutant proteins also offer promise; one could envisage both partially defective mutants and dominant-negative mutants. Techniques for generating subtle mutants rather than complete knockouts are already available (44–46), and more are being developed.

One recent advance that shows much promise is the development of methods for knocking out genes only in specific cell types or at specific times (47–49). For example, one might wish to eliminate a gene from only T or B lymphocytes, or neutrophils, or endothelial cells. While the available methods still need some refinement, it seems almost certain that such manipulations will be available soon, allowing production of a different sort of chimera in which only specific cell types are mutant.

Another way to achieve that end would be to use transgenic technology to introduce into knockout mice a copy of the ablated gene driven by a cell-type-specific promoter. Advances in transgenic technology make this an increasingly feasible option (50–55). Similar approaches could be used to introduce antisense genes or dominant-negative genes into only certain cell types to eliminate function of a given gene or its product in one cell type without affecting others. Even with some current knockout mice, one could generate such chimeras. For example, reciprocal bone marrow transplants between P-selectin-null and wild-type mice could produce mice lacking P-selectin only in their endothelial cells or only in their platelets, in order to dissect the relative importance of P-selectin in these two cell types.

Finally, we should mention the possibility of gene therapy. Certain vascular diseases, such as LAD I and Glanzmann's thrombasthemia, seem to be good prospects for gene therapy. The standard current approach to gene therapy is to introduce a transgene by transfection into some cell type, which is then reintroduced into the patient. There is always the finite possibility of introducing an undesirable mutation by the insertion of the transgene. In mice this happens in 5–10% of insertions and, although gene therapy methods would mutate only one of the two alleles in any given cell, there is the remote possibility that such a mutation could be deleterious (e.g., activate an oncogene). An alternative approach might be to use the developing methods of homologous recombination to convert one of the mutant alleles to wild-type. This is the reverse of the approach one takes to create a subtle mutation. The methods are currently far from sufficiently reproducible for this to be a useful approach. However, it seems possible, even likely, that the rapid pace of development of technology in this field will make such approaches realistic in the not-too-distant future.

CONCLUSION

The last several years have seen extraordinarily rapid progress in our understanding of the molecular basis of cell adhesion and in our ability to manipulate genes in mice. The confluence of these two rapidly advancing fields is already beginning to provide new insights into the multifarious roles of cell-adhesion molecules in vascular biology, allowing generation of animal models for human diseases and detailed dissection of the specific roles of individual genes and gene products. These advances are still accumulating rapidly, and it is certain that much will be learned by these methods in the next few years. Cell-adhesion molecules are excellent targets for therapy; given detailed understanding of the molecular interactions, one can design reagents to block them and intervene in adhesion processes that contribute to thrombosis and inflammation. Indeed, such molecules are in clinical trials and will probably soon be in the clinic. The genetically engineered animal models will aid greatly in defining the best targets for such therapeutic approaches. Thus, in addition to contributing to a deeper understanding of the basic processes, this very active area of research promises to have significant impact on the treatment of cardiovascular disease.

REFERENCES

1. Hynes RO, Lander AD. Contact and adhesive specificities in the associations, migrations, and targeting of cells and axons. Cell 1992; 68:303-322.
2. Springer TA. Traffic signals for lymphocyte recirculation and leukocyte emigration: the multistep paradigm. Cell 1994; 76:301-314.
3. Lasky LA. Selectins: interpreters of cell-specific carbohydrate information during inflammation. Science 1992; 258:964-969.
4. Bevilacqua MP, Nelson RM. Selectins. J Clin Invest 1993; 91:379-387.
5. Baumhueter S, Singer MS, Henzel W, Hemmerich S, Renz M, Rosen SD, Lasky LA. Binding of L-selectin to the vascular sialomucin CD34. Science 1993; 262:436-438.
6. Lasky LA, Singer MS, Dowbenko D, Imai Y, Henzel WJ, Grimley C, Fennie C, Gillett N, Watson SR, Rosen SD. An endothelial ligand for L-selectin is a novel mucin-like molecule. Cell 1992; 69:927-938.
7. Sako D, Chang XJ, Barone KM, Vachino G, White HM, Shaw G, Veldman GM, Bean KM, Ahern TJ, Furie B, Cumming DA, Larsen GR. Expression cloning of a functional glycoprotein ligand for P-selectin. Cell 1993; 75:1179-1186.
8. Hynes RO. Integrins: versatility, modulation, and signaling in cell adhesion. Cell 1992; 69:11-25.
9. Sánchez-Madrid F, Corbi AL. Leukocyte integrins: structure, function and regulation of their activity. Sem Cell Biol 1992; 3:199-210.
10. Phillips DR, Charo IF, Scarborough RM. GPIIb-IIIa: the responsive integrin. Cell 1991; 65:359-362.
11. Shattil SJ, Ginsberg MH, Brugge JS. Adhesive signaling in blood platelets. Curr Opin Cell Biol 1994; 6:695-704.
12. DeLisser HM, Chilkotowsky J, Yan HC, Daise ML, Buck CA, Abelda SM. Deletions in the cytoplasmic domain of platelet- endothelial cell adhesion molecule-1

(PECAM-1, CD31) result in changes in ligand binding properties. J Cell Biol 1994; 124:195–203.

13. Vaporciyan AA, DeLisser HM, Yan HC, Mendiguren II, Thom SR, Jones ML, Ward PA, Albelda SM. Involvement of platelet-endothelial cell adhesion molecule-1 in neutrophil recruitment *in vivo*. Science 1993; 262:1580–1582.

14. Takeichi M. Cadherins: a molecular familly important in selective cell-cell adhesion. Annu Rev Biochem 1990; 59:237–252.

15. Takeichi, M. Cadherin cell adhesion receptors as a morphogenetic regulator. Science 1991; 251:1451–1455.

16. Lampugnani MG, Resnati M, Riateri M, Pirott R, Pisacane A, Houen G, Ruco LP, Dejana E. A novel endothelial-specific membrane protein is a marker of cell–cell contacts. J Cell Biol 1992; 118:1511–1522.

17. Capecchi MR. Altering the genome by homologous recombination. Science 1989; 244:1288–1292.

18. George EL, Hynes RO. Gene targeting and generation of mutant mice for studies of cell–extracellular matrix interactions. Meth Enzymol 1994; 245:386–420.

19. Schnieke A, Harbers K, Jaenisch R. Embryonic lethal mutation in mice induced by retrovirus insertion into the $\alpha_1(I)$ collagen gene. Nature 1983; 304:315–320.

20. Kuivaniemi H, Tromp G, Prockop DJ. Mutations in collagen genes: causes of rare and some common diseases in humans. FASEB J 1991; 5:2052–2060.

21. George EL, Georges-Labouesse EN, Patel-King RS, Rayburn H, Hynes RO. Defects in mesoderm, neural tube and vascular development in mouse embryos lacking fibronectin. Development 1993; 119:1079–1091.

22. Hynes RO. Fibronectins. New York: Springer-Verlag, 1990.

23. Saga Y, Yagi T, Ikawa Y, Sakakura T, Aizawa S. Mice develop normally without tenascin. Genes Dev 1992; 6:1821–1831.

24. Yang JT, Rayburn H, Hynes RO. Embryonic mesodermal defects in α_5-integrin-deficient mice. Development 1993; 119:1093–1105.

25. Wilson RW, Ballantyne CM, Smith CW, Montgomery C, Bradley A, O'Brien WE, Beaudet AL. Gene targeting yields a CD18-mutant mouse for study of inflammation. J Immuno 1993; 151:1571–1578.

26. Anderson DC, Springer TA. Leukocyte adhesion deficiency: an inherited defect in the Mac-1, LFA-1, and p150,95 glycoproteins. Annu Rev Med 1987; 38:175–194.

27. Arnaout MA. Structure and function of the leukocyte adhesion molecules CD11/CD18. Blood 1990; 75:1037–1050.

28. Mayadas TN, Johnson RC, Rayburn H, Hynes RO, Wagner DD. Leukocyte rolling and extravasation are severely compromised in P-selectin-defient mice. Cell 1993; 74:541–544.

29. Larsen E, Celi A, Gilbert GE, Furie BC, Erban JK, Bonfanti R, Wagner DD, Furie B. PADGEM protein: A receptor that mediates the interaction of activated platelets with neutrophils and monocytes. Cell 1989; 59:305–312.

30. Kubes P, Kanwar S. Histamine induces leukocyte rolling in post-capillary venules: A P-selectin-mediated event. J Immunol 1994; 152:3570–3577.

31. Ley K, Gaehtgens P, Fennie C, Singer MS, Lasky LA, Rosen SD. Lectin-ike cell adhesion molecule 1 mediates leukocyte rolling in mesenteric venules in vivo. Blood 1991; 77:2553–2555.

32. von Adrian UH, Chambers JD, McEvoy LM, Barsgatze RF, Arfors KE, Butcher EC. Two-step model of leukocyte-endothelial cell interaction in inflammation. Proc Natl Acad Sci USA 1991; 88:7538–7542.

33. Arbonés ML, Ord DC, Ley K, Ratech H, Maynard-Curry C, Otten G, Capon DJ, Tedder TF. Lymphocyte homing and leukocyte rolling and migration are impaired in L-selectin (CD62L) deficient mice. Immunity 1994; 1:247–260.

34. Weller A, Isenmann S, Vestweber D. Cloning of the mouse endothelial selectins. J Biol Chem 1992; 267:15176–15183.

35. Subramaniam M, Koedam JA, Wagner DD. Divergent fates of P- and E-selectins after their expression on the plasma membrane. Molec Biol Cell 1993; 4:791–801.

36. Dustin ML. Two-way signalling through the LFA-1 lymphocyte adhesion receptor. Bioessays 1990; 12:421–427.

37. Xu H, Gonzalo JA, St Pierre Y, Williams IR, Kupper TS, Cotran RS, Springer TA, Gutierrez-Ramos JC. Leukocytosis and resistance to septic shock in intercellular adhesion molecule 1–deficient mice. J Exp Med 1994; 180:95–109.

38. Sligh JE, Ballantyne CM, Rich SS, Hawkins HK, Smith CW, Bradley A, Beaudet AL. Inflammatory and immune responses are impaired in mice deficient in intercellular adhesion molecule 1. Proc Natl Acad Sci USA 1993; 90:8529–8533.

39. Etzioni A, Frydman M, Pollack S, Avidor I, Phillips ML, Paulson JC, Gershoni-Baruch R. Brief report: Recurrent severe infections caused by a novel leukocyte ahesion deficiency. N Engl J Med 1992; 327:1789–1792.

40. Kieffer N, Phillips DR. Platelet membrane glycoproteins: functions in cellular interactions. Annu Rev Cell Biol 1990; 6:329–357.

41. Sadler JE. A revised classification of von Willebrand Disease. Thromb Haemost 1994; 71:520–525.

42. Mortensen RM, Conner DA, Chao S, Geisterfer-Lowrance AAT, Seidman JG. Production of homozygous mutant ES cells with a single targeting construct. Molec Cell Biol 1992; 12:2391–2395.

43. Chen J, Lansford P, Stewart V, Young F, Alt FW. RAG-2-deficient blastocyst complementation: an assay of gene function in lymphocyte development. Proc Natl Acad Sci USA 1993; 90:4528–4532.

44. Hasty P, Ramirez-Solis R, Krumlauf R, Bradley A. Introduction of a subtle mutation into the Hox-2.6 locus in embryonic stem cells. Nature 1991; 350:243–246.

45. Askew GR, Doetschman T, Lingrel JB. Site-directed point mutations in embryonic stem cells: a gene-targeting tag-and-exchange strategy. Molec Cell Biol 1993; 13:4115–4124.

46. Wu H, Liu X, Jaenisch R. Double replacement: strategy for efficient introduction of subtle mutations into the murine Colla-1 gene by homologous recombination in embryonic stem cells. Proc Natl Acad Sci USA 1994; 91:2819–2823.

47. Lakso M, Sauer B, Mosinger B, Lee EJ, Manning RW, Yu SH, Mulder KL, Westphal H. Targeted oncogene activation by site-specific recombination in transgenic mice. Proc Natl Acad Sci USA 1992; 89:6232–6236.

48. Orban PC, Chui D, Marth JD. Tissue- and site-specific DNA recombination in transgenic mice. Proc Natl Acad Sci USA 1992; 89:6861–6865.

49. Gu H, Marth JD, Orban PC, Mossmann H, Rajewsky K. Deletion of a DNA

polymerase β gene segment in T cells using cell type-specific gene targeting. Science 1994; 265:103–106.

50. Ornitz DM, Moreadith RW, Leder P. Binary system for regulating transgene expression in mice: targeting *int-2* gene expression with yeast *GAL4/UAS* control elements. Proc Natl Acad Sci USA 1991; 88:698–702.

51. Hyde SC, Gill DR, Higgins CF, Trezise AEO, MacVinish LJ, Cuthbert AW, Ratcliff R, Evens MJ, Colledge WH. Correction of the ion transport defect in cystic fibrosis transgenic mice by gene therapy. Nature 1993; 362:250–255.

52. Cox GA, Cole NM, Matsumura K, Phelps SF, Hauschka SD, Campbell KP, Faulkner JA, Chamberlain JS. Overexpression of dystrophin in transgenic *mdx* mice eliminates dystrophic symptoms without toxicity. Nature 1993; 364:725–729.

53. Strauss WM, Dausman J, Beard C, Johnson C, Lawrence JB, Jaenisch R. Germ line transmission of a yeast artificial chromosome spanning the murine α$_1$(I) collagen locus. Science 1993; 259:1904–1907.

54. Kozarsky KF, Wilson JM. Gene therapy: adenovirus vectors. Curr Opin Gen Devel 1993; 3:499–503.

55. Gerard RD, Meidell RS. Adenovirus-mediated gene transfer. Trends Cardiovasc Med 1993; 3:171–177.

16

Cardiogenesis and Molecular Physiology

Robert S. Ross
University of California, San Diego, School of Medicine
La Jolla
and Veterans Administration Hospital—San Diego
San Diego, California

Howard A. Rockman and Kenneth R. Chien
University of California, San Diego, School of Medicine
La Jolla, California

INTRODUCTION

Congenital heart disease currently affects one in 200 births in the United States annually. Each case is composed of a constellation of well-defined defects such as abnormalities in septation, valvular formation, vascular development, cardiac chamber growth, and right-to-left positional orientation of the heart or great vessels. These phenotypical changes are easily identified, yet few associated candidate genes or molecular insights have been recognized. Although our current understanding of cardiovascular developmental defects is itself at an embryonic stage, this area is likely to be one of the major beneficiaries of advances in mouse genetics through the generation of transgenic and gene-targeted animal model systems. Recent contributions to our understanding of the developmental regulation of genetic markers of cardiac chamber formation and specification, combined with studies on the molecular switches that regulate the expression of these markers, are beginning to provide a foundation from which to analyze the complex process of cardiogenesis. In addition, the development of microsurgical approaches, miniaturized catheterization techniques, microangiography, and high-speed video microscopy has allowed the quantitative analysis of complex physiological phenotypes in vivo, in genetically manipulated adult mice and recently in the murine embryo. Further, a number of laboratories have identified chromosomal markers that are closely linked with genetic cardiovascular developmental defects in humans, including Holt-Oram syndrome, supravalvular aortic stenosis, and the "Catch-22 syndromes" (conotruncal anomaly face, Shprintzen's syndrome, and DiGeorge's syndrome).

Utilizing molecular insights from invertebrate systems, human genetics, and epidemiology, a number of genetically based mouse models of congenital cardiovascular anomalies are now being developed. This brief review highlights a few of the recent advances in the field of cardiogenesis, and discusses currently available approaches to assess adult murine cardiac physiology in vivo, a scientific trend that is paving the way toward the fusion of cardiac molecular biology and physiology, i.e., the evolution of cardiovascular molecular physiology. Subsequently, a brief discussion is provided of the development of technology that will allow the assessment of physiology, function, and flow in the cardiovascular system of murine embryos in vivo, with placental circulation intact. Within a few years, our capabilities now available to study adult murine models of cardiovascular disease will become available for study of models of congenital anomalies in the mouse embryo.

CARDIOGENESIS

The formation of the myocardium is one of the earliest and most critical steps during vertebrate embryogenesis. Cardiac progenitor tissue arises from the lateral plate mesoderm, giving rise to a pair of tubular primordia localized symmetrically on either side of the midline, which consist of pre-endocardial cells at its inner surface, premyocardial cells at the outer surface, and an intermediate layer of extracellular material termed cardiac jelly. Subsequently, these primordia fuse to form a primitive heart tube that can be divided into various segments destined to become distinct cardiac chambers, i.e., truncus, bulboventricular, ventricular, atrial, and sinus venosus (1) (Figure 1).

One experimental approach used to begin to understand the complex nature of cardiac development has been to identify the mechanisms that regulate expression of genetic markers during a particular developmental stage of interest. The inherent assumption is that the elucidation of the mechanisms that regulate the expression of these genes will eventually lead to the identification of the molecular switches that control many aspects of the cardiac muscle cell phenotype at a certain stage of development. Thus, the genetic marker could be viewed as the final step in the molecular pathway(s) that control the complex processes of cardiac commitment, regional specification, maturation, and morphogenesis. Initial work in this area has been hampered by the lack of continuous, differentiated cardiac muscle cell lines that mimic various stages of cardiac growth and development. However, based on the availability of a panel of well-characterized cardiac muscle genes, and the recent development of transgenic, transfection, and microinjection approaches for studies in a variety of cardiac experimental

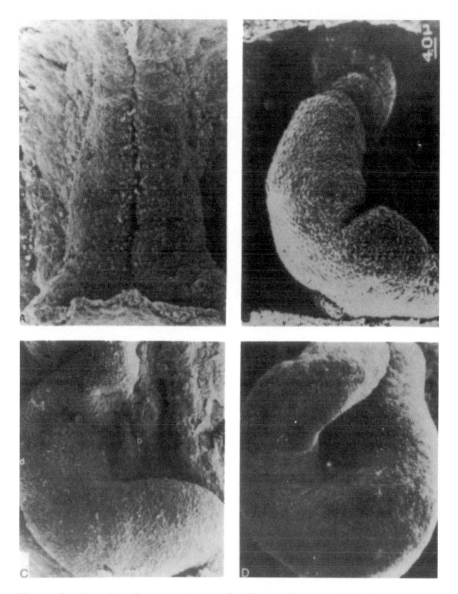

Figure 1 Scanning electron micrographs illustrate looping of the heart tube in the chick embryo. (A) Heart as the linear tube stage following fusion of the paired primordia. *a* (Center of photo) indicates the insertion of the ventral mesocardium where the paired primordia are fusing. (B) Early looping stage. (C) Late stage of looping with *b* (center of photo) indicating the dorsal mesocardium and *d* (left of photo) representing the dextrorotated portion of the loops. (D) Later stage of development with * (center of photo) indicating the primary cardiac sulcus. (From Ref. 1.)

model systems, the unraveling of the pathway(s) that control the complex physiology of cardiac development now appears to be experimentally feasible. A few of these advances are outlined below.

Cardiac Commitment

Cardiac and skeletal muscle are derived from embryonic mesoderm and express many of the same muscle-specific genes. Several transcription factors that regulate tissue-specific expression of skeletal muscle are also involved in tissue- specific expression in cardiac muscle, although several genes expressed in both skeletal and cardiac tissue appear to utilize different *cis* regulatory regions within each context. The skeletal muscle program appears to be activated in response to the expression of members of the MyoD family (for a more detailed review, see Chapter 1). The MyoD family is a member of the helix-loop-helix transcription factor family and includes MyoD, myogenin, MRF-4, and myf5 (2). Ectopic expression of each of these factors in cells of diverse embryonic origin results in activation of skeletal muscle genes, which has led to their being described as master regulators. Although many of the skeletal muscle genes that are activated in response to the expression of the MyoD family are also those that are expressed in cardiac muscle, the MyoD family does not appear to be expressed in cardiac cells. The coexpression of a single contractile protein gene in these two striated muscle subtypes has suggested the possibility that analogous pathways may be responsible for the activation of myogenesis in these two distinct striated muscle subtypes.

Prior to the cloning of MyoD, the existence of a dominant regulator of the skeletal muscle phenotype had been inferred from experiments in which genomic DNA from skeletal muscle cells was transfected into a continuous embryonic fibroblast cell line, C3H10T½ (10T½) cells, and was able to convert them to differentiating skeletal muscle at a frequency compatible with there being one or a few closely linked loci responsible for the conversion (3). Additional evidence for a dominant skeletal muscle determination gene came from cell fusion studies in which polyethylene glycol–mediated fusion of skeletal muscle myotubes with a variety of other cell types resulted in the activation of a broad range of skeletal muscle genes in nonmuscle nuclei within heterokaryons (4,5). The extent of activation was dependent on cell type, the nuclear ratio of each cell type within each heterokaryon, and the length of time after fusion. Activation of skeletal muscle genes was seen with the greatest frequency in muscle fibroblast heterokaryons and occurred at high frequencies even when fibroblast nuclei outnumbered skeletal muscle nuclei. One explanation for this facility of conversion was that both cell types are of mesodermal origin.

Although much has been learned about the determination of skeletal muscle, relatively little is known about the factors involved in cardiac muscle determination. To date, while a number of helix-loop-helix proteins have been isolated from cardiac muscle, no functional cardiac analog of MyoD has been described. To address this problem, recent studies have been performed with cardiac-fibroblast heterokaryons to investigate the existence of a dominantly acting cardiac determination factor (6), in a manner analogous to the initial studies of Blau and Baltimore (4,5). A novel experimental approach was used employing primary embryonic fibroblasts from transgenic mice as a means of assaying for the activation of a cardiac promoter-luciferase reporter transgene within fibroblast nuclei. This approach provided a potential means of genetic selection for a dominantly acting positive factor and can be generalized to other systems. Three markers of the cardiac lineage were examined: a myofibrillar protein promoter, myosin light chain-2 (MLC-2), a secreted protein, atrial natriuretic factor (ANF); and a transcription factor, myocyte-specific enhancer factor-2 (MEF-2). MEF-2 is specific to both cardiac and skeletal muscle cells. In a majority of the heterokaryons with an equal ratio of cardiac to fibroblast nuclei, none of these cardiac markers was expressed, indicating that the cardiac phenotype is not dominant over the embryonic fibroblast phenotype. The distinction from previous results of skeletal muscle is emphasized by these results with MEF-2, which is dominantly expressed in skeletal-muscle fibroblast heterokaryons but not in cardiac-fibroblast heterokaryons, supporting divergent regulation of MEF-2 expression in the two cell types. These studies provide clear evidence that divergent pathways may regulate cardiac and skeletal myogenesis and that, in contrast to a dominant skeletal muscle phenotype, the cardiac muscle phenotype may be recessive. As such, it may be difficult to directly extrapolate to cardiogenesis observations made with the dominantly acting myogenic determination genes in the MyoD family. However, a number of independent laboratories have recently identified new members of the basic helix-loop-helix protein family that are expressed in heart (7,8). It will become of interest to directly determine whether these factors can dominantly activate the cardiac muscle gene program in nonmuscle cells in a manner analogous to the MyoD family.

These studies are supported by an increasing body of evidence that suggests that divergent pathways may regulate the expression of a single contractile protein gene in the cardiac versus skeletal context. One of the clearest examples of this phenomenon has been in the studies of the cardiac/slow-twitch skeletal MLC-2 gene, which is expressed as an abundant transcript in both cardiac ventricular muscle and in slow-twitch muscle (e.g., soleus muscle). Recent studies have identified a 250 bp MLC-2v promoter fragment that can confer ventricular-specific expression at the earliest stages of cardiogenesis in the early looped heart tube (9), while levels of the reporter activity are found at background levels in slow skeletal muscle. Similarly, transient assays in cultured muscle cells have documented divergent *cis* regulatory elements in the murine creatine kinase promoter region that are required for muscle specificity in cardiac versus skeletal muscle (10). Studies of the troponin C promoter have also shown that distinct regulatory elements can mediate cardiac and skeletal muscle-specific expression in transient assays in cultured muscle cells (11). Thus, it is becoming increasingly clear that caution is warranted in extrapolating from the studies of muscle cell commitment and myogenesis in the skeletal muscle to the context of cardiogenesis.

Cardiac Mesodermal Specification

Classic embryological studies in a wide variety of organisms utilizing tissue explant studies as well as fate maps (12,13) have suggested that the initial specification of the amphibian heart in *Xenopus laevis* arises early during embryogenesis, and perhaps is underway during the blastula or gastrula stages. Explantation of late gastrula "cardiac" mesoderm with subsequent transplantation to epidermal vesicles leads to heart formation in 50% of explants, while similar experiments utilizing the more developed, early neurula tissue lead to a 100% success in formation of beating hearts. Further, if the dorsal-ventral axis of the embryo is disrupted or prevented from forming (by such maneuvers as high-pressure or ultraviolet irradiation), cardiac formation is impaired. Finally, it has been suggested that anterior or pharyngeal endoderm is an inducer of cardiac mesoderm.

With these types of studies as a basis, recent studies have been initiated to determine the molecular events that govern these phenotypical changes. Recent work in the fruitfly, *Drosophila*, has provided a vivid example of the power of genetic approaches to identify genes that regulate the complex process of cardiogenesis. Although these organisms lack a multichambered heart, they do contain a dorsal vessel that pumps hemolymph to various body tissues. In contrast to vertebrate embryos, cardiac commitment and specification of the mesoderm in *Drosophila melanogaster* appear to occur later, following completion of gastrulation. By this time, the mesoderm has spread to form a subectodermal layer. Fate analysis of individual cells in *Drosophila* embryos (14) show that they do not appeared committed during or shortly following gastrulation. One of the earliest steps in specification of the *Drosophila* mesoderm is division of this tissue into visceral (giving rise to the gut) and somatic (giving rise to the body wall) primordia. Somewhat later, further subdivision leads to specification of individual body wall musculature, and dorsal crest mesoderm are found to be specified to heart (15). With questions arising as to the molecular basis for this (and other) patterning events in *Drosophila*, a set of genes was ultimately identified that is termed the homeobox gene superfamily. [For review see McGinnis and Krumlauf (16).]

The homeobox genes function as transcriptional regulators that appear to be critical during development, acting as positional specifiers. They were first characterized in mutant *Drosophila* that were noted to have homeotic mutations, in which one body segment or part was replaced with another, structurally normal part. As an example, the *Antennapedia* mutant shows a normally formed leg in place of the antenna (Figure 2) (17). The homeobox genes are characterized by a conserved DNA-binding motif termed the homeodomain. Utilizing degenerate polymerase chain reaction (PCR) primers based on a previously identified homeobox gene, a new gene was identified in *Drosophila* that is termed *tinman* (or *msh*-2, NK-4), as mutations in this gene lead to a heartless fly (18,19). *Tinman* is initially expressed at the blastoderm stage, shortly after the appearance of *twist*, another gene required for mesoderm formation. While initially expressed in a diffuse manner in mesoderm, *tinman* soon becomes restricted to the dorsal mesoderm, from which arise the visceral musculature and heart. Analysis of the phenotype of *Drosophila* embryos deficient in the chromosomal region that contains *tinman* reveals

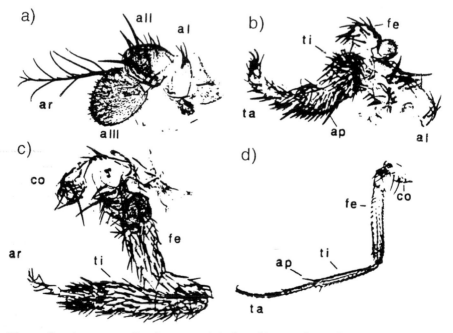

Figure 2 *Antennapedia* phenotype is induced in transformed *Drosophila*. (a) Example of a normal antenna (×68). (b and c) Two examples of antennae from mutant flies showing that the main parts of the antenna are transformed into leg structures (×22 and ×68, respectively). (d) Normal mesothoracic leg (×22). aI, antennal segment I; aII, antennal segment II; aIII, antennal segment III; ar, arista; co, coxa; fe, femur; ti, tibia; ap, apical bristle; ta, tarsus. (From Ref. 17.)

abnormal or reduced formation of somatic muscles and no differentiation of visceral or heart muscle. Thus, *tinman* appears to code for a homeobox-containing gene critical for *Drosophila* heart formation.

As is common with identification of a gene family within one organism, investigators soon extended their search for homeobox genes to other organisms and located homologous genes in *Xenopus*. Again utilizing degenerate PCR primers, Tonissen and coworkers (20) identified a 150 bp sequence from *Xenopus* adult heart mRNA that appeared to encode a homeodomain region. Subsequently, this sequence was utilized as a probe to screen a neurula stage *Xenopus* cDNA library, and the full-length sequence of this new gene was identified and termed Xnkx-2.5. Significant homology was seen between this gene and *Drosophila tinman*, as well as murine Nkx-2.5 (discussed below). Initially the gene is expressed in the front of the embryo in the presumptive cardiac mesoderm and soon after localizes to the heart region of the embryo. Although this gene is expressed at significant levels throughout the heart during development, expression is extinguished in the adult heart, revealing low-level expression elsewhere in the mature organism, in both foregut and the pharyngeal region. Indeed, this gene either may be critical in specifying the cardiac phenotype or, alternatively, is located downstream of another critical gene playing this role. A second group of investigators (21) has also isolated a series of *Xenopus* homeobox-containing genes related to the *Drosophila msh*-2 class of genes. In preliminary studies, this group has found four distinct mRNA species that are expressed in the heart. One, termed XCX2, is expressed by the gastrula stage, and is restricted to the heart after the neurula stage. Indeed, these genes may also be important in specification of the cardiac genotype in *Xenopus* and, further, may be distinctly related or identical to the Xnkx-2.5 gene discussed above.

As a means of extending this data to mammalian species, one must realize that both mouse and humans have four homeobox gene clusters spread throughout the genome, most recently termed HOXA-HOXD. Utilizing embryonic stem (ES) cells derived from the mouse blastocyst, specific disruption of a target gene can be effected (see Chapter 10). The clonally selected ES cells that contain this disruption can then be incorporated into murine blastocysts, utilizing microinjection techniques. The chimeric blastocysts can be reimplanted into a pseudopregnant host mother and ultimately generate a "line" of mice that contain targeted disruption of the specific gene of interest. Through mating of animals harboring this heterozygous mutation, animals with the -/- (or null) phenotype can ultimately be generated. This technology has alternatively been called generation of a "gene knockout." Utilizing this methodology, Chisaka and Capecchi (22) created mice with a null mutation in the hoxA-3 (hox 1.5) gene and subsequently demonstrated the importance of vertebrate homeobox genes in cardiac development. Although not restricted to cardiac developmental abnormalities, these animals with the hoxA- 3⁻/hoxA-3⁻ genotype display numerous abnormalities in the heart including atrial dilatation and hypertrophy, a bicuspid aortic valve, and abnormalities in the pulmonary valve. They survive to term but die shortly after birth, presumably of cardiac abnormalities. Still, numerous other tissues are involved in this mutant phenotype including the thymus and parathyroids, which are absent; craniofacial and pharyngeal, as well as vascular abnormalities are also seen. Although these animals indeed have pathological abnormalities akin to DiGeorge's syndrome, it is not yet clear whether this cardiac phenotype represents a primary requirement for hoxA-3 in the cardiovascular system or a secondary effect on other tissues. Interestingly, disruption of a neighboring gene, hoxA-1 (hox 1.6), shows no cardiac abnormalities but, rather, ones in the inner ear and hindbrain.

Finally, several novel murine homeobox genes that are expressed in the developing heart have recently been identified (21,23). Utilizing both mouse genomic and murine cardiac cDNA libraries, Lints et al. (23) identified two genes related to the *Drosophila msh-2* gene, which they have termed Nkx-2.5 and Nkx-2.6. To identify the tissue distribution of expression, RNase protection assays were performed with embryonic and adult tissue samples, as well as ones from cell culture lines. In the fetus, strong expression was noted only in the heart. Adult samples revealed strongest expression in the heart, with weak expression in spleen and tongue but none in skeletal muscle. Finally, cell lines showed positive expression of Nkx-2.5 in NIH3T3 fibroblasts as well as thymic and lymph node stromal cells (SCL-19 and TSL-1 cells, respectively.) Similar to several other homologous homeodomain species, Nkx-2.6 was detected in brain while Nkx-2.5 was not. In vivo, whole mount analysis of murine embryos first showed detection of Nkx-2.5 at the early headfold (late primitive streak) stage, in a crescent shape at the extreme anterior and lateral portions of the embryo. A short time later it becomes apparent that expression can be detected in both mesodermal and endodermal components of the anterior splanchnopleure, a region considered fated to be myocardium. Thus, it is apparent that this gene is expressed in regions of the embryo that may harbor the earliest known cardiac precursors. In comparison to other early cardiac markers, such as α-cardiac actin or β-myosin heavy chain, Nkx-2.5 appears to be expressed at an even earlier developmental timepoint. Further, at later embryonic stages (e.g., 12.5 p.c.) this homeodomain gene is found in both atrial and ventricular chambers (Figure 3). Interestingly, expression is also found in the pharyngeal endoderm (a potential cardiac induction site) and also in tongue myoblasts, differentiated skeletal muscle, and a small area of stomach and developing spleen. Thus, Nkx-2.5 is suggested to play an important role in cardiac commitment or differentiation both by its appearance in the regions of the murine embryo that contain both cardiac precursors and in potential cardiac inducing cells. Gene targeting experiments such as that described above could clarify its role in murine cardiac development.

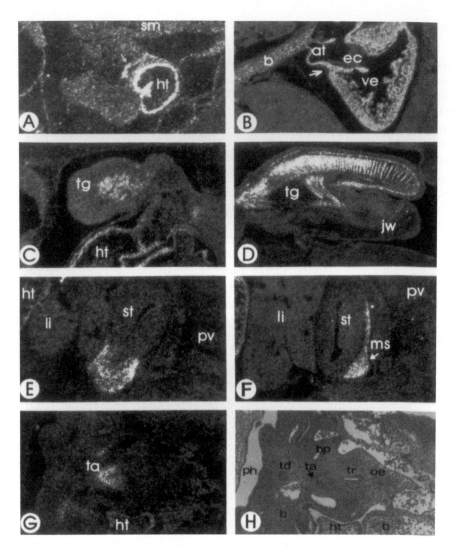

Figure 3 Expression of Nkx-2.5 as assessed by in situ hybridization analysis of murine embryos. A–G, dark microscopy; H, light-field microscopy of panel G. (A) Day 8.5 p.c. section showing positive signal in the heart and pharyngeal floor (arrow). (B) Day 12.5 p.c. section reveals hybridization to both atria and ventricle. Epicardial tissue (arrow) does not express Nkx-2.5 (C and D) Days 11.5 and 14.5 p.c. sections, respectively, showing hybridization to lingual myoblasts and differentiated muscles. (E) Day 12.5 p.c. sections shows positive signal in distal end of the stomach mesoderm. (F) Day 12.5 p.c. section at a level different from that of panel E shows positive signal in probable spleen anlage. (G and H) Dark- and light-field microscopy, respectively, of day 11.5 P.C. section revealing positive hybridization in heart and developing thyroid. at, atrium; b, blood in which a false-positive hybridization signal results from chemical reaction between erythrocytes and the photographic emulsion; bp, branchial pouch; ec, endocardial cushion; ht, heart; jw, jaw; li, liver; ms, mesentery; oe, esophagus; ph, pharynx; pv, prevertebrae; sm, somites; st, stomach; ta, thyroid anlage; tg, tongue; tr, trachea; ve, ventricle. (From Ref. 23.)

Regional Specification

Introduction

In addition to providing a scientific foundation for understanding the initial stages of cardiogenesis and commitment to the cardiac muscle lineage, molecular and genetic approaches are beginning to unravel the complex process of ventricular chamber specification, maturation, and septation (24). Transgenic approaches have led to the identification of the key regulatory elements required to maintain chamber-restricted expression of genetic markers of the ventricular phenotype (25). In turn, the nuclear factors that occupy these sites and that presumably regulate chamber-restricted expression are beginning to be identified (25). Both positive and negative markers for the ventricular phenotype have been well characterized (9,26,27), not only allowing the identification of various atrial and ventricular progenitors but also serving to define a series of maturational steps within ventricular muscle cell lineages that can be discriminated on the basis of a distinct temporal and combinatorial pattern of expression of these individual chamber-restricted markers at the single-cell level (28). The next few sections summarize these recent advances, utilizing the myosin light chain-2v as a genetic marker for the process of ventricular specification, and the down-regulation of an atrial marker (MLC-2a) as a genetic marker for the process of ventricular chamber maturation, expansion of the compact zone, and muscular septation in the developing murine embryonic heart. A recently developed mouse model for ventricular chamber defects is described, and we discuss its potential utility to dissect ventricular chamber defects via mouse genetics. Finally, the importance of the embryonic stem cell/embryod body system in the study of regional specification is discussed.

Molecular Markers for the Process of Ventricular Specification (MLC-2v) and Maturation/Septation (MLC-2a)

As noted above, to study the molecular mechanisms that control patterning of the heart tube during early cardiogenesis, we have used the ventricular myosin regulatory light chain-2 gene, which is expressed in the ventricular segment of the rodent primitive heart tube. MLC-2v is the phosphorylatable, regulatory ventricular myosin light-chain isoform, which is expressed in slow-twitch skeletal and cardiac muscle (29). Within the normal myocardium, the MLC-2v gene is expressed exclusively in the ventricular chamber and is not detectable in the atrial muscle at significant levels (29). As noted earlier, the ventricular MLC-2 gene displays a restricted pattern of expression to the ventricular portion of the primitive heart tube (9), which stands in contrast to other known chamber-specific markers that are expressed in all portions of the early embryonic heart (30). During the subsequent stages of heart development, the expression of the MLC-2v gene in the atrium remains negligible, suggesting the utility of MLC-2v as an early genetic marker of regional specification during murine cardiogenesis. In independent lines of transgenic mice, a 250 bp MLC-2 promoter fragment can confer ventricular specificity to a luciferase reporter gene in the embryonic myocardium (9), thereby providing a molecular model system for understanding positional specification of the primitive heart tube, and the acquisition of the ventricular muscle cell phenotype.

Studies have been performed in vivo to map the regulatory elements within the 250 bp MLC-2 promoter fragment that mediate the processes of cardiac and ventricular-specific expression of the MLC-2 gene. Transgenic mice have been generated that harbor MLC-2-luciferase fusion genes containing mutations in five distinct *cis* regulatory elements of the 250 bp MLC-2 promoter region (HF-1a, HF-1b, HF-2, HF-3, and E-box sites) (25). These studies document the importance of both the HF-1a and the HF-1b sites for the maintenance of ventricular muscle expression in the in vivo context, since point mutations that abolish binding of the respective cardiac muscle factors to these sites result in significant crippling of promoter activity (versus wild-type) in several independent transgenic lines. This positive regulatory effect appears to be specific for the HF-1 elements, as mutations in the HF-2 or an E-box site have little effect on decreasing the level of expression of the MLC-2-luciferase fusion gene in the ventricular myocardium of transgenic mice. The mutations in the E-box site, in fact, significantly up-regulated reporter activity in the soleus, gastrocnemius, and uterus. Mutations in another conserved regulatory element, HF-3, not only resulted in a significant up-regulation (>100-fold) of the luciferase fusion gene in all muscle tissues examined, but the luciferase reporter activity was only marginally increased in liver. Thus, both E-box and HF-3 regulatory elements appear to function in vivo as negative regulatory elements that primarily act to suppress expression in the muscle cell context. These studies in MLC-2-luciferase transgenic mice indicate that both positive (HF-1a/HF-1b) and negative (E-box and HF-3) regulatory elements may mediate ventricular muscle-specific expression. Thus, it is likely that the interaction between these discrete elements and their corresponding *trans*-acting factors is critical for determining whether the MLC-2 gene will be switched on or off in various muscle subtypes during the course of murine myogenesis.

To assess whether the atrial counterpart of the MLC-2v gene (MLC-2a) could also serve as a chamber-specific marker, we have cloned an atrial MLC-2 cDNA (554 bp), which displayed homology to the human MLC-2a cDNA at both the nucleotide (87%) and amino acid (95%) levels (26). Northern blot, reverse transcriptase-linked polymerase chain reaction, RNase protection, and

Western blot analysis revealed atrial-restricted expression in the adult mouse heart, very low levels in aorta, and no detectable expression in ventricle, skeletal muscle, uterus, or liver. In situ hybridization studies during mouse embryogenesis revealed cardiac-specific expression throughout days 8 to 16 postcoitum, with atrial restricted expression from day 12 and qualitatively greater atrial expression than ventricular from day 9 (Figure 4) (26). Thus, a preferential pattern of expression in the atria occurs prior to septation. The MLC-2a gene was differentially regulated when compared with MLC-2v expression during embryonic stem cell cardiogenesis in vitro (see below), with MLC-2a transcript levels detectable from day 6 in suspension cultures as compared to day 9 for MLC-2v. The regional-specific expression of the MLC-2a and MLC-2v genes in their respective chambers occurs at an earlier time in cardiogenesis than all other described mammalian cardiac muscle genes and provides genetic markers for chamber specification (atrial and ventricular) in both the in vitro and in vivo contexts. Based on these results, a model for the sequential maturation of ventricular muscle cell lineages has been proposed (see Table 1).

Retinoids: A Signaling Molecule Critical for Cardiac Morphogenesis

The events and molecules that are critical for cardiac morphogenesis are poorly understood at this time. The retinoids, including retinoic acid (RA) and related vitamin A derivatives, comprise a class of molecules that has been investigated for some time as to their importance in cardiac development. Since Wilson and Warkany's study over 40 years ago (31), it has been known that vitamin A deficiency can result in ventricular chamber dysmorphogenesis. Paradoxically, RA exposure can lead to numerous teratogenic effects in the vertebrate embryo, including abnormalities in the axial skeleton, cranial and cardiac neural crest–derived tissues, and the limbs, as well as the heart (32). This teratogenicity is critically dependent on the dose and timing of exposure during embryonic development.

The complexity of these effects can become to be understood in the context of the numerous retinoic acid receptors. This family of nuclear receptors, which are ligand-dependent transcription factors, is composed of six members: the three retinoic acid receptors (RARs) and three distinct retinoid X receptors (RXRs) (33). Ligand specificity overlaps in that both RARs and RXRs bind 9-*cis* RA while RARs can also bind all-*trans* RA. Further complexity is introduced in that RXRs are known to bind DNA either as homodimers or as heterodimers with thyroid hormone receptors, vitamin D receptors, or peroxisome proliferator receptors (34). Finally, each receptor subtype is known to be expressed in a specific tissue fingerprint during

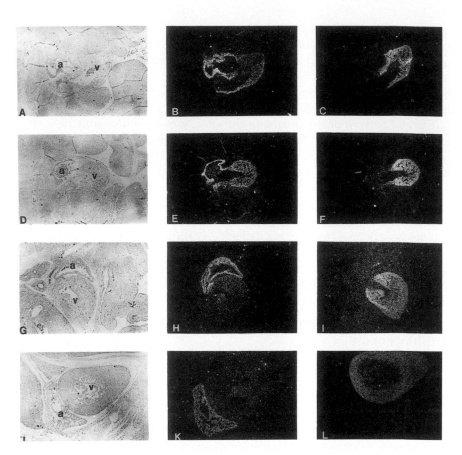

Figure 4 Myosin light chain-2 expression during murine embryonic development as assessed by in situ hybridization. Staged normal mouse sections were hybridized with either MLC-2a or MLC-2v riboprobes. a, atrial; v, ventricular. (A) Day 11 p.c. sagittal. (B) Dark-field image of A hybridized with the MLC-2a riboprobe. (C) Section adjacent to A with MLC-2v hybridization. (D) Additional day 11 p.c. sagittal. (E) Dark-field image of D hybridized with the MLC-2a riboprobe. (F) Section adjacent to D with MLC-2v. (G) Day 12 p.c. sagittal. (H) Dark-field image of G with the MLC-2a riboprobe hybridization. (I) Section adjacent to G hybridized with MLC-2v. (J) Day 14 p.c. sagittal. (K) Dark-field image of J hybridized with the MLC-2a riboprobe. (L) Section adjacent to J with MLC-2v hybridization. ×100. (From Ref. 26.)

Table 1 Stages of Atrial and Ventricular Muscle Cell Maturation as Assessed by Atrial and Ventricular Myosin Light Chain-2 and ANF Expression

Cardiac muscle cell lineage	MLC-2a	MLC-2v	ANF
Unspecified cardiac progenitor	+	−	−
Early ventricular progenitor	+	+	+
Late ventricular progenitor	+	+	+
Embryonic ventricular phenotype	−	+	+
Adult ventricular phenotype	−	+	−
Embryonic and adult atrial phenotype	+	−	+

The assignments of molecular markers are based upon in situ analysis of murine cardiogenesis in vivo (9,26) and is consistent with the temporal activation of these chamber-specific markers during in vitro chamber specification in embryonic stem cells (28).

embryonic development and in the adult. As an example, RXRα is expressed in the intestine, muscle, liver, kidney, skin, and heart of the adult. As a means to determine the function of these various receptor subtypes, a gene targeting approach has been utilized to generate mice deficient in expression of a specific receptor. Mice harboring homozygous mutations in the RARα gene do not show a profound phenotype (35). Given the expression of the RXRα gene in the heart, evaluation of targeted disruption of this gene seemed logical for evaluation of its importance in cardiac morphogenesis. In collaboration with the Evans laboratory, we have recently characterized a line of mice harboring targeted disruption of the RXRα gene (36).

The organization and importance of various domains of the RXRα gene are well known. The third exon of the RXRα gene is essential for receptor function, as it contains a portion of the DNA-binding domain. Utilizing this information, a gene targeting vector was constructed that would replace a portion of the third exon, its splice donor, as well as the 5′ end of the following intron, with an antisense neomycin resistance gene. This would provide one means of selection for homologous recombination between the targeting vector and the intact genome. The targeting vector was introduced into murine ES cells, and appropriate positive and negative selection criteria were set in 77 colonies. Of these clones, three tested positive for incorporation of the mutant, and one of the three was successfully used to colonize chimeric embryos by introduction into blastocysts. Successful transmission through the germline produced heterozygous mice that were normal in all respects. These mice were in turn bred to produce homozygotes (−/−) with a true null phenotype; i.e., they showed no detectable expression of intact RXRα transcript.

Production of RXRα homozygotes resulted in embryonic lethality. Analysis of homozygous embryos was performed at various stages of development. Embryos from day 9.5–11.5 p.c. were normal externally and histologically. It was at day 12.5 p.c. that the effects of the RXRα mutation became evident, with abnormalities in both the heart and the liver. Most easily noted, and found in every embryo examined, was an underdeveloped liver by day 12.5 p.c., approximately 30% of wild-type by mass. Histologically, the liver appeared normal at this stage. Interestingly, by day 14.5 p.c., the liver showed significant recovery in that mass was equivalent to approximately 60% of wild-type littermates. Thus, it is unlikely that this hepatic proliferative defect is responsible for embryonic death. Further analysis of the homozygotes revealed near normal appearance, yet a frequent, and sometimes dramatic, subdermal edema, particularly in the periorbital region. The edema suggested the possibility of intrauterine cardiac failure, and thus the development of the heart in these embryos was analyzed.

Through day 11.5, the (–/–) embryos appeared to have normal cardiac development. At day 12.5 p.c., the heart began to show significant abnormalities compared to wild-type or heterozygous littermates. The visceral pericardium appeared irregular and the pericardial space was prominent (although these findings returned toward normal by day 14.5 p.c.) Although the ventricular wall thickness and mass of the mutant heart were comparable to those in control animals, the contour of the ventricular surface was irregular. In comparison, it appeared that the day 12.5 p.c. mutant (–/–) hearts had matured to that equivalent to a wild-type 11.5 p.c. embryonic heart. Thus, a maturational delay in cardiac development was suggested.

By day 14.5 p.c., all homozygous RXRα embryos display ventricular hypoplasia, with absence of proliferation of the compact zone of the myocardium (Figure 5). Organized contractile activity was still evident in the mutant ventricles. Trabeculation was present but reduced, and was disorganized, particularly in the muscular septal region. Except for one (–/–) embryo, all displayed a ventricular septal defect. Aortic and pulmonary valves and outflow tracts, which have neural crest origins, appeared normal. Right ventricular dilatation and right atrial enlargement were frequently seen, and one case of mitral atresia was found. Thus, the primary defect in the RXRα knockout mice appears to be a maturational lag or arrest in the ventricular myocyte, with sporadic mutations presumably being only a secondary phenomenon. The time of embryonic lethality in the (–/–) animals is during the phase when exponential growth of the entire embryo (approximately 50% weight increase per day) is occurring. It is thus likely that the defective heart is unable to maintain adequate output relative to the needs of the embryo, demand outstrips supply, and embryonic congestive heart failure may ultimately lead to death. These results document an essential role for RXRα during embryonic

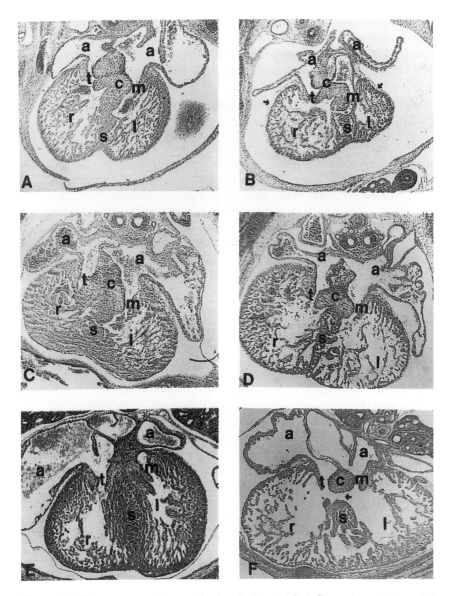

Figure 5 Transverse sections at the level of ventricle inflow valves A, C, and E are from a wild-type embryo; B, D, and F are from a homozygous RXRα-deficient (-/-) littermate. Embryos were isolated at E12.5 (A, B), E13.5 (D, D), and E14.5 (E, F) of development. a, atria; c, endocardial cushion; l, left ventricle; m, mitral valve; r, right ventricle; s, interventricular septum; t, tricuspid valve. Arrows in B point to the loosely attached pericardial layer; arrow in F indicates a ventricular septal defect at the septal–cushion fusion. (From Ref. 36.)

(and particularly cardiac) development. Indeed, an identical cardiac phenotype to that shown in these animals was seen in 1949 with classic nutritional studies of vitamin A deficiency (31). These mice should provide a unique model in which to study cardiac development at the molecular level.

Retinod Induction of the Ventricular Phenotype in ES Cell-Derived Cardiac Muscle Lineages

While each of these studies has been useful in identifying the early nature of cardiac chamber specification, the molecular cues for the regional specification and maturation of distinct atrial and ventricular cardiac muscle lineages are largely unknown. One of the major experimental difficulties has been the absence of in vitro models to permit the direct evaluation of molecules that may be involved in these processes. Further, since the formation of distinct cardiac chambers is essential for the viability of the growing embryo, in vivo studies using gene targeted mice may be complicated by lethal cardiac phenotypes and associated difficulties in distinguishing primary from secondary effects. Currently, our understanding of the pathways for cardiac chamber specification, maturation, and morphogenesis is relatively limited (37). In this regard, we have recently documented that embryonic stem cells can be driven into the ventricular-specific phenotype (e.g., the expression of a molecular marker of the ventricular phenotype, MLC-2v) during the in vitro differentiation of ES cells into beating embryoid bodies (EBs) (28). The molecular cues that control patterning of the heart tube during early cardiogenesis are largely unknown. Previous studies have explored the embryonic stem (ES) cell differentiation system to determine if this in vitro model could be useful in studying the process of regional specification of cardiac muscle cells at the earliest possible stages (28). As assessed by polymerase chain reaction, ribonuclease protection, in situ hybridization, and immunohistochemical analyses, ES cell differentiation into embryoid bodies is characterized by the transcriptional and translational activation of the ventricular myosin light chain-2 gene, demonstrating that ventricular specification occurs during ES cell cardiogenesis. The finding of a ventricular-specific marker in an in vitro system in the absence of an intact heart tube provides evidence for cardiac regional specification independent of positional cues or physiological stimuli. The temporal expression of the myogenic regulatory factors, myogenin and MyoD, suggests activation of the skeletal muscle program following cardiac myogenesis in vitro, indicating temporal fidelity to the progression of in vivo myogenesis. These data establish the mouse embryonic stem cell system as a model for cardiac chamber specification and suggest a promising approach in the study of regional specification in genetically engineered cardiac muscle cells.

Recently, we have developed a technique to purify cardiac muscle cell lineages from the EBs, on a modified Percoll gradient, which ordinarily contains <3% cardiac myocytes (38). These purified EB-derived cardiac myocyte cultures contain greater than 60–90% beating cardiac muscle cells from a given preparation. Immunofluorescence assay of these cells reveals that approximately 20–40% of the cardiac cells express MLC-2v when isolated from day 14 beating embryoid bodies. Exposure of these purified cardiac muscle cells to all-*trans* retinoic acid (which is readily converted to 9-*cis* RA, the RXR ligand) results in a greater than twofold increase in the relative proportion of cardiac cells that express the ventricular marker MLC-2v, while no significant increase is found in the number of cells that express MLC-2a. These results are also readily apparent in dual immunofluorescence studies in which the ratio of MLC-2v can be assessed and normalized to the number of cells expressing a marker found in all cardiac muscle cell lineages (myomesin or titin). The effect is not mimicked by the administration of other agents that are known to influence the cardiac muscle program, including IGF-1, and the α-adrenergic agonist phenylephrine. This effect appears to be stage-dependent, since purified cardiac muscle cell lineages from day 9 embryoid bodies are significantly less responsive and do not display a statistically significant increase in the number of MLC-2v staining cells following incubation with all-trans retinoic acid. These studies document that retinoids can have a direct effect on driving cardiac muscle lineages into the ventricular phenotypic pathway, a result that is consistent with previous observations that RXRα-dependent step(s) are required for the proliferation of ventricular muscle lineages (i.e., MLC-2v- expressing cells) during ventricular chamber development (see subsequent sections). To prove the fidelity of this in vitro assay as it relates to the in vivo context, it will be necessary to examine the RA responses of cardiac cells derived from parental ES cells that contain a double allelic knockout of the RXRα gene. If such is the case, the EB-derived cardiac muscle cell system could represent a valuable in vitro model in which to assess the subset of cardiac muscle genes influenced by RXRα deficiency and, therefore, potentially implicated in ventricular chamber dysmorphogenesis.

The Neural Crest in Cardiac Development

While most of the mature heart is derived from cells found in the primitive embryonic heart tube, it has been known for some time that the cranial neural crest (CNC) is important for formation of the outflow tract as well as postganglionic cardiac innervation (39,40). The neural crest arises from the lateral aspect of the neural plate, the neural folds. As the neural tube is formed, the neural crest cells begin migration from this area to others throughout the embryo, primarily in the face, pharyngeal apparatus, thymus, thyroid, parathyroids, and heart. Studies utilizing quail-chick chimeric embryos (41) have enabled investigators to follow the migration of CNC-derived cells. Cells from this region, which extends from the level of the midotic placode to the caudal limit of somite 3, migrate initially into pharyngeal arches 3, 4, and 6 (Figure 6). In the pharyngeal arches they contribute to the endothelium of the aortic arch arteries, while some of these cells migrate farther and populate the outflow tract to form the aorticopulmonary septum, a portion of the aortic and pulmonary valves, and portions of the cardiac ganglia.

To confirm the importance of the cardiac neural crest, ablation of portions of this embryonic region were performed in the chick embryo (42). The chick was chosen because of the relatively easy surgical manipulation of the chick embryo, the brief gestational period, and the fact that cell tracing with chimeric transplantation of quail cells (as discussed above) to chick embryos had been accomplished. These studies are generally carried out at 24–30 hours of incubation, with further incubation of the surgically manipulated embryos until days 8–11 (43). If the entire cardiac neural crest is ablated,

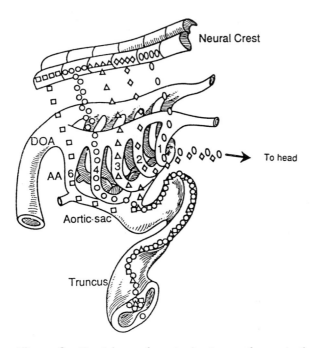

Figure 6   Cranial neural crest migratory pathways in the pharyngeal region. Neural crest cells in pharyngeal arches 3, 4, and 5 migrate from the pharyngeal region into the outflow tract of the developing heart and participate in formation of the aorticopulmonary truncal septa. (From Ref. 39.)

persistent truncus arteriosus results. While a ventricular septal defect is a constant component in these types of experiments, other abnormalities may or may not be present, depending on small variances in size of the ablation. For example, the aorticopulmonary septum may be present or absent, and the position of the single outflow vessel can be variable. With significantly smaller ablations of the cardiac neural crest or, alternatively, ablations in regions of the cranial neural crest outside the cardiac region, numerous defects may arise, including double-outlet right ventricle, tetratology of Fallot, or Eisenmenger's complex. Further, associated abnormalities are seen in the aortic arch (interrupted aortic arch is common) and the inflow tract (tricuspid atresia, tricuspid stenosis, or double-inlet LV).

The cardiac neural crest has been shown to be a distinct subset of the cranial neural crest. If substitution experiments are performed in which the cardiac neural crest is removed and replaced by other regions of the cranial or caudal neural crest, persistent truncus arteriosus results (44). Thus, this substitution is equivalent to the full ablation discussed above and suggests that the cardiac neural crest phenotype is determined early in embryonic development. Recent experiments have documented the ability to rescue the effects of cardiac neural crest ablation by backtransplantation of cells (45). Neural folds were cultured for 3 days and then cells were backtransplanted into ablated embryos. Transplantation resulted in 50% improvement in heart development and, when the cultured cells had been incubated in the presence of leukemia inhibitory factor (LIF), an almost 200% improvement was seen. Further, migratory pathways of the backtransplanted cells were in essence normal. Why the LIF results in improved rescue is at this point unknown.

The model of chick neural crest ablation has been used to bridge the gap between developmental biology and traditional physiology (46). Ablation procedures were performed in 202 chick embryos; then 105 of these, as well as 25 control embryos, were viewed via microcinephotographic technique. In this technique, the ablation procedures were performed at 24–30 hours of incubation and the eggs were resealed. At day 3, the eggs were reopened and the embryos exposed, and high-speed photography at a rate of 100 frames/second was performed. Eggs were resealed and incubation was continued till day 11. Survival rate was not altered by the filming process. Analysis of the filmed blood pool in the developing heart allowed calculation of systolic and diastolic ventricular areas, ejection fraction, stroke volume, and cardiac output. While parameters between the control and experimental groups on the whole were not different, significant differences were seen between the ablated survivors at day 11 and those ablated embryos that did not survive (Table 2). Non-survivors showed statistically significant reductions in diastolic ventricular area, stroke volume, and cardiac output. Further, film analysis could document normal or abnormal looping of the embryonic heart (Figure 7). Thus, the

Table 2 Hemodynamic Parameters of Chick Embryos

Chick embryos	Heart rate (beats/min)	Ejection fraction (%)	Systolic ventricular area (mm^2)	Diastolic ventricular area (mm^2)	Stroke volume (mm^3)	Cardiac output (mm^3/min)
Control ($n = 15$)	163 ± 1	83.5 ± 1.8	0.16 ± 0.01	0.51 ± 0.02	0.22 ± 0.01	35.5 ± 1.5
Experimental ($n = 28$)	161 ± 2	84.5 ± 1.4	0.16 ± 0.01	0.53 ± 0.02	0.23 ± 0.01	36.0 ± 1.9
Survivors ($n = 13$)	161 ± 3	87.2 ± 1.6	0.15 ± 0.02	0.56 ± 0.03[a]	0.26 ± 0.02[b]	42.4 ± 2.9[b]
Nonsurvivors ($n = 15$)	161 ± 2	82.1 ± 1.9	0.17 ± 0.01	0.50 ± 0.02[a]	0.19 ± 0.01[b]	30.4 ± 1.5[b]

Values are given as mean $\pm$ SEM.
[a] $p < 0.05$.
[b] $p < 0.001$.
Source: From Ref. 46.

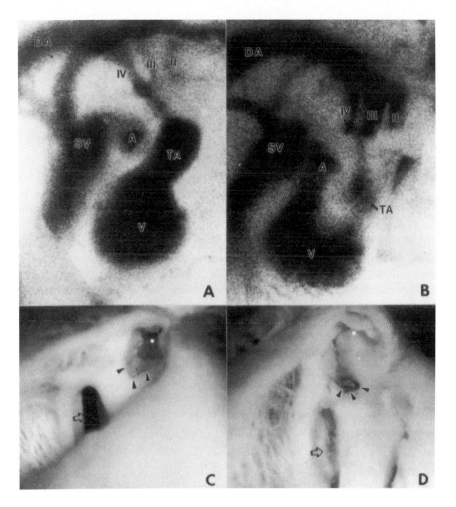

Figure 7   (A and B) Microcinephotography of stage 18 chick embryo that had undergone neural crest ablation at stages 8–10. Panel A shows normal cardiac looping while panel B shows incomplete looping. (C and D) direct photomicrographs of right ventricular view of neural crest ablated chick hearts at day 11 of incubation. Panel C shows biventricular origin of truncal artery; panel D shows truncal artery arising entirely from right ventricle. Black arrowhead, ventricular septa; white asterisk, common truncus orifice; white arrow, right ventricular inflow orifices; A, atrium; DA, dorsal aorta, SV, sinus venosus, TA, truncus arteriosus; V, embryonic ventricle, II, II, and IV represent, right second, third, and fourth aortic arch arteries, respectively. (From Ref. 46.)

ability to predict survival of defective embryos can be determined from analysis of hemodynamic parameters at an early developmental timepoint. The time has arrived for merging physiological analysis with developmental biology.

MOLECULAR PHYSIOLOGY

The generation of transgenic and gene-targeted mice is beginning to lead to new insights into the regulation of cardiac muscle gene expression and molecular determinants of cardiac development (37,47,48). Technical advances in manipulating the genome of an animal now allow investigators to achieve a level of scientific inquiry into complex physiological questions hitherto unapproachable. A major limitation to the study of physiology in the intact mouse is the small size of that laboratory animal. In addition, other factors, such as the resting conscious heart rate of 600 bpm, pose special hurtles and underscore the need for the development of specialized technology to accurately measure physiological variables. This section reviews some of the recent advances in murine cardiac physiology that have led to the application of traditional physiology to the study of genetically engineered animals and are leading to the new discipline of molecular physiology.

Assessment of Hemodynamic and Ventricular Function in the Adult Mouse

Measurement of Blood Pressure

Thirty years ago, noninvasive tail blood pressure was reported in the mouse, showing the variability between inbred strains (49). Although still a popular method for noninvasive blood pressure monitoring, direct measurement of arterial pressure is preferable for the accurate determination of blood pressure. Using fluid-filled catheters made from flame-stretched PE-50 tubing, reliable blood pressure has been recorded in both the anesthetized (50) and the conscious mouse (51). Typical systolic and diastolic aortic pressures in the 18–20-gram anesthetized mouse are 90 and 60 mm Hg, respectively. As expected, blood pressure in the conscious animal is higher, with typical values for mean aortic pressure ranging from 90 to 105 mm Hg. Some anesthetic agents will reduce resting heart rate from a range of 600–675 bpm to 300–350 bpm. Despite the miniature size of the mouse vasculature (carotid artery diameter of 0.5 mm and thoracic aortic diameter of 1 mm), accurate hemodynamic measurements can be obtained using fluid-filled catheters that have adequate frequency-response characteristics to measure blood pressure, in both the anesthetized and the conscious mouse.

In Vivo Systolic and Diastolic Function

Despite the suitability of using fluid-filled catheters for the measurement of central aortic pressure, the assessment of isovolumic phase measures of contractility, such as the maximum first derivative of the left ventricular pressure (LV dP/dt$_{max}$), requires the use of high-fidelity pressure transducers to accurately measure the change in left ventricular pressure with time at very high heart rates. Since this cannot be achieved with a fluid-filled catheter system, new miniature high-fidelity micromanometer transducers have been developed that show frequency-response curves that are flat up to 10 kHz. Figure 8 shows representative tracings of left ventricular systolic and diastolic pressure, LV dP/dt, and aortic pressure in an open-chest, anesthetized mouse using a 2F high-fidelity micromanometer. The catheter is inserted into the left atrium, advanced across the mitral valve, and secured in the left ventricle. Hemodynamic measurements can then be recorded under basal conditions and after the injection of bolus doses of isoproterenol, which results in a marked augmentation in both peak positive and peak negative dP/dt. Thus, accurate hemodynamic parameters can be obtained in the intact mouse, allowing for the hemodynamic assessment of a variety of cardiac phenotypes.

X-Ray Contrast Microangiography

To analyze in vivo ventricular function in the mouse, the technique of X-ray contrast microangiography was developed to allow quantitative assessment of ventricular volumes and ejection fraction. The technique involves injection of 100–120 µl of nonionic X-ray contrast in the jugular vein of the anesthetized intact mouse. Angiographic images are acquired on videotape under constant fluoroscopic technique in the 30° right anterior oblique and 60° left anterior oblique projections (Figure 9). X-ray images are digitized and first-pass right ventricular and levophase left ventricular video density curves are generated. Ejection fraction can then be calculated from maximal end diastolic and minimal end systolic values of each beat. Biventricular and left ventricular end diastolic volumes are obtained from trace biplane images (52–54). Figure 10 is an angiographic image of the mouse in the left anterior oblique projection following contrast injection. There is filling of both the right and left ventricle, which is clearly separated by the intraventricular septum. Also shown are a 1 × 1 cm grid for magnification correction and a lead marker to correct for scatter and veiling glare. This technique of digital contrast angiography, which allows for the quantitative assessment of in vivo cardiac function, will be useful to study the effect of gene manipulation on ventricular function.

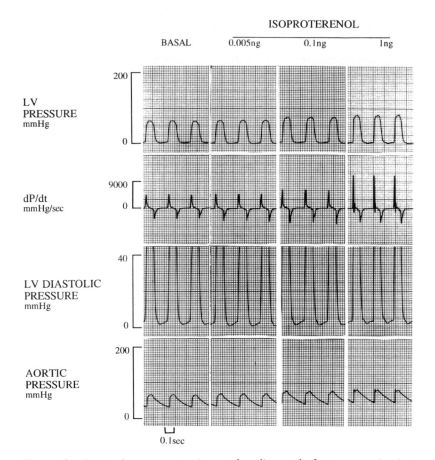

Figure 8 Original pressure tracings at baseline and after progressive intravenous doses of isoproterenol in a normal adult mouse. Simultaneous LV pressure, dP/dt, LV diastolic pressure, and aortic pressure are recorded. Administration of isoproterenol results in a marked increase in maximal positive dP/dt associated with an increase in systolic pressure due to the augmented contractility. The expected decline in LV diastolic pressure with isoproterenol is balanced by the infusion of volume resulting in minimal change in LV diastolic pressure. The dose of isoproterenol was chosen to have maximal inotropic effects with small influences on heart rate. LV, left ventricular pressure; dP/dt, first derivative of left ventricular pressure; ng, nanogram.

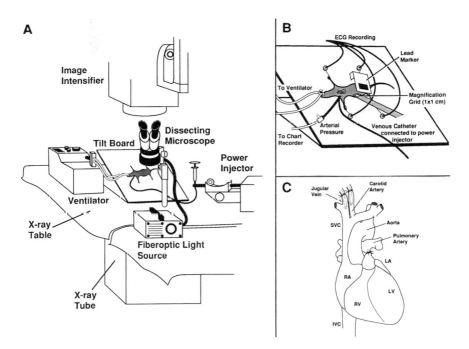

Figure 9 Diagram outlining the facility for microangiographic quantification of in vivo cardiac function. (A and B) Under fluoroscopy, angiographic images are recorded on videotape following power injection of 120 µl of X-ray contrast into the external jugular vein over a period of 1–2 seconds in 30° right anterior oblique and 60° left anterior oblique projections. X-ray images are digitized at 30 frames/second with a resolution matrix of 512×512 pixels with 256 shades of gray using a time base corrector interfaced to a computer system. First-pass RV and levophase LV video-density curves are generated using field-by-field subtraction of images, interlaced to provide a temporal resolution equivalent to 60 frames/second. Biventricular and LV end-diastolic volumes are obtained from traced biplane images using the area-length method after grid correction for magnification. RV end-diastolic volume was calculated by subtracting the LV end-diastolic volume and calculated septal volume (weight/1.05) from the biventricular end-diastolic volume. (C) Pulmonary artery banding can be performed to create a condition of RV overload by tying a suture against either a 25 gauge needle (moderate stenosis) or a 26 gauge needle (severe stenosis).

Cardiac Output and Regional Blood Flow in the Intact Mouse

A major advance in the study of murine physiology has been the development of techniques for the determination of blood flow in conscious mice (55). Using modifications of the reference microsphere and dilution technique, cardiac output, regional blood flow, and intravascular fluid volumes can be

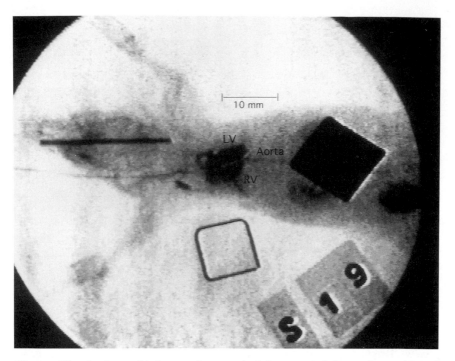

Figure 10   Angiographic image of a normal adult mouse in left anterior position after injection of X-ray contrast. In addition to both ventricles being filled, the aorta, both kidneys, and the bladder can be visualized. Also shown is the tracheal cannula for ventilation and intravenous catheter filled with X-ray contrast. (Adapted from Ref. 37. Copyright 1993 by the AAAS.)

obtained in the conscious mouse with the injection of radiolabeled microspheres directly into the left ventricle while a reference sample is withdrawn at a constant rate from the femoral artery. For 25–30-gram mice, cardiac output in the conscious mouse is 16 ml/min, with a calculated stroke volume of 25 μl/beat. The total blood volume in mice of this weight is approximately 2.3 ml. This technique has been applied to study the hemodynamic alterations of prolonged atrial natriuretic factor elevation in transgenic mice that constitutively overexpress the murine ANF gene (56). Circulating plasma levels of ANF are fourfold higher in these mice and are associated with a reduction in mean arterial pressure by 24 mm Hg. Interestingly, transgenic mice that have a reduction in central aortic pressure have no difference in cardiac output, stroke volume, and heart rate compared to nontransgenic controls. This suggests that the chronic reduction in blood pressure is due to a marked reduction in total peripheral vascular resistance and demonstrates the utility of this technique in understanding physiological processes in transgenic mice.

Physiological Measurements in the Excised Mouse Heart

Although great advances have been made in assessing in vivo ventricular function in the mouse, it clearly has limitations in the study of murine physiology because of the complex integration of the central nervous system and neurohumoral modulation in the intact animal. In this regard, important contributions have been made in the development of an isolated perfused mouse heart preparation (57-59). Adapting techniques used for larger species, instrumentation has been miniaturized so that cardiac contractile function can be studied in an isolated perfused work-performing murine heart preparation (Figure 11). The appealing feature of a working mouse heart preparation is the ability to control the hemodynamic environment (afterload, preload, and heart rate) with great precision while measuring indices of systolic and diastolic function such as intraventricular pressure, dP/dt, relaxation time, time to half relaxation, and time to peak pressure.

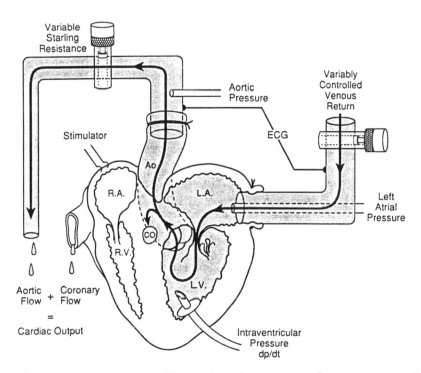

Figure 11 Flow diagram of the work-performing mouse heart preparation. RA and RV, right atrium and ventricle; ECG, electrocardiogram; CO, coronary outflow. (Adapted from Ref. 58.)

Physiological Measurements in Mice with Abnormal Cardiac Phenotypes

Overload Conditions

Applying microsurgical techniques to induce aortic constriction has resulted in a model of left ventricular pressure overload hypertrophy that results in the significant increase in heart weight associated with many of the molecular markers shown to be involved in the hypertrophic process (60,61).

In the pursuit of a murine model of ventricular hypertrophy that progresses to decompensated heart failure, lessons learned from aortic constriction have been applied to the pulmonary artery to yield a model of right ventricular hypertrophy and failure (52). This model of right ventricular overload is created by the microsurgical constriction of the pulmonary artery (Figures 9C and 12). Applying the technique of digital contrast of microangiography described earlier, right ventricular volume and ejection fraction can be quantified in normal and pressure- overloaded right ventricles. Representative angiographic images illustrate the effect of pulmonary artery banding

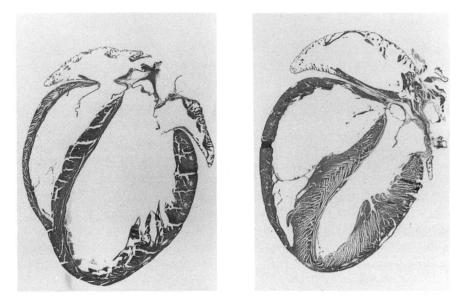

Figure 12 Representative histological sections of a sham-operated (left) and a pulmonary artery–banded (right) heart subjected to a moderate stenosis. The heart is fixed under constant venous perfusion pressure of 5 mmHg. Note the increase in wall thickness of the right ventricle and significant dilatation of both the right ventricular and atrial chambers. The left ventricle is shown as a small underfilled chamber.

on the mouse heart (Figure 13). The right ventricle dilates (end diastole) with a concomitant increase in residual volume at end systole, implying a reduction in ventricular function as shown by the diminished ejection fraction (Figure 13). To assess the transition from the phase of cardiac compensation to overt heart failure, two levels of pulmonary artery constriction can be applied to this model, resulting in a phenotype of compensatory concentric right ven-

SHAM

RV END-DIASTOLE   RV END-SYSTOLE

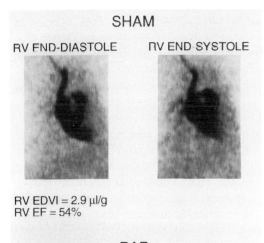

RV EDVI = 2.9 μl/g
RV EF = 54%

PAB

RV END-DIASTOLE   RV END-SYSTOLE

RV EDVI = 3.6 μl/g
RV EF = 36%

Figure 13 Representative angiographic images demonstrating RV dilatation and dysfunction following chronic pulmonary artery banding. Increased RA and RV chamber size with retrograde filling of the dilated inferior vena cava with contrast is shown. Increase residual volume at end systole indicates decreased ejection fraction and RV dysfunction. PAB, pulmonary artery banding; RV, right ventricle; RV EDVI, RV end-diastolic volume/body weight; RV EF, RV ejection fraction. (Adapted from Ref. 52.)

tricular hypertrophy following a moderate overload and decompensated right ventricular function with chamber dilatation following severe overload (52).

Heart failure is associated with an impairment of β-adrenergic receptor transduction as demonstrated by the down-regulation in β-adrenergic receptor number and a decrease in agonist-mediated inotropy in human failing myocardium (62). To investigate the consequence of overexpressing β-adrenergic receptors in the ventricular myocardium, transgenic mice were generated with the cardiac-specific overexpression of the human β_2-adrenergic receptor, resulting in a 55–200-fold increase in receptor density (63). In vivo assessment of left ventricular function in transgenic mice demonstrates a 70% increase in baseline left ventricular dP/dtmax compared to control animals (Figure 14). Interestingly, the administration of isoproterenol does not augment left ventricular dP/dtmax in the transgenic animal, demonstrating maximal activation of the β-adrenergic receptor system. In contrast, graded doses of isoproterenol result in marked increases in dP/dt_{max} in control animals (Figure 14).

Integrating mouse genetics with conventional physiological wisdom has led to the new era of molecular physiology. Molecular biologists are now embracing the small-animal physiologists to help unravel the mysteries of many cardiac diseases. One can see the potential, for example, of applying the model of heart failure, such as the right ventricular overload model described earlier, to transgenic mice that overexpress the β-adrenergic receptor, and then to assess the effect on ventricular function using high-fidelity pressure manometry and X-ray contrast microangiography. This not only will lead to important insights into the mechanisms of cardiac disease, but potentially may lead to new therapeutic strategies as the technology of in vivo gene transfer becomes refined.

Myocardial Infarction

As our understanding of the mechanisms of left ventricular remodeling following myocardial infarction improves, genetic models in the mouse that overexpress molecules that can alter ischemic damage (e.g., heat shock protein) will yield important insights into the mechanism of post–myocardial infarction ventricular dilatation and dysfunction. A model of myocardial infarction has been developed in the mouse by applying the technique of left anterior descending artery occlusion previously developed in the rat (64) (Figure 15) that can be used to test candidate molecules thought to be important in the pathogenesis of ischemic myocardial damage and postinfarction remodeling. The mouse system also offers other potential advantages, such as the availability of inbred genetic strains. In particular, the physiological consequence of genetic disorders of the extracellular matrix has been studied with passive ventricular and isolated muscle mechanics, which has led to valuable information about the influence of the extracellular matrix on cardiac mechanics (65,66).

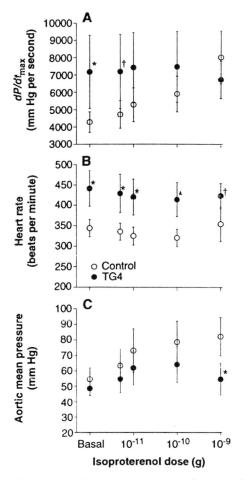

Figure 14 In vivo assessment of contractility in transgenic mice that overexpress the human β_2-adrenergic receptor (TG4). A 2F high-fidelity micromanometer transducer is placed in the left ventricle, and left ventricular pressure and the first derivative (LV dP/dt, panel A) are recorded at baseline and after doses of isoproterenol. (B) Heart rate is significantly higher in the TG4 animals at baseline and during isoproterenol. (C) At the highest dose of isoproterenol, aortic pressure falls in the transgenic group, probably due to peripheral β_2-adrenergic receptor–stimulated vasodilatation uncompensated by further increases in cardiac output since contractility is maximal. Data is mean ± SD; $n = 7$ for each group; *$p < 0.05$. (Adapted from Ref. 63. Copyright 1994 by the AAAS.)

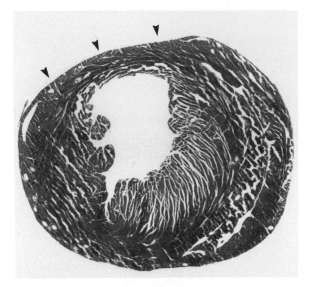

Figure 15   Representative histological section of a myocardial infarction in an adult mouse following left anterior descending artery occlusion. There is loss of muscle mass and thinning of the anterior wall (arrowheads). The right ventricle and interventricular septum are on the right side of the photomicrograph.

New Directions

Although advances have been made in developing strategies to assess in vivo ventricular function, they have required the placement of miniature catheters using microsurgical techniques. Development of noninvasive techniques to assess cardiac function in the mouse will provide the ability to temporally monitor cardiac function noninvasively. Emerging into this frontier are modifications of the techniques of echocardiography (67) and magnetic resonance imaging (68) to assess regional wall motion abnormalities and left ventricular mass in small animals. Further modifications of these techniques should allow for accurate noninvasive assessment of ventricular function in the mouse in the near future.

In Vivo Cardiac Physiology During Murine Embryogenesis: Ventricular Chamber Dysfunction in RXRα-Deficient Embryos

As discussed above, considerable progress has been made in the development and application of miniaturized technology and digitized microangiography to assess cardiac function and physiology in vivo, in adult transgenic and gene-targeted mice (60,61,63). Recently, in collaboration with Drs. John Ross and Geert Schmid-Schoenbein, we have developed the capability of quantitatively assessing the in vivo ventricular function of the RXRα-deficient murine embryos, mentioned above (36). For these studies, a pregnant mouse is anesthetized, intubated, and placed on a respirator. Each embryo is sequentially removed from the uterus and individually examined via high-speed video microscopy with the placental circulation intact. Real-time imaging of ventricular contraction can be visualized, as the embryos are relatively translucent. The injection of contrast (fluoresceinated albumin) via micropipette techniques allows the quantitation of ventricular chamber function. Analysis of a representative normal embryo studied at a heart rate of 120 bpm using bright-field imaging of the blood pool, which allows an edge-detection program to track end diastolic and end systolic areas of a single beat, yields a left ventricular ejection fraction of 0.70. Representative digitized images of the embryo heart (day 14 p.c.) in a single affected littermate are illustrated in Figure 16. The imaging system is calibrated in 50-μ increments using a 2-mm scale. In addition to these static images, survey of one of the mutant RXRα

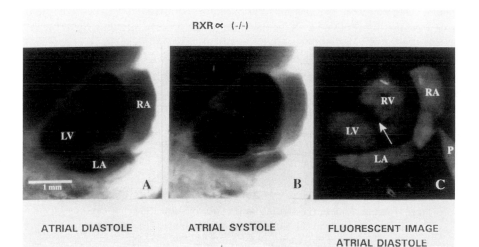

Figure 16   Digitized images of an RXRα (–/–) embryonic heart at day 14 p.c. (A and B) Static frames of blood-pool imaging using high-speed videomicroscopy. (A) Atrial diastole, (B) atrial systole. Note that the ventricles have expanded in the interval from A to B. (C) Fluorescent image using fluorescinated albumin microinjection in atrial diastole. The arrow designates the site of the ventricular chamber defect. RA, right atrial; LA, left atrial; RV, right ventricle; LV, left ventricle; P, micropipette.

embryos in real time revealed evidence of severe depression of ventricular contractile function, as well as complete heart block. This constellation suggests that there may be defective formation of the ventricular muscular septum in RXRα -/- mice, with associated severe impairment of the ventricular conduction system due to defective formation of the bundle branches. These studies would suggest that the disruption of the RXRα gene not only may result in the ventricular chamber hypoplasia and defects in the muscular septum, but also is associated with defects in the ontogeny of the ventricular conduction system (E. Dyson, G. Schmid-Schoenbein, R. M. Evans, J. Ross, Jr., K. R. Chien, unpublished observations). It remains to be seen whether this conduction system defect is more widespread and also results in defective formation of the conduction system in other cardiac compartments, including the SA and AV nodal systems. This single example illustrates the critical importance of merging physiological analysis with molecular models of cardiac development.

SUMMARY

The past two decades of cardiovascular biology and medicine have been based largely on the consideration of the heart and vasculature as an integrated physiological system, a view that has resulted in major therapeutic advances. The field is now on the threshold of a molecular therapeutic era. By allowing the molecular analysis of in vivo cardiovascular physiology, recent advances in mouse and human genetics may lead to the generation of a host of novel, biologically targeted therapeutic options. Given the multifactorial and polygenic nature of cardiovascular diseases, it is now necessary to incorporate the cardiologist, physiologist, and molecular biologist to unravel the mysteries of cardiac growth and development.

REFERENCES

1. Steding G, Seidl W. Contribution to the development of the heart. Part I: Normal development. Thorac Cardiovasc Surg 1980; 28:388–389.
2. Edmondson DG, Olson EN. Helix-loop-helix proteins as regulators of muscle-specific transcription. J Biol Chem 1993; 268:755–758.
3. Lassar AB, Paterson BM, Weintraub H. Transfection of a DNA locus that mediates the conversion of 10T1/2 fibroblasts to myoblasts. Cell 1986; 47:649–656.
4. Blau HM. Differentiation requires continuous active control. Annu Rev Biochem 1992; 61:1213–1230.
5. Blau HM, Baltimore D. Differentiation requires continuous regulation. J Cell Biol 1991; 112:781–783.
6. Evans SM, Tai L-T, Tan VP, Newton CB, Chien KR. Heterokaryons of cardiac

myocytes and fibroblasts reveal the lack of dominance of the cardiac muscle phenotype. Mol Cell Biol 1994; 14:4269–4279.

7. Kurabayashi M, Jeyaseeian R, Kedes L. Two distinct cDNA sequences encoding the human helix-loop-helix protein Id2. Gene 1993; 133(2):305–306.

8. Litvin J, Montgomery MO, Goldhamer DJ, Emerson CP Jr, Bader DM. Identification of DNA-binding protein(s) in the developing heart. Dev Biol 1993; 156(2): 409–417.

9. O'Brien TX, Lee KJ, Chien KR. Positional specification of ventricular myosin light chain-2 expression in the primitive murine heart tube. Proc Natl Acad Sci 1993; 90:5157–5161.

10. Amacher SL, Buskin JN, Hauschka SD. Multiple regulatory elements contribute differentially to muscle creatine kinase enhancer activity in skeletal and cardiac muscle. Mol Cell Biol 1993; 13(5):2753–2764.

11. Parmacek MS, Vora AJ, Shen T, Barr E, Jung F, Leiden JM. Identification and characterization of a cardiac-specific transcriptional regulatory element in the slow/cardiac troponin C gene. Mol Cell Biol 1992; 12(5):1967–1976.

12. Holtfreter J. Differenzierungspotenzen isolieerter Teile der Urodelengastrula. Wilhelm Roux Arch EntwMech Org 1938; 138:522–656.

13. Dale L, Slack JMW. Fate map for the 32-cell stage of Xenopus laevis. Development 1987; 99:527–551.

14. Beer J, Technau GM, Campos-Orteg JA. Lineage analysis of transplanted individual cells in embryos of Drosophila melanogaster. Commitment and proliferative capabilities of mesodermal cells. Wilhelm Roux Arch Dev Biol 1987; 196:220–230.

15. Hartenstein V, Jan YN. Studying Drosophila embryogenesis with P-lacZ enhancer trap lines. Wilhelm Roux Arch Dev Biol 1992; 201:194–220.

16. Mcginnis W, Krumlauf R. Homeobox genes and axial patterning. Cell 1992; 68:283–302.

17. Schneuwly S, Lemenz R, Gehring WJ. Redesigning the body plan of Drosophila by ectopic expression of the homoeotic gene Antennapedia. Nature 1987; 325:816–818.

18. Bodmer R, Jan LY, Jan YN. A new homeobox-containing gene, msh2, is transiently expressed early during mesoderm formation of Drosophila. Development 1990; 110:661–669.

19. Azpiazu N, Frasch M. Tinman and bagpipe: two homeobox genes that determine cell fates in the dorsal mesoderm of Drosophila. Genes Dev 1993; 7:1325–1340.

20. Tonissen KF, Drysdale TA, Lints TJ, Harvey RP, Krieg PA. Xnkx-2.4, a Xenopus gene related to Nkx-2.4 and tinman: Evidence for a conserved role in cardiac development. Dev Biol 1994; 162:325–328.

21. Komuro I, Izumo S. CsX: a murine homeobox-containing gene specifically expressed in the developing heart. Proc Natl Acad Sci 1993; 90(17):8145–8149.

22. Chisaka O, Capecchi MR. Regionally restricted developmental defects resulting from targeted disruption of the mouse homeobox gene hox-1.5. Nature 1991; 350:473–479.

23. Lints TJ, Parsons LM, Hartley L, Lyons I, Harvey RP. Nkx2.5: a novel murine homeobox gene expressed in early heart progenitor cells and their myogenic descendants. Development 1993; 119:419–431.

24. Chien KR, Zhu H, Knowlton KU, Miller-Hance W, van Bilsen M, O'Brien TX, Evans SM. Transcriptional regulation during cardiac growth and development. Annu Rev Physiol 1993; 55:77-95.

25. Lee K, Hickey R, Zhu H, Chien KR. Positive regulatory elements (HF-1a/HF-1b) and a novel negative element (HF-3) mediate ventricular muscle-specific expression of myosin light chain-2-luciferase fusion genes in transgenic mice. Mol Cell Biol 1994; 14:1220-1229.

26. Kubalak SW, Miller-Hance WC, O'Brien TX, Dyson E, Chien KR. Chamber-specification of atrial myosin light chain-2 expression precedes septation during murine cardiogenesis. J Biol Chem 1994; 269:16961-16970.

27. Lyons GE. In situ analysis of the cardiac muscle gene program during embryogenesis. Trends Cardiovasc Med 1994; 4:70-77.

28. Miller-Hance WC, LaCorbiere M, Fuller SJ, Evans SM, Lyons G, Schmidt C, Robbins J, Chien KR. *In vitro* chamber specification during embryonic stem cell cardiogenesis: Expression of the ventricular myosin light chain-2 gene is independent of heart tube formation. J Biol Chem 1993; 268:25244-25252.

29. Lee KJ, Ross RS, Rockman HA, Harris A, O'Brien TX, van Bilsen M, Shubeita H, Kandolf R, Brem G, Price J, Evans SM, Zhu H, Franz WM, Chien KR. Myosin light chain-2-luciferase transgenic mice reveal distinct regulatory programs for cardiac and skeletal muscle-specific expression of a single contractile protein gene. J Biol Chem 1992; 267:15875-15885.

30. Lyons GE, Schiaffin S, Sassoon D, Barton P, Buckingham M. Developmental regulation of myosin gene expression in mouse cardiac muscle. J Cell Biol 1990; 111:2427-2436.

31. Wilson JG, Warkany J. Aortic-arch and cardiac anomalies in the offspring of vitamin A deficient rats. Am J Anat 1949; 85:113-155.

32. Morriss-Kay G, ed. Retinoids in normal development and teratogenesis. Oxford: Oxford University Press, 1992.

33. Leid M, Kastner P, Chambon P. Multiplicity generates diversity in the retinoic acid signalling pathways. Trends Biochem Sci 1992; 17:427-433.

34. Kliewer SA, Umesono K, Mangelsdorf DJ, Evans RM. Retinoid X receptor interacts with nuclear receptors in retinoic acid, thyroid hormone, and vitamin D$_3$ signalling. Nature 1992; 355:446-449.

35. Li E, Sucov HM, Lee K-F, Evans RM, Jaenisch R. Normal development and growth of mice carrying a targeted disruption of the α1 retinoic acid receptor gene. Proc Natl Acad Sci USA 1993; 90:1590-1594.

36. Sucov HM, Dyson E, Gumeringer CL, Price J, Chien KR, Evans RM. RXRα mutant mice establish a genetic basis for vitamin A signaling in heart morphogenesis. Genes Dev 1994; 8:1007-1018.

37. Chien KR. Molecular advances in cardiovascular biology. Science 1993; 260:916-917.

38. Schnee J, Fuller S, Kubalak S, Chien KR. Unpublished observation.

39. Kirby ML, Waldo KL. Role of neural crest in congenital heart disease. Circulation 1990; 82:332-340.

40. Le Lievre CS, Le Douarin NM. Mesenchymal derivatives of the neural crest: analysis of chimaeric quail and chick embryos. J Embryol Exp Morphol 1984; 34:125-154.

41. Phillips MT, Kirby ML, Forbes G. Analysis of cranial neural crest distribution in the developing heart using quail-chick chimeras. Circ Res 1987; 60:27-30.
42. Kirby ML, Gale TF, Stewart DE. Neural crest cells contribute to aorticopulmonary septation. Science 1983; 220:1059-1061.
43. Kirby ML, Turnage KL, Hays BM. Characterization of conotruncal malformations following ablation of "cardiac" neural crest. Anat Rec 1985; 213:87-93.
44. Kirby ML. Plasticity and predetermination of mesencephalic and trunk neural crest transplanted into the region of the cardiac neural crest. Dev Biol 1989; 134:401-412.
45. Kirby ML, Kumiski DH, Myers T, Cerjan C, Mishima N. Backtransplatation of chick cardiac neural crest cells cultured in LIF rescues heart development. Dev Dynamics 1993; 198:296-311.
46. Tomita H, Connuck DM, Leatherbury L, Kirby ML. Relation of early hemodynamic changes to final cardiac phenotype and survival after neural crest ablation in chick embryos. Circulation 1991; 84:1289-1295.
47. Koretsky AP. Investigation of cell physiology in the animal transgenic technology. Am J Physiol 1992; 262:C261-C275.
48. Mockrin SC, Dzau VJ, Gross KW, Horan MJ. Transgenic animals: new approaches to hypertension research. Hypertension 1991; 17(3):394-399.
49. Schlager G. Systolic blood pressure in eight inbred strains of mice. Nature 1966; 6061:519-520.
50. Rockman HA, Wachhorst SP, Mao L, Ross J Jr. Angiotensin II receptor blockade prevents ventricular hypertrophy and ANF gene expression with pressure overload in mice. Am J Physiol 1994; 267:H1-H8.
51. Steinhelper ME, Cochrane KL, Field LJ. Hypotension in transgenic mice expressing atrial natriuretic factor fusion genes. Hypertension 1990; 16:301-307.
52. Rockman HA, Ono S, Ross RS, Jones LR, Karimi M, Bhargava V, Ross J Jr, Chien KR. Molecular and physiological alterations in murine ventricular dysfunction. Proc Natl Acad Sci USA 1994; 91:2694-2698.
53. Ono S, Bhargava V, Miyamoto MI, Guazzi M, Duerr RL, Ono S, Mao L, Rockman HA, Ross J Jr. *In vivo* assessment of left ventricular remodelling after myocardial infarction by digital contrast angiography in the rat. Cardiovasc Res 1994; 28:349-357.
54. Bhargava V, Hagan G, Miyamoto MI, Ono S, Rockman HA, Ross J Jr. Systolic and diastolic global right and left ventricular function: Assessment in small animals using an automated angiographic technique. In: Computers in Cardiology. Durham, NC: IEEE Computer Society, 1992:191-194.
55. Barbee RW, Perry BD, Re RN, Murgo JP. Microsphere and dilution techniques for the determination of blood flows and volumes in conscious mice. Am J Physiol 1992; 263:R728-R733.
56. Barbee RW, Perry BD, Re RN, Murgo JP, Field LJ. Hemodynamics in transgenic mice with overexpression of atrial natriuretic factor. Circ Res 1994; 74:747-751.
57. Ng WA, Grupp IL, Subramaniam A, Robbins J. Cardiac myosin heavy chain mRNA expression and myocardial function in the mouse heart. J Circ Res 1991; 69:1742-1750.
58. Grupp IL, Subramaniam A, Hewett TE, Robbins J, Grupp G. Comparison of

normal, hypodynamic, and hyperdynamic mouse hearts using isolated work-performing heart preparations. Am J Physiol 1993; 265:H1401–H1410.

59. Hewett TE, Grupp IL, Grupp G, Robbins J. α-Skeletal actin is associated with increased contractility in the mouse heart. Circ Res 1994; 74:740–746.

60. Rockman HA, Ross RS, Harris AN, Knowlton KU, Steinhelper ME, Field L, Ross J, Chien KR. Segregation of atrial specific and inducible expression of an ANF transgene in an *in vivo* murine model of cardiac hypertrophy. Proc Natl Acad Sci USA 1991; 88:8277–8281.

61. Rockman HA, Knowlton KU, Ross J Jr, Chien KR. *In vivo* murine cardiac hypertrophy: A novel model to identify genetic signaling mechanisms which activate an adaptive physiologic response. Circulation 1993; 87:14–21.

62. Bristow MR, Ginsburg R, Minobe W, Cubicciotti RS, Sageman WS, Lurie K, Billingham ME, Harrison DC, Stinson EB. Decreased catecholamine sensitivity and β-adrenergic-receptor density in failing human hearts. N Engl J Med 1982; 307:205–211.

63. Milano CA, Allen LF, Rockman HA, Dolber PC, McMinn TR, Chien KR, Johnson TD, Bond RA, Lefkowitz RJ. Enhanced myocardial function in transgenic mice overexpressing the β2-adrenergic receptor. Science 1994; 264:582–586.

64. Pfeffer MA, Pfeffer JM, Fishbein MC, Fletcher PJ, Spadaso J, Kloner RA, Braunwald E. Myocardial intact size and ventricular function in rats. Circ Res 1979; 44:503–512.

65. Omens JH, Rockman HA, Covell JW. Passive ventricular mechanics in tight-skin mice. Am J Physiol 1994; 266:H1169–H1176.

66. Capasso JM, Robinson TF, Anversa P. Alterations in collagen cross-linking impair myocardial contractility in the mouse heart. Circ Res 1989; 65:1657–1664.

67. Manning WJ, Wei JY, Katz SE, Litwin SE, Douglas PS. In vivo assessment of LV mass in mice using high frequency cardiac ultrasound: Necropsy validation. Am J Physiol 1994; 266:H1672–H1675.

68. Manning WJ, Wei JY, Fossel ET, Burstein D. Measurement of left ventricular mass in rats using electrocardiogram-gated magnetic resonance imaging. Am J Physiol 1990; 258:H1181–H1186.

17

Gene Therapy for Cardiovascular Disease

An Introduction

Fred D. Ledley
GeneMedicine, Inc.
and Baylor College of Medicine
Houston, Texas

W. French Anderson
University of Southern California School of Medicine
Los Angeles, California

INTRODUCTION

In 1992, the first clinical trial of gene therapy for cardiovascular disease began with the introduction of a gene for the LDL receptor into the liver of a patient with familial hypercholesterolemia (1). The first clinical trial of gene therapy began on September 14, 1990, when investigators at the National Institutes of Health treated a young girl to correct a genetic deficiency that crippled her immune system, by introducing genes into her lymphocytes. Since that time clinical investigation of gene therapy has expanded rapidly. By the end of 1994, more than 80 clinical trials had been approved by the NIH to treat diseases such as cancer, AIDS, familial hypercholesterolemia, cystic fibrosis, and adenosine deaminase deficiency, and more than 200 patients had participated in these trials. The encouraging results from these trials, as well as advances being made in laboratory and animal studies, have begun to validate the potential of gene therapy for cardiovascular disease in general.

The growing acceptance of gene therapy for cardiovascular disease is part of a more general realization that the clinical utility of gene therapy will not be restricted to the rare or life-threatening inherited disorders that have served as models for basic research in gene delivery and gene expression (2–4). The definition of gene therapy has broadened, from simply replacing a missing function in an inherited disorder to enhancing normal functions, providing new therapeutic functions, or interfering with pathological functions associated with many common diseases. While gene therapy provides much-needed hope for many inherited diseases, the most important clinical and social impact of gene therapy is likely to derive from the treatment of common, multifactorial disorders such as cancer, atherogenesis, and inflammation.

Gene therapy may come to play a central role not only in the treatment of disease but also in developing novel and potent prevention strategies. With the rapid progress of the Human Genome Project, there is a growing understanding of the multifaceted, genetic contribution to common, multifactorial disease. Most diseases, including cardiovascular diseases such as hyperlipidemia, atherogenesis, thrombosis, and myocardial damage, are increasingly understood to have genetic as well as environmental components. Genetic disease has historically responded poorly to conventional medical and surgical therapeutic approaches (5). It may be predicted that the genetic component of multifactorial diseases might, similarly, prove to be largely recalcitrant to conventional therapies. Unlike conventional medicines that alter the environmental component of a disease, gene therapy is capable of altering the intrinsic, genetic component of disease. Optimal therapy for multifactorial diseases may involve a combination of conventional medicines aimed at treating the environmental component of disease in conjunction with a gene-based medicine aimed at the genetic component of disease (6).

The potential for applying gene therapy to common diseases in routine clinical practice has increased the need for a clinical as well as technical perspective on these emerging technologies. In considering the role of gene therapy for cardiovascular disease, for example, it is necessary not only to consider whether gene therapy will have therapeutic effect but to understand the specific clinical needs that might be satisfied by gene therapy. It is also necessary to consider whether gene therapy will compete effectively with new and emerging pharmaceutical and surgical therapies in terms of effectiveness, safety, cost, and acceptance by patients, providers, reimbursement agencies, and regulatory authorities. Moreover, it is necessary to understand not only how gene therapy technologies will affect the practice of cardiologists and surgeons in the future, but how the patients' and providers' perspectives will, in turn, shape the development of gene therapies. This chapter describes the principles and evolving technologies for gene therapy and research aimed at applying these technologies for cardiovascular research.

WHAT IS GENE THERAPY?

The principle of somatic cell gene therapy is that genes can be introduced into selected cells in the body to treat genetic or acquired disorders. Somatic gene therapy is, in its essence, analogous to conventional therapies using pharmaceutical products, biological products, or transplantation in which novel materials are introduced into the body for therapeutic effect (7). Gene therapy differs from conventional biological or pharmaceutical therapy in that the product that is administered to the patient—the gene—is not itself therapeutic. Rather, in gene therapy the gene is introduced into certain target cells and causes these cells to express a therapeutic product encoded by the gene. Gene therapy does *not* require repairing or replacing mutant genes. While the targeted correction of gene defects has been achieved in certain cell and animal models using homologous recombination, the technical limitations of current methods make it unlikely that these techniques will find significant applications in clinical practice over the next decade.

The process of gene therapy involves the delivery of a gene to the nucleus of target cells in the body. There are several steps to gene therapy (Figure 1):

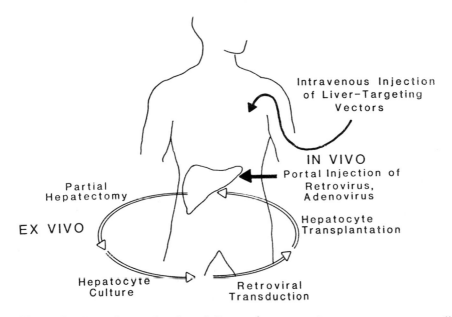

Figure 1 Gene therapy involves delivery of an expression vector to a target cell, internalization of the vector into the nucleus of the cell, transcription and translation to produce a therapeutic gene product, and trafficking of the therapeutic gene product to an intracellular or extracellular site of action. To accomplish this, gene therapy requires a sequence encoding a therapeutic gene product, a vector that directs the synthesis of this product by a target cell, and a vector-delivery system capable of introducing the vector into the appropriate cell within the patient.

1. Administration of the gene to the patient
2. Delivery of the gene to the appropriate target cell
3. Uptake into the cell and into its nucleus
4. Transcription of DNA into RNA
5. Translation of the RNA into a therapeutic protein
6. Posttranslational modification of the product
7. Transit of the product to particular compartments within the cell, or elsewhere in the body, where its therapeutic function is required

The various component technologies required for gene therapy are designed to control each of these steps, leading to safe and effective function of the therapeutic product. The challenge for gene therapy is to develop methods that effectively deliver the gene to a sufficient number of cells and achieve adequate expression of the therapeutic gene products.

A variant approach to gene therapy involves administering bioactive RNA molecules directly to patients rather than administering DNA which is capable of being transcribed into RNA. RNA itself is a potent molecule that can function in several ways: by forming a "triplex" structure with DNA, by binding to binding RNA with an "antisense" sequence, by exerting an enzymatic activity as a "ribozyme," or by binding to various determinants within the body as a "selex" structure. The challenge for using RNA as a therapeutic agent also involves the delivery of sufficient amounts of material to appropriate cells in vivo.

Gene therapy is not necessarily permanent, and may be applied to treat intercurrent illnesses as well as chronic disease. Moreover, while there is considerable appeal to the possibility of permanently curing or preventing chronic and inherited genetic disease, permanent therapies may introduce unnecessary clinical risk in light of the inevitable complications of clinical care, such as misdiagnosis, poor compliance, and adverse events. It is unlikely that even the most sophisticated molecular descriptions of disease will eliminate the human elements of variability that make the practice of medicine an art as well as a science. Thus, many gene therapies currently under development are designed to function like conventional medicines with a finite duration of action and predictable pharmacokinetic properties that allow the physician to adjust the level of therapy, or even terminate therapy, in response to a patient's immediate needs.

It should be emphasized that somatic gene therapy explicitly does not involve manipulation of the inherited genetic material in the sperm and egg. It is possible to engineer changes in the inherited germ cells of experimental animals using various nonspecific or specific targeting methods, including homologous recombination. These genetic manipulations have an important place in basic research in cardiovascular disease. The clinical application of such technologies, however, would embody significant technical, ethical, and social risk (8,9). Current federal regulations for research performed under the authority of the NIH, however, proscribe the application of these techniques in clinical practice (10). It is unlikely that these restrictions will be relaxed in the near future.

CLINICAL APPLICATIONS OF GENE THERAPY

Three types of clinical indications for gene medicines may be distinguished: those that involve gene insertion for inherited gene defects, those aimed at providing controlled expression of biological products, and those aimed at remodeling intrinsic cellular functions to make cells more resistant to disease or more susceptible to conventional therapies.

Therapy for Inherited Deficiency States

The most straightforward application of gene therapy is the treatment of autosomal or X-linked recessive diseases in which the deficiency of an essential gene product leads to a disease. In familial hypercholesterolemia, one example of such a disease, deficiency of the LDL receptor leads to hypercholesterolemia and advanced atherogenesis. There is a growing list of examples of therapy for such disorders in animal models. For example, research in the LDL receptor–deficient Watanabe rabbit has demonstrated the feasibility of lowering cholesterol by introducing a gene for the LDL receptor into hepatocytes (11). Similarly, studies in factor IX–deficient, hemophilia A dogs demonstrated that introduction of the factor IX gene into hepatocytes can correct the clotting defect (12). Clinical trials have demonstrated success in treating adenosine deaminase deficiency in rare cases of immune deficiency caused by an inherited defect in this enzyme (13), replacement of CFTR in the nasal epithelium of patients with cystic fibrosis (14), and introduction of the LDL receptor into the liver of patients with familial hypercholesterolemia (1). Cardiovascular diseases with simple autosomal genetics that would be amenable to such gene therapy are relatively rare, and it is likely that the most important applications will involve treatment of more common, multifactorial diseases.

Controlled Delivery of Biological Products

Perhaps the most important application of gene therapy will be for the controlled delivery of biological products, including proteins or bioactive RNA molecules. The limiting problem in developing clinical applications for many biological products is the lack of effective methods for controlled delivery for these molecules. Many biological products require continuous intravenous infusions or frequent injections because of their short half-lives or relatively narrow therapeutic index. The concept behind using gene therapy to deliver a biological product is that DNA vectors can be introduced into cells within the body and that these cells will then express the therapeutic product in vivo. The ability to localize expression of the product within a specific tissue may be particularly important. For example, it may be desirable to express growth factors, angiogenesis factors, growth-inhibitory factors, vasodilators, or inotropic agents at specific sites within the body for a local effect, thus minimizing unintended effects of these materials on other tissues. The ability to control the level, location, and duration of action of potent biological products may significantly enhance their clinical applicability and acceptance.

Altering the Intrinsic Characteristics of Cells

Finally, the unique property of gene therapies is their ability to alter the intrinsic characteristics of cells by targeting gene products to discrete intracellular or extracellular locations and allowing the expression of a biological product to be regulated by normal cellular pathways. For example, the therapeutic use of *trans*-acting regulatory factors that control the growth and differentiation of cardiac or smooth muscle may require localization of these proteins within the nucleus. Similarly, the therapeutic use of the LDL receptor may require localization in membrane. Proteins are normally targeted to specific intracellular locations during the process of translation and posttranslational modification. It would be difficult, if not impossible, to deliver exogenously administered proteins to specific locations within the cell. In contrast, gene therapies lead to production of proteins by normal pathways and can be designed to deliver gene products to various sites within the cell, to adjacent cells, or to secretory pathways leading to systemic circulation. It is likely that many of the genes discovered by genomic research will be nuclear, cytoplasmic, or membrane proteins that will find therapeutic applications through gene therapy.

METHODS FOR GENE THERAPY

The three components of gene therapy are a gene encoding a therapeutic product, an expression vector that contains the gene and causes the gene product to be expressed at therapeutic levels from an appropriate target cell, and a vector-delivery system that delivers the vector to the target cell.

Therapeutic Genes

With the rapid progress toward understanding the molecular basis of cardiovascular disease, many genes with putative therapeutic potential have been identified. These include both proteins and bioactive RNA molecules. Not every gene that has an interesting biological action, however, has potential for use as a gene therapy, and not every deviation from normal gene expression or every mutation that may be rectified represents a rational target for clinical therapy. Often gene therapy will be directed not at restoring normal, homeostatic functions but at creating novel biological functions to counteract disease processes. Such therapies will create novel biological situations and raise novel questions about the role of certain genes and gene products in homeostasis or disease. In many instances, it is necessary to understand the effect of a gene product at pharmacological rather than physiological concentrations, the effect of a gene product in cell types different from normal, and the effect of regulating gene expression from chimeric rather than natural promoters. Gene insertion may prove to be much like metabolic engineering in simple organisms (15), in which the effects of genetic manipulation are not always predictable from first principles alone.

Expression Vectors

The function of the expression vector is to direct transcription of the administered gene into RNA, translation of the therapeutic product from RNA, and proper posttranslational modification or compartmentalization of the product. This is achieved by constructing a DNA molecule that contains elements that control each of these processes. For example, the promoter and enhancer sequences control the rate of transcription. These sequences can restrict expression to certain cell types (tissue specificity) or provide regulation by endogenous or exogenous stimuli. The initiation and termination codons are essential for translation. Certain leader sequences within the gene direct the gene product to the cytoplasm, nucleus, membrane, certain organelles, or secretory pathways.

Sometimes it is desirable for a gene to be expressed at normal, physiologically controlled levels. In such cases the expression vector can be constructed using genetic elements that normally direct expression of the therapeutic gene product. More often, however, gene therapy is aimed at producing nonphysiological levels of expression in a cell. For example, gene therapy can involve secretion of high levels of a gene product from a relatively small number of cells. Also, gene therapy may require restoring expression of a gene product that is normally suppressed by a disease process by overriding pathogenic regulatory mechanisms or expressing gene products in novel, heterotopic locations. Such nonphysiological expression is achieved by constructing vectors that contain control elements derived from various different genes, allowing the level and location of expression to be controlled for optimal therapeutic effect.

The limiting factor in gene therapy is often achieving sufficient amounts of the product to exert a therapeutic effect. For this reason, many of the vectors currently being considered as potential therapeutics use genetic elements from viruses that are known to provide high levels of expression during viral infections. Examples of such vectors are those that incorporate the Rous sarcoma virus (RSV) or cytomegalovirus (immediate early) (CMV) promoters or enhancers. The use of cell-specific promoters to restrict gene expression to cells such as hepatocytes, smooth muscle, or cardiac muscle may provide more control over the regulation, reproducability, and pharmacokinetics of gene therapy.

It is also possible to design vectors to control the duration of gene expression. Certain vectors are intended to permanently integrate genes into the chromosomes of the host cell to achieve indefinite expression of the therapeutic product. This is usually accomplished by incorporating viral elements into the vector that allow the vector to be inserted into the chromosomes of the host cell. Viral vectors incorporating elements of murine leukemia virus or adeno-associated virus are designed for "permanent" therapy. Other vectors may be designed to maintain the therapeutic gene as an extrachromosomal (episomal) element within the nucleus that can be replicated and repaired like a normal gene without integrating into the host's chromosome. Most vectors, for example, those based on adenovirus or nonviral delivery methods, do not persist indefinitely in the target cell but are eliminated over time. Such vectors may be used like conventional medicines, with adjustable dosing and schedules of administration to treat acute disease or establish steady-state levels of the gene product. Effective therapy with such vectors will require methods for vector delivery with low toxicity, a convenient mode of administration, and infrequent dosing to achieve acceptance and compliance in the clinic.

Vector Delivery

There are several different approaches to vector delivery (Table 1). These can be generally divided into the use of genetically engineered cells, the use of viral vectors, and the use of gene-based medicines. These three approaches differ significantly in terms of their applicability to certain disease and organ targets, the manner in which they are administered in clinical practice, and their risks.

Genetically Engineered Cells

Cell-based therapy involves transplanting a cell into the body to express a therapeutic product. The cell may be a genetically engineered autologous cell as in ex vivo gene therapy, an allogeneic cell, or even a xenogeneic cell.

Table 1 Methods for Vector Delivery

Feature	Method
Cell-based[a]	
Viral	Retrovirus
	Adeno-associated virus
DNA	Electroporation
	Transfection
Cells	Xenograft
Viral[b]	
Defective	Retrovirus
	Adeno-associated virus
Attenuated	Adenovirus
	Herpesvirus
	Papilloma virus
Nucleic acid[c]	
DNA	"Naked" DNA
	Formulated DNA
	Gene gun
	Jet injection
Lipid DNA	Cationic lipids
	Liposomes
Receptor DNA	Proteins
	Small molecules
Polymers	Dendrimers
Peptides	Polylysine
	Hemagglutinin

[a]Cell-based therapies involve the genetic manipulation of cells ex vivo prior to reimplantation into the body.
[b]Viral vectors are designated as defective if they encode the therapeutic gene and no viral gene products, or attenuated if they express certain residual viral gene products in addition to the therapeutic product.
[c]Nucleic acids including DNA and RNA can be administered directly as therapeutic molecules. Various methods for enhanced delivery are shown.

Cell-based therapy is commonly performed by cultivating cells in the laboratory, introducing genes into these cells using viral vectors of DNA transfection methods, and then transplanting these cells back into the body, where the cells are intended to persist indefinitely. The limiting factor in many cell-based therapies is obtaining controlled gene expression in vivo and developing novel methods for cellular transplantation.

Viral Vectors

Most research on gene therapy to date has focused on the use of genetically engineered viral particles to introduce genes into cells. Viral infection is the archetype for the process of introducing genes into cells. The steps in such gene transfer include binding of the vector particle to specific receptors on the surface of the target cell, followed by internalization into the cell by endocytosis or fusion with cellular membranes, release of the vector into the cytoplasm, and, finally, trafficking of the vector to the nucleus. The goal of using viral vectors for gene delivery is to engineer replication-defective virus particles that retain elements required for gene transfer but lack elements required for their pathogenic effect and replication.

The prototype viral vectors are those derived from Moloney murine leukemia virus (Figure 2). These viral vectors are completely "defective" in that they encode no viral gene products, but they retain the ability to enter cells, transfer their genes to the nucleus, and integrate these genes into the chromosomes of the host cell (16). These are theoretical risks associated with retroviral vectors, including the potential for spontaneous recombination to form a replication-competent virus, and the potential for insertional mutagenesis from random integration into the genome (17). Other vectors, such as those derived from adenovirus (Figure 3) (18,19) or herpesvirus (20), are, in contrast, only replication-defective and still express many viral gene products. Such vectors sometimes exhibit residual, attenuated cytopathic effects.

Each virus exhibits certain properties that may be exploited for somatic gene therapy, as well as certain limitations. For example, retroviral vectors permanently integrate their genes into the chromosomes of the engineered cell and offer the prospect for permanently altering the characteristics and function of the cell (16). These vectors, however, will transform only dividing cells, are difficult to produce in large quantity, and are prone to recombination. Retroviral vectors have been extensively studied for a variety of clinical indications, and more than 50 clinical trials using these vectors are currently in progress (2–4). Most of these trials employ ex vivo strategies in which the patient is not directly exposed to retroviral particles. The ex vivo strategy involves removing cells or tissue from the patient, cultivating these cells in the laboratory, introducing a vector into the cells, and then introducing the genetically engineered cells back into the patient by transfusion or transplantation. In contrast, other approaches to gene therapy involve direct administration of genetic material to patients in vivo like conventional medicines.

Adeno-associated virus vectors are similarly defective and are capable of permanently inserting genes into the genome. Unlike retrovirus, adeno-associated viral vectors are capable of transferring genes into nondividing cells. These vectors, however, are less extensively characterized, and safety studies, as well as methods for efficiently producing the recombinant vector, are less

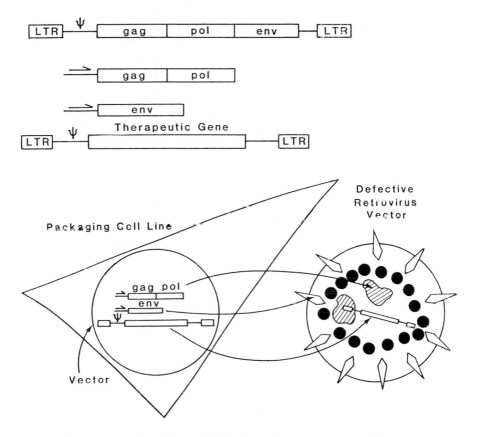

Figure 2 Construction of "defective" retroviral vectors based on Moloney murine leukemia virus, the complete genome of which (top) encodes three polyproteins: *gag*, *pol*, and *env*. The expression of these proteins in a cell will result in assembly of a retroviral particle. To generate a defective retroviral vector, several constructs are made containing portions of the complete genome. One contains only the sequences encoding the *gag* and *pol* proteins. Another contains only the sequences encoding *env*. The third is the retroviral vector containing two LTR sequences and the psi sequences, which is essential for packaging in addition to the therapeutic gene. (Bottom) A packaging cell line can be established that expresses *gag*, *pol*, and *env* from the constructs shown in the top panel. When the vector is introduced into the packaging cell line, a virus particle will be assembled that contains the *gag*, *pol*, and *env* proteins as well as the vector that contains only the therapeutic gene. These defective retroviral particles are capable of infecting cells but do not encode any viral gene products or pathogenic properties.

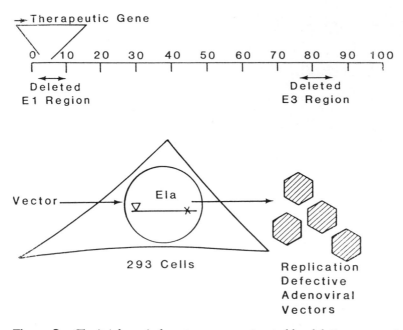

Figure 3 (Top) Adenoviral vectors are constructed by deleting two regions of the adenoviral genome: the E3 region, which is not essential for viral production or proliferation, and the E1 region, which is essential for activating the late gene of adenovirus and stimulating production of the adenovirus particle. Recombinant genes are commonly inserted in the E1 region, with the length of the insert limited by the size of the E3 and E1 deletions. (Bottom) Adenoviral particles are produced by introducing the vector into the 293 cell line, which constitutively produces E1 functions. The presence of E1 stimulates production of late proteins encoded by the vector- producing particles capable of infecting cells, but not further propagation in the absence of E1.

complete (21). Herpes viral vectors have the potential for persisting indefinitely within a cell in a latent state without the risk of integration into the host-cell chromosome. These vectors are not completely defective, and the current generation of herpesvirus vectors retain considerable cytopathic effect (20).

Adenovirus vectors are effective agents for gene transfer. These vectors can be produced in large quantities and are capable of transferring genes into nondividing cells, leading to high-level expression of recombinant gene products for several weeks (19,22). Unlike vectors that integrate into the chromosome of the target cell or persist indefinitely in an episomal location, adenoviral infection leads to transient expression of recombinant genes for a period of days to weeks. Adenoviral vectors have been important in validating the potential for gene therapy in animal models, and several clinical trials using adenoviral vectors for gene therapy are currently underway. These vectors are not completely defective and exhibit cytopathicity and immunogenicity, which may limit their clinical application, particularly since blocking antibodies are formed that may preclude repetitive dosing (19).

Gene-Based Medicines

There is increasing interest in the administration of DNA expression vectors directly to patients like conventional medicines. This is accomplished by using drug-delivery methods to deliver vectors to specific cellular targets, penetrate the periplasmic membrane, and enter the nucleus. In its simplest form, DNA-expression vectors injected into skeletal or cardiac muscle will be taken up by cells, and gene products may be expressed for prolonged periods of time (weeks to months) (23). Few other cell types beside muscle are able to take up DNA after a simple interstitial injection, and various approaches have been described for enhancing the uptake of DNA into cells by using cationic lipids (24), receptor-mediated uptake (25,26), polymeric carriers (27), liposomes (28), or particle bombardment (29). Various strategies for enhancing the efficiency of uptake have been described, including the use of endosomal release (30-32) and nuclear targeting agents to enhance trafficking to the nucleus, as well as the development of synthetic carriers rather than peptides for targeting surface receptors. The goal of these studies is to develop methods for gene delivery that are less inherently toxic than viral vectors and that may be administered by conventional routes so that gene-based medicines may come to be used alongside conventional medicines in routine clinical practice.

APPLICATIONS OF GENE THERAPY FOR CARDIOVASCULAR DISEASE

Cardiovascular disorders present a variety of challenges for gene therapy, ranging from the need to alter systemic metabolism of lipids to altering the local response to angioplasty. This section provides an introduction to the progress that has been made toward gene therapy for cardiovascular disease and the challenges that remain. Various specific experiences are described in detail in Chapters 18-21.

Hyperlipidemia

The first clinical trials of gene therapy for cardiovascular disease have been aimed at lowering cholesterol by expressing gene products that alter cholesterol metabolism. Research in several laboratories has shown that expression of an LDL-receptor gene in LDL receptor–deficient Watanabe rabbits (11), transgenic mice (33), or normal mice (34) results in lowering of cholesterol. The initial patient with familial hypercholesterolemia treated by gene therapy exhibited a lowering of LDL and an increase in the HDL and HD/LDL ratio, demonstrating that further investigation of this approach is warranted (1). Other approaches to gene therapy for hypercholesterolemia have also been described, including expression of apoA1 (35) and the recently described VLDL receptor (36). While studies to date have validated the potential for lowering cholesterol by genetic manipulation, a substantially greater reduction in cholesterol may be required to alter the progression of atherogenesis. The ultimate goal of such therapy is to prevent atherogenesis by treating asymptomatic patients, or patients with early-stage diseases, who are at risk for cerebral, coronary, or peripheral vascular disease. To achieve compliance in such patient populations, it will be necessary to develop therapies that have a relatively long duration of action and a low incidence of adverse effects, and that are noninvasive and cost-effective.

Gene Transfer to Vessel Walls

Methods for introducing genes into endothelial cells have been described that may have application in treating atherogenesis and thrombosis. Gene transfer into arterial endothelium has been described using several techniques, including implantation of genetically modified endothelial (37–39) or smooth-muscle cells (40), local instillation of liposomes (41), DNA-coated catheters (42), or adenoviral vectors (43). Various applications of this technology have been proposed, including the local expression of t-PA to treat thrombosis (t-PA expression) (44) and the expression of FGF to induce neovascularization of atherosclerotic plaques (45). This method has also been useful for generating animal models of intimal hyperplasia by expressing PDGF (46) and vasculitis by expressing an allogenic histocompatibility antigen (47). Restenosis may be a particularly appropriate target for gene therapy since local instillation of genes during angioplasty may provide localized expression of factors to prevent the proliferation of intimal tissues following the procedure. Systemic therapy targeted to endothelium would require improvements in methods for gene delivery.

Gene Transfer to Myocardium

Several methods for introducing genes into cardiac muscle have been described, including direct administration of DNA-expression vectors (48) and infection with adenoviral vectors (49). Among the various applications of gene delivery to the myocardium that have been proposed is the expression of growth factors such as FGF to induce neovascularization of infarcted tissue.

SOCIAL AND REGULATORY ISSUES IN GENE THERAPY

The current climate of acceptance of gene therapy by the public, physicians, and the pharmaceutical industry followed many years of intense debate concerning the ethical and social implications of genetic engineering in human subjects (7,50–53). It is now commonly accepted that therapeutic applications of genes as medicines are not significantly different from the therapeutic applications of conventional pharmaceutical drugs, devices, or transplants, and raise few novel clinical, regulatory, or ethical issues. Clinical trials of somatic gene therapy are governed by the same principles of medical ethics, safety, risk/benefit, and responsibility that pertain to other therapeutic trials (54–57).

Product licensing, establishment licensing, and labeling of gene-therapy products will similarly be governed by the same regulations as biological and pharmaceutical products. The licensing of gene medicines will be under the jurisdiction of the Center for Biological Evaluation and Research (CBER), which, in 1991 and 1993, promulgated the "Points to Consider in Human, Somatic Cell Therapy and Gene Therapy" (55,56). In addition to review by the FDA, clinical trials receiving federal support, or taking place at federally supported institutions, are reviewed by the Recombinant DNA Advisory Committee of the NIH (10) and require approval of the Director of the NIH. The primary function of the RAC is to provide a forum for public review of gene therapy and to be a backup for local Institutional Review Boards (bioethics) and Institutional Biosafety Committees, which may not have in-depth experience with gene therapies. Similar review panels have been established in other countries.

With progressive pharmaceuticalization of gene medicines, gene therapy will largely resemble conventional biological, pharmaceutical, and vaccine products, and the need for separate, public review of gene-therapy protocols may come to be restricted to novel issues such as germline genetic engineering or new viral vector systems. Increasingly, the regulatory guidelines that govern gene therapy, and the clinical standards for investigation, will come to resemble the standards for protection of human subjects, Good Clinical Practice, and Good Manufacturing Practice that govern conventional biological and pharmaceutical products.

CONCLUSION

Cardiovascular disease is increasingly being understood as the interaction of environmental and genetic factors. Many different genes and gene products have been described that may have a role in the therapy or prevention of disease. It is likely that gene therapy will come to play a critical role in clinical practice, as a means both for controlled delivery of biological products and for altering the inherent genetic factors contributing to disease.

The first applications of gene therapy for cardiovascular disease are now in clinical trials. Many other potential applications are being studied in preclinical animal models and may enter clinical trials in the next several years. These early experiments are only pioneering examples of the various approaches to gene therapy that may arise from a more complete understanding of the molecular biology of cardiovascular disease and homeostasis. While these experiments are establishing the validity of genetic approaches to cardiovascular disease, it is unlikely that any of the methods currently employed will find widespread application in the practice of cardiologists or cardiac surgeons. Rather, ongoing refinement of vectors for gene expression and vector-delivery systems will be necessary to produce products that will meet the criteria for general use by practitioners and patients. These criteria include not only effectively satisfying an unmet clinical need but doing so in a way that is safe; cost-effective; acceptable to patients, providers, physicians, reimbursement agencies, and regulatory authorities; and superior to novel pharmaceutical and biological products.

ACKNOWLEDGMENTS

F.D.L. is a founder with equity interest in GeneMedicine, Inc. W.F.A. is a consultant with stock options in Genetic Therapy Inc.

REFERENCES

1. Grossman M, Raper SE, Kozarsky K, Stein EA, Englehardt JF, Muller D, Lupien PJ, Wilson JM. Successful ex vivo gene therapy directed to the liver in a patient with familial hypercholesterolemia. Nat Genet 1994; 6(4):335-341.
2. Anderson WF. Human gene therapy. Science 1992; 256:808-813.
3. Miller AD. Human gene therapy comes of age. Nature 1992; 357:455-460.
4. Ledley FD. Are contemporary methods for somatic gene therapy suitable for clinical applications? Clin Invest Med 1993; 16:78-88.
5. Hayes A, Costa T, Scriver CR, Childs B. The effect of Mendelian disease on human health II: response to treatment. Am J Med Genet 1985; 21:243-255.
6. Ledley FD. The therapeutic promise of molecular genetics. J Invest Derm 1994; 103(5 suppl):2S-5S.
7. Anderson WF, Fletcher JC. Gene therapy in human beings: when is it ethical to begin? N Engl J Med 1980; 303:1293-1297.
8. Anderson WF. Human gene therapy: scientific and ethical considerations. J Med Phil 1985; 10:275-291.

9. Fletcher JC, Anderson WF. Germ-line gene therapy: a new stage of debate. Law Med Health Care 1991; 20:26-39.

10. Recombinant DNA Advisory Committee. Points to consider in the design and submission of human somatic cell gene therapy protocols. Fed Reg 54(169): 36698-36703. Reprinted in Hum Gene Ther 1990; 1:93-103.

11. Chowdhury JR, Grossman M, Gupta S, et al. Long term improvement of hypercholesterolemia after ex vivo gene therapy in LDLR deficient rabbits. Science 1991; 254:1802-1805.

12. Kay MA, Rothenberg S, Landen CN, Bellinger DA, Leland F, Toman C, Finegold MF, Thompson AR, Reed MS, Brinkhous KM, Woo SLC. In vivo gene therapy of hemophilia B, sustained partial correction in factor IX deficient dogs. Science 1992; 262:117-119.

13. Culver KW, Anderson WF, Blaese RM. Lymphocyte gene therapy. Hum Gene Ther 1991; 2:107-109.

14. Zabner J, Couture LA, Gregory RJ, Graham SM, Smith AE, Welsh MJ. Adenovirus-mediated gene transfer transiently corrects the chloride transport defect in nasal epithelia of patients with cystic fibrosis. Cell 1993; 75(2):207-216.

15. Bailey JE. Metabolic engineering. Science 1991; 252:1668-1675.

16. Miller AD. Retrovirus packaging cells. Hum Gene Ther 1990; 61:5-14.

17. Cornetta K, Morgan RA, Anderson WF. Safety issues related to retroviral mediated gene transfer in humans. Hum Gene Ther 1991; 2:5-14.

18. Berkner KL. Development of adenovirus vectors for the expression of heterologous genes. BioTechniques 1988; 6:616-629.

19. Trapnell BC. Adenoviral vectors for gene transfer. Adv Drug Del Rev 1993; 12:185-199.

20. Geller AI, Keyomarsi K, Bryan J, Pardee AB. An efficient deletion mutant packaging system for defective herpes simplex virus vectors: potential applications to human gene therapy and neuronal physiology. Proc Natl Acad Sci USA 1990; 87:8950-8954.

21. Kotin RM. Prospects for the use of adeno-associate virus as a vector for human gene therapy. Hum Gene Ther 1994; 5:793-801.

22. Sratford-Perricaudet LD, Levrero M, Chasse JF, Perricaudet M, Briand P. Evaluation of the transfer and expression in mice of an enzyme-encoding gene using a human adenovirus vector. Hum Gene Ther 1990; 1:241-256.

23. Wolff JA, Malone RW, Williams P, Chong W, Acsadi G, Jani A, Felgner PL. Direct gene transfer into mouse muscle in vivo. Science 1990; 247:1465-1468.

24. Felgner PL, Gadek TR, Holm M, et al. Lipofection: a highly efficient, lipid-mediated DNA-transfection procedure. Proc Natl Acad Sci USA 1987; 84:7413-7417.

25. Wagner E, Zenke M, Cotten M, Beug H, Birnstiel ML. Transferrin-polycation conjugates as carriers for DNA uptake into cells. Proc Natl Acad Sci USA 1990; 87:3410-3414.

26. Wu GY, Wu CH. Receptor-mediated gene delivery and expression in vivo. J Biol Chem 1988; 263:14621-14624.

27. Haensler J, Szoka FC Jr. Synthesis and characterization of a trigalactosylated bisacridine compound to target DNA to hepatocytes. Bioconj Chem 1993; 4:85-93.

28. Nicolau C, Cudd A. Liposomes as carriers of DNA. Crit Rev Ther Drug Carrier Syst 1989; 6:239–271.

29. Yang NS, Burkholder J, Roberts B, Martinell B, McCabe D. In vivo and in vitro gene transfer to mammalian somatic cells by particle bombardment. Proc Natl Acad Sci 1990; 87:9568–9572.

30. Curiel DT, Agarwal S, Wagner E, Cotten M. Adenovirus enhancement of transferrin-polylysine-mediated gene delivery. Proc Natl Acad Sci USA 1991; 88:8850–8854.

31. Curiel DT, Wagner E, Cotten M, Birnstiel ML, Agarwal S, Li CM, Loechel S, Hu PC. High efficiency gene transfer mediated by adenovirus coupled to DNA-polylysine complexes. Hum Gene Ther 1992; 3:147–154.

32. Wagner E, Plank C, Zatloukal K, Cotten M, Birnstiel ML. Influenza virus hemagglutinin HA-2N-terminal fusogenic peptides augment gene transfer by transferrin-polylysine-DNA complexes: toward a synthetic virus-like gene-transfer vehicle. Proc Natl Acad Sci USA 1992; 89:7934–7938.

33. Ishibashi S, Brown MS, Goldstein JL, et al. Hypercholesterolemia in low density lipoprotein receptor knockout mice and its reversal by adenovirus mediated delivery. J Clin Invest 1993; 92:883–893.

34. Herz J, Gerard RD. Adenovirus mediated transfer of low density lipoprotein receptor gene acutely accelerates cholesterol clearance in normal mice. Proc Natl Acad Sci USA 1993; 90(7):2812–2816.

35. Kopfler WP, Willard M, Betz T, Willard JE, Gerard RD, Meidell RS. Adenovirus-mediated transfer of a gene encoding human apoplipoprotein A-1 into normal mice increases circulating high-density lipoprotein cholesterol. Circulation 1994; 90(3):1319–1327.

36. Oka K, Tzung KW, Sullivan M, Lindsay E, Baldini A, Chan L. Human very-low density lipoprotein receptor complementary DNA and deduced amino acid sequence and localization of its gene VLDLR to chromosome band 9p24 by fluorescence in situ hybridization. Genomics 1994; 20:298–300.

37. Nabel EG, Plautz G, Boyce FM, Stanley JC, Nabel GJ. Recombinant gene expression in vivo within endothelial cells of the arterial wall. Science 1989; 244:1342–1344.

38. Wilson JM, Birinyi LK, Salomon RN, Libby P, Callow AD, Mulligan RC. Implantation of vascular grafts lined with genetically modified endothelial cells. Science 1989; 244:1344–1346.

39. Dichek DA, Neville RF, Zwiebel JA, Freeman SM, Leon MB, Anderson WF. Seeding of intravascular stents with genetically engineered endothelial cells. Circulation 1990; 80:1347–1353.

40. Plautz G, Nabel EG, Nabel GJ. Introduction of vascular smooth muscle cells expressing recombinant genes in vivo. Circulation 1991; 83:578–583.

41. Nabel EG, Gordon D, Yang ZY, et al. Gene transfer in vivo with DNA-liposome complexes: lack of autoimmunity and gonadal localization. Hum Gene Ther 1992; 3:649–656.

42. Reissen R, Rahimizadeh H, Blessing E, Takeshita S, Barry JJ, Isner JM. Arterial gene transfer using pure DNA applied directly to a hydrogel-coated angioplasty balloon. Hum Gene Ther 1993; 4:749–758.

43. Guzman RJ, Lemarchand P, Crystal RG, Epstein SE, Finkel T. Efficient and selective adenovirus mediated gene transfer into vascular neointima. Circulation 1993; 88:2838–2848.

44. Dichek DA, Nussbaum O, Degen SJF, Anderson WF. Enhancement of fibrinolytic activity of sheep endothelial cells by retroviral mediated gene transfer. Blood 1991; 77:533–541.

45. Nabel EG, Yang ZY, Plautz G, et al. Recombinant fibroblast growth factor-1 promotes intimal hyperplasia and angiogenesis in arteries in vivo. Nature 1993; 362:844–846.

46. Nabel EG, Yang Z, Liptay S. Recombinant platelet derived growth factor B gene expression in porcine arteries induce intimal hyperplasia in vivo. J Clin Invest 1993; 91:1822–1829.

47. Nabel EG, Plautz G, Nabel GJ. Transduction of a foreign histocompatibility gene into the arterial wall induced vasculitis. Proc Natl Acad Sci 1992; 89:5147–5161.

48. Ascadi G, Jiao S, Jani A, Duke D, Williams P, Chong W, Wolff JA. Direct gene transfer and expression into rat heart in vivo. New Biol 1990; 3:71–81.

50. Nelson JR. The role of religions in the analysis of the ethical issues of human gene therapy. Hum Gene Ther 1991; 1:43–48.

51. Murray TH. Human gene therapy, the public, and the public policy. Hum Gene Ther 1990; 1:49–54.

52. Walters L. The ethics of human gene therapy. Nature 1986; 320:225–227.

53. Fletcher JC. Evolution of ethical debate about human gene therapy. Hum Gene Ther 1990; 1:55–61.

54. Ledley FD. Clinical considerations in the design of protocols for somatic gene therapy. Hum Gene Ther 1991; 2:77–84.

55. Epstein SL. Regulatory concerns in human gene therapy. Hum Gene Ther 1991; 2:243–249.

56. Kessler DA, Siegel JP, Noguchi PD, Zoon KC, Feiden KL, Woodcock J. Regulation of somatic cell therapy and gene therapy by the Food and Drug Administration. N Engl J Med 1993; 329:1169–1173.

57. Ledley FD. Designing clinical trials of somatic gene therapy. Proc NY Acad Sci 1994; 75(2):207–216.

18

Gene-Transfer Approaches to Vascular Disease

Elizabeth G. Nabel and Gary J. Nabel
University of Michigan
Ann Arbor, Michigan

INTRODUCTION

Gene transfer, the introduction and expression of recombinant genes in host cells, offers novel approaches to the study and treatment of vascular diseases. Considerable interest in gene transfer has fostered the development of methodologies for the delivery of recombinant genes to vascular endothelial and smooth-muscle cells in vitro and in vivo. To date, gene transfer to the vasculature has been employed largely for two purposes: 1) the examination of biological function and the regulation of sequences that control gene expression and 2) the development of animal models of vascular disease. In these animal models, the pathophysiology of vascular disease can be studied, and potential new therapies can be developed.

Several aspects of vascular biology and pathogenesis are appropriate for study by gene transfer. Many vascular lesions are anatomically localized to specific sites in the circulation, and recombinant genes can be directly introduced to these local lesions by catheter. Genes delivered by catheter to vascular lesions could therefore potentially prevent the development of additional lesions by expression of the recombinant gene. Stable expression of a recombinant gene may not be necessary to treat some vascular proliferative disorders.

Recombinant genes can be delivered to organs, such as the liver and kidney, via the vasculature. Catheter-based gene delivery therefore offers the potential to modify tissue- or organ-specific gene expression selectively. While there have been initial successes in demonstrating the feasibility of vascular gene transfer, there are also technical challenges. These include achieving regulated and cell-specific gene expression in target cells, developing efficient gene-transfer delivery systems, and producing animal models of vascular disease that mimic human diseases. Progress in vascular gene transfer is reviewed in this chapter, including viral and nonviral vector development, ex vivo and in vivo gene-transfer methods, animal models of vascular gene transfer to study disease pathophysiology, and potential therapeutic applications.

VASCULAR CELL TARGETS FOR GENE TRANSFER

The normal artery consists of three layers: intima, media, and adventitia. The intima is lined by endothelium on the lumen and bounded by the internal elastic lamina. In addition to endothelium, the intima consists of a thin layer of connective tissue that contains occasional solitary smooth-muscle cells (depending on species and artery). With increasing age, there is a concentric increase in intimal smooth-muscle cells. Endothelial cells regulate hemostasis (1,2), vascular contractility (3,4), cellular proliferation (5,6), and inflammation (7,8) in the vessel wall. Following arterial injury, lesions develop within the intima by interaction of the endothelium with macrophages, platelets, smooth-muscle cells, and T lymphocytes (9). Endothelial mitogens produced by macrophages include platelet-derived growth factor (PDGF) (10), fibroblast growth factor (FGF) (11), and transforming growth factor (TGF) (12). Endothelial cells express genes for a number of growth-regulatory molecules including PDGF (13), basic fibroblast growth factor (bFGF) (14), TGF-β (15), interleukin-1 (IL-1) (16), and tumor necrosis factor alpha (TNF-α) (17). These factors can induce secondary changes in gene expression in endothelial and smooth-muscle cells. Endothelial cells are a principal source of mitogenic and activating factors for underlying macrophages through factors such as macrophage colony stimulating factor (M-CSF) or oxidized low-density lipoproteins (oxLDL) (18). Endothelial cells participate in leukocyte chemotaxis and adhesion through potent chemotactic factors such as oxidized LDL (19) and MCP-1 (20) and cellular adhesion molecules such as VCAM (7). In addition, endothelial cells form nitric oxide (21) and PGI2 (22), which regulate dilation or constriction of the artery. Finally, endothelial cells regulate multiple functions within the vessel wall and are a major target for gene transfer.

The media is the muscular wall of the artery bounded by the internal and external elastic lamina. The media consists of spiraling layers of smooth-muscle cells attached to one another and interspersed with collagen fibrils and proteoglycans. The primary function of the media is arterial contraction or dilation. Following vascular injury, medial smooth-muscle cells may proliferate and migrate to the intima and contribute to the development of vascular hypertrophy and hyperplasia (9). Smooth-muscle cells respond to and elaborate many chemotactic and mitogenic factors, including PDGF (6), bFGF (23,24), TGF-β (25), IL-1 (16), and TNF-α (26). Smooth-muscle cells are a large reservoir of cells that could potentially elaborate recombinant gene products. The fact that they are ideal targets for gene transfer allows us to study their role in vascular lesion pathogenesis.

The adventitia consists of a dense collagenous structure containing collagen fibrils, elastic fibers, and many fibroblasts together with smooth-muscle cells. It is a highly vascular tissue and contains nerve fibers as well. The adventitia provides the outermost portion of the media and elastic arteries with nutrition via the vaso vasorum. Direct vascular gene transfer via the lumen often results in cell transfection in the adventitia, which may be mediated by transit through the vaso vasorum or transmurally through the arterial wall. The role of the adventitia in the pathogenesis of vascular lesions has not been well defined, but these cells also represent potential targets for gene transfer.

VECTORS

Viral and nonviral vectors have been employed in vascular gene-transfer studies to overcome the natural resistance of endothelial and smooth-muscle cells to foreign DNA, thus facilitating the delivery and entry of the recombinant genetic material into these cells. Vectors showing varying degrees of effectiveness include viral vectors, such as retrovirus, adenovirus, and adenovirus-augmented, receptor-mediated conjugates, and nonviral vectors, such as cationic liposomes, liposome viral conjugates, and polymer gels.

Viral Vectors

Viral vectors were among the first employed in gene-transfer studies (27,28). To enter a cell, the virus must have access to the appropriate viral receptor on the cell surface. After cell entry, the virus uses the biosynthetic pathways of the host cell to produce viral DNA, RNA, and protein. Among the four viral vectors that have been used in preclinical gene-transfer studies, retroviruses and adenoviruses are the vectors primarily used in studies of gene transfer to vascular tissue.

Retrovirus

The life cycle of retroviruses, which are RNA viruses, commences with entry into cells by a retroviral receptor. Viral RNA is copied by reverse transcriptase into DNA in the cytoplasm; DNA migrates to the host cell nucleus and integrates into the target-cell chromosomal DNA. The viral genome reproduces with each cell division. Viral RNA and proteins are synthesized in the host cytoplasm and bud through the cell membrane.

Although their ability to stably infect many types of mammalian cells makes retroviruses valuable as vectors (29), their ability to become part of the infected cell can allow retroviruses to become dangerous to the host if the reproduction of the viral part of the vector is not controlled. Hence, retroviruses are modified for use as vectors. Removal of the viral structural genes necessary for replication while leaving the packaging signal intact will render the retroviral vector replication-incompetent so that wild-type retroviruses cannot be produced (30). The altered retrovirus is then transfected into a packaging cell line. This contains plasmids encoding all the retrovirus structural genes under control of the regulatory sequences of the viral promoter but lacks the packaging signal (31). The vector transcripts are recognized by the viral proteins involved in packaging, and a replication-deficient virion is produced. Introduction of multiple modifications of the helper provirus—so that more than one recombination event is required to generate wild-type virus—provides further protection to the host. (For a review of the construction of retroviral vectors, see Refs. 27, 32, and 33.)

Retroviral vectors were employed in many early in vitro and in vivo demonstrations of vascular gene transfer (34–42). Despite the relative ease with which stable retroviral infection is accomplished in many cell types, the efficiency of gene transfer into vascular cells via retroviral vectors is relatively low. Approximately 5 to 10% of rat, porcine, or bovine endothelial cells are routinely infected in vitro, while 0.1 to 1% of endothelial cells have been infected in vivo (43,44). In addition, retroviral vectors are effective only in replicating cells (45), thus limiting the utility of this vector in quiescent vascular cells. Retroviral particles are also labile, causing the virus to be inactivated in vivo by a complement-mediated process in primates and humans. In addition, the cDNA insert that the retroviral vector can accommodate is relatively small, usually less than 5 kb.

Finally, the safety of retroviral vectors for us in long-term gene transfer in vivo is not yet known. There has been one report of wild-type virus producing lymphomas in a study employing retroviral vectors introduced into hematopoietic stem cells in primate bone-marrow transplantation; this complication arose from contamination by replication-competent virus from packaging cell lines (46). In contrast, 5-month studies of porcine serum and peripheral blood lymphocytes derived from direct retroviral gene-transfer studies into porcine arteries via retroviral vectors have revealed no wild-type retroviral recombination (47). Additional safety assessment may be provided by a large number of human studies employing retroviral vector-mediated gene transfer (27,28) and more recent trials of direct transfection by retroviral injection (48).

Generally, retroviral vectors have been associated with lower rates of transfection. Other vectors, such as the adenovirus and cationic liposomes, have provided higher rates of gene transfer into the vasculature. Such deficiencies of retrovirus vectors have stimulated pursuit of other vectors for vascular gene transfer.

Adenovirus

Adenoviruses—nonenveloped, double-stranded DNA viruses—have received considerable attention recently as in vivo gene-delivery vectors (49–51). The life cycle of the adenovirus is characterized by attachment to an adenoviral glycoprotein receptor on mammalian target cells and entry into cells by receptor-mediated endocytosis (52). Adenoviruses escape lysosomal degradation by adenoviral capsid proteins, and translocate to the cell nucleus. These viruses integrate with low frequency in dividing cells and do not require replication for transfer and expression of recombinant genes. Expression of viral genes depends on cellular transcription factors and expression of the adenoviral E1 gene (early region), which encodes a transactivator of viral gene expression (53). During lytic infection, the viral genome replicates to several thousand copies per cell. The viral genome associates with core proteins and is packaged into capsids by self-assembly of major capsid proteins.

The dependence of viral replication on the E1 gene is the basis for construction of replication-incompetent adenoviral vectors (54). The insertion of foreign genes into the adenoviral genome as a replacement for the E1 region results in a replication-defective adenovirus. Adenovirus vectors are produced by homologous recombination in 293 cells, a human embryonic kidney cell line, or certain cell lines that contain an integrated copy of the adenovirus E1 region (55). Typically, the foreign gene is introduced into a bacterial plasmid that contains a small region of the left adenovirus type 5 genome from which the native E1 gene has been deleted. The bacterial plasmid is cotransfected into 293 cells with an incomplete adenoviral genome; homologous recombination between the two DNAs generates a recombinant genome in which the E1 region is replaced by the foreign gene. Viral stock can be propagated in 293 cells to high titers, generally 10^9–10^{10} plaque-forming units/ml.

While adenoviruses have a number of favorable properties, they also have significant limitations. Gene expression in vascular cells following adenoviral infection usually persists for only several weeks (56,57). In addition, adenoviral vectors are not incorporated stably into the genome: Evidence suggests that in actively dividing cells, recombinant adenovirus is maintained transiently as an episomal genetic element (52). Although transient gene expression may be desirable for some vascular therapies, such as smooth-muscle cell proliferation following angioplasty, other vascular diseases may require long-term gene expression. Stratford-Perricaudet and colleagues (58) observed expression of reporter genes in muscle and liver tissue of neonatal mice for at least 12 months following intravenous injection, while gene expression persisted in adult mice for less than 1 month (58). Loss of foreign gene sequences have been observed in liver (59), artery (56), muscle (58), and other tissues following adenoviral infection. Transient expression may be secondary to loss of DNA, promoter extinction, and host immune responses.

Because adenoviruses are ubiquitous, host adults have neutralizing antibodies to one or more adenoviral serotypes. In addition, adenoviral infection of experimental animals can result in neutralizing antibody titers directed against adenovirus capsid proteins (60,61). Serum-neutralizing antibodies have also been detected following adenovirus-mediated gene transfer to rabbit synovium (62). While low levels of neutralizing antibodies do not appear to have adverse clinical effects, it remains to be determined whether host immune responses to the adenovirus will inhibit repeated infection and preclude repeated administrations of adenovirus (63). Repeat administration of E1-deleted vectors results in successful gene transfer and expression in the lung, but the efficiency is reduced in relation to the presence of human adenovirus-neutralizing antibody (64). While multiple administrations of adenoviral vectors may be important for the treatment of inherited diseases such as cystic fibrosis, single treatments may be required for vascular diseases such as restenosis. Nonetheless, vector modifications to limit host immune responses are being investigated.

First-generation recombinant adenoviruses (deletion of E1 genes and partial deletion of E3 genes) have been useful for in vivo applications of gene transfer; however, gene expression has been transient and associated with tissue inflammation, particularly in the liver and lung (65–67). These inflammatory effects in the mouse lung are diminished when recombinant adenovirus is employed from which the E2A gene, in addition to E1 sequences, has been deleted. Inactivation of the E2A gene, in addition to E1, is associated with longer gene expression and less airway inflammation in the lung (68) and liver (69). No inflammatory response was detected in the coronary vasculature following intra-arterial adenovirus infusion in rabbits (70). Mononuclear cells were observed in the adventitia of porcine peripheral arteries following local arterial gene transfer, but no arterial necrosis or vasculitis was present (71). Infection of pulmonary arteries with adenovirus was also associated with mild degrees of perivascular inflammation; however, pulmonary arteries instilled with saline or liposomes also exhibited similar mild accumulation of perivascular mononuclear cells (72). Further investigations are under way to determine the effects of vascular infection with adenovirus vectors.

Recombinant adenoviral vectors have been employed to introduce genes into multiple cell lines in vitro and tissues in vivo. These include cardiac and skeletal muscle (58), arteries (56,57,70,73,74), brain (75–77), lung (63, 78–83), synovium (62), and liver (59,69,84). Adenoviral vectors have several advantages, including efficient infection of mammalian cells and expression in nondividing cells in vivo. This vector is relatively stable; can be purified, grown, and concentrated to high titer; and is able to accommodate large cDNA inserts. Extrachromosomal replication of the adenoviral vector greatly reduces the chances of mutation by random integration and dysregulation of important host cellular genes. Adenoviral infections have not been associated with human malignancies.

Adenovirus-Augmented, Receptor-Mediated Gene Delivery

Receptor-mediated gene delivery using ligand DNA complexes linked to inactivated adenovirus is another potential vector for the vasculature. These vectors consist of two components: DNA condensed with polylysine, which allows binding to the virus, and a ligand, most commonly transferrin, conjugated to the virus complex (85–87). The DNA polylysine adenovirus conjugate enters cells by transferrin ligand binding to transferrin receptors. The complex is internalized by receptor- mediated endocytosis. The adenovirus functions to disrupt lysosomes in the cytoplasm, diminishing DNA degradation and enhancing DNA release into the cytoplasm (88). DNA is maintained as an episome, and integration is infrequent. Adenoviral conjugate vectors have been employed in gene transfer to hepatocytes (89) and hematopoietic cells (90). Another potential advantage of this vector includes the use of other synthetic or natural ligands to specifically target DNA to cells. The size of DNA that can be delivered is not limited by packaging constraints, thus increasing the number of genes that might be delivered with this vector. Current investigations are evaluating the utility of this vector in the vasculature as well as the duration and stability of gene expression in host cells.

Nonviral Vectors

Nonviral vectors offer potential safety advantages by avoiding possible infectious complications of viruses. Nonviral vectors employed in the vasculature include cationic liposomes, liposome viral conjugates, and polymer gels.

Cationic Liposomes

Cationic liposomes are positively charged vehicles that interact spontaneously and rapidly with polyanions such as DNA, mRNA, and antisense oligonucleotides. This interaction results in a liposome polynucleotide complex that captures or incorporates all of the polynucleotide (91,92). This complex is taken up by cells with a 10 to 100 times greater efficiency of negatively charged or neutral liposomes (93).

Cationic lipid reagents are a growing class of compounds that are useful for transfection of cells in vitro and for gene delivery in vivo. The prototype cationic lipid is DOTMA (N[1- (2,3-dioleyloxy)propyl]-N,N,N-trimethylammonium chloride) (93). When mixed with an equimolar amount of DOPE (dioleoyl phosphatidylethanolamine), DOTMA forms a cationic liposome, commercially available as Lipofectin (GIBCO-BRL, Gaithersburg, Maryland). Another cationic liposome, composed of DC-cholesterol (3b[N-(N'N'-dimethylaminoethane)- carbamoyl] cholesterol) mixed with DOPE, has proven to be nontoxic to cells over a wide range of concentrations in vitro and is metabolized in vivo (94). This vector has provided efficient gene transfer into malignant tumors in vivo (95). A limitation to these liposome preparations was liposome aggregation at high concentrations, limiting the amount of DNA delivered to cells. Modifications have been made to increase the concentrations of DNA that can be delivered to cells. Examples of second-generation cationic liposomes include DOSPA (2,3-dioleyoxy-N-[2(sperminecarboxamido)ethyl]-N,N-dimethyl-1-propanaminiumtrifluoracetate) mixed with DOPE (Lipofectamine, GIBCO-BRL).

The lipofection transfection methodology is notable for its simplicity. An aliquot of sterile aqueous cationic liposome is introduced into a polynucleotide solution, and the mixture is added to tissue culture cells or introduced into tissues in vivo. These features result in a transfection methodology that is convenient, reproducible, and effective in a wide variety of cell types. The effectiveness of liposome transfection depends in part on the molar ratio between the polynucleotide and liposome. For each nucleic acid and liposome preparation, optimal molar ratios are determined. For example, differences in target cell, plasmid and liposome preparation require adjustments to optimize transfection efficiency. Liposomal transfection has been used effectively and efficiently in multiple cell types in vitro and in vivo, including endothelial cells (43,47,96), lung epithelial cells (97), and tumor cells (95). Cationic liposomes have been employed in a human gene therapy trial to treat malignant melanoma and for catheter-based gene delivery in animals and humans (98). Studies in mice, rabbits, and pigs have demonstrated the safety and minimal toxicity of DNA liposome transfection in vivo by intra-arterial, intravenous, or intratumor injections (99,100).

Advantages of liposome vectors include safety for transfection of human tissues, minimal risk of mutagenesis, and acceptance of large amounts of DNA, which permits transfection of varying nucleotides. Following transfection with cationic liposomes, plasmid DNA is maintained in an extrachromosomal form. Therefore, expression can be relatively transient, from weeks to months. In nondividing tissues, however, these extrachromosomal plasmids can be maintained stably, particularly if they do not synthesize immunogenic gene products. Transfection efficiency depends on target tissue, nucleotide, and liposome preparation. Studies are ongoing to improve cationic liposome formulations to enhance delivery and improve transfection efficiency. In addition, the use of cationic liposomes to deliver antisense nucleotides (101,102) and ribozymes to vascular and other cells is under investigation.

Liposome Viral Conjugates

Another form of liposome gene transfer is the encapsulation of plasmid DNA and nuclear proteins in liposomes that are introduced into cells by membrane fusion mediated by hemagglutinating virus of Japan (HVJ). The nuclear protein, high-mobility group 1 (HMG1), also has been used to facilitate migration of plasmid DNA to the nucleus.

Purified, heat-inactivated HVJ is used to fuse DNA-loaded liposome vesicles with cell membranes. HVJ liposome transfection has increased expression of DNA in adult rat liver (103,104) and kidney (105). HVJ liposome-mediated gene transfer has been employed in a rat arterial model to investigate the in vivo function of angiotensin converting enzyme and renin genes. This method yields an efficiency of ≥30% of vascular cells transfected without local toxicity. These investigators have also developed HVJ liposome delivery of antisense oligonucleotides to rat carotid arteries (106,107).

Polymer Gels

Polynucleotides have been applied to polymer gels that are used to coat catheter balloons or directly applied to arteries. One example is a hydrogel catheter, which consists of an angioplasty balloon coated with a hydrophilic polyacrylic acid polymer (108). Polynucleotides are applied to the gel-coated balloon. When the balloon is inflated in an artery, the polynucleotide is pressed into the arterial wall. This method has been successfully used to deliver recombinant DNA to rabbit arteries in vivo (109). One disadvantage of the hydrogel catheter is that polynucleotide can be rapidly washed off the balloon after exposure to the bloodstream, so protective sheaths may be required during transport in the arterial circulation. In addition, the quantity of polynucleotide that can be deposited externally or internally in arteries has not been determined. Pluronic gel has been employed to deliver antisense oligonucleotides to the external (110) and internal (111) surfaces of arteries.

GENE TRANSFER INTO THE VASCULATURE

Major research efforts have been devoted to developing methods for introducing and expressing recombinant DNA and other polynucleotides in the vasculature in vivo. The goals of these studies have been to define gene function and to develop novel therapeutic approaches to vascular diseases. Innovative techniques of gene transfer have included seeding of cells transduced with recombinant genes to arteries or vascular prostheses, injection of recombinant genes into cardiac muscle, and direct gene transfer into peripheral, pulmonary, and coronary arteries.

Ex Vivo Gene Transfer

Ex vivo or cell-mediated gene transfer of endothelial and smooth muscle cells has been demonstrated in several animal models. Transfected autologous endothelial cells have been seeded onto denuded porcine (112) and rabbit (113) iliofemoral arteries. Genetically modified vascular smooth muscle cells also have been seeded onto denuded porcine iliofemoral arteries (114) and rat carotid arteries (115). Seeding of autologous transfected endothelial cells onto rat skeletal muscle capillaries has been shown to be feasible (116).

Vascular ex vivo gene transfer may have an important role to play in device technology as a means of optimizing fibrinolysis of vascular devices highly predisposed to thrombosis (117). Genetically modified canine endothelial cells seeded onto prosthetic grafts and reimplanted into canine carotid arteries produced reporter gene expression for at least 5 weeks (118), and stents have been seeded with genetically modified sheep endothelial cells (119).

Seeding of genetically modified smooth muscle cells has been achieved in both a porcine model (114) and a rat model, as denuded rat arteries have been seeded with the human adenosine deaminase gene (115). The latter study addresses the issue of the quantity of recombinant protein that may be produced by seeded cells and suggests that such protein can exert effects locally. Additional studies have improved conditions for increasing protein expression from retrovirally transduced endothelial cells (120), the efficiency of retroviral gene transfer (113), and the number and longevity of seeded cells (121).

Direct Gene Transfer

Direct gene transfer has been employed to deliver recombinant DNA to normal, injured, and atherosclerotic arteries of several animal species. Variation in the efficiency of gene transfer, in the duration of gene expression, and among the cell types expressing the recombinant gene exists between the different animal models. This variation is dependent on the vector, mode of delivery, artery, and species. Initial studies demonstrated the feasibility of retroviral and liposome-mediated gene transfer into peripheral porcine arteries (47), canine coronary and peripheral arteries (96,122), and normal (44) and balloon-dilated atherosclerotic (43) rabbit arteries, using reporter genes *E. coli lacZ* or firefly luciferase. These studies demonstrated the feasibility of direct in vivo gene transfer into coronary and peripheral arteries. Gene expression was observed in endothelial and smooth muscle cells and in 0.1 to 10% of intimal and medial cells. Gene expression was transient following liposomal transfection but was more prolonged—up to 5 months—following retroviral infection. Both vectors appeared to be safe for in vivo gene delivery. Local and systemic toxicities were not reported, including a lack of wild-type virus or mutagenesis following direct retroviral infection.

To further improve the transfection efficiencies, adenoviral vectors have been used in recent direct gene transfer studies. These vectors produce higher rates of transient gene expression in the vasculature. Adenoviral vectors encoding reporter genes have been employed in direct gene transfer experiments in the peripheral and coronary vasculature in several animals, including rats (57,74), rabbits (70,123), and sheep (56). Ex vivo infusion of a reporter gene and a human alpha-1 antitripsin gene into intact human umbilical veins produced gene expression in the endothelium 24 hours later (73). Direct in vivo transfection of sheep carotid arteries in veins with replication-incompetent adenoviral vectors encoding a *lacZ* gene and a human cystic fibrosis transmembrane regulator gene resulted in gene expression for up to 14 days after infection (56). Gene expression was not detectable at 28 days. Adenoviral infection of injured rat carotid arteries resulted in reporter gene expression in smooth muscle cells in the intima (74) and media (57). The efficiency of gene transfer in intimal smooth muscle cells of injured rat arteries was reported to be from 10% to more than 75% (74). Findings of other investigators demonstrated that approximately 30% of smooth muscle cells in the media of injured vessels were transduced.

Catheter infusion of adenoviral vectors encoding reporter genes has been shown to be an efficient method for induction of gene expression in coronary arteries in the myocardium of rabbits (70). Single intracoronary infusion of adenovirus in rabbits resulted in recombinant gene expression in coronary arteries and surrounding myocardium 2 weeks later. Approximately 10 to 100% of coronary endothelial and vascular smooth muscle cells exhibited reporter gene expression, while gene expression was observed in 60 to 80% of surrounding cardiac myocytes. Inflammatory cells and necrosis were not observed in the vasculature or myocardium. Gene expression was transient, and reporter gene expression was lost in coronary arteries and myocardium within 1 month. In this system, adenovirus was infused through the endhole of a coronary catheter, permitting possible systemic spread of adenovirus to other organs. Adenovirus DNA was detected by PCR in the liver, kidney, lung, brain, and testes of rabbits 5 days after viral infusion; however, it is not known whether this is associated with organ pathology. Nonetheless, percutaneous translumenal gene transfer is a promising approach to coronary and myocyte transduction.

Many factors influence the cell types transduced, the duration of gene expression, and the efficiency of gene transfer following arterial delivery. An important parameter is the delivery catheter. The efficiency in types of cells transduced may depend on the catheter delivery system. Several catheters have been employed in direct gene transfer in vivo, including a double balloon catheter (USCI, C. R. Bard), a perforated catheter (USCI, C. R. Bard), a hydrogel-coated balloon catheter (Boston Scientific), and a spiral catheter (SCIMED). Depending on the animal species and artery studied, these catheters may produce different levels of gene expression, transduce different cell types, and produce different tissue injury even within the same artery. Instillation pressure can also influence the depth of vector into the artery and the transduction of cells in the intima, media, or adventitia (109,124,125). Viral titer and vector dose will also influence levels of gene transfer and the duration of gene expression. Local transfection into a focal arterial segment with removal of vector after the transfection will limit the systemic spread of the vector and cDNA. Alternatively, infusion of vector through the endhole of a balloon catheter into the arterial circulation may result in systemic spread and transfection of the microcirculation of organs distant to the transfection site. These factors should be carefully considered when executing gene transfer experiments.

Animal Models of Vascular Gene Transfer

Direct gene transfer can be used as a somatic model to define gene function in the vessel wall. Local in vivo gene transfer has several advantages over systemic administration of a protein factor. First, the target gene is transfected into a local specific site in the vasculature, which avoids systemic effects. Second, gene transfer results in gene expression by the cell, which permits analysis of gene function over a longer period of time, unlike a single administration of protein that is removed rapidly from the circulation. Third, a transfected vascular segment can be compared with adjacent nontransfected segments or with contralateral control arteries. Fourth, the biological function of a recombinant gene in vivo can be analyzed. For example, transfection of renin or ACE genes into rat smooth muscle cells using HVJ liposomes produces an increase in rat angiotensin II levels and smooth muscle cell proliferation (126). Transduction of a class I major histocompatibility gene, HLA-B7, into porcine arteries with retroviral or liposome vectors produces vascular inflammation mediated by CD8+ T cells generated against the recombinant HLA-B7 protein (124). Adenovirus-mediated gene transfer has been used to transiently reconstitute LDL receptor function in homozygous mice that lack LDL receptor produced by homologous recombination (59). Inactivation of low-density lipoprotein receptor-related protein (LRP) by adenoviral transfer of a dominant negative regulator of LRP function into mice lacking LDL receptors is associated with accumulation of chylomicrons (127).

Growth factors and cytokines stimulate vascular cell proliferation and vessel formation in vivo. Although the genes encoding many factors have been cloned and their mechanism of action has been defined in vitro, definition of their role in vivo has been much more difficult to analyze. To define the effect of recombinant growth factors and cytokines in vivo, plasmid expression vectors encoding the cDNAs for three human growth factor genes—platelet derived growth factor B (PDGF B) (128), a secreted form of acidic fibroblast growth factor (FGF-1) (129), and an active, secreted form of transfecting growth factor β1 (TGF-β1) (130)—were transfected into porcine peripheral arteries using liposome vectors. For each growth factor, expression of mRNA was confirmed by reverse transcriptase PCR, and protein expression in transduced cells was identified by monoclonal antibodies. Arteries transfected with PDGF B demonstrated severe intimal thickening, characterized by increased cellularity and smooth muscle cell proliferation, in contrast to control arteries transduced with a reporter gene (128). Smooth muscle cell proliferation was observed within the first 2 weeks following gene transfer. Later, 2 to 3 weeks after gene transfer, type I collagen and other extracellular matrix synthesis were evident. Macrophages and T cells were rarely observed in the intima. These data support the hypothesis that PDGF B gene expression in vivo stimulates smooth muscle cell proliferation early, after direct gene transfer, while collagen synthesis is a later feature of these lesions and may limit further expansion of the intima.

The heparin binding fibroblast growth factor family has proliferative and angiogenic properties in vivo (14). To investigate the role of FGF-1 in vascular pathology, a eukaryotic expression vector encoding a secreted form of FGF-1 was introduced into porcine arteries (129). Using an affinity-purified antibody to human FGF-1, FGF-1 protein was identified in intimal endothelial and smooth muscle cells. Intimal thickening was observed in FGF-1-transduced arteries 3 weeks following gene transfer. In addition, in vessels with expanded intima, intimal angiogenesis was observed. This angiogenesis was not present in vessels transduced with other growth factors or cytokines. These data support a role for FGF-1 in stimulating vascular angiogenesis in vivo.

TGF-β1 is a secreted protein that plays an important role in embryonal development and in repair following tissue injury. However, the function of TGF-β in vascular cell growth in vivo has been difficult to discern. The role of TGF-β1 in the pathophysiology of intimal and medial hyperplasia was investigated by gene transfer of an expression plasmid encoding active TGF-β1 in porcine arteries. Expression of TGF-β1 in normal porcine arteries resulted in substantial extracellular matrix production accompanied by intimal and medial hyperplasia. Increased procollagen and collagen synthesis was ob-

served in the neointima relative to transfected control arteries. Different
patterns of gene expression were observed in vessels transduced by TGF-β1
and PDGF B. Procollagen synthesis in TGF-β1 arteries occurred at 4 days after
gene transfer compared with later timepoints in PDGF B–transfected vessels.
In contrast, smooth muscle cell proliferation was a minor component of
intimal formation in TGF-β1 vessels, compared with PDGF B vessels. Although
all of these three recombinant growth factor genes (PDGF B, FGF-1, and TGF-
β1) stimulate vascular cell proliferation in vivo, each exerts distinct effects on
smooth muscle cell proliferation, angiogenesis, and extracellular matrix formation. These findings suggest that the generation of intimal hyperplasia
represents a common response to gene expression in multiple growth factors,
which in turn exert different effects on vessel repair. Investigations are
ongoing to examine the role of cytokine genes in early atherosclerosis.

Gene Transfer to Inhibit Vascular Smooth Muscle Cell Proliferation

Vascular injury induces gene products that stimulate smooth muscle cell
migration and proliferation, which results in intimal hyperplasia (9). Two
common vascular disorders in which this process is observed are atherosclerosis and restenosis. The pathophysiology of restenosis is complex, and
multiple factors contribute, including thrombus formation, smooth muscle
cell proliferation, extracellular matrix synthesis, and recoil of the vessel wall
(131–133). Molecular genetic interventions to limit smooth muscle cell proliferation or extracellular matrix synthesis at sites of vascular injury might
provide insight into the pathophysiology and provide possible treatments for
vascular proliferative disorders. Local delivery of an antiproliferative agent
during the peak of smooth muscle cell proliferation or collagen matrix
synthesis after balloon injury might limit expansion of the intima.

Several molecular interventions have been explored as possible treatment strategies to limit restenosis, including recombinant chimeric toxins
(134), antisense strategies (106,107,110,111), and gene transfer. We (71) and
others (135) reasoned that one approach to the selective elimination of
dividing cells is to express the herpesvirus thymidine kinase gene in dividing
smooth muscle cells after vascular injury. Thymidine kinase, when expressed
in transduced cells, phosphorylates the nucleoside analog ganciclovir into an
active toxic form (136–139). Incorporation of phosphorylated ganciclovir into
cellular DNA induces chain termination in dividing cells, which causes cell
death (140,141). A bystander effect, demonstrated in a variety of malignancies
and in smooth muscle cells, inhibits cell growth in nontransduced cells as well
(77,142–144). A low-molecular-weight metabolite of phosphorylated ganciclovir is diffusable, and neighboring nontransfected cells are rendered sensi-

tive to ganciclovir. This bystander effect amplifies the antiproliferative action of the thymidine kinase gene and ganciclovir treatment. Such strategies have been employed in different gene transfer approaches to tumor therapy, including glioma (77), mesothelioma (145), and adenocarcinoma (144).

Adenoviral vectors encoding the herpesvirus thymidine kinase gene (ADV-tk) were introduced into balloon-injured porcine iliofemoral arteries at the time of injury (71). Ganciclovir was administered 36 hours after infection for 6 days. A control group infected with the ADV-tk vector was treated with saline. In additional control studies, an E1-deleted adenoviral vector (ADV-ΔE1) was introduced into injured iliofemoral arteries, and animals were treated with ganciclovir or saline. Three weeks after balloon injury and adenoviral infection, the areas of the intima and media in each artery were measured by quantitative morphometry, and the intimal to medial (I/M) area ratios were determined. A significant reduction in the I/M area ratio was observed in vessels subject to mild or severe injury in animals transduced with ADV-tk and treated with ganciclovir, compared with animals infected with ADV-tk or ADV-ΔE1 treated with saline. A 40% reduction in intimal BrdC incorporation was observed in ADV-tk ganciclovir-treated animals compared with ADV-tk saline-treated animals 7 days after gene transfer, suggesting inhibition of smooth muscle cell proliferation. This reduction in intimal hyperplasia was stable, as a significant reduction was observed 6 weeks after treatment.

The potential for systemic toxicities following adenoviral infection was investigated by analysis of serum biochemistries and other organs. In transfected arteries, occasional mononuclear infiltrates were observed in the adventitia without evidence of necrosis. Tissues from other major organs, including nontransfected carotid artery, heart, lung, kidney, liver, spleen, skeletal muscle, and ovary, showed no significant pathological lesions, suggesting the potential safety of these vectors for clinical use. In similar experiments, as in a rat carotid balloon-injury model, infection of injured arteries with an ADV-tk vector 7 days after balloon injury followed by 2 weeks of ganciclovir, significantly reduced I/M area ratios in treated groups, compared with uninjured carotid artery, injured arteries infected with a reporter gene, and injured arteries transfected with tk and treated with saline (135). While the thymidine kinase gene is selective for dividing cells, it is nonselective for endothelial cells and smooth muscle cells. In the porcine studies, reendothelialization was apparent at 7 days, and was complete between 3 and 6 weeks. The time course of endothelial cell regrowth was delayed, suggesting that reendothelialization occurred after ganciclovir treatment. These studies demonstrate the potential utility of a molecular intervention targeted to proliferating smooth muscle cells after balloon injury. Whether this type of strategy might be effective in treating vascular proliferative disorders awaits further confirmation.

CLINICAL TRIALS

More than 100 clinical human gene transfer or gene therapy trials have been approved by the National Institutes of Health Recombinant DNA Advisory Committee (146). The majority of these are directed against inherited single-gene disorders, cancer, or AIDS. Several trials bear upon potential future trials for vascular diseases.

A completed trial of direct DNA liposome transfection for immunotherapy against melanoma has established preliminary feasibility, safety and therapeutic potential of DNA introduction into humans by both direct injection and catheter infusion into the vasculature (98). In this trial, the gene encoding a foreign major histocompatibility complex protein, HLA-B7, conveyed by a liposome vector, was introduced into HLA-B7-negative patients with advanced melanoma by direct injection into subcutaneous tumor nodules and by catheter to pulmonary melanoma metastases in the lung. Plasmid DNA and mRNA expression were confirmed by PCR, recombinant HLA-B7 protein was demonstrated in tumor biopsy tissue in all patients, and immune responses to HLA-B7 and autologous tumors were detected. One patient demonstrated regression of injected nodules after two independent treatments and one was accompanied by regression at distant sites. One patient received treatment of pulmonary metastases by catheter delivery of DNA liposomes in the right posterior basal segment artery (147). Transcatheter delivery of HLA-B7 DNA liposomes into a right posterior basal segment artery was safely performed and well tolerated in one patient with melanoma metastatic to the right lower lobe.

An ex vivo trial for homozygous familial hypercholesterolemia has been performed, and findings have been reported in one patient (148). A patient with homozygous familial hypercholesterolemia was treated by ex vivo gene transfer of a retroviral vector encoding the low-density lipoprotein (LDL) receptor into autologous hepatocytes. In situ hybridization of liver tissue 4 months after therapy demonstrated LDL receptor mRNA expression in scattered hepatocytes. Gene expression in a small number of hepatocytes was associated with an approximately 30% lowering of serum LDL levels. The patient was started on cholesterol-lowering therapy 4 months after gene therapy, and LDL lowering has remained stable for the duration of treatment. This report demonstrates the feasibility, safety, and potential efficacy of ex vivo liver-directed gene therapy for familial hypercholesterolemia. The effects of this lipid lowering on vascular disease await further study. Several human clinical trials to treat cystic fibrosis are also in progress. In these studies, adenoviral vectors are employed to deliver recombinant CFTR gene to lung epithelium (149) or nasal mucosa (83). Chloride channel function has been restored in the nasal epithelium in cystic fibrosis patients infected with CFTR-expressing adenovirus (83).

CONCLUSION

Our understanding of the molecular and cellular biology of the vascular system has grown rapidly during the past decade, and development of the components necessary for the therapeutic application of gene transfer is also proceeding at a rapid rate. Despite these advances, human gene therapy to treat vascular diseases faces many challenges. Identification of candidate genes and target cells and improvement in vectors will increase efficacy of transfer, extend duration of gene expression, and minimize toxicity, and approaches to regulation of gene expression are important goals. New approaches to the delivery of recombinant genes into vascular cells will facilitate these efforts. Further refinement of catheter-based delivery systems will also contribute to improved efficacy in clinical settings. Gene transfer in the vasculature is a useful approach to study gene function in vivo. Animal models of human vascular diseases will be employed to investigate disease pathophysiology and test gene therapy approaches. Progress in the clinical application of gene transfer systems to treat vascular diseases will be linked to the availability of vectors to transfer and maintain recombinant gene expression in vivo, to the identification of target genes, and to methods to achieve site-specific gene expression. Considerable progress has been made in refining the transfer of recombinant genes to the vasculature, and it is likely that genetic interventions will contribute to both the understanding and the treatment of vascular disorders in the future.

REFERENCES

1. Rosenberg RD. The biochemistry and pathophysiology of the prethrombotic state. Annu Rev Med 1987; 38:493-508.
2. Loskutoff DJ, Curriden SA. The fibrinolytic system of the vessel wall and its role in the control of thrombosis. Ann NY Acad Sci 1990; 598:238-247.
3. Furchgott RF, Zawadzki JV. The obligatory role of endothelial cells in the relaxation of arterial smooth muscle by acetylcholine. Nature 1980; 288:373-376.
4. Shimokawa H, Vanhoutte PM. Impaired endothelium-dependent relaxation to aggregating platelets and related vasoactive substances in porcine coronary arteries in hypercholesterolemia and atherosclerosis. Circ Res 1989; 64:900-914.
5. DiCorleto PE, Bowen-Pope DF. Cultured endothelial cells produce a platelet-derived growth factor-like protein. Proc Natl Acad Sci USA 1983; 80:1919-1923.
6. Ross R, Raines EW, Bowen-Pope DF. The biology of platelet-derived growth factor. Cell 1986; 46:155-169.
7. Cybulsky MI, Gimbrone MA Jr. Endothelial expression of a mononuclear leukocyte adhesion molecule during atherogenesis. Science 1991; 251:788-791.
8. Munro JM, Cotran RS. The pathogenesis of atherosclerosis: atherogenesis and inflammation. Lab Invest 1988; 58:249-261.

9. Ross R. The pathogenesis of atherosclerosis: a perspective for the 1990s. Nature 1993; 362:801–809.

10. Ross R, Masuda J, Raines EW, Gown AM, Katsuda S, Sasahara M, Malden LT, Masuko H, Sato H. Localization of PDGF-B protein in macrophages in all phases of atherogenesis. Science 1990; 248:1009–1012.

11. Folkman J. A heparin-binding angiogenic protein-basic fibroblast growth factor is stored within basement membrane. Am J Pathol 1988; 130:393–400.

12. Moses HL, Yang EY, Pietenpol JA. TGF-β stimulation and inhibition of cell proliferation: new mechanistic insights. Cell 1990; 63:245–247.

13. DiCorleto PE, Gajdusek CM, Schwartz SM, Ross R. Biochemical properties of the endothelium-derived growth factor: Comparison to other growth factors. J Cell Physiol 1983; 114:339.

14. Burgess WH, Maciag T. The heparin-binding (fibroblast) growth factor family of proteins. Annu Rev Biochem 1989; 58:575–606.

15. Botney MD, Bahadori L, Gold LI. Vascular remodeling in primary pulmonary hypertension: Potential role for transforming growth factor-β. Am J Pathol 1994; 144:286–295.

16. Raines EW, Dower SK, Ross R. Interleukin-1 mitogenic activity for fibroblasts and smooth muscle cells is due to PDGF-AA. Science 1989; 243:393–396.

17. Hajjar KA, Hajjar DP. Tumor necrosis factor-mediated release of platelet-derived growth factor from cultured endothelial cells. J Exp Med 1987; 166:235–245.

18. Steinberg D. Antioxidants and atherosclerosis. A current assessment (editorial). Circulation 1991; 84:1420–1425.

19. Yla-Herttuala S. Evidence for the presence of oxidatively modified low density lipoprotein in atherosclerotic lesions of rabbit and man. J Clin Invest 1989; 84:1086–1095.

20. Yla-Herttuala S, Lipton BA, Rosenfeld ME, Sarkioja T, Yoshimura T, Leonard EJ, Witztum JL, Steinberg D. Expression of monocyte chemoattractant protein 1 in macrophage-rich areas of human and rabbit atherosclerotic lesions. Proc Natl Acad Sci USA 1991; 88:5252–5256.

21. Minor RL, Myers PR, Guerra R Jr, Bates JN, Harrison DG. Diet-induced atherosclerosis increases the release of nitrogen oxides from rabbit aorta. J Clin Invest 1990; 86:2109–2116.

22. Botting R, Vane JR. Vasoactive mediators derived from the endothelium. Arch Mal Coeur 1989; 82:11–14.

23. Klagsbrun M, Edelman ER. Biological and biochemical properties of fibroblast growth factors: Implications for the pathogenesis of arteriosclerosis. Arteriosclerosis 1989; 9:269–278.

24. Lindner V, Reidy MA. Expression of basic fibroblast growth factor and its receptor by smooth muscle cells and endothelium in injured rat arteries. Circ Res 1993; 73:589–595.

25. Majesky MW, Lindner V, Twardzik DR, Schwartz SM, Reidy MA. Production of transforming growth factor beta 1 during repair of arterial injury. J Clin Invest 1991; 88:904–910.

26. Old LJ. Tumor necrosis factor (TNF). Science 1985; 230:630–632.

27. Miller AD. Retroviral vectors. Curr Top Microbiol 1992; 158:1–24.

28. Mulligan RC. The basic science of gene therapy. Science 1993; 260:926-932.
29. Varmus H. Retroviruses. Science 1988; 240:1427-1435.
30. Danos O, Mulligan RC. Safe and efficient generation of recombinant retroviruses with amphotropic and ecotropic host ranges. Proc Natl Acad Sci USA 1988; 85:6460-6464.
31. Cone RD, Mulligan RC. High-efficiency gene transfer into mammalian cells: generation of helper-free recombinant retrovirus with broad mammalian host range. Proc Natl Acad Sci USA 1984; 81:6349-6353.
32. Lever AM. Retroviral vectors. Biochem Soc Trans 1991; 19:379-383.
33. Tolstoshev P. Retroviral-mediated gene therapy-safety considerations and pre-clinical studies. Bone Marrow Transplant 1992; 1:148-150.
34. Eglitis MA, Kantoff PW, Gilboa E, Anderson WF. Gene expression in mice after high efficiency retroviral-mediated gene transfer. Science 1985; 230:1395-1398.
35. Eglitis MA, Kantoff PW, Jolly JD, Jones JB, Anderson WF, Lothrop CD. Gene transfer into hematopoietic progenitor cells from normal and cyclic hematopoiesis dogs using retroviral vectors. Blood 1988; 71:717-722.
36. Hock RA, Miller AD. Retrovirus-mediated transfer and expression of drug resistance genes in human haematopoietic progenitor cells. Nature 1986; 320:275-277.
37. Hogge DE, Humphries RK. Gene transfer to primary normal and malignant human hemopoietic progenitors using recombinant retroviruses. Blood 1987; 69:611-617.
38. Wilson JM, Jefferson DM, Chowdhury JR, Novikoff PM, Johnston DE, Mulligan RC. Retrovirus-mediated transduction of adult hepatocytes. Proc Natl Acad Sci USA 1988; 85:3014-3018.
39. Ledley FD, Darlington GJ, Tahn T, Woo SLC. Retrovirus gene transduction into primary hepatocytes: Implications for genetic therapy of liver-specific functions. Proc Natl Acad Sci USA 1987; 84:5335-5339.
40. Morgan JR, Barrandon Y, Green II, Mulligan RC. Expression of an exogenous growth hormone gene by transplantable epidermal cells. Science 1987; 237: 1476-1479.
41. Keller G, Paige C, Gilboa E, Wagner EF. Expression of a foreign gene in myeloid and lymphoid cells derived from multipotent haematopoietic precursors. Nature 1985; 318:149-154.
42. Kwok WW, Scheuning F, Stead RB, Miller AD. Retroviral transfer of genes into canine hemopoietic progenitor cells in culture: A model for human gene therapy. Proc Natl Acad Sci USA 1986; 83:4552-4555.
43. Leclerc G, Gal D, Takeshita S, Nikol S, Weir L, Isner JM. Percutaneous arterial gene transfer in a rabbit model. Efficiency in normal and balloon-dilated atherosclerotic arteries. J Clin Invest 1992; 90:936-944.
44. Flugelman MY, Jaklitsch MT, Newman KD, Casscells W, Bratthauer GL, Dichek DA. Low level in vivo gene transfer into the arterial wall through a perforated balloon catheter. Circulation 1992; 3:1110-1117.
45. Miller DG, Adam MA, Miller AD. Gene transfer by retrovirus vectors occurs only in cells that are actively replicating at the time of infection. Mol Cell Biol 1990; 10:4239-4242.

46. Recombinant NIH Advisory Committee. Bethesda, MD, 1993:1-2.
47. Nabel EG, Plautz G, Nabel GJ. Site-specific gene expression in vivo by direct gene transfer into the arterial wall. Science 1990; 249:1285-1288.
48. Nabel GJ, Fox BA, Post L, Thompson CB, Woffendin C. Clinical protocol: a molecular genetic intervention for AIDS—effects of a transdominant negative form of Rev. Hum Gene Ther 1994; 5:79-92.
49. Gerard RD, Meidell RS. Adenovirus-mediated gene transfer. Trends Cardiovasc Med 1993; 3:171-176.
50. Berkner KL. Development of adenovirus vectors for the expression of heterologous genes. Bio Techniques 1988; 6:616-629.
51. Kozarsky KF, Wilson JM. Gene therapy: adenovirus vectors. Curr Opin Genet Devel 1993; 3:499-503.
52. Berkner KL. Expression of heterologous sequences in adenoviral vectors. Curr Top Microbiol 1992; 58:39-66.
53. Nevins JR. Adenovirus E1A-dependent transactivation of transcription. Cancer Biol 1990; 1:59-68.
54. Graham FL, Prevec L. Adenovirus-based expression vectors and recombinant vaccines. In: Ellis RW, ed. Vaccines: New Approaches to Immunological Problems. Boston: Butterworth-Heinemann, 1992:363-390.
55. Graham FL, Smiley J, Russel WC, Nairu R. Characteristics of a human cell line transformed by DNA from human adenovirus type 5. J Gen Virol 1977; 36:59-72.
56. Lemarchand P, Jones M, Yamada I, Crystal RG. In vivo gene transfer and expression in normal uninjured blood vessels using replication-deficient recombinant adenovirus vectors. Circ Res 1993; 72:1132-1138.
57. Lee SW, Trapnell BC, Rade JJ, Virmani R, Dichek DA. In vivo adenoviral vector-mediated gene transfer into balloon-injured rat carotid arteries. Circ Res 1993; 73:797-807.
58. Stratford-Perricaudet LD, Makeh I, Perricaudet M, Briand P. Widespread long-term gene transfer to mouse skeletal muscles and heart. J Clin Invest 1992; 90:626-630.
59. Ishibashi S, Brown MS, Goldstein JL, Gerard RD, Hammer RE, Herz J. Hypercholesterolemia in low density lipoprotein receptor knockout mice and its reversal by adenovirus-mediated gene delivery. J Clin Invest 1993; 92:883-893.
60. Prevec L, Schneider M, Rosenthal KL, Belbeck LW, Derbyshire JB, Graham FL. Use of human adenovirus-based vectors for antigen expression in animals. J Gen Virol 1989; 70:429-434.
61. Natuk RJ, Chanda PK, Lubeck MD, Davis AR, Wilhelm J, Hjorth R, Wade MS, Bhat BM, Mizutani S, Lee S, Eichberg J, Gallo RC, Hung PP, Robert-Guroff M. Adenovirus-human immunodeficiency virus (HIV) envelope recombinant vaccines elicit high-titered HIV-neutralizing antibodies in the dog model. Proc Natl Acad Sci USA 1992; 89:7777-7781.
62. Roessler BJ, Allen ED, Wilson JM, Hartman JW, Davidson BL. Adenoviral-mediated gene transfer to rabbit synovium in vivo. J Clin Invest 1993; 92:1085-1092.
63. Zabner J, Petersen DM, Puga AP, Graham SM, Couture LA, Keyes LD, Lukason MJ, St George JA, Gregory RJ, Smith AE, Welsh MJ. Safety and efficacy of repetitive

adenovirus-mediated transfer of CFTR cDNA to airway epithelia of primates and cotton rats. Nature Genet 1994; 6:75-83.

64. Yei S, Mittereder N, Tang K, O'Sullivan C, Trapnell BC. Adenovirus-mediated gene transfer for cystic fibrosis: quantitative evaluation of repeated in vivo vector administration to the lung. Gene Therapy 1994; 1:192-200.

65. Ginsberg HS, Moldawer LL, Sehgal PB, Redington M, Kilian PL, Chanock RM, Prince GA. A mouse model for investigating the molecular pathogenesis of adenovirus pneumonia. Proc Natl Acad Sci USA 1991; 88:1651-1655.

66. Prince GA, Porter DD, Jenson AB, Horswood RL, Chanock RM, Ginsberg HS. Pathogenesis of adenovirus type 5 pneumonia in cotton rats (*Sigmodon hispidus*). J Virol 1993; 67:101-111.

67. Yang Y, Raper SE, Cohn JA, Engelhardt JF, Wilson JM. An approach for treating the hepatobiliary disease of cystic fibrosis by somatic gene transfer. Proc Natl Acad Sci USA 1993; 90:4601-4605.

68. Yang Y, Nunes FA, Berencsi K, Gonczol E, Engelhardt JF, Wilson JM. Inactivation of E2A in recombinant adenoviruses improves the prospect for gene therapy in cystic fibrosis. Nature Genet 1994; 7:362-369.

69. Engelhardt JF, Ye X, Doranz B, Wilson JM. Ablation of E2A in recombinant adenoviruses improves transgene persistence and decreases inflammatory response in mouse liver. Proc Natl Acad Sci USA 1994; 91:6196-6200.

70. Barr E, Carroll J, Kalynych AM, Tripathy SK, Kozarsky K, Wilson JM, Leiden JM. Efficient catheter-mediated gene transfer into the heart using replication-defective adenovirus. Gene Ther 1994; 1:51-58.

71. Ohno T, Gordon D, San H, Pompili VJ, Imperiale MJ, Nabel GJ, Nabel EG. Gene therapy for vascular smooth muscle cell proliferation after arterial injury. Science 1994; 265:781-784.

72. Muller DWM, Gordon D, San H, Yang ZY, Pompili VJ, Nabel GJ, Nabel EG. Catheter-mediated pulmonary vascular gene transfer and expression. Circ Res 1994; 75:1039-1049.

73. Lemarchand P, Jaffe HA, Danel C, Cid MC, Kleinman HK, Stratford-Perricaudet LD, Perricaudet M, Pavirani A, Lecocq JP, Crystal RG. Adenovirus-mediated transfer of a recombinant human a1-antitrypsin cDNA to human endothelial cells. Proc Natl Acad Sci USA 1992; 89:6482-6486.

74. Guzman RJ, Lemarchand P, Crystal RG, Epstein SE, Finkel T. Efficient and selective adenovirus-mediated gene transfer into vascular neointima. Circulation 1993; 88:2838-2848.

75. Akli S, Caillaud C, Vigne E, Stratford-Perricaudet LD, Poenaru L, Perricaudet M, Kahn A, Peschanski MR. Transfer of a foreign gene into the brain using adenovirus vectors. Nature Genet 1993; 3:224-228.

76. Davidson BL, Allen ED, Kozarsky KF, Wilson JM, Roessler BJ. A model system for in vivo gene transfer into the central nervous system using an adenoviral vector. Nature Genet 1993; 3:219-223.

77. Chen S-H, Shine HD, Goodman JC, Grossman RG, Woo SLC. Gene therapy for brain tumors: Regression of experimental gliomas by adenovirus-mediated gene transfer in vivo. Proc Natl Acad Sci USA 1994; 91:3054-3057.

78. Rosenfeld MA, Siegfried W, Yoshimura K, Yoneyama K, Fukayama M, Stier LE,

Paakko P, Gilardi P, Stratford-Perricaudet LD, Perricaudet M, Jallat WS, Pavarani A, Lecocq J-P, Crystal RG. Adenovirus-mediated transfer of a recombinant a 1-antitrypsin gene to the lung epithelium in vivo. Science 1991; 252:431–434.

79. Rosenfeld MA, Yoshimura K, Trapnell BC, Yoneyama K, Rosenthal ER, Dalemans W, Fukayama M, Bargon J, Stier LE, Stratford-Perricaudet L, Perricaudet M, Guggino WE, Pavarani A, Lecocq J-P, Crystal RG. In vivo transfer of the human cystic fibrosis transmembrane conductance regulator gene to the airway epithelium. Cell 1992; 68:143–155.

80. Mastrangeli A, Danel C, Rosenfeld MA, Stratford-Perricaudet L, Perricaudet M, Pavirani A, Lecocq J-P, Crystal RG. Diversity of airway epithelial cell targets for in vivo recombinant adenovirus-mediated gene transfer. J Clin Invest 1993; 91:225–234.

81. Engelhardt JF, Yang Y, Stratford-Perricaudet LD, Allen ED, Kozarsky K, Perricaudet M, Yankaskas JR, Wilson JM. Direct gene transfer of human CFTR into human bronchial epithelia of xenografts with E1-deleted adenoviruses. Nature Genet 1993; 4:27–34.

82. Engelhardt JF, Zepeda M, Cohn JA, Yankaskas JR, Wilson JM. Expression of the cystic fibrosis gene in adult human lung. J Clin Invest 1994; 93:737–749.

83. Zabner J, Couture LA, Gregory RJ, Graham SM, Smith AE, Welsh MJ. Adenovirus-mediated gene transfer transiently corrects the chloride transport defect in nasal epithelia of patients with cystic fibrosis. Cell 1993; 75:207–216.

84. Jaffe HA, Danel C, Longenecker G, Metzger M, Setoguchi Y, Rosenfeld MA, Gant TW, Thorgeirsson SS, Stratford-Perricaudet LD, Perricaudet M, Pavirani A, Lecocq JP, Crystal RG. Adenovirus-mediated in vivo gene transfer and expression in normal rat liver. Nature Genet 1992; 1:372–378.

85. Wagner E, Zenke M, Cotten M, Beug H, Birnstiel ML. Transferrin-polycation conjugates as carriers for DNA uptake into cells. Proc Natl Acad Sci USA 1990; 87:3410–3414.

86. Wagner E, Zatloukal K, Cotten M, Kirlappos H, Mechtler K, Curiel DT, Birnstiel ML. Coupling of adenovirus to transferrin-polylysine/DNA complexes greatly enhances receptor-mediated gene delivery and expression of transfected genes. Proc Natl Acad Sci USA 1992; 89:6099–6103.

87. Curiel DT, Agarwal S, Wagner E, Cotten M. Adenovirus enhancement of transferrin-polylysine-mediated gene delivery. Proc Natl Acad Sci USA 1991; 88:8850–8854.

88. Cotten M, Wagner E, Zatloukal K, Phillips S, Curiel DT, Birnstiel ML. High-efficiency receptor-mediated delivery of small and large (48 kilobase) gene constructs using the endosome-disruption activity of defective or chemically inactivated adenovirus particles. Proc Natl Acad Sci USA 1992; 89:6094–6098.

89. Wu GY, Wilson JM, Shalaby F, Grossman M, Shafritz DA, Wu CH. Receptor-mediated gene delivery in vivo: Partial correction of genetic analbuminemia in Nagase rats. J Biol Chem 1991; 266:14338–14342.

90. Chen S-J, Wilson JM, Muller DWM. Adenovirus-mediated gene transfer of soluble vascular cell adhesion molecule to porcine interposition vein grafts. Circulation 1994; 89:1992–1928.

91. Felgner PL, Gadek TR, Holm M, Roman R, Chan HW, Wenz M, Northrop JP,

Ringold GM, Danielsen M. Lipofection: a highly efficient, lipid-mediated DNA-transfection procedure. Proc Natl Acad Sci USA 1987; 84:7413–7417.

92. Felgner PL, Holm M, Chan H. Cationic liposome mediated transfection. Proc West Pharmacol Soc 1989; 32:115–121.

93. Felgner PL, Ringold GM. Cationic liposome-mediated transfection. Nature 1989; 337:387–388.

94. Gao X, Huang L. A novel cationic liposome reagent for efficient transfection of mammalian cells. Biochem Biophys Res Commun 1991; 179:280–285.

95. Plautz GE, Yang ZY, Wu B, Gao X, Huang L, Nabel GJ. Immunotherapy of malignancy by in vivo gene transfer into tumors. Proc Natl Acad Sci USA 1993; 90:4645–4649.

96. Lim CS, Chapman GD, Gammon RS, Muhlestein JB, Bauman RP, Stack RS, Swain JL. Direct in vivo gene transfer into the coronary and peripheral vasculatures of the intact dog. Circulation 1991; 83:2007–2011.

97. Coutelle C, Caplen N, Hart S, Huxley C, Williamson R. Gene therapy for cystic fibrosis. Arch Dis Child 1993; 68:437–440.

98. Nabel GJ, Nabel EG, Yang Z, Fox B, Plautz G, Gao X, Huang L, Shu S, Gordon D, Chang AE. Direct gene transfer with DNA liposome complexes in melanoma: Expression, biologic activity and lack of toxicity in humans. Proc Natl Acad Sci USA 1993; 90:11307–11311.

99. Nabel EG, Gordon D, Xang ZY, Xu L, San H, Plautz GE, Gao X, Huang L, Nabel GJ. Gene transfer in vivo with DNA-liposome complexes: lack of autoimmunity and gonadal localization. Hum Gene Ther 1992; 3:649–656.

100. San H, Yang ZY, Pompili VJ, Jaffe ML, Plautz GE, Xu L, Felgner JH, Wheeler CJ, Felgner PL, Gao X, Huang L, Gordon D, Nabel GJ, Nabel EG. Safety and short-term toxicity of a novel cationic lipid formulation for human gene therapy. Hum Gene Ther 1993; 4:781–788.

101. Chiang M-Y, Chang H, Zounes MA, Freier SM, Lima WF, Bennett CF. Antisense oligonucleotides inhibit intercellular adhesion molecule 1 expression by two distinct mechanisms. J Biol Chem 1991; 266:18162–18171.

102. Bennett CF, Chiang MY, Chan H, Shoemaker JE, Mirabelli CK. Cationic lipids enhance cellular uptake and activity of phosphorothioate antisense oligonucleotides. Mol Pharm 1992; 41:1023–1033.

103. Kaneda Y, Iwai K, Uchida T. Increased expression of DNA cointroduced with nuclear protein in adult rat liver. Science 1989; 243:375–378.

104. Kato K, Nakanishi M, Kaneda Y, Uchida T, Okada Y. Expression of hepatitis B virus surface antigen in adult rat liver. J Biol Chem 1991; 266:3361–3364.

105. Tomita N, Higaki J, Morishita R, Kato K, Mikami H, Kaneda Y, Ogihara T. Direct in vivo gene introduction into rat kidney. Biochem Biophys Res Commun 1992; 186:129–134.

106. Morishita R, Gibbons GJ, Ellison KE, Nakajima M, von der Leyen H, Zhang L, Kaneda Y, Ogihara T, Dzau VJ. Intimal hyperplasia after vascular injury is inhibited by antisense cdk 2 kinse oligonucleotides. J Clin Invest 1994; 93:1458–1464.

107. Morishita R, Gibbons GH, Ellison KE, Nakajima M, Zhang L, Kaneda Y, Ogihara T, Dzau VJ. Single intraluminal delivery of antisense cdc2 kinase and proliferat-

ing-cell nuclear antigen oligonucleotides results in chronic inhibition of neointimal hyperplasia. Proc Natl Acad Sci USA 1993; 90:8474–8478.

108. Riessen R, Isner JM. Prospects for site-specific delivery of pharmacologic and molecular therapies. JACC 1994; 23:1234–1244.

109. Riessen R, Rahimizadeh H, Blessing E, Takeshita S, Barry JJ, Isner JM. Arterial gene transfer using pure DNA applied directly to a hydrogel-coated angioplasty balloon. Hum Gene Ther 1993; 4:749–758.

110. Simons M, Edelman ER, DeKeyser JL, Langer R, Rosenberg RD. Antisense c-myb oligonucleotides inhibit intimal arterial smooth muscle cell accumulation in vivo. Nature 1992; 359:67–70.

111. Simons M, Edelman ER, Rosenberg RD. Antisense proliferating cell nuclear antigen oligonucleotides inhibit intimal hyperplasia in a rat carotid artery injury model. J Clin Invest 1994; 93:2351–2356.

112. Nabel EG, Plautz G, Boyce FM, Stanley JC, Nabel GJ. Recombinant gene expression in vivo within endothelial cells of the arterial wall. Science 1989; 244:1342–1344.

113. Conte MS, Birinyi LK, Miyata T, Fallon JT, Gold HK, Whittemore AD, Mulligan RC. Efficient repopulation of denuded rabbit arteries with autologous genetically modified endothelial cells. Circulation 1994; 89:2161–2169.

114. Plautz G, Nabel EG, Nabel GJ. Introduction of vascular smooth muscle cells expressing recombinant genes in vivo. Circulation 1991; 83:578–583.

115. Lynch CM, Clowes MM, Osborne WR, Clowes AW, Miller AD. Long-term expression of human adenosine deaminase in vascular smooth muscle cells of rats: a model for gene therapy. Proc Natl Acad Sci USA 1992; 89:1138–1142.

116. Messina LM, Podrazik RM, Whitehill TA, Ekhterae D, Brothers TE, Wilson JM, Burkel WE, Stanley JC. Adhesion and incorporation of lac-Z-transduced endothelial cells into the intact capillary wall in the rat. Proc Natl Acad Sci USA 1992; 89:12018–12022.

117. Dichek DA. Gene transfer in the treatment of thrombosis. Thromb Haemost 1993; 70:198–201.

118. Wilson JM, Birinyi LK, Salomon RN, Libby P, Callow AD, Mulligan RC. Implantation of vascular grafts lined with genetically modified endothlial cells. Science 1989; 244:1344–1346.

119. Dichek DA, Neville RF, Zwiebel JA, Freeman SM, Leon MB, Anderson WF. Seeding of intravascular stents with genetically engineered endothelial cells. Circulation 1989; 80:1347–1353.

120. Kahn ML, Lee SW, Dichek DA. Optimization of retroviral vector-mediated gene transfer into endothelial cells in vitro. Circ Res 1992; 71:1508–1517.

121. Podrazik RM, Whitehill TA, Ekhterae D, Williams WD, Messina LM, Stanley JC. High-level expression of recombinant human tPA in cultivated canine endothelial cells under varying conditions of retroviral gene transfer. Ann Surg 1992; 216: 233–240.

122. Chapman GD, Lim CS, Gammon RS, Culp SC, Desper S, Bauman RP, Swain JL, Stack RS. Gene transfer into coronary arteries of intact animals with a percutaneous balloon catheter. Circ Res 1992; 71:27–33.

123. Willard JE, Landau C, Glaniann B, Burns D, Jessen ME, Pirwitz MJ, Gerard RD,

Meidell RS. Genetic modification of the vessel wall: Comparison of surgical and catheter-based techniques for delivery of recombinant adenovirus. Circulation 1994; 89:2190–2197.

124. Nabel EG, Plautz G, Nabel GJ. Transduction of a foreign histocompatibility gene into the arterial wall induces vasculitis. Proc Natl Acad Sci USA 1992; 89:5157–5161.

125. Rome JJ, Shayani V, Flugelman MY, Newman KD, Farb A, Virmani R, Dichek DA. Anatomic barriers influence the distribution of in vivo gene transfer into the arterial wall: Modeling with microscopic tracer particles and verification with a recombinant adenoviral vector. Arterio Thromb 1994; 14:148–161.

126. Morishita R, Gibbons GH, Kaneda Y, Ogihara T, Dzau VJ. Novel and effective gene transfer technique for study of vascular renin angiotensin system. J Clin Invest 1993; 91:2580–2585.

127. Willnow TE, Sheng Z, Ishibashi S, Herz J. Inhibition of hepatic chylomicron remnant uptake by gene transfer of a receptor antagonist. Science 1994; 264:1471–1474.

128. Nabel EG, Yang Z, Liptay S, San H, Gordon D, Haudenschild CC, Nabel GJ. Recombinant platelet-derived growth factor B gene expression in porcine arteries induces intimal hyperplasia in vivo. J Clin Invest 1993; 91:1822–1829.

129. Nabel EG, Yang Z, Plautz G, Forough R, Zhan X, Haudenschild CC, Maciag T, Nabel GJ. Recombinant fibroblast growth factor-1 promotes intimal hyperplasia and angiogenesis in arteries in vivo. Nature 1993; 362:844–846.

130. Nabel EG, Shum L, Pompili VJ, Yang ZY, San H, Shu HB, Liptay S, Gordon D, Derynck R, Nabel GJ. Direct gene transfer of transforming growth factor β1 into arteries stimulates fibrocellular hyperplasia. Proc Natl Acad Sci USA 1993; 90:10759–10763.

131. Landau C, Lange RA, Hillis LD. Percutaneous transluminal coronary angioplasty. N Engl J Med 1994; 330:981–993.

132. Willerson JT, Yao S-K, McNatt J, Cui K, Anderson HV, Ostro M, Buja LM. Liposome-bound prostaglandin E1 often prevents cyclic flow variations in stenosed and endothelium-injured canine coronary arteries. Circulation 1994; 89:1786–1791.

133. Jackson CL. Animal models of restenosis. Trends Cardiovasc Med 1994; 4:122–130.

134. Epstein SE, Siegall CB, Biro S, Fu YM, FitzGerald D, Pastan I. Cytoxic effects of a recombinant chimeric toxin on rapidly proliferating vascular smooth muscle cells. Circulation 1991; 84:778–787.

135. Guzman RJ, Hirschowitz EA, Brody SL, Crystal RG, Epstein SE, Finkel T. In vivo suppression of injury-induced vascular smooth muscle cell accumulation using adenovirus-mediated transfer of herpes simplex thymidine kinase gene. Proc Natl Acad Sci USA 1994; 91:10732–10736.

136. Borrelli E, Heyman R, Hsi M, Evans RM. Targeting of an inducible toxic phenotype in animal cells. Proc Natl Acad Sci USA 1988; 85:7572–7576.

137. Gordon JW, Scangos GA, Plotkin DJ, Barbosa JA, Ruddle FH. Genetic transformation of mouse embryos by microinjection of purified DNA. Proc Natl Acad Sci USA 1980; 77:7380–7384.

138. Heyman RA, Borrelli E, Lesley J, Anderson D, Richman DD, Baird SM, Hyman R, Evans RM. Thymidine kinase obliteration: creation of transgenic mice with controlled immune deficiency. Proc Natl Acad Sci USA 1989; 86:2698–2702.

139. Breakefield XO, DeLuca NA. Herpes simplex virus for gene delivery to neurons. New Biol 1991; 3:203–218.

140. Smith KO, Galloway KS, Kennell WL, Ogilvie KK, Radatus BK. A new nucleoside analog, 9-[[2-hydroxy-1-(hydroxymethyl)ethoxyl]methyl]guanine, highly active in vitro against herpes simplex virus types 1 and 2. Antimicrob Agents Chemother 1982; 22:55–61.

141. Field AK, Davies ME, DeWitt C, Perry HC, Liou R, Germershausen J, Karkas JD, Ashton WT, Johnston DB, Tolman RL. 9-([2-hydroxy-1-(hydroxymethyl)ethoxy]-methyl)guanine: a selective inhibitor of herpes group virus replication. Proc Natl Acad Sci USA 1983; 80:4139–4143.

142. Moolten FL, Wells JM. Curability of tumors bearing herpes thymidine kinase genes tranferred by retroviral vectors. J Natl Cancer Inst 1990; 82:297–300.

143. Culver KW, Ram Z, Wallbridge S, Ishii H, Oldfield EH, Blaese RM. In vivo gene transfer with retroviral vector-producer cells for treatment of experimental brain tumors. Science 1992; 256:1550–1552.

144. Plautz G, Nabel EG, Nabel GJ. Selective elimination of recombinant genes in vivo with a suicide retroviral vector. New Biol 1991; 3:709–715.

145. Smythe WR, Hwang HC, Amin KM, Eck SL, Wilson JM, Kaiser LR, Albelda SM. Successful adenovirus-mediated gene transfer in an in vivo model of human malignant mesothelioma. Ann Thorac Surgery 1994; 57:1395–1401.

146. National Institutes of Health Recombinant DNA Advisory Committee. Human gene marker/therapy clinical protocol. Hum Gene Ther 1994; 5:787–789.

147. Nabel EG, Yang ZY, Muller D, Chang AE, Gao X, Huang L, Cho KJ, Nabel GJ. Safety and toxicity of catheter gene delivery to the pulmonary vasculature in a patient with metastatic melanoma. Hum Gene Ther 1994; 5:1089–1094.

148. Grossman M, Raper SE, Kozarsky K, Stein EA, Engelhardt JF, Muller D, Lupien PJ, Wilson JM. Successful ex vivo gene therapy directed to liver in a patient with familial hypercholesterolemia. Nature Genet 1994; 6:335–341.

149. Wilson JM. Vehicles for gene therapy. Nature 1993; 365:691–692.

19

Gene-Transfer Approaches to Myocardial Diseases

Leslie A. Leinwand
Albert Einstein College of Medicine
Bronx, New York

Jeffrey M. Leiden
University of Chicago
Chicago, Illinois

The ability to introduce genes efficiently into a wide variety of tissues and cell types in vivo, including those of the myocardium, has been an intense area of recent investigation for both basic research purposes and gene therapy (1–5). From a basic science perspective, it has become increasingly apparent that the study of gene regulation in cultured cells is not always equivalent to that in the intact animal. Because there are no established cardiac myocyte cell lines, and because successful transfection of adult cardiocytes has not been reported, the majority of studies of cardiac gene expression have been performed in primary cultures of fetal and neonatal cardiocytes. Therefore, the relationship between the developmental stage of the cultured cells and the intact adult cardiocytes must also be brought into question. In addition to facilitating in vivo studies of cardiac gene expression, the ability to stably overexpress specific gene products in cardiac myocytes in vivo would represent a powerful tool with which to study the effects of those gene products on cardiac physiology. Although basic studies of cardiac gene expression and physiology can be performed in transgenic mice, the generation of transgenic animals is costly and time-consuming (6). For these reasons, studies of cardiac gene expression and the physiology of genetically modified cardiac myocytes would be greatly enhanced by a simple and efficient means of in vivo gene transfer. In addition to its usefulness as a basic research tool, in vivo gene transfer into the myocardium might also have important therapeutic implications for a wide variety of cardiovascular diseases, including acquired and inherited cardiomyopathies as well as coronary artery disease.

When contemplating gene transfer into the myocardium, the terminally differentiated state of the cardiac myocyte, as well as the absence of stem cells, needs to be considered (7,8). These properties preclude the use of ex vivo gene-transfer approaches, and greatly limit the utility of such vector systems as the replication-defective retroviruses that require cell proliferation for efficient gene transduction (9). A simple means of achieving gene transfer into muscle was discovered by Wolff and colleagues (10), who demonstrated that skeletal muscle could take up and express naked plasmid DNA administered via intramuscular injection. This observation was extended to include cardiac muscle (11–13), and this method has proven to be useful for the analysis of muscle-specific gene expression (see below) in vivo (13,14). More recent advances in myocardial gene transfer have included the development of diverse gene-delivery systems ranging from DNA complexed to cationic liposomes (15) to recombinant adenoviruses (16–18). This review focuses on two methods for the introduction of genes into the cardiovascular system: naked DNA injection and recombinant adenoviruses. In addition, we describe the potential uses of in vivo gene transfer for studies of basic cardiac gene expression and physiology and for the somatic gene therapy of human cardiovascular disease. The reader is referred to several recent reviews (2,4,5) for additional discussions of cardiovascular gene-delivery systems and the possible therapeutic applications of gene transfer into the myocardium and vasculature.

DIRECT INJECTION OF DNA

Following the intriguing observation by Wolff and colleagues that mouse skeletal muscle could take up and express naked plasmid DNA administered via a hypodermic needle injection (10), several groups reported that cardiac muscle shared this property (11–13). In fact, cardiac muscle appears to be significantly more efficient than skeletal muscle in taking up and/or expressing naked DNA (13). The uptake of naked DNA appears to be unique to striated muscle (12). Despite the widespread use of this technique, there is currently no understanding of the mechanism by which the DNA enters the muscle cell. The basic protocol for myocardial injection involves exteriorization of the heart through a thoracotomy, a single injection of closed circular DNA into the cardiac apex, replacement of the heart into the chest, and assays for recombinant gene expression days to months later (19,20). The requirements for effective gene transfer are simple in that the DNA should be in a closed circular state (linear DNA is much less efficient) and delivered in normal saline or sucrose (19,20). It appears that DNA injected in larger volumes may result in more effective gene transfer. Once administered, the plasmid DNA appears to remain episomal, and expression is very stable, with reports of gene expression 1.5 years after DNA injection (21). Following gene injection, the vast majority of gene expression is restricted to the area surrounding the injection site (22). There is very little pathology that accompanies this procedure except for an inflammatory response to the needle injury itself (19,20). A point of some discussion is whether an immune response to the transferred gene product may develop (12).

Gene injection provides a means of both determining the relative transcription rates of genes in adult heart cells in vivo and identifying the elements of those genes that mediate their responses to complex physiological stimuli (13,14). The general approach is the same as that used in cultured cell transfection analyses. That is, a potential regulatory DNA sequence is cloned upstream of a reporter gene and then coinjected into the myocardium with a second plasmid to serve as an internal standard. This reference plasmid encodes a second reporter gene drive by a constitutive promoter. The tissue is harvested 5–7 days after DNA injection and both reporter gene products are quantitated. A major consideration in these experiments is that care must be taken to remain within the linear range of input DNA. It seems likely that at high DNA concentrations the cellular transcription machinery becomes limiting, resulting in competition between active promoters. This phenomenon is important when designing and interpreting experiments that ask quantitative questions concerning gene regulation. Therefore, it is essential that a dose-response curve be generated for each promoter under study. In general, the stronger the promoter, the less DNA should be injected to avoid competition for the basal transcriptional machinery.

In summary, naked DNA injection is a simple means of achieving stable gene transfer into cardiac myocytes. Safety concerns are minimized by the observation that the DNA remains episomal. The disadvantages of this technology are that relatively few cells in the heart take up and express DNA and that the mode of administration is quite invasive.

REPLICATION-DEFECTIVE ADENOVIRUSES TO PROGRAM RECOMBINANT GENE EXPRESSION IN THE MYOCARDIUM

Although extremely useful for basic studies of cardiac gene expression, the therapeutic utility of direct DNA injection into the myocardium has been limited both by the invasive nature of the injection technique and by the low efficiency of in vivo gene transfer produced by this method. In contrast, recent studies (16–18) have suggested that replication-defective adenoviruses may represent an extremely useful delivery system for efficiently programming therapeutic levels of recombinant proteins in the heart. Adenoviruses are double-stranded linear DNA viruses that cause respiratory-tract infections in humans (23,24). The adenovirus genome is a 36 kD DNA molecule that is divided evenly into 100 map units (mu). Adenoviruses efficiently infect a large number of replicating and nonreplicating cell types in vitro and in vivo (23,24). The ability to infect nonreplicating cells is of particular importance for cardiac gene transfer because cardiac myocytes are terminally differentiated cells that lack the ability to divide in vivo (7). Unlike retroviruses, which can be produced at titers of 10^5-10^6 pfu/ml, replication-defective adenovirus vectors can be produced as high-titer stocks (10^{10}-10^{11} pfu/ml) (23,24). In addition, these vectors can accommodate cDNA or genome inserts of up to 7–9 kb (23,24). Finally, adenovirus vectors display a favorable safety profile; they do not integrate into the host genome and have not been associated with human malignancies. Moreover, wild-type adenoviruses have been used safely in large-scale human vaccination trials (25).

Most of the replication-defective adenoviruses used for in vivo gene transfer have been derived from adenovirus serotypes 2 or 5 and have been made replication-defective by deletion of the E1 region of the viral genome (mu 1–9) (23,24). This region encodes the E1A and E1B proteins, which, in the wild-type virus, are responsible for regulating the induction of late viral gene expression and for the transforming activity of adenoviruses in vitro (26). Some adenovirus vectors also carry deletions in the nonessential E3 region of the viral genome (mu 78.5–84.7) to facilitate the incorporation of large promoter-cDNA inserts (27). Because these viruses lack the E1 genes, they are replication-defective, and can be grown only in packaging cell lines such as human 293 cells, which constitutively express the adenovirus E1 proteins in *trans*. (28) Such E1-deleted adenovirus vectors are generated by homologous recombination following cotransfection of 293 cells with a plasmid containing the left end of the viral genome (mu 0–1 and 9–16) and a recombinant cDNA-promoter cassette and wild-type adenovirus DNA from the right end of the viral genome (mu 9–100) (23,24). High-titer stocks of these viral vectors are prepared by growth in 293 cells followed by CsCl density gradient purification and concentration. Such vectors are highly infectious for a wide variety of animal and human cells in vivo, but are unable to regenerate infectious virus following in vivo infection due to their inability to express E1 gene products. During the last several years, a number of groups have demonstrated efficient in vivo gene transfer into the heart following direct injection or intracoronary infusions of such E1-deleted replication-defective adenoviruses (16–18).

In Vivo Gene Transfer into Cardiac Myocytes Following Direct Injection of Adenovirus Vectors into the Myocardium

Prior to direct injection into the myocardium, recombinant adenovirus (type 5) bearing reporter genes were tested for their ability to infect fetal and adult cardiac myocytes in vitro (18). Both cell types were extremely efficiently infected, such that the definition of a plaque-forming unit, as defined on an optimally permissive cell line, applied to cardiocytes in vitro. Therefore, cardiac myocytes have adenovirus receptors that are not limiting in number and are fully functional. A recombinant adenovirus bearing the CAT reporter gene driven by the cytomegalovirus LTR was then tested for its ability to infect the myocardium in vivo following intracardiac injection (18). 6×10^7 pfu of virus were injected into the apex of the rat heart and compared with the amount of CAT activity that could be obtained from injection of a plasmid DNA bearing the same CAT construct. As compared to DNA injection, virus injection resulted in levels of CAT activity several orders of magnitude higher when assayed 5 days following injection.

Following intracardiac muscle injection, immunohistochemical staining demonstrated that the efficiency of gene transfer achieved in cardiac muscle was extremely high, reaching about 15% of the myocytes in the heart (18). The amount of CAT protein in the left ventricle 5 days after infection was determined to be approximately 200 mg, indicating that it may be possible to achieve phenotypic modification of target organs with adenovirus vectors (29). In addition to the myocytes, all cell types in the heart were seen to be infected, including the vasculature. Substantial CAT activity was also found in all tissues tested following intramyocardial adenovirus injection. The efficient gene transfer that occurred in other tissues was somewhat surprising. The basis for this is unknown, but it is not due solely to the coronary circulatory system since a deliberate injection into the coronary cavity did not result in as efficient expression in other tissues as the injection of virus into the myocardium (29). Despite the extremely high levels of gene expression found following an intracardiac muscle injection of adenovirus, the majority of expression was restricted to the area around the injection site. The duration of expression of recombinant gene expression was found to be quite short, with an absence of expression and viral DNA 80 days after injection. As described in detail below, the limited duration of recombinant gene expression observed following in vivo gene transfer using adenoviruses is most likely due to an immune response directed against the adenovirus-infected cells.

In Vivo Gene Transfer into Cardiac Myocytes Following Intravascular Administration of Replication-Defective Adenoviruses

Although direct injection of adenovirus vectors into the heart yields efficient recombinant gene expression in cardiac myocytes directly adjacent to the site of injection, from a therapeutic standpoint it would be desirable to develop a noninvasive method for the delivery of adenovirus vectors into large regions of the myocardium. In this regard, several groups have demonstrated efficient in vivo gene transfer into the myocardium following intravenous (16) or intra-arterial infusions (17) of replication-defective adenovirus vectors. Stratford-Perricaudet and coworkers (16) administered 10^9 pfu of the AdRSVβgal virus containing the bacterial lacZ gene under the transcriptional control of the Rous sarcoma virus LTR intravenously into neonatal mice. Although the bulk of the virus infected liver, recombinant lacZ gene expression was also observed in approximately 0.2% of cardiac myocytes. Recombinant gene expression appeared to be stable in cardiac myocytes in these neonatal animals for periods of as long as 1 year following injection.

Barr et al. (17) used an intracoronary catheter to infuse 2×10^9–10^{10} pfu of AdCMV.lacZ virus into the coronary arteries of adult rabbits. Five days after catheter-mediated virus infusion, recombinant lacZ expression was observed in between 10 and 60% of cardiac myocytes in the area of distribution to the infused coronary artery. Interestingly, recombinant gene expression was seen in both atrial and ventricular myocytes, and was evenly distributed from the epicardial to the endocardial surfaces of the ventricular wall. In addition, lacZ gene expression was observed in nonmyocytic connective tissue cells in the myocardium. Of note, the efficiency of catheter-mediated in vivo gene transfer using this approach was 50–100 times higher than that observed following direct DNA injections into the heart. However, in contrast to direct DNA injections, recombinant gene expression in cardiac myocytes following adenovirus-mediated gene transfer in adult immunocompetent animals was short-lived. In the studies of Stratford-Perricaudet and coworkers (16), significant decreases in recombinant gene expression were seen 21 days after intravenous infusions of adenovirus into adult immunocompetent mice. Barr et al. (17) demonstrated the complete loss of recombinant gene expression 4 weeks after intracoronary infusions of AdCMV.lacZ into adult immunocompetent rabbits. Using a sensitive PCR-based assay, they were able to show that the loss of recombinant gene expression observed in these rabbits was due to a loss of the viral genome. Moreover, histological analyses of these hearts demonstrated a mild to moderate mononuclear cell inflammatory response.

As described below, recent studies (30,31) have suggested that the loss of recombinant gene expression following adenovirus-mediated in vivo gene transfer results from a cytotoxic T-cell immune response directed against the adenovirus-infected cells. It is also important to note that Barr et al. demonstrated viral DNA in a wide variety of organs, including testis, liver, and brain, 5 days after intracoronary infusion of adenoviruses (17). The extent of viral infection of these organs was not quantitated. However, these results suggest that it will be important to develop methods for limiting viral infection and/or recombinant gene expression to the heart following intravascular delivery of these vectors. Current approaches to this problem include the use of local delivery systems to limit the extent of viral infection, and the use of cardiac myocyte-specific transcriptional regulatory elements to restrict recombinant gene expression to cardiac myocytes in vivo.

In summary, these studies demonstrated that replication defective adenoviruses infused intravenously or into the coronary arteries can cross the coronary vasculature and infect both cardiac myocytes and nonmyocytic connective tissue cells in the myocardium. It is possible to infect as many as 60% of the cardiac myocytes in the area of distribution of a coronary artery following a single intracoronary infusion of virus. The uniform distribution of infection of cardiac myocytes suggests that virus traverses the microvasculature. However, the mechanism of viral egress from the coronary circulation remains unclear. Although promising, the use of this method for human therapy is currently limited by the shortterm nature of recombinant gene expression observed in adult immunocompetent animals, and by the widespread infection observed following intracoronary delivery of virus.

Immune Responses to Replication-Defective Adenoviruses

As described above, work from a number of groups has demonstrated that recombinant gene expression programmed by the in vivo administration of replication-defective adenoviruses is transient in adult immunocompetent animals (16–18, 29, 30, 32–34). There are several possible explanations for the short-lived nature of adenovirus-mediated recombinant gene expression in vivo. First, it is possible that the viral genome that is maintained as a linear episome is degraded, and thereby lost from infected cells. Alternatively, it is possible that adenovirus-infected cells are eliminated as a result of an immune or inflammatory response. Finally, it is possible that the viral genome is maintained stably in an episomal state, but that recombinant gene expression is eliminated due to promoter shutoff in vivo.

Recent experiments have strongly suggested that a cytotoxic T cell response directed against adenovirus-infected cells is responsible for the transient gene expression observed in vivo following administration of replication-defective adenoviruses. First, a prominent mononuclear cell inflammatory response has been observed in liver (30), skeletal muscle (31), and myocardium (17) following the in vivo administration of E1-deleted replication-defective adenoviruses. This infiltrate is composed of CD4$^+$ and CD8$^+$ T lymphocytes as well as macrophages. Second, in contrast to the results observed in adult immunocompetent animals, several groups have demonstrated stable (>6 months) recombinant gene expression following adenovirus administration to a variety of immunocompromised adult animals, including SCID (31) and nude mice (30), RAG-2-deficient mice, and mice lacking expression of class I MHC (30). Each of these strains lacks cytotoxic T lymphocytes, which are known to be important mediators of antiviral responses. These results suggest that immunosuppression may facilitate long-term adenovirus-mediated recombinant gene expression in adult animals. However, a recent study (29) demonstrated that cyclosporine treatment resulted in only a partial improvement in the duration of gene expression in adult rats.

In contrast to the results seen in adult animals, several groups have demonstrated long-term recombinant gene expression following intravenous or intramuscular administration of replication-defective adenoviruses to neonatal rats, mice, and rabbits (29,31). In an elegant series of experiments, Wilson and coworkers (30) have demonstrated that CD8$^+$ T cells from adenovirus-inoculated animals can eliminate adenovirus-mediated recombinant gene expression in SCID mice using classic adoptive transfer techniques. Taken together, these experiments provide strong evidence that in adult immunocompetent animals, a CD8$^+$ cytotoxic T-cell response eliminates adenovirus-infected cells, thereby abrogating recombinant gene expression. In contrast, infection of immunocompromised animals that lack the ability to mount an effective CTL response results in stable recombinant gene expression. Similarly, administration of adenoviruses to neonatal animals appears to result in the induction of tolerance and the concomitant stable recombinant gene expression seen in these animals.

The antigenic determinants of the antiadenovirus immune response remain somewhat unclear. However, recent studies (35,36) have suggested that the current generation of E1-deleted viruses express low levels of late viral gene products that are known to be highly immunogenic in vivo. This late viral gene expression may reflect the presence of E1-replacing activities in a variety of primary cell types in vivo.

There are at least two approaches that could be used to circumvent the antiadenovirus immune responses seen in immunocompetent animals following infection with the first generation of E1-deleted adenovirus vectors. First, it may be possible to devise immunosuppressive or tolerizing regimens that might abrogate the antiadenovirus immune responses. Alternatively, it may be possible to construct new generations of adenovirus vectors that lack the expression of late viral gene products. In particular, there is great interest in the construction of E1-plus-E2- and/or E4-deleted viruses, which should produce much lower levels of late viral proteins (37). Despite the technical difficulties associated with these approaches, the finding of stable long-term recombinant gene expression following adenovirus infection of immunocompromised or tolerized animals strongly suggests that abrogation of the anti-adenovirus immune response will result in stable recombinant gene expression in adult immunocompetent animals in vivo.

THERAPEUTIC POTENTIAL OF IN VIVO GENE TRANSFER IN THE HEART

Direct DNA injection and adenovirus-mediated in vivo gene transfer into the myocardium hold great promise as basic scientific tools with which to better understand cardiac-specific gene expression and myocardial function. As described above, direct DNA injection has been shown to be a rapid and powerful in vivo assay of both basal and inducible cardiac-specific promoter activity. Similarly, catheter-mediated in vivo gene transfer using replication-defective adenoviruses can be used to overexpress wild-type or mutant proteins in specific regions of the heart in order to study the effects of overexpression on heart rate, contractility, or diastolic function in vivo.

In addition to its importance as a basic scientific tool, in vivo gene transfer holds great promise for the treatment of a number of cardiovascular diseases. In this review, we describe three potential therapeutic applications of in vivo gene transfer into the myocardium: 1) the expression of recombinant angiogenesis factors to stimulate collateral vessel formation in the myocardium, 2) the expression of structural and contractile proteins for the correction of inherited cardiomyopathies, and 3) the expression of cell-surface receptors and sarcomeric proteins to improve the contractile function of cardiac myocytes. These examples in no way represent an exhaustive list of the potential therapeutic applications of this technology, but instead are illustrative of the potential of these approaches for the therapy of human cardiovascular diseases.

Myocardial Angiogenesis

Coronary atherosclerosis and myocardial infarction remain the leading cause of mortality in the United States. Despite significant advances in the surgical and catheter-based approaches to coronary artery disease, there are many patients with extensive disease that is not amenable to either surgical or catheter-mediated revascularization. Previous clinical studies have suggested that patients with severe epicardial coronary atherosclerosis who develop spontaneous collateral blood flow fare better than those patients who lack such collateral vessels (38). This finding suggests that the ability to program the growth of new capillaries in areas of ischemic myocardium might prove useful for the treatment of patients with severe epicardial coronary athero- sclerosis. During the last decade, genes encoding a large number of angiogene- sis factors have been identified and cloned (39). These include members of the fibroblast growth factor family (FGF-1-FGF-6) (40,41), VEGF (42,43), angiogenin (44), and scatter factor (45). Recent studies have demonstrated that intravascular or intracoronary artery infusions of bFGF (46,47) or VEGF (48) result in increased collateral blood flow in areas of ischemic muscle or myocardium. The availability of these cloned angiogenesis factors and their demonstrated efficacy in producing angiogenesis suggest the possibility of using direct DNA injection or adenovirus-mediated gene transfer to continu- ously express these proteins in the myocardium.

In a recent series of experiments, we (E. Barr and J. Leiden, unpublished results) have shown that direct injection of plasmid DNA encoding the secreted angiogenesis factor FGF-5 (49) results in 30–40% increases in capil- lary density, 3 weeks after injection into normal rat left ventricle. These preliminary studies have suggested the feasibility of using in vivo gene transfer to program neovascularization of the myocardium. However, many questions remain to be addressed before this approach could be considered feasible for human therapy. In particular, it will be important to determine the stability of neovascularization following in vivo gene transfer. In addition, because many angiogenesis factors have displayed oncogenic potential (39–41), it will be important to rule out the possibility that the long-term expression of these factors will cause tumors in vivo. Similarly, a number of angiogenesis factors have been shown to cause neointimal proliferation. Thus, it will be important to determine the effects of overexpression of specific angiogenesis factors in the coronary vasculature in vivo. Finally, it will be important to demonstrate that neovascularization programmed by in vivo gene transfer provides protec- tion against myocardial ischemia and infarction in large animal models. Studies designed to address these issues are currently in progress.

In Vivo Gene Transfer for the Therapy of Hypertrophic Cardiomyopathies

The problem of gene therapy for the hypertrophic cardiomyopathies is confounded by the fact that this disorder is typically inherited in an autosomal dominant fashion (50). At least 30–50% of hypertrophic cardiomyopathies are caused by point mutations in proteins of the sarcomere (51–53) (see also Chapter 5). Biochemical and cell biological analyses have suggested that the mutant gene products behave as dominant negative proteins. This means that the mutant protein interferes with the function of the wild-type protein. In many cases, dominant negative mutations exert their effects in the context of a vast excess of wild-type protein. Therefore, the prospects for gene therapy for these autosomal dominant disorders are difficult to contemplate. At this point, it seems probable that an inactivation of the mutant allele is most likely to be successful.

In Vivo Gene Transfer for the Treatment of Acquired Defects in Myocardial Contractility

Primary and secondary abnormalities of cardiac contractility are important contributors to the pathophysiology of a number of cardiovascular diseases, including the idiopathic dilated and viral cardiomyopathies and acute myocardial ischemia. Although these diseases are not associated with known single-gene defects, it may be possible to develop generic gene-therapy approaches that can be used to treat the abnormalities in cardiac contractility that are associated with these cardiomyopathies. Recent transgenic mouse models have suggested that the contractile phenotype of cardiac myocytes can be modified by the overexpression or ectopic expression of specific cell-surface receptors or contractile proteins (54,55). For example, Metzger et al. (54) reported that the ectopic expression of the skeletal troponin C protein in cardiac myocytes of transgenic mice (which normally express only the cardiac isoform of troponin C) renders those myocytes resistant to the negative inotropic effects of intracellular acidosis. Because intracellular acidosis appears to be the major cause of decreased contractility following acute ischemia, these results suggested that overexpression of sTnC in the heart by in vivo gene transfer might significantly ameliorate the loss of contractility resulting from acute ischemic insults. In a second series of experiments Milano et al. (55) produced a transgenic mouse that overexpressed the β-2adrenergic receptor in cardiac myocytes. Hearts from these transgenic animals demonstrated increased contractility even in the absence of exogenous β-adrenergic stimulation. Thus, it is tempting to speculate that overexpression of β-adrenergic receptors by in vivo gene transfer might similarly increase myocardial contractility in patients with cardiomyopathies and CHF.

Despite the promise of these in vivo gene-transfer approaches for the treatment of congestive heart failure associated with decreased systolic function, it will be important to show that such approaches are able to alter contractility in diseased as well as normal cardiac myocytes and, more importantly, that the alterations in contractility produced by these approaches decreases the morbidity and mortality associated with the acquired cardiomyopathies. This last consideration is particularly important because clinical trials of several positive inotropic drugs in patients with cardiomyopathies have demonstrated an increased incidence of cardiovascular mortality (56). Finally, it is important to emphasize that the two reports described above represent only the first examples of the feasibility of gene therapy for the treatment of abnormal cardiac contractility. Ongoing experiments examining the effects of overexpression of a wide variety of genes in the heart promise to yield important new genetic approaches to this difficult clinical problem.

SUMMARY AND FUTURE DIRECTIONS

In recent years, we have witnessed dramatic and surprising improvements in our ability to program recombinant gene expression in the myocardium by in vivo gene transfer. Direct DNA injection represents a simple and powerful tool for studies of transcriptional regulation in vivo. However, the invasive nature of this procedure and the relatively low efficiency of in vivo gene transfer currently limit its usefulness for human therapy. In contrast, replication-defective adenoviruses represent a safe and efficient method for delivering genes to myocardium. In particular, the ability to efficiently transduce cardiac myocytes in vivo following catheter-mediated intra-coronary artery adenovirus infusion makes this virus the vector of choice for human therapy. The feasibility of using the current generation of E1-deleted adenovirus vectors is limited by the fact that these vectors generate a CTL response in vivo that leads to the elimination of virus-infected cells in adult immunocompetent hosts. Nevertheless, the finding of longterm recombinant gene expression following adenovirus-mediated in vivo gene transfer into immunocompromised animals strongly suggests that it will be possible to develop new generations of adenoviruses (or appropriate immunosuppressive regimens) that will allow these viruses to be used to program stable recombinant gene expression in the heart.

The recent elucidation of the mutant genes responsible for a number of inherited cardiomyopathies, as well as the results of animal experiments designed to study the effects of overexpression of specific genes on cardiac structure and function, has suggested several novel approaches to the gene therapy of cardiac disease. These include the use of cloned angiogenesis factors for the treatment of ischemic cardiomyopathies, the expression of sarcomeric proteins for the treatment of inherited and acquired cardiomyopathies, and the overexpression of cell-surface receptors for the treatment of systolic dysfunction and heart failure. Ongoing studies of the genetics of human cardiovascular disease and animal studies designed to elucidate the effects of overexpression of specific genes on cardiac function promise to rapidly expand the repertoire of potential gene-therapy approaches for the treatment of human cardiac disease.

REFERENCES

1. Verma IM. Gene therapy. Scientific American 1990; 68–84.
2. Swain JL. Gene therapy: A new approach to the treatment of cardiovascular disease. Circulation 1989; 80:1495–1496.
3. Mulligan RC. The basic science of gene therapy. Science 1993; 260:926–932.
4. Leinwand L, Leiden JM. Gene transfer into cardiac myocytes *in vivo*. Trends Cardiovasc Med 1991; 1:271–276.
5. Barr E, Leiden JM. Somatic gene therapy for cardiovascular disease. Trends Cardiovasc Med 1994; 4:57–63.
6. Palmiter RD, Brinster RL. Germ-line transformation of mice. Annu Rev Genet 1986; 20:465–499.
7. Zak R. Development and proliferation capacity of cardiac muscle cells. Circ Res 1974; 34–35:11–17.
8. Watanabe A, Green F, Farmer BB. The Heart and Cardiovascular System. New York: Raven Press, 1986.(#
9. Adam MA, Miller AD. Gene transfer by retrovirus vectors occurs only in cells that are actively replicating at the time of infection. Molec Cell Biol 1992; 10:4239–4242.
10. Wolff JA, Malone RW, Williams P. Direct gene transfer into mouse muscle *in vivo*. Science 1990; 247:1465–1468.
11. Lin H, Parmacek MS, Morle G, Bolling S, Leiden M. Expression of recombinant genes in myocardium *in vivo* after direct injection of DNA. Circulation 1990; 82:2217–2221.
12. Ascadi G, Jiao S, Jani A, Duke D, Williams P, Chong W, Wolff JA. Direct gene transfer and expression into rat heart *in vivo*. New Biologist 1991; 3:71–81.
13. Kitsis R, Buttrick P, McNally E, Kaplan M, Leinwand LA. Hormonal modulation of a gene injected into rat heart *in vivo*. Proc Natl Acad Sci USA 1991; 88:4138–4142.
14. Parmacek MS, Vora AJ, Shen T, Barr E, Jung F, Leiden JM. Identification and characterization of a cardiac-specific transcriptional regulatory element in the slow/cardiac troponin C gene. Molec Cell Biol 1992; 12:1967–1976.
15. Zhu N, Liggitt D, Liu Y, Debs R. Systemic gene expression after intravenous DNA delivery into adult mice. Science 1993; 261:209–211.
16. Stratford Perricaudet L, Makeh I, Perricauden M. Widespread long-term gene transfer to mouse skeletal muscles and heart. J Clin Invest 1992; 90:626–630.
17. Barr E, Carroll J, Tripathy Sk, Kozarsky K, Wilson J, Leiden JM. Efficient catheter-mediated gene transfer into the heart using replication-defective adenovirus. Gene Ther 1994; 1:51–58.
18. Eisler A, Falck-Pedersen E, Alvira M, et al. Quantitative determination of adeno-virus-mediated gene delivery to rat cardiac myocytes *in vitro* and *in vivo* Proc Natl Acad Sci 1993; 90:11498–11502.
19. Barr E, Lin H, Parmacek MS, Leiden JM. Direct gene transfer into cardiac myocytes *in vivo*. Methods 1992; 4:169–176.
20. Wolff JA, Williams P, Acsadi G, Jiao S, Jani A, Chong W. Conditions affecting direct gene transfer into rodent muscle *in vivo* Biotechniques 1991; 11:474–485.
21. Wolff JA, Ludtke JJ, Acsadi G, Williams P, Jani A. Longterm persistence of plasmid

DNA and foreign gene expression in mouse muscle. Hum Molec Genet 1992; 1:363-369.

22. Kitsis RN, Buttrick PM, Kass AA, Kaplam M, Leinwand LA. Gene transfer into adult rat heart *in vivo*. Meth Molec Genet 1993; 1:374-392.

23. Berkner K. Development of adenovirus vectors for the expression of heterologous genes. Biotechniques 1988; 6:616-629.

24. Kozarsky KF, Wilson JM. Gene therapy: adenovirus vectors. Curr Opin Genet Devel 1993; 3:499-503.

25. Chanock R, Ludwig W, Huebner RJ, et al. Immunization by selective infection with type 4 adenovirus grown in human diploid tissue cultures. I. Safety and lack of oncogenicity and tests for potency in volunteers. JAMA 1966; 195:445-452.

26. Graham F, van der Eb A. Transformation of rat cells by DNA of human adenovirus. Virology 1973; 54:536-539.

27. Graham F, Smiley J, Russell WC. Characteristics of a human cell line transformed by DNA from adenovirus type 5. J Gen Virol 1977; 36:59-74.

28. Haj-Ahmad Y, Graham F. Development of a helper-independent human adenovirus vector and its use in the transfer of the herpes simplex virus thymidine kinase gene. J Virol 1986; 57:267-274.

29. Kass-Eisler A, Falk-Pedersen E, Elfenbein D, Alivira M, Buttrick PM, Leinwand LA. The impact of developmental stage, route of administration and the immune system on adenovirus-mediated gene transfer. Gene Ther 1994. In press.

30. Engelhardt JF, Ye X, Doranz B, Wilson JM. Ablation of E2a in recombinant adenovirus improves transgene persistence and decreases immune response in mouse liver. Proc Natl Acad Sci USA 1994. In press.

31. Tripathy Sk, Goldwasser E, Barr E, Leiden JM. Stable delivery of physiologic levels of recombinant erythropoietin to the systemic circulation by intramuscular injection of replication-defective adenovirus. 1994. In press.

32. Engelhardt JF. Adenovirus-mediated transfer of the CFTR gene to lung of non-human primates: biological efficacy study. Hum Gene Ther 1993; 4:759-769.

33. Zabner J, et al. Safety and efficacy of repetitive adenovirus-mediated transfer of CFTR cDNA to airway epithelia of primates and cotton rats. Nat Genet 1994; 6:75-83.

34. Zabner J, et al. Adenovirus-mediated gene transfer transiently corrects chloride transport defect in nasal epithelia of patients with cystic fibrosis. Cell 1993; 75:207-216.

35. Ginsberg HS, et al. A mouse model for investigating the molecular pathogenesis of adenovirus pneumonia. Proc Natl Acad Sci USA 1991; 88:1651-1655.

36. Berencsi K, et al. Pathogenicity of early region 3-replacement adenovirus recombinants in cotton rat and mouse lung. J Gen Virol 1994. In press.

37. Yang Y, Nunes FA, Berencsi K, Gonczol E, Engelhardt JF, Wilson JM. Inactivation of E2a in recombinant adenoviruses improves the prospect for gene therapy in cystic fibrosis. Nat Genet 1994; 7:362-369.

38. Sabia PJ, Powers ER, Ragosta M, et al. An association between collateral blood flow and myocardial viability in patients with recent myocardial infarction. N Engl J Med 1992; 327:1825-1831.

39. Folkman J, Klagsbrun M. Angiogenic factors. Science 1987; 235:442-447.

40. Burgess WH, Maciag T. The herapin-binding (fibroblast) growth factor family of proteins. Annu Rev Biochem 1989; 58:575-606.
41. Gospodarowicz D. Fibroblast growth factor. Crit Rev Oncogene 1989; 1:1-26.
42. Keck PJ, Hauser SD, Krivi G, et al. Vascular permeability factor, an endothelial cell mitogen related to PDGF. Science 1989; 246:1306-1312.
43. Leung DW, Cachianes G, Kuang WJ, Goeddel DV, Ferrara N. Vascular endothelial growth factor is a secreted angiogenic mitogen. Science 1989; 246:1306-1309.
44. Vallee BL, Riordan JF. Chemical and biochemical properties of human angiogenin. Adv Exp Med Biol 1988; 234:41-53.
45. Rubin JS, Chan AM-L, Bottaro DP, et al. A broad-spectrum human lung fibroblast-derived mitogen is a variant of hepatocyte growth factor. Proc Natl Acad Sci USA 1991; 88:415-419.
46. Yanagisawa-Miwa A, Uchida Y, Nakamura F, et al. Salvage of infarcted myocardium by angiogenic action of basic fibroblast growth factor. Science 1992; 257:1401-1403.
47. Battler A, Scheinowitz M, Bor A, et al. Intracoronary injection of basic fibroblast growth factor enhances angiogenesis in infarcted swine myocardium. J Am Coll Cardiol 1993; 22:2001-2006.
48. Banai S, Jaklitsch MT, Shou M, et al. Angiogenic-induced enhancement of collateral blood flow to ischemic myocardium by vascular endothelial growth factor in dogs. Circulation 1994; 89:1-7.
49. Zhan X, Bates B, Hu X, Goldfarb M. The human FGF-5 oncogene encodes a novel protein related to fibroblast growth factors. Molec Cell Biol 1988; 8:3487-3495.
50. Maron BJ, Bonow RO, Cannon RO III, Leon MB, Epstein SE. Hypertrophic cardiomyopathy: interrelations of clinical manifestations, pathophysiology, and therapy. N Engl J Med 1987; 316:780-789, 844-852.
51. Geisterfer-Lowrance AA, Kass S, Tanigawa G, et al. A molecular basis for familial hypertrophic cardiomyopathy: a beta cardiac myosin heavy chain gene missense mutation. Cell 1990; 62:999-1006.
52. Seidman CE, Seidman JG. Mutations in cardiac myosin heavy chain genes cause familial hypertrophic cardiomyopathy. Molec Biol Med 1991; 8:159-166.
53. Thierfelder L, Watkins H, MacRae C, et al. α Tropomyosin and cardiac troponin T mutations cause familiar hypertrophic cardiomyopathy: a disease of the sarcomere. Cell 1994; 77:701-712.
54. Metzger JM, Parmacek MS, Barr E, Cochrane KL, Field LJ, Leiden(0*0*0*JM. Skeletal troponin C confers contractile sensitivity to acidosis in cardiac myocytes from transgenic mice. Proc Natl Acad Sci USA 1993; 90:9036-9040.
55. Milano CA, Allen LF, Rockman HA, et al. Enhanced myocardial function in transgenic mice overexpressing the β2-adrenergic receptor. Science 1994; 264: 582-586.
56. Packer M. Is activation of the sympathetic nervous system beneficial or detrimental to the patient with chronic heart failure? Lessons learned from clinical trials with βadrenergic agonists and antagonists. J Cardiovasc Pharmacol 1989; 14:538-543.

20

Gene Therapy Approaches to Metabolic and Cardiovascular Disorders

Louis C. Smith and Randy C. Eisensmith
Baylor College of Medicine
Houston, Texas

Savio L. C. Woo
Howard Hughes Medical Institute
Baylor College of Medicine
Houston, Texas

INTRODUCTION

The basic-science investment of this society in the study of human diseases has yielded an understanding of the molecular pathologies of many inherited metabolic disorders. By 1993, molecular defects in catalytic, receptor, transport, or structural proteins had been identified for more than 200 inborn errors of metabolism (1). This information permits the unambiguous diagnosis of these disorders, either prenatally or shortly after the onset of symptoms early in life. Some of these disorders involve defects in only a single gene product, whose expression is often limited to a single tissue type. For these disorders, previous biochemical-based studies often led to the development of conventional therapies that have been relatively successful in treating the more serious consequences of these diseases. Two examples of disorders of this type include the monogenic diseases phenylketonuria (PKU), caused by a deficiency of the hepatic enzyme phenylalanine hydroxylase (PAH), and hemophilia B, caused by a functional deficiency of factor IX (FIX), an essential cofactor in the intrinsic coagulation pathway.

Unfortunately, for many other debilitating hereditary disorders there are no such specific therapies, only complex strategies to minimize the progression and complications of the particular disease. A prime example is familial hypercholesterolemia (FH), which is caused by defects in the receptor that mediates the internalization of low-density lipoprotein (LDL) (2). This disease is associated with severe hypercholesterolemia and premature coronary artery disease, and, in many cases, is resistant to conventional drug therapy.

One alternative to conventional therapies for these diseases is somatic gene therapy, whereby a functional gene is introduced into the cells in which the expression of that gene has been compromised, and to express the gene product in amounts sufficient to affect the course of the disease. Thus, rather than the disease being treated, the underlying genetic defect would be corrected and thereby prevent, reverse, stabilize, or at least slow the progression of the disease, depending on the nature of the disease and when in its course treatment is initiated. At present, gene therapy offers the potential of a complete cure for metabolic or cardiovascular disorders in which the mechanism and progress of the disease process are well understood.

In general, three components are necessary for the implementation of somatic gene therapy: the potentially therapeutic gene must be cloned; vectors must be developed for the transfer of the gene into the appropriate somatic cell targets; and an animal model should be available for establishing the relative efficiencies of different methods of gene transfer. Two approaches have been examined for the delivery of therapeutic genes. The first involves the removal of the target cells from the patient, the introduction of therapeutic genes into these cells using viral vectors or other transfecting agents, and the return of the transduced cells into the patient by autologous transplantation to reconstitute the missing function or functions (3–5). This ex vivo approach, while conceptually straightforward, has the disadvantages that tissue removal may require a major surgical procedure, and that the culturing of the cells is a labor-intensive procedure. The second approach would involve the direct introduction of genetic material into specific tissue sites using a targeting mechanism. The in vivo approach, although conceptually simple, requires the ability to direct the injected genetic material into the proper tissues or organs. This chapter presents and discusses experiments in which these two approaches have been used to treat metabolic and cardiovascular disorders in animal models and in human patients, using PKU and hemophilia B as examples of metabolic disorders and familial hypercholesterolemia as a prototypical cardiovascular disorder. Since all three of these disorders stem from hepatic deficiencies, the discussion deals primarily with methods of gene transfer into the liver.

GENE THERAPY FOR METABOLIC DISORDERS

This section focuses on two monogenic disorders, phenylketonuria (PKU) and hemophilia B, as models for the application of somatic gene therapy in the treatment of metabolic disorders. Classic PKU is an autosomal recessive disorder caused by a deficiency of hepatic phenylalanine hydroxylase (PAH).

While the conventional therapy of dietary restriction significantly reduces serum phenylalanine levels and largely reduces or prevents the mental impairment characteristic of this disease if initiated early in the neonatal period (6–8), there are still some limitations associated with this form of treatment. Reduced compliance or complete cessation of therapy, even in adolescence or early adulthood, can often be accompanied by a decline in mental or behavioral performance (9). Women with PKU must resume diet prior to or early in pregnancy to prevent the occurrence of developmental abnormalities and mental impairment in their offspring, the "maternal PKU" syndrome (10).

Hemophilia B is an X-linked disorder of hemostasis caused by a deficiency of factor IX (FIX) production in the liver. This disease occurs at a frequency of about 1 in 30,000 males and can result in severe bleeding episodes. The conventional therapy for hemophilia B involves the infusion of FIX derived from human plasma pooled from large numbers of donors. Although protein replacement therapy has increased both the life expectancy and the quality of life in patients with hemophilia B, there are significant problems associated with the intravenous infusion of FIX. The first problem is the transmission of potentially lethal human viruses, such as hepatitis B or HIV, that can be present in plasma-derived FIX preparations. The second problem is that the high costs and limited availability of FIX preparations severely limit the prophylactic use of these products.

Somatic gene therapy provides an attractive alternative for overcoming the limitations of conventional therapies for these two disorders. For PKU, the therapeutic human phenylalanine hydroxylase gene has been cloned (11) and a PAH deficient mouse model created by ethylnitrosourea (ENU) mutagenesis (12,13) is available for testing various methods of transfer of the human or mouse gene (14). Likewise, for hemophilia B, the human factor IX gene (15,16) is available for use in humans after the efficiency of various vectors is established through the delivery of the canine factor IX gene (17) in a naturally occurring FIX-deficient dog strain (18). Both ex vivo and in vivo methods have been applied to the correction of these two genetic disorders.

Ex Vivo Approaches to Gene Therapy for Metabolic Disorders

Various recombinant vectors have been used to deliver therapeutic or reporter genes to a number of explanted cell types, including hematopoietic progenitor cells, hepatocytes, fibroblasts, peripheral lymphocytes, skeletal muscle cells, endothelial cells, and various cancer cells. Ex vivo approaches to gene therapy for metabolic disorders have focused on the use of recombinant retroviral vectors to deliver therapeutic genes to explanted hepatocytes, hematopoietic stem cells, and peripheral lymphocytes. Recombinant retroviral vectors have been used to transfer the PAH cDNA into mouse fibroblast or hepatoma cell lines (19) and into primary hepatocytes from normal mice (20). More recently, retroviral transduction has been used to restore PAH activity in primary hepatocytes isolated from the PAH-deficient mouse strain, Pahenu[1] (12). These animals have a mutation in their PAH gene that reduces hepatic enzyme activity in homozygotes to less than 10% of normal levels. After transduction of primary hepatocytes isolated from these animals with a variant of the LNCX retrovirus (21) containing the mouse PAH cDNA, significant amounts of PAH mRNA, immunoreactivity, and pterin-dependent enzyme activity were introduced into the previously deficient hepatocytes (22).

DNA/protein complexes have also been used to introduce the PAH cDNA into primary hepatocytes. These complexes are synthetic DNA delivery systems containing a negatively charged DNA molecule that is noncovalently attached to a positively charged template such as poly-L-lysine (23) or to ethidium homodimer (24). The template molecule is then covalently attached to a protein that is bound and internalized by a specific receptor that is present on the surface of the target cell. A number of receptors expressed on the surface of the hepatocyte may be suitable for receptor-mediated transgene delivery to the liver. DNA/protein complexes have been introduced into cells in vitro through receptor-mediated endocytosis via the asialoglycoprotein receptor (23,25–28), the transferrin receptor (24,29–35), and, more recently, the folate receptor (36–39). The transduction of cells by DNA/protein complexes in vitro is greatly augmented by the coadministration of a replication-defective adenovirus (25,28,30,32–35,40). The adenovirus acts to lyse the endosome, releasing the endocytosed DNA/protein complexes before they are destroyed in the lysosome, or before the endosome is recycled to the cell surface. To reduce adenovirus-induced cell death, adenovirus has been chemically (28) or enzymatically (30) coupled to poly-L-lysine. Transfection efficiencies of 100% can be achieved when replication-defective adenovirus is present either as free virus or as a component of the DNA complex (25,28,30,32–35,40). Coadministration of DNA/protein complexes targeting the asialoglycoprotein receptor and a replication defective adenovirus has been used to introduce a plasmid expressing human PAH into primary hepatocytes isolated from PAHenu1 mice (25). Near-normal levels of PAH activity were produced in hepatocytes from these PAH-deficient animals.

These studies and numerous others have demonstrated that hepatocytes could be explanted and successfully transduced by reporter or therapeutic genes in vivo. Transduced hepatocytes can also be reimplanted into the liver (5,41). However, the clinical utility of this ex vivo approach in the treatment of metabolic or other disorders via hepatocellular transplantation may be limited by the low efficiency that is primarily a consequence of the low numbers of transduced hepatocytes that can be successfully restored to the liver. Nevertheless, the ex vivo approach may have greater promise when applied to other cell types, especially to hematopoietic or other stem cells if they can be readily identified, isolated, and transduced in vitro.

In Vivo Approaches to Gene Therapy for Metabolic Disorders

A more promising alternative to the ex vivo approach is the in vivo approach, in which the recombinant vector is administered directly into the target tissue through existing anatomical structures, such as the vasculature, bronchial system, or intestinal lumen, or through artificial ports. A growing number of studies have demonstrated efficient hepatocyte transduction following the in vivo administration of recombinant retroviral vectors (42–45). One such study demonstrated the successful transduction of 1–2% of all hepatocytes after intraportal infusion of a recombinant retroviral vector into partially hepatec-tomized mice (44). Although the transduction efficiency of this method was still relatively low, this approach has been successful in sustained partial correction of the hemophilia B phenotype in factor IX–deficient dogs (45). Whether the in vivo transduction efficiency of recombinant retroviral vectors is sufficient for future clinical applications in humans remains to be seen. Clearly, though, recombinant retroviruses have many desirable features as vectors for somatic gene therapy, including their broad host cell range, their established safety record, and their ability to stably integrate into the host cell genome, and may see some use in the treatment of metabolic disorders that require only small amounts of the therapeutic gene product to be expressed. The principal limitations of recombinant retroviral vectors are their inability to integrate in nondividing cells and their relatively low transduction efficiency in vivo.

A second viral vector that has a significantly greater transduction efficiency in in vivo applications is the recombinant adenoviral vector. Some adenoviral vectors have been created by replacing the adenoviral E1A gene, which is essential for replication, with an expression cassette consisting of a viral or cellular promoter followed by the therapeutic gene and a polyadenyla-tion signal. The resulting E1A-deleted recombinant adenoviral vector is there-fore replication-defective. Other adenoviral vectors may contain additional deletions or other alterations of the adenoviral genome. Although adenovi-ruses exhibit a natural tropism for the respiratory epithelium, recombinant adenoviral vectors can infect not only lung and other epithelial-derived tissues (46–52) but also nonproliferating cell types such as hepatocytes (53–57) and neuronal and glial cells of the central nervous system (58–60). In liver, 100% of all mouse hepatocytes could be transduced following intraportal infusion of recombinant adenoviral vectors (57). The extremely high transduction efficiencies in liver suggested that recombinant adenoviral vectors would be good candidates for the in vivo delivery of therapeutic genes to the liver in animal models for diseases caused by hepatic deficiencies, such as PKU or hemophilia B. In both cases, infusion of a recombinant adenoviral vector containing either the human PAH cDNA or the cFIX cDNA under the transcriptional control of the Rous sarcoma virus long terminal repeat (RSV-LTR) into the portal vasculature of the deficient animal model resulted in normal or supranormal levels of PAH or cFIX expression that completely normalized the disease phenotype (61,62).

Although completely successful in the short term, the therapeutic effect obtained in these studies did not persist beyond a few weeks in the case of the PAH-deficient mouse or a few months in the case of the FIX-deficient dog. This difference in persistence is probably not species-related; rather, it reflects the differences in the absolute amounts of protein activity necessary to prevent the phenotypic manifestations of these two distinct metabolic diseases. Repeated administration of the recombinant adenovirus expressing human PAH could not duplicate the original effect in the mouse model, suggesting that an immune response against the adenoviral vector was responsible for the failure of repeated treatment. This hypothesis was supported by the detection of high titers of adenovirus-neutralizing antibodies 4 weeks after infusion of an adenoviral vector (61). In addition to this humoral response, a T-cell-mediated immune response is also elicited in the host organism that gradually destroys cells transduced by recombinant adenoviruses (63,64), significantly limiting the utility of this vector system for long-term gene expression. Adenoviral vectors are currently being modified to reduce or eliminate expression of the adenoviral genes responsible for evoking this T-cell-mediated response, and these vectors are being examined for increased persistence.

Despite the fact that expression of the therapeutic gene did not persist indefinitely in the treated animals, these experiments demonstrate the principal advantages of current recombinant adenoviral vectors in gene transfer: extremely high efficiency, broad host cell range, and an ability to transduce nondividing cell types. In addition, the findings from these studies that only 10 to 20% of normal levels of PAH, or perhaps even less for cFIX, are sufficient to correct the disease phenotype in these animal models establish realistically attainable goals for future applications of somatic gene therapy in similar metabolic diseases in humans that may be achieved as more persistent recombinant adenoviral vectors or other gene delivery systems are developed.

Little data are presently available regarding the in vivo application of DNA/protein complexes in the treatment of metabolic or cardiovascular disorders. The few studies that do exist report limited transduction efficiency of hepatocytes or other target cells in vivo (40,65–68). Since most metabolic diseases will probably only be treated by these vectors in vivo, significant improvement will be required in several areas before clinical applications can proceed. However, even at their present state of development, synthetic DNA delivery systems have several significant advantages over other vector systems. These include the absence of a viral backbone, which removes any restrictions on the size of the DNA fragment that can be incorporated and precludes any of the possible cytopathic effects that can be caused by viruses, and the presence of specific receptor ligands, which provides the possibility to target specific cell types. Their chief limitations are their low transduction efficiencies in vivo and their lack of persistence.

GENE THERAPY OF CARDIOVASCULAR DISORDERS

Several features of FH make it an ideal candidate for gene therapy. The homozygous form of FH is lethal at an early age and is refractory to conventional therapy (2). The Watanabe Heritable Hyperlipidemic (WHHL) rabbits are well characterized (69,70) and widely available as the animal model of FH. Serum lipid and lipoprotein values are informative and serve as clinically relevant endpoints for measuring the response to therapy. The defects in the function of the mutant LDL receptors are clearly understood. Orthotopic liver transplantation in a human subject has shown that a functional LDL receptor in the liver is adequate therapy for the hypercholesterolemia (71). Most of circulating LDL is degraded in the liver. It is also the only organ responsible for conversion of cholesterol to bile acids, which are the excreted end product of cholesterol metabolism.

Gene therapy to correct the LDL receptor defect in FH provides specific examples of the two general approaches. In ex vivo studies, recombinant retroviral vectors have been utilized in WHHL rabbits (72-74), which are homozygous for an LDL receptor mutation and develop severe vascular disease. More recently, the ex vivo approach has been used in a human subject (75). The in vivo application of recombinant adenoviral vectors in LDL receptor knockout mice (55,56) and WHHL rabbits (76) demonstrates the technical feasibility and the present limitations of these approaches.

Ex Vivo Gene Therapy of LDL Receptor Deficiency

The first report of a correction of LDL receptor deficiency in WHHL rabbits utilized a recombinant virus that contained a functional human LDL receptor gene, as well as Moloney enhancer and promoter sequences (72). In these experiments, hepatocytes harvested from WHHL rabbits were transduced by recombinant retroviruses containing a functional human LDL receptor gene, harvested, and infused into the portal vein of WHHL recipients. The primary measure of efficacy was total serum cholesterol, which decreased significantly

2–6 days after transplantation, with an eventual return to pretreatment levels. Although proviral DNA sequences and virus-directed transcripts could be detected in liver tissue 24 hours after transplantation and in situ hybridization demonstrated provirus expression in a small population of hepatocytes distributed in periportal sections of the liver, there were no functional tests to demonstrate that a metabolically active LDL receptor was expressed.

Dichek et al. (73) also reported retroviral vector-mediated in vivo expression of human LDL receptors in the WHHL rabbit. The recombinant gene was transduced into rabbit primary skin fibroblasts, which were reimplanted into donor rabbits. In vivo LDL receptor expression and the survival of the transduced cells could be documented, in addition to significant decrease in total and LDL cholesterol levels after implantation of the transduced cells. Control experiments indicated, however, that the decreases were not mediated through the recombinant LDL receptor. These authors note that changes in lipid levels after gene therapy must be interpreted with caution.

To address the possibility that immune recognition of the human LDL receptor plays a role in the elimination of the transduced cells, the original ex vivo treatment of hypercholesterolemia in WHHL rabbits was repeated with a recombinant retrovirus containing a rabbit LDL receptor gene with expression controlled by sequences from the chicken β-actin gene (74). Changes in serum cholesterol were monitored. There was long-term improvement of the hypercholesterolemia—a 30 to 50% reduction—that persisted for as long as 4 months. In these experiments, the use of *lacZ*-transduced cells in control animals demonstrated that the experimental procedures per se did not produce the transient decrease in serum cholesterol observed previously. Recombinant-derived LDL receptor RNA was harvested from tissues with no diminution for up to 6.5 months after transplantation. In these experiments, the transduced cells in the liver were estimated to be 1–4%. Although the response to ex vivo retroviral transduction of the LDL receptor was modest, these experimental findings provided the necessary background that allowed the development of the clinical protocol for ex vivo gene therapy directed to the liver of a patient with familial hypercholesterolemia (77).

A 29-year-old woman with homozygous FH was treated by injection of her hepatocytes, which had been cultured with a recombinant retrovirus containing a normal copy of the human LDL receptor gene (75). The liver resection of the left lateral segment, about 15% of the total liver mass, was tolerated well. After 48 hours in culture, the hepatocytes were exposed to the recombinant retrovirus for 12–18 hours before harvesting for transplantation. At the time of reimplantation, about 20% of the cells were successfully transduced with the LDL receptor, as indicated by the binding of fluorescently labeled LDL. The cell infusions were unremarkable. There was no apparent pathology in a liver biopsy sample at 4 months, and in situ hybridization showed single cells expressing the gene. The baseline level of serum LDL after gene therapy was 17% lower than the pretreatment baseline. The patient also became more responsive to a cholesterol-lowering agent after gene therapy. The most significant change is the LDL:HDL ratio, which was reduced by this treatment from 10–13 to 5–8. The interpretation of these results is that the normal LDL receptor gene has been targeted to liver cells, thereby eliciting normal LDL receptor expression. The lowering of the elevated LDL-cholesterol level in plasma is ascribed to an acceleration of the receptor-mediated uptake of LDL into the liver.

Subsequent to the report of the human trial, Brown et al. (78) identified and discussed at length the minimal criteria for the conduct and evaluation of gene-therapy experiments to correct a genetic deficiency in LDL receptors. First, animal experiments should unambiguously document persistent activity of the exogenously supplied LDL receptor gene in mediating the removal of LDL from plasma. Second, any observed reduction in the level of plasma LDL cholesterol in patients receiving gene therapy must be shown to result from an increased activity of LDL receptors. Previous studies have shown that manipulation of the liver can lower LDL levels by diminishing VLDL secretion and/or by enhancing LDL removal. Third, the increased receptor activity should be shown to result from expression of the exogenously derived gene rather than from activation of the patient's own receptors in response to surgical or other manipulation of the liver. This is a difficult criterion to satisfy in receptor-defective patients because LDL receptor activity is subject to tight metabolic regulation. At least initially, clinical trials need to focus on receptor-negative patients who have no functional LDL receptors in the cells.

These criteria are important, particularly at this early stage of development of gene therapy. The mutation of the LDL receptor in the patient in this study, W66G, is known as the French Canadian-4 mutation (79). The defective protein had 25–100% of the normal amount of functional activity when studied in cultured fibroblasts from three affected homozygotes (80). For unknown reasons, the activity of this mutant protein appears to be lower in liver than it is in fibroblasts. Nevertheless, the liver's mutant protein clearly has the intrinsic capacity to bind LDL, and it may well increase its activity following the performance of a major surgical procedure. Grossman et al. (75) found a 17% fall in plasma LDL cholesterol after a 25% hepatectomy and reinfusion of hepatocytes infected with a retrovirus encoding the normal LDL receptor. No pre- and posttreatment studies of LDL turnover were performed. Because this decline could be due either to diminished lipoprotein production or to enhanced activity of the patient's own receptors, no conclusions can be reached about the efficacy of the ex vivo gene-transfer procedure performed on this patient. Thus, it is critically important that the criteria described by

Brown et al. (78) be met in subsequent studies of gene therapy or lipid disorders in humans.

In Vivo Gene Therapy of LDL Receptor Deficiency

Thus far, the major problem with retrovirus-mediated approaches has been the low efficiency of therapeutic gene delivery to the target organs, and the requirement for dividing cells. By contrast, adenovirus-mediated gene delivery is highly efficient and can be used to transduce nondividing cells, and recombinant adenoviral vectors can be conveniently produced and administered to experimental animals. In vivo administration of a recombinant adenovirus containing the LDL receptor gene has recently been shown to increase the clearance of labeled VLDL in LDL receptor knockout mice (56) and in WHHL rabbits (76).

The LDL receptor knockout mice were produced by homologous recombination in embryonic stem cells (56). In the homozygous animals, the total plasma cholesterol levels were twice those of wild-type littermates. There was a seven- to ninefold increase in intermediate-density lipoproteins (IDL) and LDL without a significant change in HDL, documented by FPLC profiles of plasma lipoproteins. Plasma triglyceride levels were normal. The half-lives for intravenously administered ^{125}I-VLDL and ^{125}I-LDL were increased by 30-fold and 2.5-fold, respectively. The clearance of ^{125}I-HDL was normal in the LDLR −/− mice. In contrast to wild-type mice, moderate amounts of dietary cholesterol (0.2% cholesterol/10% coconut oil) caused a major increase in the cholesterol content of IDL and LDL in LDLR −/−. The elevated IDL/LDL level of LDLR −/− mice was reduced to normal 4 days after the intravenous injection of a recombinant replication-defective adenovirus encoding the human LDL receptor driven by the cytomegalovirus promoter. An adenovirus containing a luciferase insert was used in control experiments. The LDL virus restored expression of LDL receptor protein in the liver receptors and increased the clearance of rabbit ^{125}I-VLDL, determined 4 days after administration of the virus. The presence of the receptor was documented by immunoblots of liver membranes. An estimated 90% of the parenchymal cells in liver expressed the adenovirus-transferred genes as judged by immunofluorescence of LDL. The time course of receptor functioning or persistence of either the vector DNA or the LDL receptor protein was not reported. The data show that the LDL receptor is responsible in part for the low levels of VLDL, IDL, and LDL in wild-type mice and that adenovirus-encoded LDL receptors can acutely reverse the hypercholesterolemic effects of LDL receptor deficiency.

Recombinant, replication-defective adenoviruses expressing the LDL using a chicken β-actin promoter was infused into the portal vein of WHHL rabbits (76). Analysis of liver tissues harvested 3 days after virus infusion demonstrated human LDL receptor protein in the majority of hepatocytes that exceeded the levels found in human liver by at least 10-fold. Transgene expression was stable for 7–10 days and diminished to undetectable levels within 3 weeks. Infusion of LDL receptor–expressing virus led to substantial reductions in serum cholesterol that returned to baseline within 3 weeks; this acute reduction in serum cholesterol was associated with accumulations of lipid in hepatocytes. The development of neutralizing antibodies to the recombinant adenovirus markedly diminished the effectiveness of a second dose. These studies illustrate the advantages of recombinant adenoviruses for the treatment of liver metabolic diseases and define issues, such as viral genome instability and blocking immune response, that need to be overcome before the promise of this technology can be fully realized.

RISK FACTORS FOR CARDIOVASCULAR DISEASE

The scientific background that establishes the currently documented risk factors for cardiovascular disease has been summarized by an international task force (81). Clinical guidelines for prevention of coronary heart disease were developed for Western populations. The known risk factors include the following lipid disorders and encompass a much wider range of clinical situations than exist with single gene defects.

Elevated Plasma Cholesterol

Cardiovascular risk is conferred by elevated plasma cholesterol levels in the form of LDL and IDL cholesterol. This risk factor is most often the result of a diet rich in saturated fat and cholesterol and poor in fiber, rather than the presence of a monogenic trait. There are individual differences in response to this diet that are caused by multiple genes (common or polygenic hypercholesterolemia). The complex interplay of diet and genetic variation is poorly understood. Intensive basic and clinical research is needed before the appropriate targets for genetic intervention can be identified. By contrast, primarily genetic diseases such as familial hypercholesterolemia—LDL receptor deficiency—are important but comparatively uncommon.

Reduced Plasma HDL Cholesterol

HDL cholesterol is also an independent and powerful predictor of CVD incidence; low levels of HDL cholesterol are associated with high risk. This inverse association is particularly powerful in women. Low HDL cholesterol often reflects obesity, cigarette smoking, lack of physical exercise, and impaired glucose tolerance, and may also be inherited.

Elevated Plasma Triglyceride Levels

Elevated plasma triglyceride levels, when accompanied by a low HDL cholesterol, are predictive of increased CVD incidence. It is not clear as yet whether
the triglyceride level is independently predictive or whether the association
is dependent on coexisting factors. Recent studies show that a high triglyceride level with a low HDL cholesterol level, even without markedly elevated
LDL cholesterol, reflects a cluster of findings that is predictive of high risk.
This cluster often includes truncal obesity, hypertension, impaired glucose
tolerance, and hyperinsulinemia with insulin resistance. Obesity, especially
when it is of truncal distribution, is a risk factor for CVD, at least in persons
aged less than 50.

Elevated Plasma LP(a) Levels

Elevated plasma Lp(a) levels are common in patients with CVD and have been
found to be predictive of increased risk in a recent longitudinal study.

Many animal models have been used in studies of diet-induced atherosclerosis (82). These model systems, together with the increasingly powerful
ability to produce transgenic (83) and knockout animals (56), will most likely
provide the critical information about the responsiveness of atherosclerosis
and cardiovascular disorders to high-level expression of specific genes
thought to be important in counteracting or arresting the disease process. For
example, significant and persistent overexpression of cholesterol 7–α-hydrox-
ylase in the liver of diet-induced hypercholesterolemic rabbits might not only
reduce the levels of βVLDL but also decrease the extent and severity of lesions
in their vascular walls.

Studies with WHHL rabbits show that vascular lesions may also be
almost completely prevented when treatment is initiated at an early age.
Lowering cholesterol retards progression of atherosclerosis, induces a degree
of regression, and may restore vasodilator responses that are lost in the
presence of hyperlipidemia and atherosclerosis. It is well accepted that the
atherosclerotic process begins in children and progresses in adult life. Moreover, Modifying plasma lipids in adults decreases morbidity and mortality from
CVD in trials lasting 3–7 years. It is plausible, therefore, that when plasma
cholesterol is lowered for longer periods, by commencing in childhood, the
benefits will be considerably greater. Once the technical procedures become
mature, because of grossly elevated cholesterol levels and lack of response to
conventional treatment, gene therapy for familial hypercholesterolemia
would ideally be employed in children to prevent progression of disease in
relatively undamaged individuals.

The highly effective value of secondary prevention is increasingly being
recognized. Treatment of risk factors in persons with CVD is of considerable

value in reducing further coronary events. This approach is a particularly efficient and cost-effective way of reducing the total burden of coronary disease in the community. In persons who have had a myocardial infarction, high plasma cholesterol, high LDL cholesterol, and low HDL cholesterol levels continue to predict increased risk of another heart attack. Other recognized risk factors—cigarette smoking, diabetes, and hypertension—also retain their predictive power after myocardial infarction. A large potential for reducing subsequent coronary events by correction of risk factors applies to all patients with prior manifestations of coronary disease unless they have major myocardial damage. The risk in the first 6 months is related largely to cardiac function and to the extent and localization of coronary disease. Subsequently, the risk of recurrent events remains high and is related to high plasma cholesterol, low HDL cholesterol, high blood pressure, smoking, and diabetes—i.e., to all the "classic" risk factors. Since 50% of patients who have a myocardial infarction have had previous manifestations of coronary disease, effective secondary prevention would contribute importantly to reducing the total number of heart attacks in the population.

PERSPECTIVES AND FUTURE DIRECTIONS

The routine use of gene therapy to treat metabolic and cardiovascular disease requires additional scientific and technological achievements. The refinement of existing viral-based methods of gene delivery and the development of novel nonviral delivery systems are occurring at an ever-increasing rate. The examples highlighted in this chapter illustrate approaches to adding a gene to replace the function of a defective gene. An important question for the future is to determine if genes can be added to produce a more favorable lipid profile, even when the primary defect is unknown. Furthermore, there are alternative strategies in which genetic material is added to enhance, augment, or prevent the function of endogenous genes to achieve a therapeutic benefit or to prevent the onset of a disorder. Such approaches are considered in greater detail in other chapters in this book. Finally, much of the excitement and potential for gene therapy lie not just in discovering cures for various illnesses, but also in devising ways to prevent the onset of disease.

REFERENCES

1. McKusick VA. Mendelian Inheritance in Man, 9th ed. Baltimore: Johns Hopkins University Press, 1990.
2. Goldstein JL, Brown MS. Familial hypercholesterolemia. In: Scriver CR, Beaudet AL, Sly WS, Valle D, eds. The Metabolic Basis of Inherited Disease. 6th ed. New York: McGraw-Hill, 1989:1215–1250.

3. Wolff JA, Yee JK, Skelly H, Moores J, Respess J, Friedmann T, Leffert H. Adult mammalian hepatocyte as target cell for retroviral gene transfer: a model for gene therapy. Somat Cell Mol Genet 1987; 13:423–428.

4. Anderson KD, Thompson JA, Dipietro JM, Montgomery KT, Reid LM, Anderson WF. Gene expression in implanted rat hepatocytes following retroviral-mediated gene transfer. Somat Cell Mol Genet 1989; 15:215–227.

5. Ponder K, Gupta S, Leland F, Darlington G, Finegold M, DeMayo J, et al. Mouse hepatocytes migrate to liver parenchyma and function indefinitely after intrasplenic transplantation. Proc Natl Acad Sci USA 1991; 88:1217–1221.

6. Bickel H, Gerrard J, Hickmans EM. The influence of phenylalanine intake on the chemistry and behavior of a phenylketonuria child. Acta Paediatr Scand 1954; 43:64–77.

7. Armstrong MD, Tyler FH. Studies on phenylketonuria. I. Restricted phenylalanine intake in phenylketonuria. J Clin Invest 1955; 34:565–580.

8. Woolf LI, Griffiths R, Moncrieff A. Treatment of phenylketonuria with a diet low in phenylalanine. Br Med J 1955; 1:57–64.

9. Smith I, Lobascher ME, Stevenson JE, Woolf OH, Schmidt H, Grubel-Kaiser S, Bickel H. Effect of stopping low-phenylalanine diet on intellectual progress of children with phenylketonuria. Br Med J 1978; 2:723–726.

10. Lenke RR, Levy HL. Maternal phenylketonuria and hyperphenylalaninemia: an international survey of the outcome of untreated and treated pregnancies. N Engl J Med 1980; 303:1202–1208.

11. Ledley FD, Grenett HE, DiLella AG, Kwok SCM, Woo SLC. Gene transfer and expression of human phenylalanine hydroxylase. Science 1985; 228:77–79.

12. McDonald D, Bode V, Dove W, Shedlovsky A. Pah^{hph-5}: a mouse mutant deficient in phenylalanine hydroxylase. Proc Natl Acad Sci USA 1990; 87:1965–1967.

13. Shedlovsky A, McDonald JD, Symula D, Dove WF. Mouse models of human phenylketonuria. Genetics 1993; 134:1205–1210.

14. Ledley FD, Grenett HE, Dunbar BS, Woo SLC. Mouse phenylalanine hydroxylase: homology and divergence from human phenylalanine hydroxylase. Biochem J 1990; 267:399–405.

15. Kurachi K, Davie EW. Isolation and characterization of a cDNA coding for human factor IX. Proc Natl Acad Sci USA 1982; 79:6461–6464.

16. Choo KH, Gould KG, Rees DJ, Brownlee GG. Molecular cloning of the gene for human anti-haemophilic factor IX. Nature 1982; 299:178–180.

17. Evans JP, Watzke HH, Ware JL, Stafford DW, High KA. Molecular cloning of a cDNA encoding canine factor IX. Blood 1989; 74:207–212.

18. Brinkhous KM, Davis PD, Graham JB, Dodds WJ. Expression and linkage of genes for X-linked hemophilia A and B in the dog. Blood 1973; 41:577–585.

19. Ledley FD, Grenett H, McGinnis-Shelnutt M, Woo, SLC. Retroviral-mediated gene transfer of human phenylalanine hydroxylase into NIH 3T3 and hepatoma cells. Proc Natl Acad Sci USA 1986; 83:409–413.

20. Peng H, Armetano D, MacKenzie-Graham L, Shen R-F, Darlington G, Ledley FD, Woo SLC. Retroviral-mediated gene transfer and expression of human phenylalanine hydroxylase in primary mouse hepatocytes. Proc Natl Acad Sci USA 1985; 85:8146–8150.

21. Miller AD, Rosman GJ. Improved retroviral vectors for gene transfer and expression. BioTechniques 1989; 7:980–990.

22. Liu T-J, Kay MA, Darlington GJ, Woo SLC. Reconstitution of enzymatic activity in hepatocytes of phenylalanine hydroxylase-deficient mice. Somat Cell Mol Genet 1992; 18:89–96.

23. Wu GY, Wu CH. Receptor-mediated in vitro gene transformation by a soluble DNA carrier system. J Biol Chem 1987; 262:4429–4432.

24. Wagner E, Cotten M, Mechtler K, Kirlappos H, Birnstiel ML. DNA-binding transferrin conjugates as functional gene-delivery agents: synthesis by linkage of polylysine or ethidium homodimer to the transferrin carbohydrate moiety. Bioconj Chem 1991; 2:226–231.

25. Cristiano R, Smith LC, Woo SLC. Hepatic gene therapy: adenovirus enhancement of receptor-mediated gene delivery and expression in primary hepatocytes. Proc Natl Acad Sci USA 1993; 90:2122–2127.

26. Markwell MK, Portner A, Schwartz AL. Alternative route of infection for viruses: entry by the asialoglycoprotein receptor of sendai virus mutant lacking its attachment protein. Proc Natl Acad Sci USA 1985; 82:978–982.

27. Neda H, Wu CH, Wu GY. Chemical modification of an ecotropic murine leukemia virus results in redirection of its target cell specificity. J Biol Chem 1991; 266:14143–14146.

28. Cristiano RJ, Smith LC, Kay MA, Brinkley B, Woo SLC. Hepatic gene therapy: efficient gene delivery and expression in primary hepatocytes utilizing a conjugated adenovirus/DNA complex. Proc Natl Acad Sci USA 1993; 90:11548–11552.

29. Wagner E, Zatloukal K, Cotten M, Kirlappos H, Mechtler K, Curiel DT, et al. Coupling of adenovirus to transferrin-polylysine/DNA complexes greatly enhances receptormediated gene delivery and expression of transfected genes. Proc Natl Acad Sci USA 1990; 89:6099–6103.

30. Wagner E, Zenke M, Cotten M, Beug H, Birnstiel ML. Transferrin-polycation conjugates as carriers for DNA uptake into cells. Proc Natl Acad Sci USA 1992; 87:3410–3414.

31. Zenke M, Steinlein P, Wagner E, Cotten M, Beug H, Birnstiel ML. Receptor-mediated endocytosis of transferrin polycation conjugates: an efficient way to introduce DNA into hematopoietic cells. Proc Natl Acad Sci USA 1990; 87:3655–3659.

32. Curiel DT, Agarwal S, Wagner E, Cotten M. Adenovirus enhancement of transferrin/polylysine-mediated gene delivery. Proc Natl Acad Sci USA 1991; 88:8850–8854.

33. Cotten M, Wagner E, Zatloukal K, Phillips S, Curiel DT, Birnstiel ML. High-efficiency receptor-mediated delivery of small and large (48 kilobase) gene constructs using the endosome-disruption activity of defective or chemically inactivated adenovirus particles. Proc Natl Acad Sci USA 1992; 89:6094–6098.

34. Plank C, Zatloukal K, Cotten M, Mechtler K, Wagner E. Gene transfer into hepatocytes using asialoglycoprotein receptor mediated endocytosis of DNA complexes with an artificial tetra-antennary galactose ligand. Bioconj Chem 1992; 3:533–539.

35. Michael SI, Huang C, Romer MU, Wagner E, Curiel DT. Bindingincompetent

adenovirus facilitates molecular conjugate-mediated gene transfer by the receptor-mediated endocytosis pathway. J Biol Chem 1993; 268:6866–6869.

36. Leamon CP, Low PS. Delivery of macromolecules into living cells: a method that exploits folate receptor endocytosis. Proc Natl Acad Sci USA 1991; 88:5572–5576.

37. Leamon CP, Low PS. Cytoxicity of momordin-folate conjugates in cultured human cells. J Biol Chem 1993; 267:24966–24971.

38. Turek JJ, Leamon CP, Low PS. Endocytosis of folate-protein conjugates: ultrastructural localization in KB cells. J Cell Sci 1933; 106:4223–4230.

39. Gottschalk S, Cristiano RJ, Smith LC, Woo SLC. Folate-mediated DNA delivery into tumor cells: potosomal disruption results in enhanced gene expression. Gene Ther 1994; 1:185–191.

40. Curiel DT, Agarwal S, Romer MU, Wagner E, Cotten M, Birnstiel ML. Gene transfer to respiratory epithelial cells via the receptor-mediated endocytosis pathway. Am J Respir Cell Mol Biol 1992; 6:247–252.

41. Gupta S, Aragona E, Vemuru RP, Bhargava KK, Burk RD, Chowdhury JR. Permanent engraftment and function of hepatocytes delivered to the liver: implications for gene therapy and liver repopulation. Hepatology 1991; 14:144–149.

42. Ferry N, Duplessis O, Houssin D, Danos O, Heard J-M. Retroviral-mediated gene transfer into hepatocytes in vivo. Proc Natl Acad Sci USA 1991; 88:8377–8381.

43. Kaleko M, Garcia JV, Miller AD. Persistent gene expression after retroviral gene transfer into liver cells in vivo. Hum Gene Ther 1991; 2:27–32.

44. Kay MA, Li Q, Liu T-J, Leland F, Toman C, Finegold M, et al. Hepatic gene therapy: persistent expression of human α-antitrypsin in mice after direct gene delivery in vivo. Hum Gene Ther 1992; 3:641–647.

45. Kay MA, Rothenberg S, Landen C, Bellinger D, Leland F, Toman C, et al. In vivo gene therapy of hemophilia B: sustained partial correction in factor IX deficient dogs. Science 1993; 262:117–119.

46. Gilardi P, Courtney M, Pairani A, Perricaudet M. Expression of human alpha-1-antitrypsin using a recombinant adenovirus vector. FEBS 1990; 267:60–62.

47. Stratford-Perricaudet LD, Levrero M, Chasse J-F, Perricaudet M, Briand P. Evaluation of the transfer and expression in mice of an enzyme-encoding gene using a human adenovirus vector. Hum Gene Ther 1990; 1:241–256.

48. Rosenfeld MA, Siegried W, Yoshimura K, Yoneyama K, Fukayama M, Stier LE, et al. Adenovirus-mediated transfer of a recombinant alpha-1-antitrypsin gene to the lung epithelium in vivo. Science 1991; 252:431–434.

49. Rosenfeld MA, Yoshimura K, Trapnell BC, Yoneyama K, Rosenthal ER, Dalemans W, et al. In vivo transfer of the human CFTR gene to the airway epithelium. Cell 1992; 68:143–155.

50. Quantin B, Perricaudet LD, Tajbakhsh S, Mandel J-L. Adenovirus as an expression vector in muscle cells in vivo. Proc Natl Acad Sci USA 1992; 89:2581–2584.

51. Bajocchi G, Feldman SH, Crystal RG, Mastrangeli A. Direct in vivo gene transfer to ependymal cells in the central nervous system using recombinant adenovirus vectors. Nature Genet 1993; 3:229–234.

52. Yang Y, Raper SE, Cohn JA, Engelhardt JF, Wilson JM. An approach for treating the hepatobiliary disease of cystic fibrosis by somatic gene transfer. Proc Natl Acad Sci USA 1993; 90:4601–4605.

53. Levrero M, Barban V, Manteca S, Ballay A, Balsamo C, Avantaggiati ML, Natoli G, Skellekens H, Tiollais P, Perricaudet M. Defective and nondefective adenovirus vectors for expressing foreign genes in vitro and in vivo. Gene 1991; 101:195–202.

54. Jaffe HA, Daniel C, Longenecker M, Metzger M, Setoguchi Y, Rosenfeld MA. Adenovirus-mediated in vivo gene transfer and expression in normal rat liver. Nature Genet 1992; 1:372–378.

55. Herz J, Gerard RD. Adenovirus-mediated transfer of low density lipoprotein receptor gene acutely accelerates cholesterol clearance in normal mice. Proc Natl Acad Sci USA 1993; 90:2812–2816.

56. Ishibashi S, Brown MS, Goldstein JL, Gerard RD, Hammer RE, Herz J. Hypercholesterolemia in LDL receptor knockout mice and its reversal by adenovirus-mediated gene delivery. J Clin Invest 1993; 92:883–893.

57. Li QT, Kay M, Finegold M, Stratford-Perricaudet L, Woo SLC. Assessment of recombinant adenoviral vectors for hepatic gene therapy. Hum Gene Ther 1993; 4:403–409.

58. Akli S, Caillaud C, Vigne E, Stratford-Perricaudet LD, Poenaru L, Perricaudet M, et al. Transfer of a foreign gene into the brain using adenovirus vectors. Nature Genet 1993; 3:224–228.

59. Davidson BL, Allen ED, Kozarsky KF, Wilson JM, Roessler BJ. A model system for in vivo gene transfer into the central nervous system using an adenoviral vector. Nature Genet 1993; 3:219–223.

60. Le Gal La Salle G, Robert JJ, Berrard S, Ridoux V, Stratford-Perricaudet LD, Perricaudet M. An adenovirus vector for gene transfer into neurons and glia in the brain. Science 1993; 259:988–990.

61. Fang B, Eisensmith RC, Li XHC, Finegold MJ, Shedlovsky A, Dove W, Woo SLC. Gene therapy for phenylketonuria: phenotypic correction in a genetically deficient mouse model by adenovirus-mediated hepatic gene transfer. Gene Ther 1994. In press.

62. Kay MA, Landen CN, Rothenberg SR, Taylor LI, Leland F, Wiehle S, Fang B, Bellinger D, Finegold M, Thompson AR, Read M, Brinkhous KM, Woo SLC. In vivo hepatic gene therapy: complete albeit transient correction of factor IX deficiency in hemophilia B dogs. Proc Natl Acad Sci USA 1993; 91:2353–2357.

63. Müllbacher A, Bellett AJD, Hla RT. The murine cellular immune response to adenovirus type 5. Immunol Cell Biol 1989; 67:31–39.

64. Yang Y, Nunes FA, Berencsi K, Furth EE, Gönczöl E, Wilson JM. Cellular immunity to viral antigens limits E1-deleted adenovirus for gene therapy. Proc Natl Acad Sci USA 1994; 91:4407–4411.

65. Wu GY, Wu CH. Receptor-mediated gene delivery and expression in vivo. J Biol Chem 1988; 263:14621–14624.

66. Wu CH, Wilson JM, Wu GY. Targeting genes: delivery and persistent expression of a foreign gene driven by mammalian regulatory elements in vivo. J Biol Chem 1989; 264:16985–16987.

67. Wu GY, Wilson JM, Shalaby F, Grossman M, Shafritz DA, Wu CH. Receptor-mediated gene delivery in vivo. J Biol Chem 1991; 266:14338–14342.

68. Chowdhury NR, Wu CH, Wu GY, Yerneni PC, Bommineni VR, Chowdhury JR.

Fate of DNA targeted to the liver by asialoglycoprotein receptor-mediated endocytosis in vivo. J Biol Chem 1993; 268:11265-11271.

69. Buja LM, Kita T, Goldstein JL, Watanabe Y, Brown MS. Cellular pathology of progressive atherosclerosis in the WHHL rabbit, an animal model of familial hypercholesterolemia. Arteriosclerosis 1983; 3:87-99.

70. Yamamoto T, Bishop RW, Brown MS, Goldstein JL, Russell DW. Deletion in cysteine-rich region of LDL receptor impedes transport to cell surface in WHHL rabbit. Science 1986; 232:1230-1237.

71. Bilheimer DM, Goldstein JL, Grundy SC, Starzl TE, Brown MS. Liver transplantation provides low density lipoprotein receptors and lowers plasma cholesterol in a child with homozygous familial hypercholesterolemia. N Engl J Med 1984; 311:1658-1664.

72. Wilson JM, Chowdhury NR, Gossman M, Wajsman R, Epstein A, Mulligan RC, Chowdhury JR. Temporary amelioration of hyperlipidemia in low density lipoprotein receptor-deficient rabbits transplanted with genetically modified hepatocytes. Proc Natl Acad Sci USA 1990; 87:8437-8441.

73. Dichek DA, Bratthauer GL, Beg ZH, Anderson KD, Newman KD, Zwiebel JA, Hoeg JM, Anderson WF. Retroviral vector-mediated in vivo expression of low-density-lipoprotein receptors in the Watanabe heritable hyperlipidemic rabbit. Somat Cell Mol Genet 1991; 17:287-301.

74. Chowdhury JR, Gossman M, Gupta S, Chowdhury NR, Baker JR Jr, Wilson JM. Long term improvement of hypercholesterolemia after *ex vivo* gene therapy in LDLR-deficient rabbits. Science 1991; 254:1802-1805.

75. Grossman M, Raper SE, Kozarsky K, Stein EA, Engelhardt JF, Muller D, Lupien PJ, Wilson JM. Successful *ex vivo* gene therapy directed to liver in a patient with familial hypercholesterolemia. Nature Genet 1994; 6:335-341.

76. Kozarsky KF, McKinley DR, Austin LL, Raper SE, Stratford-Perricaudet LD, Wilson JM. In vivo correction of low density lipoprotein receptor deficiency in the Watanabe heritable hyperlipidemic rabbit with recombinant adenoviruses. J Biol Chem 1994; 269:13695-13702.

77. Wilson JM, Grossman M, Raper SE, Baker JR Jr, Newton RS, Thoene JG. Clinical protocol for *ex vivo* therapy of familial hypercholesterolemia. Hum Gene Ther 1993; 3:179-222.

78. Brown MS, Goldstein JL, Havel RJ, Steinberg D. Gene therapy for cholesterol. Nature Genet 1994; 7:349-350.

79. Hobbs HH, Brown MS, Goldstein JL. Molecular genetics of the LDL receptor gene in familial hypercholesterolemia. Hum Mut 1992; 1:445-466.

80. Leitersdorf E, Tobin EJ, Davignon J, Hobbs HH. Common low-density lipoprotein receptor mutations in the French Canadian population. J Clin Invest 1990; 85:1014-1023.

81. Lewis B, Stein Y. Prevention of coronary heart disease: scientific background and new clinical guidelines. Nutr Metab Cardiovas Dis 1992; 2:113-156.

82. Jokinen M, Clarkson T, Prichard R. Animal models in atherosclerosis research. Exp Mol Path 1985; 42:1-28.

83. Field LJ. Transgenic mice in cardiovascular research. Annu Rev Physiol 1993; 55:97-114.

21

Vascular Smooth-Muscle-Cell Proliferation

Basic Investigations and New Therapeutic Approaches

Robert D. Rosenberg
Harvard Medical School
and Beth Israel Hospital
Boston
and Massachusetts Institute of Technology
Cambridge, Massachusetts

Michael Simons
Harvard Medical School
and Beth Israel Hospital
Boston, Massachusetts

The smooth-muscle cells (SMCs) of the blood vessel wall are normally in a relatively quiescent state. However, arterial injury or vascular disease induces a proliferative response that serves as an important mechanism in atherogenesis and also provides the basis for angioplasty-induced coronary artery restenosis, as well as bypass vein graft failure (1–4). The best known mitogen involved in inducing SMC growth is platelet-derived growth factor (PDGF), which is released from circulating platelets or synthesized by growing SMCs or endothelial cells (5). The SMCs are endowed with PDGF-α as well as PDGF-β receptors and respond to all isoforms of the mitogen. Several investigations have identified a synergistic interaction between PDGF and insulin growth factor (IGF-I), which stimulates replication of SMCs under in vitro conditions (6). Basic fibroblast growth factor (bFGF) as well as epidermal growth factor (EGF) also bind to specific receptors and induce SMC growth. Interleukin-1 (IL-1) enhances SMC proliferation in vitro by initiating production of endogenous PDGF and basic FGF (7,8), whereas transforming growth factor (TGF)-β possesses both stimulatory and inhibitory effects on cell growth (9). The end product of the blood-pressure regulating system angiotensin II is also known to affect smooth-muscle-cell proliferation and hypertrophy (10–13). The blood-clotting enzyme thrombin is generated by activation of the coagulation cascade induced by arterial injury, interacts with specific SMC receptors, and thereby stimulates growth of this cell type (14). Thus, it is readily apparent

547

that multiple growth factors and biological modulators can initiate SMC proliferation, and it is highly likely that additional undefined components play a critical role in the growth of this cell type (Figure 1). The in vivo biological potency of several of these growth factors and biological modulators has been evaluated in a rat carotid artery or pig coronary artery injury model (15–17). In these systems, the endothelium is denuded and the subsequent migration/proliferation of SMCs is quantitated. The kinetics of the overall process has been carefully characterized in the rat carotid artery injury model and, to a lesser extent, in the pig coronary artery injury model (17,18).

The interactions of the various growth factors with their specific SMC receptors induce a complex series of biochemical reactions involving protein kinase C, receptor-linked tyrosine phosphorylation, and cAMP-dependent kinases, which culminates in the activation of DNA-binding proteins and the initiation of DNA replication as well as cell division (Figure 1). The common

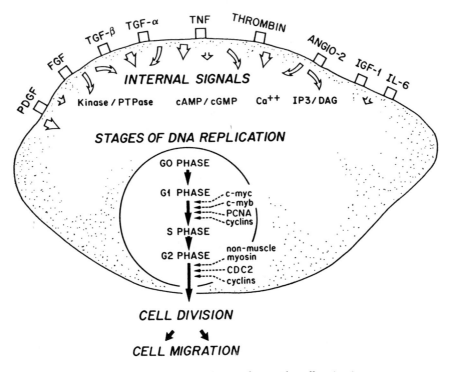

Figure 1   Common final pathway of smooth-muscle-cell activation.

signal transduction pathway that links activation of the above receptors with SMC proliferation probably includes induction of c-myc and c-fos mRNA at 30 minutes to 2 hours, with a decline to normal by 12 hours; expression of cdc 2 kinase at 8–12 hours, with a decline to normal by 16 hours; and the synthesis of c-myb protein and PCNA protein at 16–20 hours, with a decline to normal by 24 hours (20–23).

The proto-oncogene c-myb is a DNA-binding protein that may play an important role in SMC growth. The oncogene is homologous to the transforming gene of the avian myeloblastosis virus and was originally thought to be present only in hematopoietic cells. However, c-myb is synthesized by chick embryo fibroblasts as well as proliferating SMCs and may also be produced by other cell types (23,24). The expression of the proto-oncogene is growth-dependent: it occurs at low levels in quiescent cells, increases rapidly as cells begin to proliferate, and peaks in the late G_1 phase of the cell cycle (23). The oncogene appears to regulate cellular growth. Thus, expression of the oncogene in fibroblasts directly or indirectly induces the appearance of DNA polymerase alpha, histone H3, and PCNA mRNAs by a posttranscriptional mechanism and permits the cells to enter the S phase (25). Heparin, as well as the closely related heparan sulfate proteoglycans, suppresses SMC proliferation in vivo as well as in vitro (26–29). The block occurs in the late G_1 phase of the cell cycle and is associated with a decrease in c-myb levels (30) and a partial return of expression of SMC-specific contractile proteins (3). The observation that heparin's suppression of SMC growth is associated with a decrease in c-myb message level (30) and the known effects of the proto-oncogene suggest an important role for c-myb in SMC proliferation.

A number of other cell cycle-specific genes have been implicated in control of SMC proliferation. Thus, proto-oncogene c-myc, a DNA-binding protein, is expressed immediately (30 minutes to 2 hours) after arterial injury in vivo (31). After that, the proto-oncogene's expression declines but reappears once again 7 days later (32). Heparin's administration has also been associated with a decrease in c-myc as well as c-fos expression (21). The expression of PCNA (a subunit of DNA polymerase Δ) and cdc-2 kinase has also been linked to cell proliferation in a number of cell-culture systems (33,34). However, it remains to be rigorously demonstrated that the above components are members of the final common pathway that links activation of specific SMC receptors and the migration/proliferation of SMCs within the blood-vessel wall. The use of antisense oligonucleotides to inhibit expression of specific intracellular components offers the potential for defining key regulatory points in the above pathway, establishing the detailed molecular mechanisms of these gene products under in vitro and in vivo conditions and developing a novel therapeutic approach for intervening in vessel-wall disorders.

This chapter focuses mainly on c-myb because investigations of this proto-oncogene have progressed very rapidly, but available data on other regulatory components are also reviewed. In the following sections, we provide a brief primer on antisense oligonucleotide technology, describe data that reveal the critical nature of c-myb as an intracellular mediator of SMC proliferation under in vitro conditions, discuss the mechanism of action of the proto-oncogene as a regulator of intracellular calcium concentrations at the G_1/S interface of the cell cycle, demonstrate that antisense oligonucleotides directed against c-myb suppress SMC growth and migration under in vivo conditions, and summarize the available information on inhibition of growth and migration of SMC by antisense oligonucleotides directed against cdc 2 kinase, c-myc, and PCNA under both in vitro and in vivo conditions.

ANTISENSE OLIGONUCLEOTIDE TECHNIQUES

The antisense oligonucleotides employed to inhibit expression of specific mRNAs in studies described below are usually 15–20 residues in length and represent complementary copies of various regions of specific mRNAs. Previous investigations have demonstrated the biological activity of antisense oligonucleotides targeted against 5′ untranslated domains, the site of initiation of translation, the exon/intron boundaries, 3′ untranslated regions, and other areas of mRNAs (35). Although theoretical criteria have been proposed for selecting optimally effective antisense sequences (36–38), most investigators have concluded that no accurate method exists for predicting which oligonucleotide will inhibit expression of a given mRNA. Indeed, a shift of three residues in the starting position of the antisense oligonucleotide can generate major differences in biological activity (39). Thus, it is usually necessary to test a number of antisense oligonucleotides directed against different sequences of a particular mRNA. However, a good initial choice is to synthesize an antisense oligonucleotide starting at +4, which is usually quite effective (see below).

The optimal length of the antisense oligonucleotide represents a compromise between maximizing thermodynamic stability of the antisense oligonucleotide–mRNA hybrid, which increases as a function of the number and composition of bases utilized, and the need to exclude interactions with other mRNAs, which also increases as a function of the length of the antisense oligonucleotide (40). Theoretical calculations suggest that a single mismatch in the nucleotide composition between the oligonucleotide and the target mRNA sequence decreases stability of association by an average of 3.7 kcal/mol (increase in ΔG, free energy of association), while a two-nucleotide mismatch would produce an even larger decrease in duplex stability (40). These computations assume that antisense oligonucleotides must exhibit a perfect sequence match with its target mRNA to inhibit expression. However, it is widely acknowledged that antisense oligonucleotides with partial mismatches may also inhibit expression of a specific mRNA (see below). Tests on cRNA expression in *Xenopus* oocytes of the inhibitory action of antisense oligonucleotides of different lengths show that a 50% biological effect occurred at oligonucleotide lengths of 7.6 to 9.9 residues and a 95% biological effect took place at a oligonucleotide length of 12 residues (41). Interestingly, in the above system, the location of the target sequence had little effect on the biological activity of the antisense oligonucleotide (41).

The antisense oligonucleotides with a native phosphodiester linkage exhibit a relatively short half-life of about 1 to 2 hours in cell culture or under in vivo conditions because of degradation by existing nucleases. While this short time period may be sufficient for some applications, in most instances prolonged suppression of gene expression is required. Therefore, the antisense oligonucleotides have been chemically modified to reduce the rate of cleavage of the sequence by nucleases by 10–50-fold. The various modifications of the phosphodiester bond commonly utilized are listed in Table 1, with phosphorothioate and methylphosphonate oligonucleotides constituting the most frequently used modifications.

The antisense methylphosphonate oligonucleotides have previously been shown to inhibit expression of target mRNAs such as HSV-1, HIV, c-Ha-Ras, and c-myc (42–45). These antisense oligonucleotides are believed to function primarily by blocking translation of message, since this modification appears to prevent RNase H–dependent cleavage of the targeted mRNA (see below for mechanism of action) (45). Therefore, a 10–50-fold higher molar concentration of methylphosphonate oligonucleotides, as compared to phosphorothioate oligonucleotides, is often required to achieve the same magnitude of effect. The mixed chirality of methylphosphonate oligonucleotides has also been raised as an important issue because of its effects on hybridiza-

Table 1 Chemical Modifications of Synthetic Oligonucleotides

Name	Backbone modifiers[a]
Phosphodiester (normal)	$O-$
Formacetal	$-O-CH-O-$
Methylphosphonate	CH_3-
Phosphoramidate	R_2N-
Phosphorothioate	$S-$
Phosphotriester	$R-$
Phosphoroselenoate	$Se-$

[a]Backbone modification of phosphodiester bond in synthetic oligonucleotides.

tion and solubility. These products are often poorly soluble in water and other polar solvents, which limits their usefulness.

The antisense phosphorothioate oligonucleotides are also quite resistant to the action of nucleases and have been utilized to inhibit expression of many specific mRNAs, including all those discussed in the current review (46). These antisense oligonucleotides exhibit a small reduction of about 0.3 kcal/mol of free energy per residue with regard to binding to target mRNA, probably function mainly via the RNase H–dependent mechanism, and are quite soluble in water (see below for mechanism of action). These properties make the above antisense phosphorothioate oligonucleotides relatively easy to use in biological experiments. However, it has been suggested that antisense phosphorothioate oligonucleotides, as compared to native oligonucleotides, facilitate RNase H–dependent cleavage, which could allow destruction of target mRNAs with larger degrees of mismatch (see the discussion below of mechanism of action). Finally, it has also been suggested that nonspecific toxicity with regard to cell growth occurs more frequently with phosphorothioate oligonucleotides than with other modifications, although scant information on this matter is available.

Several other modifications have been employed, including 2'-O-alkyl groups on purine and pyrimidine residues (47), C-5 propene groups on pyrimidine residues (48), and the capping of oligonucleotides with 5'/3' phosphorothioate residues (49), but insufficient data are available to evaluate their properties. Perhaps the most interesting modification technique involves the production of antisense chimeric methylphosphonate/phosphorothioate oligonucleotides that exhibit significant uptake by cells, and RNase H–dependent cleavage of target mRNAs. These chimeric molecules also possess greatly improved specificity under in vitro conditions and can recognize the difference between mRNAs that differ by only a single base (50-52).

The uptake of antisense oligonucleotides by cells is poorly understood. Some investigators claim that phosphorothioate oligonucleotides bind to specific protein receptors, and are then transported into endosomes via receptor-mediated endocytosis (53,54). However, these studies are bedeviled by numerous experimental problems and the postulated receptors have not been isolated. It has also been suggested that methylphosphonate oligonucleotides enter cells via adsorptive or fluid-phase endocytosis (55). The different cell types exhibit great variations in the uptake of antisense oligonucleotides by either of the above mechanisms. If the extent of entry of antisense oligonucleotides into particular cells is limited, then the antisense oligonucleotides can be encapsulated in liposomes with or without a coating of targeting/membrane-fusion proteins, and the process of cellular uptake can be dramatically augmented. The antisense oligonucleotides that enter cells have been located in the endosomes, the lysosomes, the cytoplasm, and the nucleus

(56). It remains unclear whether chemical modifications of antisense oligonucleotides alter the above subcellular distributions, and whether the presence of antisense oligonucleotides at a particular subcellular site is of importance to its biological activity.

The antisense oligonucleotides inhibit expression of specific mRNAs by a variety of potential mechanisms. It is clear that they could potentially interfere with splicing of transcripts, suppress nuclear to cytoplasmic transport of mRNAs, initiate destruction of mRNAs via RNase H–dependent cleavage, and prevent translation of mRNAs by ribosomes (Figure 2). However, the relative importance of the various potential mechanisms of action for a particular antisense oligonucleotide is usually ill defined. Furthermore, antisense oligonucleotides can also inhibit gene transcription by several mechanisms. It is apparent that antisense oligonucleotides can form a region of triple helix with the DNA duplex when certain steric constraints are satisfied (57). The triple helix interferes with the action of RNA polymerase by altering DNA conformation, preventing binding of transcriptional factors, or inhibiting the movement of the RNA polymerase complex (57). The inhibition of binding of transcription factors is a particularly promising area. Cooney et al. (58) have described the binding of a 28 base oligonucleotide to a purine-rich sequence in the c-myc promoter that blocks transcription under in vitro conditions. Similar results have also been reported with the IL-2 receptor alpha promoter (59,60). The effects of antisense oligonucleotides on transcription are not considered in this chapter because sequences with special characteristics are usually required to observe this activity (57).

In principle, it should be possible to employ antisense oligonucleotides to inhibit expression of any specific mRNA. However, this approach is most likely to be successful with mRNAs that are present at very low copy numbers,

RNA TRANSLATION

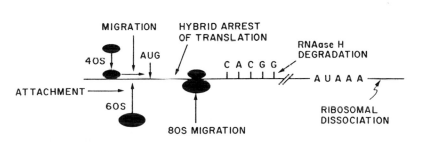

Figure 2 Mechanisms of antisense effect. Proposed oligonucleotide sites of action are indicated with arrows.

and encode proteins with relatively short half-lives. Given these considerations, it is not surprising that antisense techniques have been utilized to inhibit expression of cell-cycle mRNAs, specific growth factor/cytokine receptor mRNAs, or specific signaling molecule mRNAs (see below). In the ideal case, the antisense oligonucleotide inhibits expression of a specific mRNA without affecting other mRNAs or cellular proteins and thereby produces a specific phenotypical state. However, the antisense oligonucleotides can inhibit expression of other mRNAs because of the existence of partial sequence matches. Thus, a 13 base antisense oligonucleotide should recognize a unique sequence in the eukaryotic genome (the probability of more than one such sequence occurring in the genome, 4^{13}, is close to zero when compared to the size of the genome), but partial matches to other mRNAs are likely. The effect of weak interactions with nontarget mRNAs could be magnified with antisense phosphorothioate oligonucleotides because even partial sequence matches might induce RNase H–dependent cleavage of unrelated mRNAs. The degradation of mRNAs other than target mRNA has been observed under in vitro conditions and in *Xenopus* oocytes (61–63). In a similar fashion, the antisense oligonucleotides might also bind to critical intracellular proteins, which could then inhibit their function and produce the observed phenotypical state (64). Finally, the generation of an antisense oligonucleotide–mRNA hybrid might activate intracellular enzymes, such as P68, that would induce the synthesis of macromolecules, including γ interferon, that could produce the observed phenotypical state (65).

The above considerations suggest that four types of experimental controls should be carried out before an antisense oligonucleotide is thought to inhibit expression of a specific mRNA and thereby produce a specific phenotypical state:

1. Multiple nonoverlapping antisense oligonucleotides directed at the same mRNA sequence should inhibit expression of the same mRNA/protein and produce the same phenotypical state. This control minimizes the chance that spurious interactions with other mRNAs or proteins could be responsible for inducing the specific phenotypical state.

2. Scrambled oligonucleotides (the same base composition as the antisense oligonucleotide but with a completely different sequence) or minimal-mismatch antisense oligonucleotides (antisense oligonucleotide with two to four nucleotide mismatches) should not affect expression of the specific mRNA/protein and should not induce the phenotypical state. This control assures that the observed effect of antisense oligonucleotides is not due to nonspecific toxicity.

3. Antisense oligonucleotides directed against a specific mRNA not

involved in the phenotypical state should inhibit expression of the specific mRNA/protein, but should not induce the observed phenotypical state. This control proves that formation of an antisense oligonucleotide–mRNA hybrid is insufficient to produce the specific phenotypical state.

4. Constitute overexpression of the targeted mRNA within the cell should dramatically increase the amount of antisense oligonucleotide needed to inhibit expression of the mRNA/protein. This control shows that the antisense oligonucleotides must be interacting with the specific mRNA within the cell. These controls minimize the likelihood that antisense oligonucleotides function via interactions with a target other than the specific mRNA. In the investigations described below, these controls have been utilized to a greater or lesser extent.

THE PROTO-ONCOGENE c-myb IS A CRITICAL REGULATOR OF SMC PROLIFERATION

We initially showed that exposure of SMCs to antisense phosphorothioate c-myb oligonucleotides (+4 to +21), but not sense phosphorothioate c-myb oligonucleotides, dramatically reduced the levels of c-myb mRNA and oncoprotein (66). To this end, the well-characterized SV40–large T–transformed (SV40LT) rat aortic SMC line was growth-arrested in media containing low concentrations of growth factors and exposed to media containing high levels of growth factors as well as antisense c-myb oligonucleotides or sense c-myb oligonucleotides from the same region, and the levels of c-myb mRNA were assessed by dot blot hybridization. Figure 3 presents individual normalized dot blot counts for synchronized proliferating SMCs, antisense or sense c-myb oligonucleotide–treated SMCs, heparin-arrested SMCs, and growth-arrested SMCs. As expected, the growth-arrested cells exhibit minimal amounts of c-myb message, while serum stimulation induces a 10-fold increase in the above gene product over 24 hours. Treatment with antisense c-myb oligonucleotides produces a dramatic reduction in specific message as compared to sense oligonucleotides or control cultures. Exposure to heparin also triggers a significant decrease in c-myb message that is consistent with results previously reported from our laboratory (30).

The reduction in the concentrations of c-myb mRNA should lead to a decrease in the levels of the oncoprotein. To show this effect, SV40LT SMCs were plated at low density on glass slides, growth-arrested, and then shifted to growth factor–containing media to which were added antisense or sense c-myb oligonucleotides. After 24 hours SMCs were examined by indirect

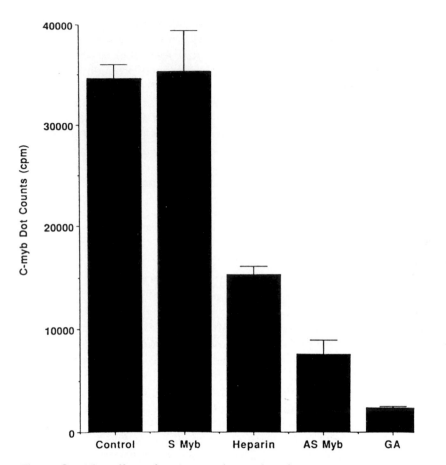

Figure 3 The effect of antisense oligonucleotides on mRNA levels. Bar graph showing results of myb RNA dot blot. The levels of c-myb message are displayed as numbers of counts per minute per dot normalized for large T RNA level. Control: untreated cells; S myb: cells treated with 25 μM sense c-myb phosphorothioate oligonucleotide; Heparin: cells treated with 100 μg/ml heparin; AS myb: cells treated with 25 μM antisense c-myb phosphorothioate oligonucleotide; GA: growth-arrested cells (96 hours in 0.5% FBS-DMEM).

immunofluorescence microscopy utilizing specific antisera against c-myb that revealed that the levels of oncoprotein in most cells treated with antisense oligonucleotides are dramatically reduced as compared to those exposed to sense oligonucleotides (66). The importance of c-myb to SMC growth was then evaluated using SV40LT SMCs as well as early-passage primary rat and mouse aortic SMCs. The SV40LT SMCs were growth-arrested in media containing low concentrations of growth factors and exposed to media containing high levels of growth factors as well as varying amounts of antisense phosphorothioate c-myb oligonucleotides (+4 to +21) or sense phosphorothioate oligonucleotides from the same region. The extent of cell proliferation was determined at 72 hours. The data shows a concentration-dependent suppression of SMC growth with antisense c-myb oligonucleotides, but not sense c-myb oligonucleotides (Figure 4).

The specificity of this effect is documented by four additional types of controls. First, the above studies were carried out with a second nonoverlapping antisense c-myb phosphorothioate oligonucleotide (+22 to +39) and a control scrambled c-myb phosphorothioate oligonucleotide that showed the same extent of inhibition of expression of c-myb mRNA and the same specific degree of suppression of SMC proliferation as outlined above (66). Second, the same studies repeated with a minimal-mismatch antisense phosphorothioate c-myb oligonucleotide in which two nucleotides of the 18 base sequence were randomly altered revealed no reduction in the levels of c-myb mRNA and a complete loss of antiproliferative activity (66). Third, experiments were conducted with antisense and sense phosphorothioate thrombomodulin (TM) oligonucleotides (+4 to +21). This cell-surface receptor has no known function in the growth of SMCs, and antisense oligonucleotides should have no effect on the proliferation of these cells. The data show that SMC growth was unaltered by the addition of the above antisense oligonucleotides despite a 90% reduction of TM mRNA levels as determined by Northern analysis (66). Fourth, we also showed that stable constitutive overexpression of c-myb in SV40LT SMCs increased the concentrations of antisense c-myb oligonucleotide needed to inhibit expression of the c-myb mRNA (Figure 5).

The antiproliferative effects of the antisense and sense oligonucleotides were also evaluated with primary rat and mouse aortic SMCs. The data show that growth of the two cell types is greatly suppressed with antisense, but not sense, phosphorothioate c-myb oligonucleotides (66). The inhibition of cell proliferation for the various cell types appears to be due to a growth arrest in the late G_1 phase of the cell cycle. In a different set of experiments, Brown et al. (67) showed that antisense c-myb oligonucleotides suppressed proliferation of primary bovine aortic SMCs.

We next established the minimal time required for exposure of SV40LT SMCs to antisense c-myb oligonucleotides to achieve maximal growth inhibition. In the studies cited above, cells were continuously exposed to oligonucleotides from the time of shift from growth arrest to the measurement of antiproliferative effect by cell count at 72 hours. In the experiments cited below, SV40LT SMCs were treated with antisense oligonucleotides for stated periods after the shift from growth arrest, washed twice with phosphate-buffered saline, placed in fresh oligonucleotide-free growth factor–containing

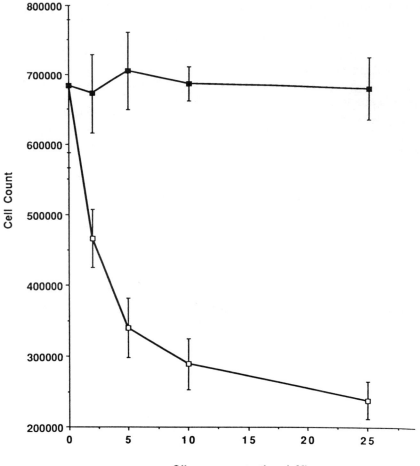

Figure 4   Line graphs showing effects of antisense c-myb on proliferation of SV40–large T–transformed rat aortic smooth-muscle cells. The antisense (light boxes) and sense (dark boxes) c-myb phosphorothioate oligonucleotides have been added at varying micromolar concentrations to growth-arrested SV40–large T–transformed rat aortic smooth-muscle cells before serum stimulation, and cell counts were obtained 72 hours later. The data are displayed as mean ± SD.

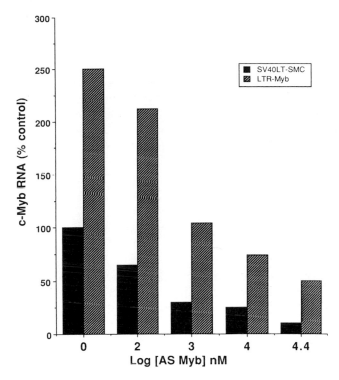

Figure 5   The effect of c-myb overexpression in SV40LT SMCs on the amount of antisense oligonucleotides needed to suppress growth in vitro. Dot-blot-determined c-myb mRNA levels (normalized for GAPDH) in the wildtype SV40LT SMCs and the c-myb-transfected LTR-Myb cells in the presence of variable amounts of antisense c-myb oligonucleotides.

media, and assessed for growth-inhibitory effect by cell count at 72 hours. The addition of antisense c-myb oligonucleotides, for as little as 1 hour, generated a significant antiproliferative effect whereas exposure for 4 hours produced a growth-inhibitory effect identical to that observed with continuous exposure for 72 hours (66). Therefore, a very brief contact of cells with antisense oligonucleotide produces a suppression of proliferation. We also ascertained whether the growth-inhibitory effects of antisense oligonucleotides are readily reversible. To this end, SV40LT SMCs were exposed for 4 hours after release from growth arrest to antisense or sense phosphorothioate c-myb oligonucleotides. The cells were subsequently washed twice and placed in fresh oligonucleotide-free growth factor–containing media. Cell counts were determined at days 3 and 5. The data revealed that SV40LT SMCs treated with antisense c-myb oligonucleotides, as compared to the corresponding sense oligonucleotides, demonstrate a significant initial suppression of growth at day 3. However, the doubling times, between days 3 and 5 with antisense c-myb, as compared to the corresponding sense oligonucleotides, were identical (66).

In summary, the above investigations reveal that SMCs express c-myb during in vitro proliferation; addition of antisense oligonucleotides for short periods of time leads to a specific destruction of the proto-oncogene mRNA as well as inhibition of oncoprotein synthesis; and inhibition of c-myb expression leads to a reversible suppression of cell growth with a block in the late G_1 phase of the cell cycle (30,66).

THE PROTO-ONCOGENE c-myb MEDIATES AN INTRACELLULAR CALCIUM RISE DURING THE LATE G_1 PHASE OF THE CELL CYCLE

The intracellular levels of ionizable calcium ($[Ca^{2+}]_i$) are critically involved in controlling cell-cycle progression and cell growth. It is widely appreciated that transient increases of $[Ca_{2+}]_i$ occur early in mitosis and during anaphase (68,69). These elevations appear to be required for disappearance of the nuclear envelope, condensation of chromosomes, breakdown of mitotic spindles, and activation of the contractile ring (70–73). The regulation by $[Ca^{2+}]_i$ of the G_1 to S transition is less thoroughly documented. However, earlier investigations have suggested that the divalent cation may be required for entry of cells into the S phase (74,75). For the above reasons, we measured the $[Ca^{2+}]_i$ of SMCs early in the cell cycle and then determined whether c-myb, which is differentially expressed at this time, might be responsible for the observed variations (76).

We first employed SV40LT SMCs to investigate changes in $[Ca^{2+}]_i$ during cell-cycle progression. To this end, SMCs were growth-arrested and then stimulated by addition of growth factor–containing media. The concentrations of $[Ca^{2+}]_i$ were determined at intervals of 8 hours. The data obtained by full-field image analysis of representative cells using Fura-2 demonstrate that $[Ca^{2+}]_i$ are unchanged for the first 8 hours, decline at 16 hours, rise to the initial levels at 24 hours, and drop back again at 32 hours (Figure 6, upper panel). Flow cytometric analysis of cellular DNA shows partial cell-cycle synchronization and reveals that $[Ca^{2+}]_i$ at 24 hours increases as the cell population enters the S phase (Figure 6, lower panel). Identical experiments carried out with primary rat aortic SMCs showed similar changes in intracellular divalent cation (76).

The concentration of c-myb mRNA in SV40LT SMCs were then determined by dot blot analysis with normalization to GAPDH mRNA, which

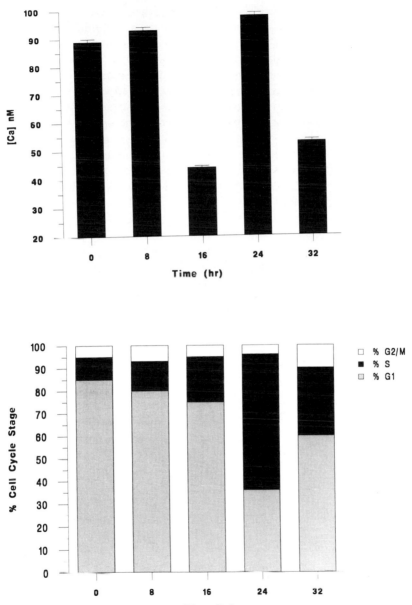

Figure 6 Intracellular calcium levels during cellcycle progression in vascular smooth-muscle cells. (Top) concentrations of intracellular calcium in SV40LT SMCs determined with Fura-2 following serum stimulation (mean ± SE). (Bottom) cell-cycle distribution determined in parallel samples.

demonstrated that message levels are low in growth-arrested cells (0 hours), increase significantly at 16 hours (late G_1), and reach a maximum at 24 hours (G_1/S interface) (Figure 7). Thus, the increased concentrations of proto-oncogene mRNA precede the elevation in $[Ca^{2+}]_i$. The measurements of $[Ca^{2+}]_i$ in SV40LT SMCs were repeated after addition of antisense or two-base-pair-mismatch antisense c-myb oligonucleotides prior to serum stimulation. The concentrations of antisense oligonucleotide selected had previously been shown to inhibit proto-oncogene expression and block S-phase progression in this cell-culture system as outlined above (66). These investigations demonstrate that addition of antisense c-myb oligonucleotide, as compared to two-base-pair-mismatch antisense c-myb oligonucleotide, almost completely suppresses elevated $[Ca^{2+}]_i$ levels observed at 24 hours but has little effect at earlier time points (Figure 7) (76).

In preliminary studies, we have employed a second technique to show the tight linkage between expression of c-myb and the elevation of $[Ca^{2+}]_i$ at the G_1/S interface. To this end, we have constructed a SV40 promoter-driven version of c-myb that lacks the DNA-binding region of the proto-oncogene but can dimerize with endogenous c-myb and thereby act as a dominant negative

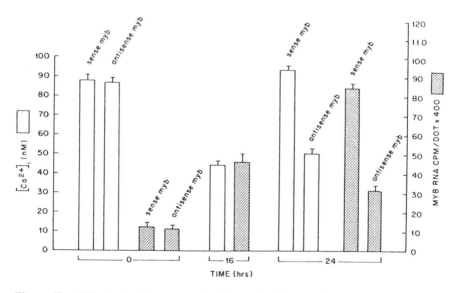

Figure 7   Effect of antisense c-myb oligonucleotide on elevated calcium levels at the G_1/S interface. The concentrations of intracellular calcium (light bars) and the levels of c-myb mRNA (dark bars) in SV40LT SMCs at 0, 16, and 24 hours after serum stimulation. The missense or antisense phosphorothioate c-myb oligonucleotides were added to cultures at a concentration of 25 μM at the time of serum stimulation (24-hour time point). The data are displayed as mean ± SE.

effector of this transacting molecule (77). The stable expression of this construct in SMCs has little effect on the resting $[Ca^{2+}]_i$ but completely suppresses the elevation of divalent cation that occurs at the G_1/S interface (Housain et al., preliminary results). Therefore, the increased $[Ca^{2+}]_i$ at the G_1/S interface appears to be under the control of c-myb, whereas the molecular events responsible for regulating intracellular calcium levels at other stages of the cell cycle are independent of the proto-oncogene. However, c-myb might function in a very indirect manner to raise $[Ca^{2+}]_i$ by allowing cells to progress to a point in the cell cycle that is associated with elevations of divalent cation.

To exclude this possibility, we determined whether expression of c-myb could directly elevate $[Ca^{2+}]_i$ independent of growth state. This was accomplished by stably transfecting SV40LT SMCs with the proto-oncogene, demonstrating that the levels of c-myb mRNA as well as the oncoprotein were elevated by two- to fourfold, and then measuring $[Ca^{2+}]_i$ (76). Full-field imaging with Fura-2 shows that the growth-arrested transfected cells, as compared to growth-arrested untransfected cells, exhibit homogeneously elevated levels of $[Ca^{2+}]_i$. The measurements reveal a substantial increase in Fura-2 ratios that corresponds to an average 1.8-fold increase in $[Ca^{2+}]_i$ (158nM ± 15.6 nM vs. 92 nM ± 9.6 nM; $p < 0.05$) (76).

We have also exposed the growth-arrested transfected and untransfected cells to antisense c-myb oligonucleotide or two-base-pair-mismatch antisense c-myb oligonucleotide. The treatment of growth-arrested transfected cells with antisense c-myb oligonucleotide, but not the minimal-mismatch antisense c-myb oligonucleotide, completely abolishes elevated $[Ca^{2+}]_i$ while the treatment of similarly designated untransfected control cells with antisense c-myb oligonucleotide had no effect on $[Ca^{2+}]_i$ (76). Furthermore, Northern analysis showed a 2.5-fold reduction in the concentrations of c-myb message in growth-arrested transfected cells after exposure to antisense c-myb oligonucleotide, as well as the lack of effect on proto-oncogene message with the minimal-mismatch antisense c-myb oligonucleotide.

We then attempted to determine whether elevated concentrations of intracellular calcium induced by c-myb are secondary to altered influx from extracellular sources. To resolve this issue, we measured the levels of $[Ca^{2+}]_i$ in growth-arrested transfected and similarly designated untransfected control cells placed for 5 minutes in 1mM calcium-containing buffer, calcium-free buffer, and phosphate-buffered saline supplemented with 2 mM EGTA. The growth-arrested transfected and untransfected control cells exhibited a decline in $[Ca^{2+}]_i$, to the same final levels. Thus, the c-myb-dependent elevations of $[Ca^{2+}]_i$ are likely to be generated by increased influx of divalent cation from the external environment rather than increased efflux of the divalent cation from an internal compartment. However, the increased entry of calcium does not appear to occur via L-type channels since nifedipine is unable to suppress the observed alterations and is unlikely to take place via T-type channels because these structures are not usually present in vascular SMCs.

We have conducted additional investigations to more precisely define the molecular mechanism by which c-myb is able to raise $[Ca^{2+}]_i$. In these studies, we have demonstrated that growth-arrested SV40LT SMCs stably transfected with c-myb exhibit a fourfold increase of IGF-I receptors and secrete significantly augmented amounts of IFG-I. This stably transfected cell type also exhibits a twofold increased extent of calcium influx and a twofold decreased extent of calcium efflux (78). The calcium influx and efflux rates were measured with radioactive divalent cation as previously outlined (78,79). The elevated calcium influx rate of transfected cells can be decreased to that of wild-type cells with addition of IGF-I-neutralizing antibody, but not control antibody. The decreased calcium-efflux rate of stably transfected cells can be increased to that of wild-type cells with antisense c-myb oligonucleotides, but not scrambled or sense c-myb oligonucleotides. Proliferating wild-type SMCs also show an increased influx rate of calcium in late G_1 that is dependent on augmented amounts of IGF-I activity as shown by inhibition of this parameter with IGF-I-neutralizing antibody, but not control antibody. However, increased levels of IFG-I receptor were not detected at any stage during the cell cycle, suggesting that significant overexpression of c-myb in stably transfected cells as outlined above may have artifactually raised the numbers of these growth-factor receptors. The wild-type SMCs also show a decreased calcium-efflux rate in late G_1 that is dependent on expression of c-myb as revealed by suppression of this parameter with antisense c-myb oligonucleotides, but not scrambled or minimal-mismatch antisense oligonucleotides (78). Thus, the concentration of $[Ca^{2+}]_i$ in SMCs appear to be regulated by the IGF-I/IGF-I receptor–dependent influx mechanism as well as an efflux mechanism that is tightly linked to expression of cmyb (Figure 8).

We have also shown that treatment of wild-type SMCs with either antisense IGF-I receptor oligonucleotides or antisense c-myb oligonucleotides induces a late G_1 block in cell proliferation that can be overcome by exposure to a calcium ionophore in amounts sufficient to raise $[Ca^{2+}]_i$ to levels of divalent cation normally observed at the G_1/S interface (78). The effect appears to be due to increased $[Ca^{2+}]_i$, rather than potential nonspecific effects of the ionophore since reduction in the extracellular concentrations of divalent cation to levels that do not allow ionophore to increase $[Ca^{2+}]_i$ inhibit the stimulatory effect of this manipulation on S-phase progression. However, the above data do not exclude the importance of other c-myb-dependent gene products in the regulation of the G_1/S transition. Indeed, it appears likely that c-myb-dependent alterations in $[Ca^{2+}]_i$ act in concert with other c-myb-dependent gene products to allow entry into the S phase.

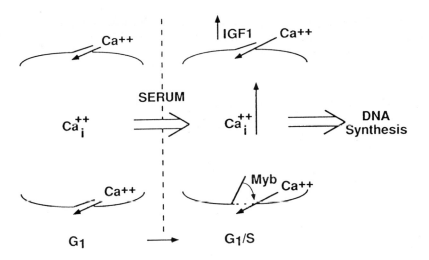

Figure 8 A theoretical model of c-myb-dependent IGF-I/IGF-I receptor expression and regulation of Ca efflux in smooth-muscle cells.

In summary, we have provided evidence for a cell-cycle-dependent alteration in SMCs of the rates of calcium influx under the control of the IGF-I/IGF-I receptor system and the rates of calcium efflux under the control of c-myb that determine $[Ca^{2+}]_i$ at the G_1/S interface, and thereby helps to regulate entry of cells into the S phase. Thus, we have uncovered a novel mechanism of action for the c-myb gene product, whose function was obscure before our investigation. However, we have no knowledge of the molecular events that elevate calcium efflux in late G_1 or the manner by which c-myb is able to subvert this process. Indeed, the pathway of calcium efflux could be located in an intracellular organelle or the plasma membrane. The effect of c-myb could be quite direct in altering transcription of an ion-pump mRNA, such as calcium ATPase, or extremely indirect in modulating transcription of an ion transporter, such as the Na^+/H^+ transporter, which might preciprocally increase the activity of the Na^+/Ca^{2+} exchanger.

The regulation of SMC proliferation as outlined above may provide a new paradigm by which different growth factors might interact to stimulate DNA synthesis and cell division. Growth factors and/or biological modulators such as IGF-I, angiotensin II, and endothelin appear to augment calcium influx independent of c-myb (80–82). PDGF, and perhaps other serum-derived growth factors, appears to enhance expression of c-myb in a posttranscriptional manner and thereby decrease calcium efflux. Therefore, it is reasonable to propose that different groups of growth factors may individually regulate the rates of calcium influx and efflux at the G_1/S interface and thereby collaborate to establish a $[Ca^{2+}]_i$ that permits a given extent of cell proliferation. The interplay between growth factors and biological modulators that raise $[Ca^{2+}]_i$ could represent an important common pathway of atherogenesis.

ANTISENSE c-myb OLIGONUCLEOTIDES INHIBIT INTIMAL ARTERIAL SMOOTH-MUSCLE-CELL PROLIFERATION IN VIVO

As outlined above, we employed antisense phosphorothioate oligonucleotides to dissect c-myb gene function in vitro. However, technical difficulties have previously prevented the use of this approach for investigating the effect of gene products in vivo. We have utilized a rat carotid artery injury model to show that local delivery of antisense phosphorothioate c-myb oligonucleotide suppresses SMC migration and growth (the term *accumulation* is used to designate the sum total of migration and proliferation observed under in vivo conditions). To this end, the left common carotid arteries of male Sprague-Daley rats were subjected to balloon angioplasty, creating a highly reproducible intimal accumulation of SMCs over the entire length of the affected blood vessel. Antisense c-myb oligonucleotides or the corresponding sense oligonucleotides were added to solutions of F127 pluronic gel at a concentration of 1 mg of oligonucleotide/ml and kept at 4°C. Immediately after balloon angioplasty, 200 µl of solution was applied to the exposed segment of the carotid artery in the neck area from which the adventitia had been stripped. The treated area of the blood vessel constituted about half the portion of the carotid artery that lies within the neck. Upon contact with tissues at 39°C, the solution gelled instantaneously, generating a translucent layer that enveloped the treated region of the carotid artery. The wounds were closed immediately after application of gel, and the rats were returned to their cages. Inspection of several additional animals revealed that the pluronic gel completely disappears over 1–2 hours.

We initially ascertained the effect of antisense c-myb oligonucleotide in suppressing oncogene mRNA levels within the rat carotid artery 2 weeks after injury. This was accomplished by surgically removing the entire treated portion of the blood vessel. Northern blot analysis shows that injured carotid artery treated with antisense oligonucleotide exhibits essentially no c-myb mRNA, whereas injured carotid artery treated with sense oligonucleotide possesses significant amounts of oncogene mRNA. The equivalent loading of mRNA was documented by hybridization of the same blot against a GAPDH cDNA probe (15). To determine the effect of antisense oligonucleotide on neointimal SMC accumulation, we performed balloon angioplasty on rats, some of which were subsequently treated with F127 pluronic gels containing antisense oligonucleotide, sense oligonucleotide, or no oligonucleotide. The rest were left untreated. Morphological examination 2 weeks after angioplasty revealed that minimal intimal accumulation occurred upon application of antisense oligonucleotide–containing gel, whereas extensive SMC accumulation was observed in the other treatment or control regimens (Figure 9). The calculated areas and ratios of the arterial segments document suppression of intimal SMC accumulation by antisense oligonucleotide without any apparent effect on medial SMC viability (15).

The specificity of the above effect was documented by two additional types of investigations. First, we conducted an additional set of studies with a nonoverlapping antisense c-myb oligonucleotide (+22 to +39) and a scrambled control oligonucleotide from the same region. The data showed the same extent of SMC growth suppression as outlined above (15). Second, we carried out an additional set of studies using antisense two-base-pair-mismatch cmyb oligonucleotide, antisense c-myb oligonucleotide, and sense oligonucleotide as outlined above. Previous in vitro results show that antisense mismatch oligonucleotide exerts no effect on c-myb mRNA levels and SMC growth (66). The in vivo data reveal that antisense mismatch oligonucleotide is not able to inhibit intimal SMC accumulation (15). The rat carotid arteries were also examined to evaluate the lengthwise extent of antisense oligonucleotide–mediated suppression of SMC accumulation. Figure 10 (right) shows a representative treated region of a rat carotid artery within the neck and an untreated segment of the same blood vessel within the chest just distal to the aortic arch. Similar patterns of SMC accumulation were noted with other antisense oligonucleotide–treated rats. The action of antisense oligonucleotide is apparently limited to the portion of carotid artery immediately in contact with gel. These animals exhibit a transition from minimal to extensive intimal accumulation over 3–4 mm of arterial length starting just proximal to the application of gel (Figure 10, left).

Based on in vitro studies, we suspect that the synthesis and release of numerous mitogens allows interaction with a correspondingly large set of specific receptors to initiate a variety of secondary messages that eventually lead to SMC proliferation and migration (83–87). Unfortunately, the experimental tools available for studying these critical biological systems under in vivo conditions are quite limited. Previous investigators have partially elucidated the growth factors responsible for intimal accumulation by infusing specific antibodies against individual mitogens such as PDGF or basic FGF (88,89). However, no general approach was available to identify critical intracellular intermediates of the above pathway. We thought that antisense oligonucleotides might be useful in pinpointing these key components by specifically inhibiting their expression and then determining the effects of the suppression on SMC response in vivo. Data summarized above strongly imply

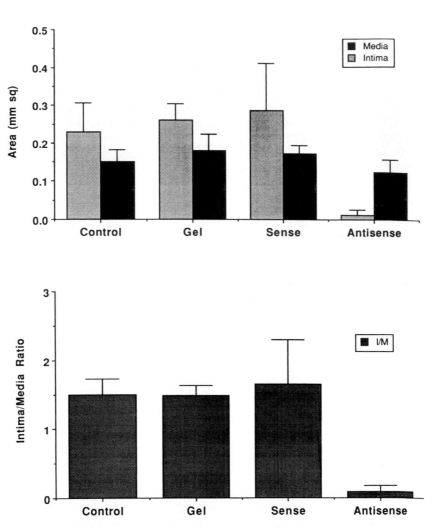

Figure 9   Effect of antisense and sense c-myb oligonucleotides on neointimal formation in rat carotid arteries subjected to balloon angioplasty. The upper panel shows mean cross-sectional areas of the intimal and medial regions of rat carotid arteries that were untreated or treated with pluronic gel, pluronic gel containing 200 μg of sense oligonucleotide, or pluronic gel containing 200 μg of antisense oligonucleotide. The lower panel depicts the ratio of intimal to medial areas for the same four groups of animals. The data are provided as mean ± SD. (From Ref. 15.)

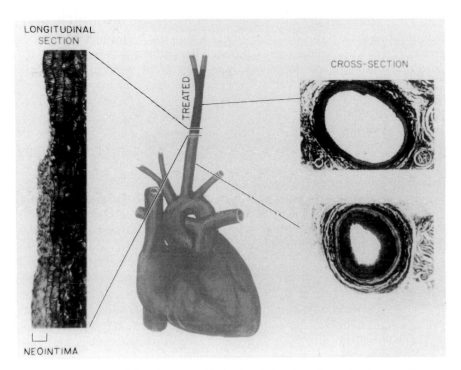

Figure 10 Spatial distribution of the antiproliferative effect of antisense oligonucle-otides. (Right) typical cross-sections (Mason trichrome, ×80) from treated and un-treated segments of the left carotid artery of a rat. (Left) longitudinal section of the transitional area between these zones (Mason trichrome, ×200). The shaded area represents the extent of application of pluronic gel.

that c-myb is a critical intermediate of a common mitogenic pathway required by all growth factors involved in SMC migration and growth. Given the previously documented in vitro effect of antisense oligonucleotide, we believe that the proto-oncogene is involved in proliferation. The drug heparin and blood-vessel-wall heparan sulfates probably also inhibit SMC growth via sup-pression of c-myb (30). Thus, endothelial cell and SMC heparan sulfates appear to constitute a natural blood-vessel-wall antiproliferative pathway.

To the best of our knowledge, these studies constitute the first reported use of antisense oligonucleotide to inhibit synthesis of a normal gene product under in vivo conditions in higher eukaryotic species with a subsequent effect on a cellular process. We suspect that our positive results are due to the local delivery of high concentrations of an antisense oligonucleotide modified to reduce the rate of in vivo degradation. The success of the above strategy suggests that antisense oligonucleotides can be used in vivo to suppress other low-abundance gene products segmentally within the blood-vessel wall and thereby delineate in vivo the critical intracellular pathways essential for SMC migration and growth. It is also readily apparent that the above approach can be utilized to dissect the critical molecular events that control other mechanisms of the blood-vessel wall. Furthermore, intimal SMC accumulation constitutes a major pathological event responsible for the long-term failure of coronary and peripheral arterial bypass grafts as well as the development of stenosis following coronary artery angioplasty (1,90). Several promising approaches have been described for suppressing this process, but none has proven clinically useful as yet (91–93). The antisense oligonucleotides constitute a new class of therapeutic agents that can be locally targeted to inhibit specific gene products that result in suppression of intimal SMC accumulation, and possibly other acute pathological processes such as thrombosis.

It is clear that the action of antisense oligonucleotides on in vivo SMC accumulation might be dramatically altered by several experimental variables. These factors include the severity of arterial injury that stimulates neointimal accumulation, the levels of the intracellular target mRNA/protein and its kinetics of expression after arterial injury, the technique used to carry out oligonucleotide delivery, the amount of oligonucleotide deposited in the blood-vessel wall, the precise sequence of oligonucleotide used, and the persistence of the oligonucleotide at the injury site. In the sections below, we briefly review six investigations recently completed by four laboratories of the effects of antisense phosphorothioate oligonucleotides on in vivo SMC accumulation. These studies employed antisense phosphorothioate oligonucleotides of varying sequence directed against three intracellular targets, adventitial and intra-arterial methods of oligonucleotide delivery, and two animal models of arterial injury. These investigations emphasize the importance of several critical variables listed above and suggest that much experimental work needs to be completed to optimize results obtained with this approach. The interrelationship between the delivery technique and the amounts of oligonucleotide needed to inhibit neointimal SMC accumulation is elucidated by three investigations carried out with the rat carotid artery injury model.

Morishita et al. (94) employed a mixture of two separate antisense cdc2 kinase and PCNA oligonucleotides (–9 to +9 of the mouse sequence of cdc2 kinase and +4 to +21 of the rat sequence of PCNA) at levels of 4 nmoles to suppress neointimal accumulation in the rat carotid artery model. The two types of antisense oligonucleotides were encapsulated in liposomes complexed with the protein coat of a Sendai virus and delivered with an intra-arterial catheter system. The above approach significantly enhances the cellular uptake of oligonucleotides by blood-vessel-wall SMCs. The single delivery of both oligonucleotides after arterial injury induced a marked decrease in the levels of cdc2 and PCNA mRNAs at 24 hours and suppressed neointimal accumulation by 60 to 70% at 2 weeks, with no loss of effect at 8 weeks. However, the single delivery of either oligonucleotide by the above approach had no effect on SMC accumulation at 2 weeks. The specificity of antisense oligonucleotides is indicated by the absence of any biological effect with sense oligonucleotides and four-base-pair-mismatch antisense oligonucleotides from the same region under in vitro and in vivo conditions.

Abe et al. (95) utilized a single antisense cdc2 and cdk2 kinase oligonucleotides (−9 to +9 of the mouse cdc2 and cdk2 sequences) at a level of 40 nmoles to suppress neointimal accumulation in the rat carotid artery injury model. The antisense oligonucleotide was delivered by the adventitial route with Pluronic 127 as outlined above. The single delivery of either of the above oligonucleotides after arterial injury induced a 50% reduction in the activity of the corresponding kinase in the blood-vessel wall at 2 weeks, and suppressed neointimal accumulation by about 60% at 2 weeks. The specificity of antisense oligonucleotides is suggested by the lack of any biological effect with sense oligonucleotides from the same region under in vivo conditions.

Simons et al. (96) used a single antisense PCNA oligonucleotide (+4 to +21 of the rat sequence) at a level of 250 nmoles to inhibit SMC accumulation in the rat carotid artery injury model. The antisense oligonucleotide was delivered by the adventitial route with Pluronic 127 as outlined above. The single delivery of the above oligonucleotide after arterial injury induced a 77% reduction in PCNA mRNA levels in the blood-vessel wall at 24 hours, and suppressed neointimal accumulation by about 80% at 2 weeks. The specificity of the antisense oligonucleotide is indicated by two additional types of investigations. On the one hand, a second nonoverlapping antisense PCNA oligonucleotide (+22 to +39 of the rat sequence) was also able to inhibit expression of the target mRNA and suppress neointimal accumulation as outlined above. On the other hand, sense and scrambled oligonucleotides from the same regions used to construct the two antisense PCNA oligonucleotides described above had no effect on target mRNA or neointimal accumulation.

The above investigations graphically demonstrate that the in vivo action of antisense oligonucleotides is dependent on the nature of the delivery system and the amount of antisense oligonucleotide used. The antisense oligonucleotides directed against cdc2/cdk2 kinases or PCNA exhibit a potent in vivo effect on neointimal accumulation when delivered in large amounts by an adventitial route. The same antisense oligonucleotides must be used together to achieve the same extent of biological effect when delivered at 10–50-fold lower levels by a more efficient intra-arterial delivery approach. In the latter case, the two antisense oligonucleotides appear to function in a synergistic fashion. However, it remains unclear whether the amounts of oligonucleotide delivered completely explain the observed differences. It is possible that the two delivery methods are responsible for different patterns of distribution and persistence of antisense oligonucleotides within the blood-vessel wall and therefore lead to significant variations in the biological potency of these components.

The importance of the antisense oligonucleotide sequence needed to inhibit neointimal accumulation is documented by three investigations carried out with the rat carotid artery and the pig coronary artery injury models.

Bennett et al. (97) employed local delivery of a single antisense c-myc phosphorothioate oligonucleotide (+1 to +16 of the human sequence) at a level of 40 nmoles to suppress neointimal accumulation in the rat carotid artery injury model. The antisense oligonucleotide was delivered by the adventitial route with Pluronic 127 as previously described (83). The single delivery of the above oligonucleotide after arterial injury induced a 75% reduction in the levels of c-myc mRNA at 2 hours in the blood-vessel wall, and suppressed neointimal accumulation by about 60% at 2 weeks. The above inhibitory effect was localized to the distal treated area of the carotid artery, but was absent from the proximal untreated area of the same carotid artery as previously depicted for antisense c-myb oligonucleotides (see Figure 10). The specificity of antisense oligonucleotides is suggested by the lack of any in vivo effect on neointimal accumulation with sense oligonucleotides from the same region. The specificity of the antisense oligonucleotides is also supported by in vitro experiments. On the one hand, no growth inhibitory effect is observed with two-base-pair-mismatch antisense c-myc oligonucleotides as well as antisense α-SMC actin or antisense GAPDH oligonucleotides. In addition, treatment of an SMC clone constitutively overexpressing c-myc with levels of antisense c-myc oligonucleotides previously used to inhibit expression of the proto-oncogene mRNA showed no suppression of the message.

Edelman et al. (98) utilized delivery of either of two antisense phosphorothioate c-myc oligonucleotides (+4 to +21 or +22 to +39 of the human sequence) at levels of 200 nmoles to suppress neointimal accumulation in a rat carotid artery injury model. Both of the antisense oligonucleotides had previously been shown to exhibit antiproliferative but not antimigratory effects under in vitro conditions. The antisense c-myc oligonucleotides were delivered by the adventitial route for short periods of 1-2 hours with Pluronic 127 or for long periods of 14 days with ethylene vinyl acetate copolymer (EVAc). The single delivery of the antisense c-myc oligonucleotide for short periods after arterial injury induced a 60 to 75% reduction in the levels of c-myc mRNA and oncoprotein at 24 hours and an 84% reduction in SMC proliferation at 72 hours as measured with BudR. However, 1 week after arterial injury, the effect of antisense oligonucleotides was completely lost with regard to expression of oncoprotein and neointimal accumulation. The administration of the same antisense c-myc phosphorothioate oligonucleotides for 14 days using an EVAc-based delivery system after arterial injury resulted in a virtually identical extent of inhibition of the levels of c-myc mRNA and oncoprotein at 24 hours as noted above, and a suppression of neointimal accumulation of 75% at 2 weeks. A similar set of experiments carried out with antisense c-myb oligonucleotides using both short and long times of delivery showed a suppression of neointimal accumulation of about 85% at 2 weeks. The specificity of antisense oligonucleotides is suggested by the lack of any in vivo effect on c-myc mRNA or neointimal accumulation with sense or scrambled oligonucleotides from the same region.

Shi et al. (17) used local delivery of a single antisense phosphorothioate c-myc oligonucleotide (+1 to +16 of the human sequence) at a level of 200 nmoles to suppress neointimal accumulation in a pig coronary artery injury model. The antisense oligonucleotides were delivered by the intra-arterial route with a perfusion catheter. The delivery was accomplished by expanding the catheter balloon against the inner surface of the blood vessel and then injecting the oligonucleotides into the blood-vessel wall via lateral sideports at a pressure of 4 atmospheres over 30 seconds. This method of delivery resulted in approximately 30-fold higher concentrations of oligonucleotides at the site of injection as compared to other organs, which persisted for at least 3 days. The single delivery of the above oligonucleotide after arterial injury suppressed neointimal accumulation in the injured coronary artery by about 70% at 4 weeks. The specificity of the antisense oligonucleotides is suggested by the lack of any in vivo effect on neointimal accumulation with sense oligonucleotides from the same region. The specificity of the antisense oligonucleotides is also supported by in vitro experiments in which no growth-inhibitory effect is observed with scrambled c-myc oligonucleotides or four-base-pair-mismatch antisense c-myc oligonucleotides from the same region.

The above investigations vividly show that the in vivo action of antisense oligonucleotides on neointimal SMC accumulation is dependent on the precise sequence used. The antisense c-myc oligonucleotides (+1 to +16 of the human sequence) exhibit a potent in vivo effect on SMC accumulations when delivered over short time periods either by an adventitial route or with a perfusion catheter in the rat carotid artery or pig coronary artery injury models. However, the antisense c-myc oligonucleotides with a different sequence (+4 to +21 or +22 to +39 of the human sequence) exhibit an in vivo effect on SMC accumulation for only about 3 days when delivered over a short time period in the rat carotid artery injury model. The prolonged administration of these two antisense oligonucleotide sequences is required to achieve an in vivo effect on neointimal accumulation at 2 weeks. The available experimental data suggest that the biological potency of the three antisense c-myc oligonucleotides is not significantly different under in vitro conditions, which implies that some other property such as persistence of these components at the site of injury might be responsible for the differences observed under in vivo conditions. The latter possibility is particularly attractive because c-myc is expressed for a short time immediately after arterial injury and then again at 7 days. Thus, the inhibition of expression of c-myc mRNA at early time points, as compared to other proto-oncogenes' messages, may not be sufficient to exert a long-term effect on SMC accumulation.

CONCLUSIONS

We have described recent investigations employing antisense oligonucleotides that attempt to define a common intracellular pathway required for growth-factor-dependent stimulation of SMC proliferation and/or migration. These studies have revealed a novel mechanism of action of one of these intracellular components, and have also demonstrated that antisense techniques can be utilized to inhibit specific mRNAs under in vivo conditions. These investigations have sparked interest in this technology, and stimulated others to use this approach under in vivo conditions in other areas, especially neurobiology (99–101). Indeed, clinical trials with antisense oligonucleotides are currently underway to determine their efficacy in several human disorders including acute myeloblastic leukemia and post-angioplasty restenosis. However, it should be apparent from the above discussion that numerous basic issues about antisense techniques, including their specificity, require resolution before this approach can be used in an optimally effective manner.

REFERENCES

1. Dilley RJ, McGeachie JK, Prendergast FJ. A review of proliferative behavior, morphology and phenotypes of vascular smooth muscle. Atherosclerosis 1987; 63:99–107.
2. Libby P, Warner SJC, Solomon RN, Birinyi LK. Production of platelet-derived growth factor-like mitogen by smooth muscle cells from atheroma. N Engl J Med 1988; 318:1493–1498.
3. Clowes AW, Clowes MM, Kocher O, Ropraz P, Chaponnier C, Gabbiani G. Arterial smooth muscle cells in vivo: relationship between actin isoform expression and mitogenesis and their regulation by heparin. J Cell Biol 1988; 107:1939–1945.
4. Schwartz SM, Reidy MA. Common mechanisms of proliferation of smooth muscle in atherosclerosis and hypertension. Hum Pathol 1987; 18:240–247.
5. Ross R, Raines EW, Bowen Pope DF. The biology of platelet-derived growth factor. Cell 1986; 46:155–169.
6. Banscota NK, Taub R, Zellner K, King GL. Insulin, insulin-like growth factor I and

platelet derived growth factor interact additively in the induction of the proto oncogene c-myc and cellular proliferation in cultured bovine aortic smooth muscle cells. Mol Endocrinol 1989; 3:1183–1190.

7. Libby P, Warner SJC, Friedman GB. Interleukin 1: a mitogen for human vascular smooth muscle cells that induces the release of growth-inhibitory prostanoids. J Clin Invest 1988; 81:487–498.

8. Gay CG, Winkles JA. Interleukin 1 regulates heparin-binding growth factor 2 gene expression in vascular smooth muscle cells. Proc Natl Acad Sci USA 1991; 88:296–300.

9. Majack RA. Beta-type transforming growth factor specifies organizational behavior in vascular smooth muscle cell cultures. J Cell Biol 1987; 105:465–471

10. Lam JY, Lacoste L, Bourassa MG. Cilazapril and early atherosclerotic changes after balloon injury of porcine carotid arteries. Circulation 1992; 85:1542–1547.

11. Powell JS, Clozel J-P, Muller RKM, Kuhn H, Hefti F, Hosang M, Baumgartner HR. Inhibitors of angiotensin-converting enzyme prevent myointimal proliferation after vascular injury. Science 1989; 245:186–188.

12. Daemen MJAP, Lombardi DM, Bosman FT, Schwartz SM. Angiotensin II induces smooth muscle cell proliferation in the normal and injured rat arterial wall. Circ Res 1991; 68:450–456.

13. Owens GK. Control of hypertrophic versus hyperplastic growth of vascular smooth muscle cells. Am J Physiol 1989; 257:H1755–1765.

14. McNamara CA, Sarembock IJ, Gimple LW, Fenton JW 2d, Coughlin SR, Owens GK. Thrombin stimulates proliferation of cultured rat aortic smooth muscle cells by a proteolytically activated receptor. J Clin Invest 1993; 91:94–98.

15. Simons M, Edelman ER, DeKeyser JL, Langer R, Rosenberg RD. Antisense cmyb oligonucleotides inhibit intimal arterial smooth muscle cell accumulation in vivo. Nature 1992; 359:67–70.

16. Azrin MA, Mitchel JF, Pedersen C, Curley T, Bow LM, Alberghini TV, Waters DD, McKay RG. Inhibition of smooth muscle cell proliferation *in vivo* following local delivery of antisense c-myb oligonucleotides during angioplasty. J Am Coll Cardiol 1994; 23:396A.

17. Shi Y, Fard A, Galeo A, Hutchinson HG, Vermani P, Dodge GR, Hall DJ, Shaheen F, Zalewski A. Transcatheter delivery of c-myc antisense oligomers reduces neointimal formation in a porcine model of coronary artery balloon injury. Circulation 1994; 90:944–951.

18. Clowes AW, Reidy MA, Clowes MM. Kinetics of cellular proliferation after arterial injury. I. Smooth muscle growth in the absence of endothelium. Lab Invest 1983; 49:327–333.

19. Carter AJ, Laird JR, Farb A, Kufs W, Wortham DC, Virmani R. Morphologic characteristics of lesion formation and time course of smooth muscle cell proliferation in a porcine proliferative restenosis model. J Am Coll Cardiol 1994; 24:1398–1405.

20. Kindy MS, Sonenshein GE. Regulation of oncogene expression in cultured aortic smooth muscle cells. J Biol Chem 1986; 261:12865–12868.

21. Pukac LA, Castellot JJ Jr, Wright TC Jr, Caleb BL, Karnovsky MJ. Heparin inhibits c-fos and c-myc mRNA expression in vascular smooth muscle cells. Cell Regul 1990; 1:435–443.

22. Murray A, Hunt T. The Cell Cycle. New York: WH Freeman, 1993.
23. Thompson CB, Challoner PB, Neiman PE, Groudine M. Expression of the c-myb proto-oncogene during cellular proliferation. Nature 1986; 319:374-380.
24. Luscher B, Eisenman RN. New light on myc and myb. Part II. Myb Genes Dev 1990; 4:2235-2241.
25. Travali S, Ferber A, Reiss K, Sell C, Koniecki J, Calabretta B, Baserga R. Effect of the myb gene product on the expression of the PCNA gene in fibroblasts. Oncogene 1991; 6:887-894.
26. Clowes AW, Karnovsky MJ. Suppression by heparin of smooth muscle cell proliferation in injured arteries. Nature 1977; 265:625-626.
27. Reilly CF, Fritze LMS, Rosenberg RD. Antiproliferative effects of heparin on vascular smooth muscle cells are reversed by epiderman growth factor. J Cell Physiol 1986; 131:149-157.
28. Guyton JR, Rosenberg RD, Clowes AW, Karnovsky MJ. Inhibition of rat arterial smooth muscle proliferation by heparin. Circ Res 1980; 46:625-634.
29. Fritze LMS, Reilly CF, Rosenberg RD. An antiproliferative heparan sulfate species produced by postconfluent smooth muscle cells. J Cell Biol 1985; 100:1041-1049.
30. Reilly CF, Kindy MS, Brown KE, Rosenberg RD, Sonenshein GE. Heparin prevents vascular smooth muscle cell progression through the G_1 phase of the cell cycle. J Biol Chem 1989; 264:6990-6995.
31. Gadeau A-P, Campna M, Dsgranges C. Induction of cell-cycle dependent genes during cell cycle progression of arterial smooth muscle cells in culture. J Cell Physiol 1991; 146:356-361.
32. Miano J, Vlasic N, Robert R, Stemerman M. Smooth muscle cell immediate-early gene and growth factor activation follows vascular injury. Arterioscl Thromb 1993; 13:211-219.
33. Hunter T, Karin M. The regulation of transcription by phosphorylation. Cell 1992; 70:375-387.
34. Furukawa Y, Piwnica Worms H, Ernst TJ, Kanakura Y, Griffin JD. Cdc2 gene expression at the G1 to S transition in human T lymphocytes. Science, 1990; 250:805-808.
35. Neckers L, Whitesell L, Rosolen A, Geselowitz DA. Antisense inhibition of oncogene expression. Crit Rev Oncog 1992; 3:175-231.
36. Murray JAH, Crockett N. Antisense techniques: An overview. In: Murray JAH, ed. Antisense RNA and DNA. New York: Wiley-Liss, 1992.
37. Wang S, Dolnick BJ. Quantitative evaluation of intracellular sense: antisense RNA hybrid complexes. Nucl Acid Res 1993; 21:4383-4391.
38. Stull RA, Taylor LA, Szoka FC Jr. Predicting antisense oligonucleotide inhibitory efficacy: a computational approach using histograms and thermodynamic indices. Nucl Acid Res 1992; 20:3501-3508.
39. Speir E, Epstein SE. Inhibition of smooth muscle cell proliferation by an antisense oligodeoxynucleotide targeting the messenger RNA encoding proliferating cell nuclear antigen. Circ 1992; 86:538-547.
40. Freier SM, Lima WF, Sanghvi YS, Vickers T, Zounes M, Cook PD, Ecker DJ. Thermodynamics of antisense oligonucleotide hybridization. In: Erickson RP,

Izant JG, eds. Gene Regulation: Biology of Antisense RNA and DNA. New York: Raven Press, 1992: 95–108.

41. Fakler B, Herlitze S, Amthor B, Zenner HP, Ruppersberg JP. Short antisense oligonucleotide-mediated inhibition is strongly dependent on oligo length and concentration but almost independent of location of the target sequence. J Biol Chem 1994; 269:16187–16194.

42. Smith CC, Aurelian L, Reddy MP, Miller PS, Ts'o POP. Antiviral effect of an oligo (nucleoside methylphosphonate) complimentary to the splice junction of herpes simplex virus type 1 immediate early pre-mRNAs 4 and 5. Proc Natl Acad Sci USA 1986; 83:2787–2791.

43. Sarin PS, Agrawal S, Civeira MP, Goodchild J, Ikeuchi T, Zamecnik PC. Inhibition of acquired immunodeficiency syndrome virus by olidodeoxynucleotide methylphophonates. Proc Natl Acad Sci USA 1988; 85:7448–7451.

44. Gray GD, Herandez OM, Hebel D, Root M, Pow-Sang JM, Wickstrom E. Antisense DNA inhibition of tumor growth by c-Ha-ras oncogene in nude mice. Cancer Res 1993; 53:577–580.

45. Winstrom E, Bacon TA, Wickstrom EL. Down regulation of c-myc antigen expression in lymphocytes of Emu-c-myc transgenic mice treated with anti-c-myc DNA methylphosphonates. Cancer Res 1992; 52:6741–6745.

46. Stein CA, Cheng YC. Antisense oligonucleotides as therapeutic agents—is the bullet really magical? Science 1993; 261:1004–1012.

47. Iribarren AM, Sproat BS, Neuner P, Sulston I, Ryder U, Lamond AI. 2'-O-alkyl oligoribuncleotides as antisense probes. Proc Natl Acad Sci USA 1990; 87:7747–7751.

48. Wagner RW, Matteucci MD, Lewis JG, Gutierrez AJ, Moulds C, Froehler BC. Antisense gene inhibition by oligonucleotides containing C-5 propyne pyrimidines. Science 1993; 260:1510–1513.

49. Hoke GD, Draper K, Freier SM, Gonzalez C, Driver VB, Zounes MC, Ecker DJ. Effects of phosphorothioate capping on antisense oligonucleotide stability, hybridization and antiviral efficacy versus herpes simplex virus infection. Nucl Acid Res 1991; 19:5743–5748.

50. Giles RV, Spiller DG, Tidd DM. Chimeric oligodeoxynucleotide analogs: enhanced cell uptake of structures which direct ribonuclease H with high specificity.

51. Giles RV, Tidd DM. Enhanced Rnase H activity with methylphosphonodiester/phosphodiester chimeric antisense oligonucleotides.

52. Giles RV, Tidd DM. Increased specificity of antisense oligodeoxynucleotide targeting of RNA cleavage by Rnase H using chimeric methylphosphonodiester/phosphodiester structures.

53. Yakubov LA, Deeva EA, Zarytova VF, Ivanova EM, Ryte AS, Yurchenko LV, Vlassov VV. Mechanism of oligonucleotide uptake by cells: involvement of specific receptors? Proc Natl Acad Sci USA 1989; 86:6454–6458.

54. Loke SL, Stein CA, Zhang XH, Mori K, Nakanishi M, Subainghe C, Cohen JS, Neckers LM. Characterization of oligonucleotide transport into living cells. Proc Natl Acad Sci USA 1989; 86:3474–3478.

55. Akhtar S, Shoji Y, Juliano RL. Pharmaceutical aspects of the biological stability

and membrane transport characteristics of antisense oligonucleotides. In: Erickson RP, Izant JG, eds. Gene Regulation: Biology of Antisense RNA and DNA. New York: Raven Press, 1992: 133–146.

56. Leonetti JP, Mechti N, Degols G, Gagnor C, Lebleu B. Intracellular distribution of microinjected antisense oligonucleotides. Proc Natl Acad Sci USA 1991; 88:2702–2706.

57. Helene C, Thuong NT, Harel-Bellan A. Control of gene expression by triple helix-forming oligonucleotides. In: Basega R, Denhardt DT, eds. Antisense Strategies. Ann NY Acad Sci 1992; 600:27–36.

58. Cooney M, Czernuszewicz G, Postel EH, Flint SJ, Hogan ME. Site-specific oligonucleotide binding represses transcription of the human c-myc gene in vitro. Science 1988; 241:456–459.

59. Orson FM, Thomas DW, McShan WM, Kessler DJ, Hogan ME. Oligonucleotide inhibition of IL2R alpha mRNA transcription by promoter region collinear triplex formation in lymphocytes. Nucl Acid Res 1991; 19:3435–3441.

60. Grigoriev M, Praseuth D, Robin P, Guesse AL, Thuong NT, Helene C, Harel-Bellan A. A triple-helix forming oligonucleotideintercalator conjugate acts as a transcriptional repressor via inhibition of NF-k-B binding to interleukin2receptor alpha regulatory sequence. J Biol Chem 1992; 267:3389–3395.

61. Mirabelli CK, Bennett CF, Anderson K, Crooke ST. In vitro and in vivo pharmacologic activities of antisense oligonucleotides. Anti-Cancer Drug Design 1991; 6:647–661.

62. Woolf TM, Jennings CG, Rebagliati M, Melton DA. The stability, toxicity and effectiveness of unmodified and phosphorothioate antisense oligodeoxynucleotides in Xenopus oocytes and embryos. Nucl Acid Res 1990; 18:1763–1769.

63. Woolf TM, Melton DA, Jennings CG. Specificity of antisense oligonucleotides in vivo. Proc Natl Acad Sci USA 1992; 89:7305–7309.

64. Block LC, Griffin LC, Latham JA, Vermass EH, Toole JJ. Selection of single-stranded DNA molecules that bind and inhibit human thrombin. Nature 1992; 355:564–566.

65. Offerman MK, Medford RM. Induction of VCAM-1 gene expression by double-stranded RNA occurs by a p68 kinase-dependent pathway in endothelial cells (abstr). Clin Res 1993; 41:262a.

66. Simons M, Rosenberg RD. Antisense nonmuscle myosin heavy chain and c-myb oligonucleotides suppress smooth muscle cell proliferation in vitro. Circ Res 1992; 70:835–843.

67. Brown KE, Kindy MS, Sonenshein GE. Expression of the c-myb proto-oncogene in bovine vascular smooth muscle cells. J Biol Chem 1992; 267:4625–4630.

68. Keith CH, Ratan R, Maxfield FR, Bajer A, Shelanski Z. Local cytoplasmic calcium gradients in living mitotic cells. Nature 1985; 316:848–850.

69. Poenie M, Alderton J, Steinhardt R, Tsien R. Calcium rises abruptly and briefly throughout the cell at the onset of anaphase. Science 1986; 233:886–889.

70. Steinhardt RA, Alderton J. Intracellular free calcium rise triggers nuclear envelope breakdown in the sea urchin embryo. Nature 1988; 332:364–366.

71. Twigg J, Patel R, Whitaker M. Translational control of InsP3-induced chromatin

condensation during the early cell cycles of sea urchin embryos. Nature 1988; 332:366–369.

72. Sisken JE, Silver RB, Barrows GH, Grasch SD. In: Advances in Microscopy. New York: Alan R Liss, 1985: 73–87.

73. McIntosh JR, Koonce MP. Mitosis. Science 1989; 246:622–628.

74. Paul D, Ristom HJJ. Cell cycle control by Ca^{++} ions in mouse 3T3 cells and in transformed 3T3 cells. J Cell Physiol 1979; 98:31–40.

75. Pardee AB, Dubrow R, Hamlin JL, Kletzien RF. Animal cell cycle. Annu Rev Biochem 1978; 47:715–750.

76. Simons M, Morgan KG, Parker C, Collins E, Rosenberg RD. The proto-oncogene c-myb mediates an intracellular calcium rise during the late G_1 phase of the cell cycle. J Biol Chem 1993; 268:627–632.

77. Badiani P, Corbella P, Kioussis D, Marvel J, Weston K. Dominant interfering alleles define a role for c-myb in T-cell development. Genes Dev 1994; 8:770–782.

78. Simons M, Hideao A, Salzman EW, Rosenberg RD. c-Myb affects intracellular calcium handling in vascular smooth muscle cells. Am J Physiol: Cell Regul Physiol, 1995; 37:C856–C868.

79. Takada K, Amino N, Tada H, Miyai K. Relationship between proliferation and cell cycle-dependent Ca^{2+} influx induced by a combination of thyrotropin and insulin-like growth factor-I in rat thyroid cells. J Clin Invest 1990; 86:1548–1555.

80. Kojima I, Matsunaga H, Kurokawa K, Ogata E, Nishimoto I. Calcium influx: an intracellular message of the mitogenic action of insulin-like growth factor-I. J Biol Chem 1988; 263:16561–16567.

81. Gonzalez E, Salomonsson M, Kornfeld M, Gutierrez AM, Morsing P, Persson AE. Different action of angiotensin II and noradrenaline on cystolic calcium concentration in isolated and perfused afferent arterioles. Acta Physiol Scand 1992; 145:299–300.

82. Gardner JP, Tokudome G, Tomonari H, Maher E, Hollander D, Aviv A. Endothelin-induced calcium responses in human vascular smooth muscle cells. Am J Physiol 1992; 262.

83. Banskota NK, Taub R, Zellner K, Olsen P, King GL. Characterization of induction of protooncogene c-myc and cellular growth in human vascular smooth muscle cells by insulin and IGF-I. Diabetes 1989; 38:123–129.

84. Deuel TF, Huang JS. Platelet-derived growth factor: structure, function, and roles in normal and transformed cells. J Clin Invest 1984; 74:669–676.

85. Lindner V, Lappi DA, Baird A, Majack RA, Reidy MA. Role of basic fibroblasts growth factor in vascular lesion formation. Circ Res 1991; 68:106–113.

86. Jawien A, Bowen-Pope DF, Lindner V, Schwartz SM, Clowes AM. Platelet-derived growth factor promotes smooth muscle migration and intimal thickening in a rat model of balloon angioplasty. J Clin Invest 1992; 89:507–511.

87. Hansson GK, Jonasson L, Holm J, Clowes MM, Clowes AW. Alpha-interferon regulates vascular smooth muscle proliferation and Ia antigen expression in vivo and in vitro. Circ Res 1988; 63:712–719.

88. Lindner V, Reidy MA. Proliferation of smooth muscle cells after vascular injury is inhibited by an antibody against basic fibroblast growth factor. Proc Natl Acad Sci USA 1991; 88:3739–3743.

89. Ferns GA, Raines EW, Sprugel KH, Motani AS, Reidy MA, Ross R. Inhibition of neointimal smooth muscle accumulation after angioplasty by an antibody to PDGF. Science 1991; 253:1129-1132.
90. Nobuyoshi M, Kimura T, Ohishi H, Horiuchi H, Nosaka H, Hamasaki N, Yokoi H, Kim KJ. Restenosis after percutaneous transluminal coronary angioplasty: pathology observations in twenty patients. Am Coll Cardiol 1991; 17:433-439.
91. Jonasson L, Holm J, Hannson GK. Cyclosporin A inhibits smooth muscle proliferation in the vascular response to injury. Proc Natl Acad Sci USA 1988; 85:2303-2306.
92. Powell JS, Clozel JP, Muller RK, Kuhn H, Hefti F, Hosang M, Baumgartner HP. Inhibitors of angiotensin-converting enzyme prevent myointimal proliferation after vascular injury. Science 1989; 245:186-188.
93. Pukac LA, Hirsch GM, Lormeau J-C, Petitou M, Choay J, Karnovsky M. Antiproliferative effects of novel, nonanticoagulant heparin derivatives on vascular smooth muscle cell in vitro and in vivo. Am J Pathol 1991; 139:1501-1509.
94. Morishita R, Gibbons GH, Ellison KE, Nakajima M, Zhang L, Kaneda Y, Ogihara T, Dzau VJ. Single intraluminal delivery of antisense cdc2 kinase and proliferating-cell nuclear antigen oligonucleotides results in chronic inhibition of neointimal hyperplasia. Proc Natl Acad Sci USA 1993; 90:8474-8478.
95. Abe J, Zhou W, Taguchi J, Takuwa N, Miki K, Okazaki H, Kurokawa K, Kumada M, Takuwa Y. Suppression of neointimal smooth muscle cell accumulation in vivo by antisense cdc2 and cdk2 oligonucleotides in rat carotid artery. Biochem Biophys Res Comm 1994; 198:16-24.
96. Simons M, Edelman ER, Rosenberg RD, Antisense PCNA oligonucleotides inhibit intimal hyperplasia in a rat carotid injury model. J Clin Invest 1994; 93:2351-2356.
97. Bennett MR, Anglin S, McEwan JR, Jagoe R, Newby AC, Evan GI. Inhibition of vascular smooth muscle cell proliferation in vitro and in vivo by c-myc antisense oligodeoxynucleotides. J Clin Invest 1994; 93:820-828.
98. Edelman ER, Simons M, Sirois MG, Rosenberg RD. c-Myc in vasculo-proliferative disease. Circ Res. In press.
99. Wahlestedt C, Golanov E, Yamamoto S, Yee F, Ericson H, Yoo H, Inturrisi CE, Reis DJ. Antisense oligodeoxynucleotides to NMDA-R1 receptor channel protect cortical neurons from excitotoxicity and reduce focal ischaemic infarctions. Nature 1993; 363:260-263.
100. Gyurko R, Wielbo D, Phillips MI, Antisense inhibition of AT1 receptor mRNA and angiotensinogen mRNA in the brain of spontaneously hypertensive rats reduces hypertension of neurogenic origin. Regulatory Peptides 1993; 49:167-174.
101. Harrison P. Antisense: into the brain. Lancet 1993; 342:254-255.

Index

581

About the Editor

Stephen C. Mockrin is Deputy Director, Division of Heart and Vascular Diseases, National Heart, Lung and Blood Institute, National Institutes of Health, Bethesda, Maryland. The author or coauthor of numerous professional papers that reflect his research interests in genetics, gene therapy, hypertension, molecular biology, biotechnology, and cellular motility, he is a Fellow of the American Heart Association's Council for High Blood Pressure Research and a member of the American Society for Cell Biology, among other organizations. Dr. Mockrin received the B.S. degree (1968) in chemistry from the University of Michigan, Ann Arbor, and the Ph.D. degree (1973) in biochemistry from the University of California, Berkeley.